Birds of Prey

José Hernán Sarasola
Juan Manuel Grande • Juan José Negro
Editors

Birds of Prey

Biology and conservation in the XXI century

Springer

Editors
José Hernán Sarasola
Centro para el Estudio y Conservación de las Aves Rapaces en Argentina (CECARA)
Universidad Nacional de La Pampa (UNLPam) & Consejo Nacional de Investigaciones Científicas y Técnicas de Argentina (CONICET)
Santa Rosa, La Pampa, Argentina

Juan Manuel Grande
Centro para el Estudio y Conservación de las Aves Rapaces en Argentina (CECARA)
Universidad Nacional de La Pampa (UNLPam) & Consejo Nacional de Investigaciones Científicas y Técnicas de Argentina (CONICET)
Santa Rosa, La Pampa, Argentina

Juan José Negro
Department of Evolutionary Ecology
Doñana Biological Station (EBD-CSIC)
Seville, Spain

ISBN 978-3-030-08850-7 ISBN 978-3-319-73745-4 (eBook)
https://doi.org/10.1007/978-3-319-73745-4

Softcover re-print of the Hardcover 1st edition 2018

Printed on acid-free paper

This Springer imprint is published by the registered company Springer International Publishing AG part of Springer Nature.
The registered company address is: Gewerbestrasse 11, 6330 Cham, Switzerland

Preface

Birds of prey have fascinated humans through time perhaps more than any other group of birds. Proof of the respect, admiration, and veneration that raptors gained in ancient but also in modern human civilizations is found in the diversity of symbolisms, from hieroglyphs to country flags and emblems, condensing the ideals that raptors transmitted to these cultures, such as power, vitality, sacredness, and nobility. From the eighteenth century to modern societies, however, and although raptors are still admired as singular and as the most majestic species among birds, some divorce occurred with part of the society, and a majority of raptor species have suffered and still suffer from active human persecution.

From an ecological perspective, birds of prey are the main avian top predators found in almost all mainland terrestrial ecosystems all around the world, playing a key role in regulating prey populations and structuring natural communities. However, such features also make raptors very susceptible to environmental changes, being species that have suffered the most from the impacts of human modifications of natural habitats. Changes, by the way that are jeopardizing biodiversity conservation at an unprecedented global scale. The industrial revolution beginning in the mid-nineteenth century allowed the development of technical advances that lay the foundations for the current well-being of modern societies, including energy and food production. However, these advances, along with the expansion of communication and transportation infrastructures, needed to fulfill the increasing demands of a growing world human population. This resulted in the expansion of agricultural areas and urban sprawl, along with all the infrastructures needed for the supply of food, water, and energy, and thus permitted the transformation of enormous areas of natural habitats and the exploitation of natural resources at rates never recorded before. In this context, many raptor species around the world have become severely threatened and their populations reduced in numbers to levels that compromise their conservation.

This book is aimed to provide readers, either students, researchers, or plain raptor enthusiasts, with the most exhaustive up-to-date review of topics related to raptor biology and conservation. Even though we are facing a severe environmental crisis, new technologies and accumulated knowledge have permitted the development of a number of tools that may traduce in the improvement of methodologies applied in the field of avian biology and conservation. In this sense, this book intends to gather exhaustive revisions of some of the latest technical and methodological advances on the study of raptors. Most of these advances were unforeseen even just a few years ago and currently represent important research opportunities for the study of birds of prey. In addition, novel ways in which raptors are threatened by human activities are recorded in conjunction with alternatives to mitigate such impacts.

This book, focused mainly in diurnal raptors but with some references and examples to nocturnal ones in general topics, is divided into three main sections. The first section deals with some of the most important aspects on the general biology and ecology of raptors: from phylogeny and taxonomy to breeding, behavioral ecology, migration, and dispersion, including a review of one of the particular functions that birds of prey play in natural communities as dispersers of plant seeds.

The second section is concerned with the interactions between raptors and humans, human activities, and the resulting human-modified environments. In this section, chapters cover issues from the protagonism of diurnal raptors since prehistory, in ancient human civilizations and up to the present day. This section deals, for instance, with the main threats that raptors currently face when living in urban and human-altered environments. This section also includes a review of the state of the art in some of the most important anthropogenic threats for raptors, such as avian electrocution in powerlines and lead poisoning.

The third and last section is concerned with raptor conservation worldwide. Species conservation is strongly influenced by their particular habits and behavior, but also by the singularities of the ecosystems or regions they live in. For that reason, chapters in this part include both biogeographical and taxonomical approaches. Thus, some particular groups, such as the Old World vultures, are analyzed in detail, while biogeographical-based assessments include some of the regions holding a greater diversity of raptors and also some of the most threatened species. Two of the chapters, as an example, deal with the development of novel tools applied to the study, monitoring, and hands-on conservation of birds of prey, including the latest available molecular techniques in conservation genetics, or the use of unmanned aerial vehicles (UAVs) for nest surveying minimizing disturbance.

As editors, we wish to acknowledge and heartily thank all our coauthors in this book. Their expertise and love for raptors have made possible this joint effort with contributions from all over the world. If this book helps, even minimally, to provide directions for preserving raptors and their habitats, our goals when we embarked in this endeavor will be fulfilled.

Santa Rosa, Argentina — José Hernán Sarasola
Santa Rosa, Argentina — Juan Manuel Grande
Seville, Spain — Juan José Negro
25 January 2018

Contents

Part I
General Biology

Chapter 1
Phylogeny, Taxonomy, and Geographic Diversity of Diurnal Raptors: Falconiformes, Accipitriformes, and Cathartiformes

David P. Mindell, Jérôme Fuchs, and Jeff A. Johnson

Introduction

An ideal taxonomy for organisms incorporates comprehensive knowledge of existing species diversity and their phylogenetic relationships. This knowledge is used in developing consistent criteria for recognizing and naming species as well as monophyletic groups (clades) above the species level, including genera, families, and orders. This provides well-justified, consensus names for taxa that can be used globally in studying and managing the health of species and their populations. Having the taxonomic hierarchy of names reflects evolutionary history and advances our understanding of the origins and causes of change over time in biological diversity.

Ideal classifications also include information on the relative distinctiveness of clades, as indicated by their assigned rank. However, placement of taxa into categories of genus, family, order, and others is contentious because decisions are subjective, varying with the relative priority given to different kinds of traits (e.g., morphological, molecular, ecological) both within and among organismal groups.

It has long been suggested that geological age for clade origins be used as a standardized criterion for assigning taxonomic rank across life forms (e.g., Bigelow 1956; Hennig 1966; Sibley and Ahlquist 1990), as time (age) is an objective metric that can be applied and compared across all life forms. There is no such objective

D. P. Mindell (✉)
Museum of Vertebrate Zoology, University of California, Berkeley, CA, USA

J. Fuchs
Institut Systématique Evolution Biodiversité (ISYEB), Muséum national d'Histoire naturelle, CNRS, Sorbonne Université, EPHE 57 rue Cuvier, CP51, 75005 Paris, France

J. A. Johnson
Department of Biological Sciences, Institute of Applied Sciences, University of North Texas, Denton, TX, USA
e-mail: Jeff.Johnson@unt.edu

J. H. Sarasola et al. (eds.), *Birds of Prey*,
https://doi.org/10.1007/978-3-319-73745-4_1

metric regarding the degree of difference in morphology, physiology, life history, behavior, or niche breadth that can be broadly and objectively applied to guide taxonomic category delineation across disparate groups in general and in diurnal raptors in particular. Thus, current classifications for birds vary considerably in the criteria, if any, that have been used in assignment of taxonomic categories. Further, passage of time can be measured throughout life's history, whereas the histories of particular organismal or avian traits, which might inform the ranking of categories, are less easily discovered, due to convergence, natural selection, and chance events.

Although reliable age estimates remain unavailable for many groups, this is changing. As methods improve for dating phylogenies using molecular and fossil data in combination, this approach, in which taxonomic categories reflect approximate ages of named taxa, is becoming more common (see Avise and Johns 1999; Holt and Jønsson 2014).

The value of classifications that reflect phylogeny and relative age for named clades stems from their use in conservation (see in this book Chap. 15) and studies of evolution. Conservation of biodiversity requires availability of widely accepted names for species that correspond to distinctive evolutionary units or lineages, and, correspondingly, species are most often the focus of monitoring and management plans. Conservation efforts also benefit from understanding the degree of differentiation among species, as relative distinctiveness for taxa, including phylogenetic diversity (e.g., Faith 1992), may factor in prioritizing efforts. Relative distinctiveness is often based on phylogenetic tree branch lengths and/or estimates of taxon ages.

Higher-level taxonomic categories are sometimes used as units in studies of ecology and evolution, including, for example, studies of diversification and extinction rates. Such studies generally assume taxonomic categories are comparable in age. However, several analyses have noted the disparity or inconsistency in ages of higher-level taxa for vertebrates (Avise and Johns 1999; Avise and Liu 2011). Holt and Jønsson (2014, p. 1013) found that avian genera are "*particularly inconsistent* [in age] *and only slightly better than a randomly delineated taxonomy.*" Similarly, Jønsson et al. (2016) found that within the *Corvides*, a large group of oscine passeriform birds, genera were generally not consistent in age. Having classifications where taxonomic categories were similar in age would promote understanding of biodiversity's evolution by providing units across organismal groups that could be used in comparative studies.

Despite the benefits, taxonomic classification is often deferred, as comprehensive phylogenies with age estimates for phylogenetic nodes remain scarce or preliminary for many groups. Recent advances in our understanding of raptor evolution have raised the potential for new classification efforts, particularly for orders and families.

We have two primary objectives here. First, we provide a phylogenetically informed classification for diurnal raptors (i.e., orders Accipitriformes, Cathartiformes, Falconiformes), including an initial effort to equilibrate orders and families by age. This includes review of the literature as well as new analyses of published Accipitriformes DNA sequences. Second, we provide geographic analyses of species richness (SR), phylogenetic diversity (PD), mean pairwise distances (MPD), and mean nearest taxon distances (MNTD) measures across and within eco-regions for Falconiformes and Accipitriformes. As part of this, we evaluate the impact on these values of the potential extinction for species currently deemed at risk by the IUCN (International Union for Conservation of Nature).

Diurnal Raptor Phylogeny and Classification

Recent analyses are congruent in showing that the traditional raptor families Accipitridae (hawks, eagles, kites, Old World vultures) and Falconidae (falcons, caracaras) are not sister taxa (Cracraft et al. 2004; Ericson et al. 2006; Hackett et al. 2008; McCormack et al. 2013; Jarvis et al. 2014; Mahmoud et al. 2014; Prum et al. 2015), and the traditional order Falconiformes (e.g.. Peters 1931) has been revised. Currently, Accipitriformes is the order including hawks, eagles, and kites, and Falconiformes are reserved for the falcons, forest falcons, and caracaras (see Gill and Donsker 2017). Recent analyses also agree in showing that the New World vulture clade (Cathartidae) is sister to current Accipitriformes, not closely related to storks in the family Ciconiidae, as hypothesized earlier (Sibley and Ahlquist 1990) (Fig. 1.1), and similar in age (arising ca. 60 mya) to orders Accipitriformes and Falconiformes (Table 1.1). Correspondingly, we support placement of Cathartidae as the sole family within the order Cathartiformes (Jarvis et al. 2014; Chesser et al. 2016). The secretarybird and osprey species are placed in the families Sagittariidae and Pandionidae, respectively, and it is reasonable to ask if these warrant separate families based on their ages (see below). Current classification for the three orders of diurnal raptors (Falconiformes, Accipitriformes, Cathartiformes) is provided in Table 1.1. This reflects the synthesis by Gill and Donsker (2017), and a small number of changes suggested here is noted with asterisks in Table 1.1.

Falconiformes

We have recently published phylogenies for all Falconiformes species (Fuchs et al. 2011, 2012, 2015), and these are largely congruent with studies based on fewer species by Griffiths (1999) and Griffiths et al. (2004). In seeking to keep classification consistent with phylogeny, our best estimate for Falconidae phylogeny leads to three recommended changes to traditional taxonomy. One is changing *Polihierax insignis* to *Neohierax insignis* to maintain monophyly for genera as used by Swann (1922). A second is to recognize *Falco columbarius aesalon*, the Old World Merlin, as a distinct species, *F. aesalon*. This is based on its reciprocal monophyly with *F. columbarius*, the New World Merlin, and its genetic distinctiveness being similar or greater to that found between other distinct *Falco* species (Fuchs et al. 2015). The third recommendation is expanding the genus *Daptrius* to include all species

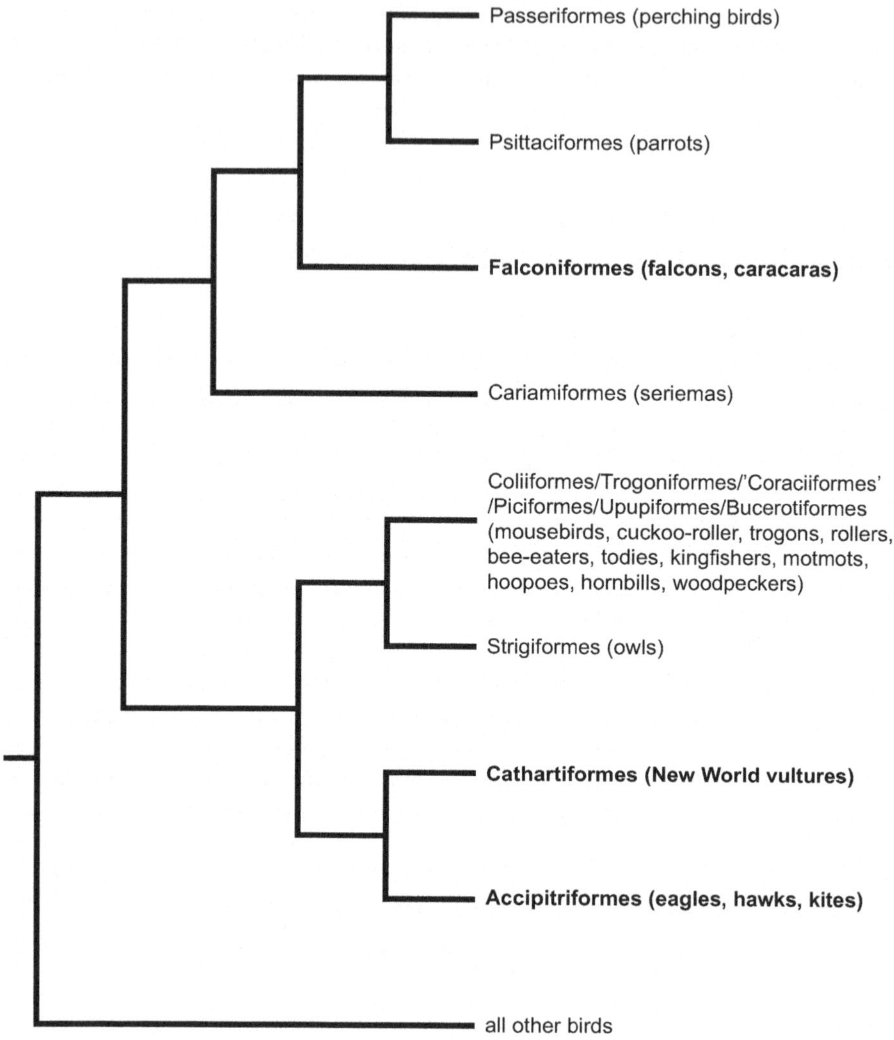

Fig. 1.1 Avian phylogeny showing placement of three diurnal raptor orders (in bold) among other birds. (see McCormack et al. (2013), Jarvis et al. (2014), Prum et al. (2015))

currently in *Milvago*, *Phalcoboenus*, and *Ibycter*. The traditional falconid taxonomy, with the changes noted above, does yield genera of similar age.

Falconiformes are estimated to be about 60 MY old, and this is similar to estimated ages for other orders, including Cathartiformes and Accipitriformes (Jarvis et al. 2014; Prum et al. 2015; Table 1.1). Regarding the naming of clades within Falconiformes, we suggest recognizing families Falconidae (falcons, caracaras, *Polihierax* and *Neohierax*) and Herpetotheridae (laughing falcon and forest falcons) with respective age estimates of 24 MY each (Fuchs et al. 2015). This would provide

Table 1.1 Classification for diurnal raptor orders Falconiformes, Accipitriformes, and Cathartiformes

Order Cathartiformes* (60[1]/58[2] mya)
Family Cathartidae, new world vultures (7 species; 60[1]/58[2] mya)
***Cathartes*:** Turkey vulture, *C. aura*; lesser yellow-headed vulture, *C. burrovianus*;
Greater yellow-headed vulture, *C. melambrotus*
***Coragyps*:** Black vulture, *C. atratus*
***Sarcoramphus*:** King vulture, *S. papa*
***Gymnogyps*:** California condor, *G. californianus*
***Vultur*:** Andean condor, *V. gryphus*
Order Accipitriformes (60[1]/58[2])
Family Sagittariidae, secretarybird (1 species; 40[2] mya)
***Sagittarius*:** Secretarybird, *S. serpentarius*
Family Pandionidae, ospreys (2 species; 27[2] mya)
***Pandion*:** Western osprey, *P. haliaetus*; eastern osprey, *P. cristatus*
Family Accipitridae, kites, hawks, and eagles (255 species; 27[2] mya)
***Elanus*:** Black-winged kite, *E. caeruleus*; black-shouldered kite, *E. axillaris*;
White-tailed kite, *E. leucurus*; letter-winged kite, *E. scriptus*
***Gampsonyx*:** Pearl kite, *G. swainsonii*
***Chelictinia*:** Scissor-tailed kite, *C. riocourii*
***Polyboroides*:** African harrier-hawk, *P. typus*; Madagascan harrier-hawk, *P. radiatus*
***Gypohierax*:** Palm-nut vulture, *G. angolensis*
***Gypaetus*:** Bearded vulture, *G. barbatus*
***Neophron*:** Egyptian vulture, *N. percnopterus*
***Eutriorchis*:** Madagascan serpent eagle, *E. astur*
***Leptodon*:** Grey-headed kite, *L. cayanensis*; white-collared kite, *L. forbesi*
***Chondrohierax*:** Hook-billed kite, *C. uncinatus*; Cuban kite, *C. wilsonii*
***Pernis*:** European honey buzzard, *P. apivorus*; crested honey buzzard, *P. ptilorhynchus*;
Barred honey buzzard, *P. celebensis*; Philippine honey buzzard, *P. steerei*
***Elanoides*:** Swallow-tailed kite, *E. forficatus*
***Lophoictinia*:** Square-tailed kite, *L. isura*
***Hamirostra*:** Black-breasted buzzard, *H. melanosternon*
***Aviceda*:** African cuckoo-hawk, *A. cuculoides*; Madagascan cuckoo-hawk, *A. madagascariensis*;
Jerdon's Baza, *A. jerdoni*; Pacific Baza, *A. subcristata*; black Baza, *A. leuphotes*
***Henicopernis*:** Long-tailed honey buzzard, *H. longicauda*; black honey buzzard, *H. infuscatus*
***Necrosyrtes*:** Hooded vulture, *N. monachus*
***Gyps*:** White-backed vulture, *G. africanus*; white-rumped vulture, *G. bengalensis*;
Indian vulture, *G. indicus*; slender-billed vulture, *G. tenuirostris*; Rüppell's vulture, *G. rueppelli*;
Himalayan vulture, *G. himalayensis*; griffon vulture, *G. fulvus*; cape vulture, *G. coprotheres*
***Sarcogyps*:** Red-headed vulture, *S. calvus*
***Trigonoceps*:** White-headed vulture, *T. occipitalis*
***Aegypius*:** Cinereous vulture, *A. monachus*
***Torgos*:** Lappet-faced vulture, *T. tracheliotos*
***Spilornis*:** Crested serpent eagle, *S. cheela*; great Nicobar serpent eagle, *S. klossi*;
Mountain serpent eagle, *S. kinabaluensis*; Sulawesi serpent eagle, *S. rufipectus*;

(continued)

Table 1.1 (continued)

Philippine serpent eagle, *S. holospilus*; Andaman serpent eagle, *S. elgini*
Pithecophaga**:** Philippine eagle, *P. jefferyi*
Circaetus**:** Short-toed snake eagle, *C. gallicus*; Beaudouin's snake eagle, *C. beaudouini*;
Black-chested snake eagle, *C. pectoralis*; brown snake eagle, *C. cinereus*;
Southern-banded snake eagle, *C. fasciolatus*; western banded snake eagle, *C. cinerascens*;
Congo serpent eagle, *C. spectabilis*
Terathopius**:** Bateleur, *T. ecaudatus*
Macheiramphus**:** Bat hawk, *M. alcinus*
Harpyopsis**:** Papuan eagle, *H. novaeguineae*
Morphnus**:** Crested eagle, *M. guianensis*
Harpia**:** Harpy eagle, *H. harpyja*
Nisaetus**:** Changeable hawk-eagle, *N. cirrhatus*; Flores hawk-eagle, *N. floris*;
Mountain hawk-eagle, *N. nipalensis*; Legge's hawk-eagle, *N. kelaarti*;
Blyth's hawk-eagle, *N. alboniger*; Javan hawk-eagle, *N. bartelsi*;
Sulawesi hawk-eagle, *N. lanceolatus*; Philippine hawk-eagle, *N. philippensis*;
Pinsker's hawk-eagle, *N. pinskeri*; Wallace's hawk-eagle, *N. nanus*
Spizaetus**:** Black hawk-eagle, *S. tyrannus*; black-and-white hawk-eagle, *S. melanoleucus*;
Ornate hawk-eagle, *S. ornatus*; black-and-chestnut hawk-eagle, *S. isidori*
Stephanoaetus**:** Crowned eagle, *S. coronatus*
Lophotriorchis: Rufous-bellied eagle, *L. kienerii*
Polemaetus: Martial eagle, *P. bellicosus*
Aquila: Tawny eagle, *A. rapax*; Steppe eagle, *A. nipalensis*; Spanish Imperial eagle, *A. adalberti*;
Eastern imperial eagle, *A. heliaca*; Gurney's eagle, *A. gurneyi*; Golden eagle, *A. chrysaetos*;
Wedge-tailed eagle, *A. audax*; Verreaux's eagle, *A. verreauxii*; Cassin's eagle, *A. africana*;
Bonelli's eagle, *A. fasciata*; Bonaparte's eagle, *A. spilogaster*
Hieraaetus: Wahlberg's eagle, *H. wahlbergi*; Booted eagle, *H. pennatus*;
Little eagle, *H. morphnoides*; Pygmy eagle, *H. weiskei*; Ayres's eagle, *H. ayresii*
Lophaetus: Long-crested eagle, *L. occipitalis*
Ictinaetus: Black eagle, *I. malaiensis*
Clanga: Lesser spotted eagle, *C. pomarina;* Indian spotted eagle, *C. hastata;*
Greater spotted eagle, *C. clanga*
Harpagus: Double-toothed Kite, *H. bidentatus*; Rufous-thighed Kite, *H. diodon*
Kaupifalco: Lizard buzzard, *K. monogrammicus*
Micronisus: Gabar goshawk, *M. gabar*
Melierax: Dark chanting goshawk, *M. metabates*;
Eastern chanting goshawk, *M. poliopterus*; Pale chanting goshawk, *M. canorus*
Urotriorchis: Long-tailed hawk, *Urotriorchis macrourus*
Erythrotriorchis: Chestnut-shouldered goshawk, *E. buergersi*;
Red goshawk, *E. radiatus*
Megatriorchis: Doria's goshawk, *Megatriorchis doriae*
Accipiter: Tiny hawk, *A. superciliosus*; Semicollared hawk, *A. collaris*;
Crested goshawk, *A. trivirgatus*; Sulawesi goshawk, *A. griseiceps*;

(continued)

Table 1.1 (continued)

Grey-bellied hawk, *A. poliogaster;* Red-chested goshawk, *A. toussenelii*;
African goshawk, *A. tachiro*; Chestnut-flanked Sparrowhawk, *A. castanilius*;
Shikra, *A. badius*; Nicobar sparrowhawk, *A. butleri*; Levant sparrowhawk, *A. brevipes*;
Chinese sparrowhawk, *A. soloensis*; Frances's sparrowhawk, *A. francesiae*;
Spot-tailed sparrowhawk, *A. trinotatus*; Grey goshawk, *A. novaehollandiae*;
Variable goshawk, *A. hiogaster*; Brown goshawk, *A. fasciatus*;
Black-mantled goshawk, *A. melanochlamys*; Pied goshawk, *A. albogularis*;
White-bellied goshawk, *A. haplochrous*; Fiji goshawk, *A. rufitorques*;
Moluccan goshawk, *A. henicogrammus*; Slaty-mantled goshawk, *A. luteoschistaceus*;
Imitator goshawk, *A. imitator*; Grey-headed goshawk, *A. poliocephalus*;
New Britain goshawk, *A. princeps*; Red-thighed Sparrowhawk, *A. erythropus*;
Little sparrowhawk, *A. minullus*; Japanese sparrowhawk, *A. gularis*; Besra, *A. virgatus*;
Dwarf sparrowhawk, *A. nanus*; Rufous-necked sparrowhawk, *A. erythrauchen*;
Collared sparrowhawk, *A. cirrocephalus*; Bismarck sparrowhawk*, *A. brachyurus*;
Vinous-breasted sparrowhawk, *A. rhodogaster*; Madagascan sparrowhawk, *A. madagascariensis*;
Ovambo sparrowhawk, *A. ovampensis*; Eurasian sparrowhawk, *A. nisus*;
Rufous-breasted sparrowhawk, *A. rufiventris*; Sharp-shinned hawk, *A. striatus*;
White-breasted hawk, *A. chionogaster*; Plain-breasted hawk, *A. ventralis*;
Rufous-thighed hawk, *A. erythronemius*; Cooper's hawk, *A. cooperii*;
Gundlach's hawk, *A. gundlachi*; Bicolored hawk, *A. bicolor*; Chilean hawk, *A. chilensis*;
Black sparrowhawk, *A. melanoleucus*; Henst's goshawk, *A. henstii*;
Northern goshawk, *A. gentilis*; Meyer's goshawk, *A. meyerianus*
Circus: **Western** marsh harrier, *C. aeruginosus*; Eastern marsh harrier, *C. spilonotus*;
Papuan harrier, *C. spilothorax*; Swamp harrier, *C. approximans*;
African marsh harrier, *C. ranivorus*; Reunion harrier, *C. maillardi*;
Malagasy harrier, *C. macrosceles*; Long-winged harrier, *C. buffoni*;
Spotted harrier, *C. assimilis*; Black harrier, *C. maurus*; Hen harrier, *C. cyaneus*;
Northern harrier, *C. hudsonius*; Cinereous harrier, *C. cinereus*; Pallid harrier, *C. macrourus*;
Pied harrier, *C. melanoleucos*; Montagu's harrier, *C. pygargus*
Milvus: Red kite, *M. milvus*; Black kite, *M. migrans*;Yellow-billed kite, *M. aegyptius*;
Haliastur: Whistling kite, *H. sphenurus*; Brahminy kite, *H. indus*
Haliaeetus: White-bellied Sea eagle, *H. leucogaster*; Sanford's Sea eagle, *H. sanfordi*;
African fish eagle, *H. vocifer*; Madagascan fish eagle, *H. vociferoides*;
Pallas's fish eagle, *H. leucoryphus*;White-tailed eagle, *H. albicilla*; Bald eagle, *H. leucocephalus*;
Steller's sea eagle, *H. pelagicus*; Lesser fish eagle, *H. humilis*;
Grey-headed fish eagle, *H. ichthyaetus*
Butastur: Grasshopper buzzard, *B. rufipennis*; White-eyed buzzard, *B. teesa*;
Rufous-winged buzzard, *B. liventer*; Grey-faced buzzard, *B. indicus*;
Ictinia: Mississippi kite, *I. mississippiensis*; Plumbeous Kite, *I. plumbea*
Busarellus: Black-collared hawk, *Busarellus nigricollis*
Rostrhamus: Snail kite, *Rostrhamus sociabilis*
Helicolestes: Slender-billed Kite, *Helicolestes hamatus*

(continued)

Table 1.1 (continued)

Geranospiza: Crane hawk, *Geranospiza caerulescens*
Cryptoleucopteryx: Plumbeous hawk, *Cryptoleucopteryx plumbea*
Buteogallus: Slate-colored hawk, *B. schistaceus*; common black hawk, *B. anthracinus*;
Cuban black hawk, *B. gundlachii*; rufous crab hawk, *B. aequinoctialis*;
Savanna hawk, *B. meridionalis*; white-necked hawk, *B. lacernulatus*;
Great black hawk, *B. urubitinga*; solitary eagle, *B. solitarius*; Chaco eagle, *B. coronatus*
***Morphnarchus*:** Barred hawk, *Morphnarchus princeps*
***Rupornis*:** Roadside hawk, *Rupornis magnirostris*
***Parabuteo*:** Harris's hawk, *P. unicinctus*; white-rumped hawk, *P. leucorrhous*
***Geranoaetus*:** White-tailed hawk, *G. albicaudatus*; variable hawk, *G. polyosoma*;
Black-chested buzzard-eagle, *G. melanoleucus*
***Pseudastur*:** Mantled hawk, *P. polionotus*; white hawk, *P. albicollis;*
Grey-backed hawk, *P. occidentalis*
***Leucopternis*:** Semiplumbeous hawk, *L. semiplumbeus*; black-faced hawk, *L. melanops*;
White-browed hawk, *L. kuhli*
***Buteo*:** Grey hawk, *B. plagiatus*; grey-lined hawk, *B. nitidus*;
Red-shouldered hawk, *B. lineatus*; Ridgway's hawk, *B. ridgwayi*;
Broad-winged hawk, *B. platypterus*; white-throated hawk, *B. albigula*;
Short-tailed hawk, *B. brachyurus*; Hawaiian hawk, *B. solitarius*; Swainson's hawk, *B. swainsoni*;
Galapagos hawk, *B. galapagoensis*; zone-tailed hawk, *B. albonotatus*;
Red-tailed hawk, *B. jamaicensis*; rufous-tailed hawk, *B. ventralis*; ferruginous hawk, *B. regalis*;
Rough-legged buzzard, *B. lagopus*; upland buzzard, *B. hemilasius*; eastern buzzard, *B. japonicus*;
Himalayan buzzard, *B. burmanicus*; long-legged buzzard, *B. rufinus*;
Cape Verde buzzard, *B. bannermani*; Socotra buzzard, *B. socotraensis*;
Common buzzard, *B. buteo*; forest buzzard, *B. trizonatus*; mountain buzzard, *B. oreophilus*;
Archer's buzzard, *B. archeri*; red-necked buzzard, *B. auguralis*;
Madagascar buzzard, *B. brachypterus*; augur buzzard, *B. augur*; jackal buzzard, *B. rufofuscus*
Order Falconiformes (60[1]/58[2] mya)
Family Herpetotheridae,* laughing and forest falcons (8 species; 30.2 mya)
***Herpetotheres*:** Laughing falcon, *H. cachinnans*
***Micrastur*:** Barred forest falcon, *M. ruficollis*; Plumbeous forest falcon, *M. plumbeus*;
Lined forest falcon, *M. gilvicollis*; cryptic forest falcon, *M. mintoni*;
Slaty-backed forest falcon, *M. mirandollei*; collared forest falcon, *M. semitorquatus*;
Buckley's forest falcon, *M. buckleyi*
Family Falconidae, caracaras, falcons (66 species; 30.2 mya)
***Daptrius*:** Black caracara, *D. ater*; red-throated caracara, *D. americanus**;
Carunculated caracara, *D. carunculatus**; mountain caracara, *D. megalopterus**;
White-throated caracara, *D. albogularis**; striated caracara, *D. australis**;
Yellow-headed caracara, *D. chimachima**; chimango caracara, *D. chimango**
***Caracara*:** Northern crested caracara, *C. cheriway*;
Southern crested caracara, *C. plancus*
***Spiziapteryx*:** Spot-winged falconet, *S. circumcincta*

(continued)

Table 1.1 (continued)

***Polihierax*:** Pygmy falcon, *P. semitorquatus*
***Neohierax*:** White-rumped falcon, *N. insignis**
***Microhierax*:** Collared falconet, *M. caerulescens*; black-thighed falconet, *M. fringillarius*;
White-fronted falconet, *M. latifrons*; Philippine falconet, *M. erythrogenys*;
Pied falconet, *M. melanoleucos*
***Falco*:** Lesser kestrel, *F. naumanni*; common kestrel, *F. tinnunculus*; rock kestrel, *F. rupicolus*;
Malagasy kestrel, *F. newtoni*; Mauritius kestrel, *F. punctatus*; Seychelles kestrel, *F. araeus*;
Spotted kestrel, *F. moluccensis*; nankeen kestrel, *F. cenchroides*;
American kestrel, *F. sparverius*; greater kestrel, *F. rupicoloides*; fox kestrel, *F. alopex*;
Grey kestrel, *F. ardosiaceus*; Dickinson's kestrel, *F. dickinsoni*; banded kestrel, *F. zoniventris*;
Red-necked falcon, *F. chicquera*; red-footed falcon, *F. vespertinus*; Amur falcon, *F. amurensis*;
Eleonora's falcon, *F. eleonorae*; sooty falcon, *F. concolor*; aplomado falcon, *F. femoralis*;
American Merlin*, *F. columbarius*; Eurasian Merlin*, *F. aesalon**; bat falcon, *F. rufigularis*;
Orange-breasted falcon, *F. deiroleucus*; Eurasian hobby, *F. subbuteo*; African hobby, *F. cuvierii*;
Oriental hobby, *F. severus*; Australian hobby, *F. longipennis*;
New Zealand falcon, *F. novaeseelandiae*; brown falcon, *F. berigora*; grey falcon, *F. hypoleucos*;
Black falcon, *F. subniger*; lanner falcon, *F. biarmicus*; laggar falcon, *F. jugger*;
Saker falcon, *F. cherrug*; gyrfalcon, *F. rusticolus*; prairie falcon, *F. mexicanus*;
Peregrine falcon, *F. peregrinus*; barbary falcon, *F. pelegrinoides*; Taita falcon, *F. fasciinucha*

Differences relative to Gill and Donsker (2017) are based on recent analyses (see text) and are marked by asterisks (*). Age estimates in millions of years ago (mya) are given for orders and families; superscript numbers denote references: [1]Jarvis et al. 2014; [2]Prum et al. 2015

family-level taxa within the broad range of recent estimates for other birds (e.g., Prum et al. 2015). Point estimates for ages of Falconiformes genera range from about 2 to 19 million years old (Fuchs et al. 2015).

Cathartiformes

We have also recently published a phylogeny for all seven Cathartiformes species (Johnson et al. 2016). Two primary clades were identified: (1) black vulture (*Coragyps atratus*) together with the three *Cathartes* species (lesser yellow-headed vulture *C. burrovianus*, greater yellow-headed vulture *C. melambrotus*, and turkey vulture *C. aura*) and (2) king vulture (*Sarcoramphus papa*), California condor (*Gymnogyps californianus*), and Andean (*Vultur gryphus*) condor. Support values for taxon relationships within the two basal clades were inconsistent between analyses with the exception of black vulture being sister to a monophyletic Cathartes clade. Increased support for a yellow-headed vulture clade was recovered in the species tree analysis using additional nuclear loci.

These two primary Cathartiformes clades are estimated to have first diverged about 14 MY ago, with genera ranging in age from about 3 to 12 mya (Johnson et al. 2016). Age estimates for Cathartiformes range from about 69 MY old (Johnson et al. 2016) old to about 60 MY old (Jarvis et al. 2014; Prum et al. 2015). Ninety-five percent credibility intervals for these estimates overlap, and the latter two estimates are based on much larger datasets.

Accipitriformes

Accipitriformes, currently with 255 species and 70 genera in 3 families, are the largest diurnal raptor order and correspondingly presents the greatest challenge for systematists. No single analysis including all species with estimates for divergence dates has yet been conducted. However, a variety of species-level analyses have been conducted, and they provide compelling support for many relationships (see Riesing et al. 2003; Gamauf and Haring 2004; Wink and Sauer-Gürth 2004; Helbig et al. 2005; Lerner and Mindell 2005; Griffiths et al. 2007; Lerner et al. 2008, 2017; Breman et al. 2013; Barrowclough et al. 2014; Oatley et al. 2015).

One of the most striking findings of the analyses above is non-monophyly for traditional subfamilies (e.g., Peters 1931; Jollie 1977; Amadon and Bull 1988) and many of the traditional genera including: *Buteo*, *Leucopternis*, *Aquila*, *Hieraaetus*, *Spizaetus*, *Accipiter*, and *Circaetus*. The findings of non-monophyly for genera have been addressed to some degree in recent changes to the IOC listing (Gill and Donsker 2017). However, a clear understanding of the relationships among Accipitriformes genera and their timing of diversification is still lacking as most of these studies focus on particular genera or subfamilies.

We provide new analyses here, compiling published sequence data (GenBank; ncbi.nlm.nih.gov) from the studies referenced above and making a supermatrix with sequence data for ten selected loci (mitochondrial ND1, ND2, CO1, Cytochrome b, nuclear MB intron-2, FGB intron-5, FGB intron-7, TGFb2 intron-5, c-myc, and RAG1). The use of supermatrices, combining data from multiple individuals for a taxon, requires careful vetting of data and some empty (missing) data cells. However, its phylogenetic utility is well established (see Philippe et al. 2017). We conducted preliminary analyses to detect low-quality sequences (e.g., having uncertainty codes or insertion-deletion events in protein-coding genes that are not multiples of three), misidentifications, or contaminations. When several sequences of equal quality were available for the same taxon, we retained the longest ones. Whenever possible, we used sequences that were derived from the same individual. The best fitting models for each locus were selected using the BIC criterion, as implemented in TOPALi (Milne et al. 2009). The concatenated supermatrix was analyzed using Beast (Drummond et al. 2012), assuming a strict molecular clock. All loci were allowed to have their own molecular clock and substitution models.

Divergence times were estimated using three independent calibration points. The first is the fossil *Circaetus rhodopensis* (Boev 2012), which represents the first appearance of the genus *Circaetus* in the fossil record. This fossil was dated to the Late Miocene (Turolian = Late Meotian), ca 7.5 mya. This fossil was used to calibrate the split between the genera *Circaetus* and *Terathopius*. We modeled this calibration point using a lognormal distribution (stdev = 0.25 offset = 7.25). The second is the fossil *Mioneophron longirostris* (Li et al. 2016) for the first appearance of the Gypaetinae in the fossil record. The fossil was dated to the Upper Miocene Liushu Formation, Linxia Basin, and was used to calibrate the split between *Polyboroides* and the three Gypaetinae genera *(Gypaetus, Neophron, Gypohierax)*. This calibration point was also modeled using a lognormal distribution (stdev = 0.25 offset = 7.25). The third is the fossil *Aegypius varswaterensis* (Manegold et al. 2014) for the first appearance of the genus *Aegypius* in the fossil record. This fossil was dated to the Early Pliocene of the Upper Varswater Formation at Langebaanweg, South Africa. This fossil was used to calibrate the split between *Aegypius* and its sister genus *Torgos*. We modeled this calibration point using a lognormal distribution (stdev = 0.25 offset = 3.6).

We ran three independent analyses for 100 million iterations on the CIPRES 3.1 gateway server (www.cipres.org; Miller et al. 2010) before combining the log and tree files. We used TRACER v1.7 (Rambaut et al. 2014) to ensure that our effective sample size for all Bayesian analyses of the underlying posterior distribution were large enough (>200) for meaningful estimation of parameters.

Three Accipitriformes families (Accipitridae, Pandionidae, and Sagittariidae) have long been in use (Peters 1931). Sagittariidae has been estimated to be about 40 million years old, and Accipitridae and Pandionidae have been estimated to be about 27 mya (Prum et al. 2015). These ages fall within the range estimated for other diurnal raptor family ages, including Falconidae and Herpetotheridae, and within the range for many other avian families recently estimated in Jarvis et al. 2014 and Prum et al. 2015. However, that range is broad (50–10 MYA), and estimates are still based on only a limited sampling of confamilial taxa. Our analyses of Accipitriformes (Fig. 1.2a) did not include non-Accipitriformes outgroups, and we consider the sampling of multiple related families and orders to provide better family-level age estimates.

Our analyses combining published DNA sequences data into a new supermatrix provide new phylogenetic and temporal hypotheses for several clades and suggest that divergences within Accipitridae began around the Eocene/Oligocene boundary about 34 mya, with the split of the *Elanus/Gampsonyx* clade from the other Accipitridae genera, as recovered by Lerner and Mindell (2005), Griffiths et al. (2007), and Barrowclough et al. (2014), with many of the subfamily-level lineages having appeared in the mid-Miocene (Fig. 1.2a–d). Point estimates for ages of Accipitridae genera range from about 2 to 20 million years old.

Peters (1931) recognized eight Accipitridae subfamilies (Buteoninae, Circinae, Accipitrinae, Circaetinae, Milvinae, Perninae, Aegypiinae, Elaninae), although, as noted above, all have been found to be polyphyletic to some degree. As our understanding of the phylogeny and age of Accipitridae genera is improving, it makes sense to reconsider those traditional subfamily names, because they have been, and can continue to be, useful in discussing the macroevolution and ecology of more inclusive groups (groups of genera). For this reason, we include subfamily names

a

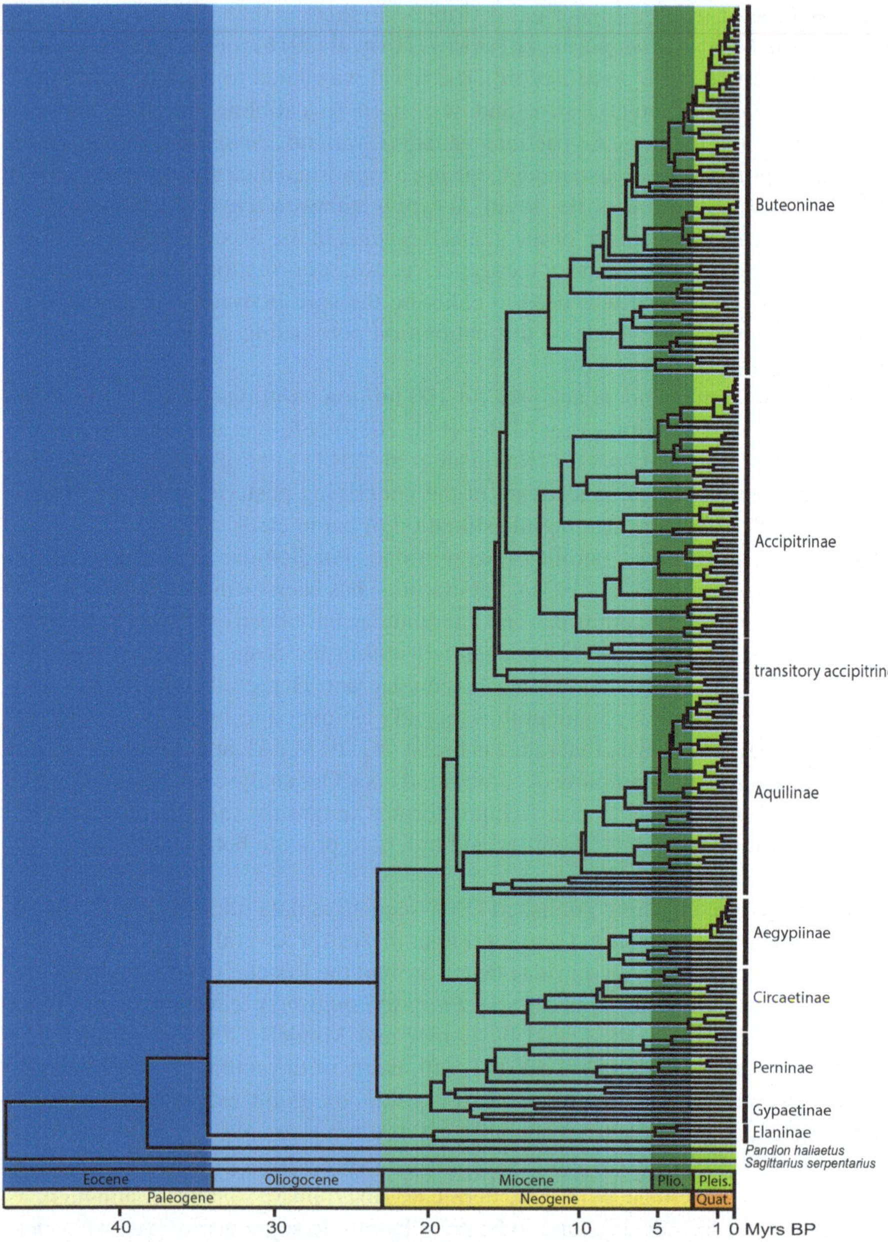

Fig. 1.2 Phylogeny for Accipitriformes based on newly assembled datasets from published DNA sequences, for: (**a**) all Accipitriformes study taxa; (**b**) Accipitrinae and relatives; (**c**) Aquilinae, Aegypiinae, Circaetinae, Perninae, Gypaetinae, *Elaninae*, and relatives; and (**d**) Buteoninae and relatives. To help visualize non-monophyly in (**b**) *Accipiter* species are shown in red; (**c**) *Aquila* species in red, *Circaetus* in blue, and *Aviceda* in green. * indicates posterior probability of 0.95–0.99. ** indicates posterior probability of 1.0. † designates the location of the three calibration points used in estimating divergence times in (**a**)

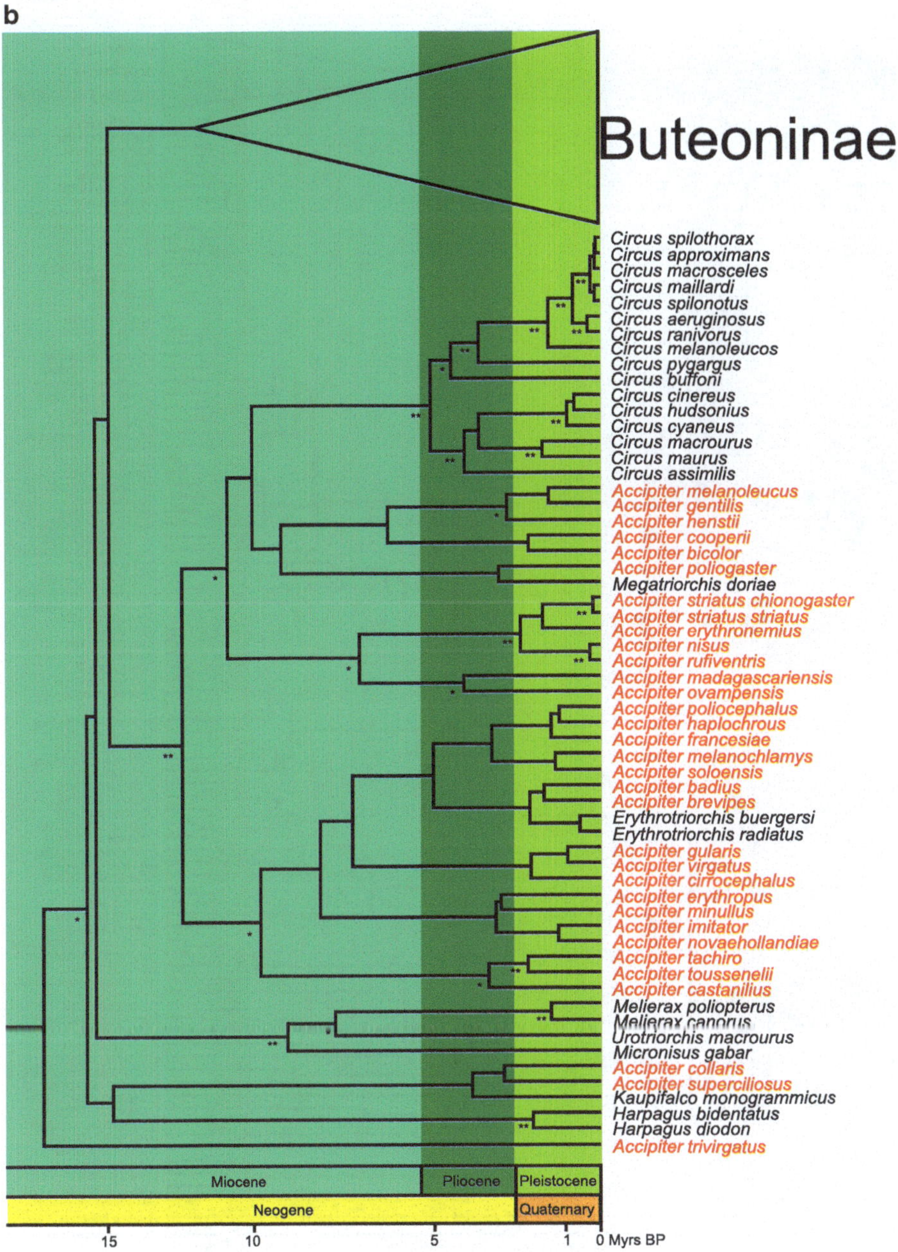

Fig. 1.2 (continued)

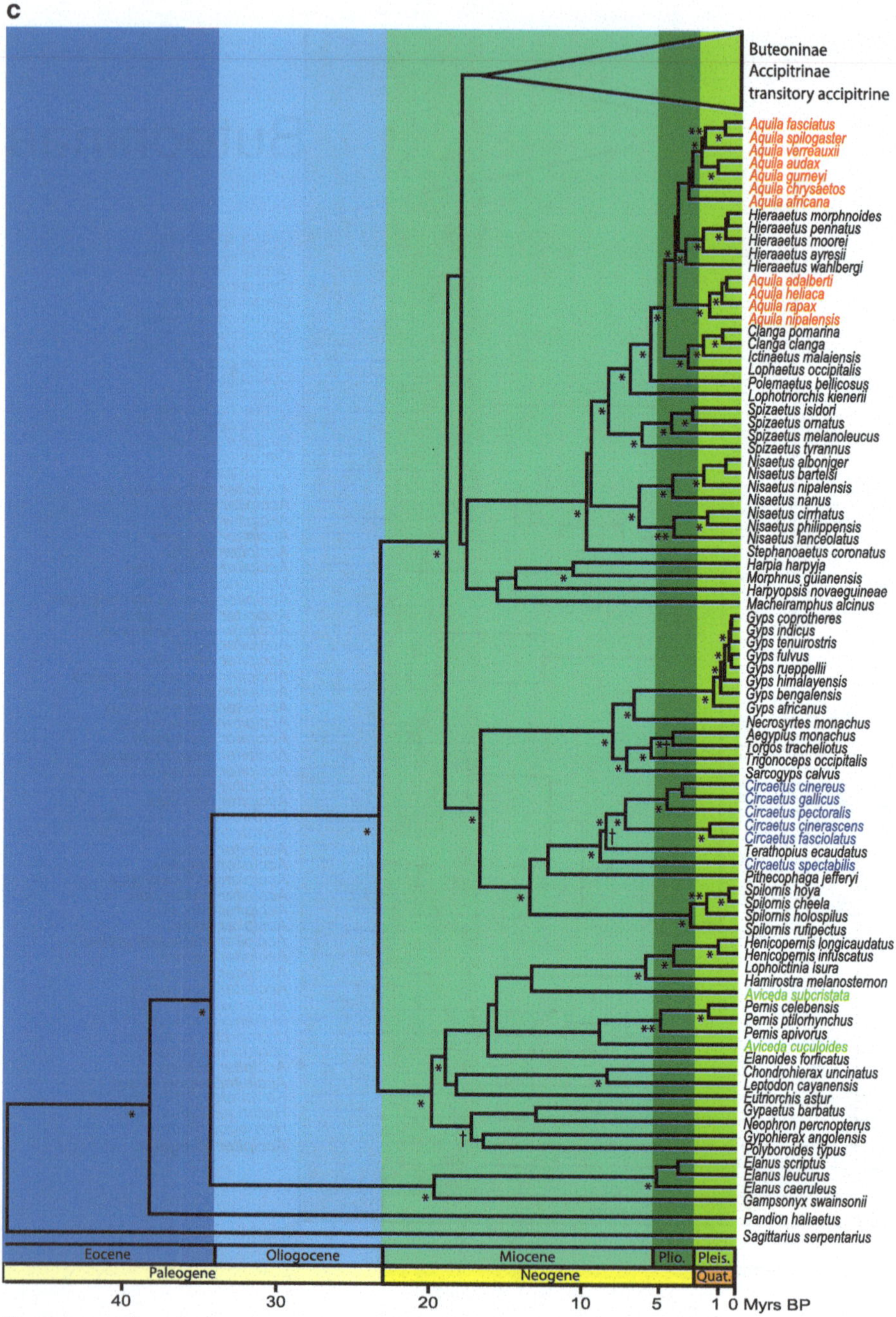

Fig. 1.2 (continued)

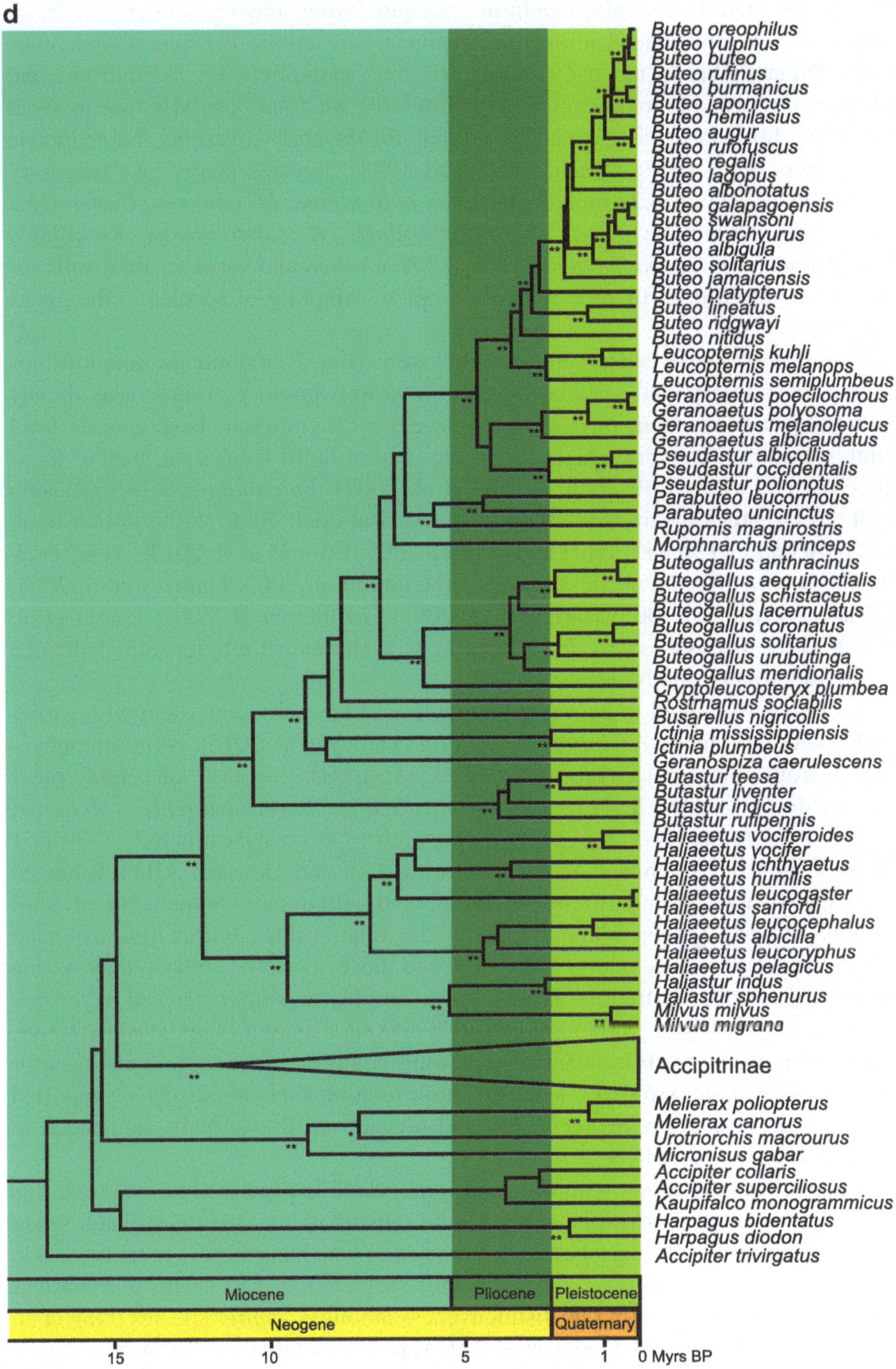

Fig. 1.2 (continued)

for relevant clades of similar age in the Accipitriformes phylogenies (Fig. 1.2a–d; Buteoninae, Accipitrinae, Aquilinae, Aegypiinae, Circaetinae, Perninae, Gypaetinae, and Elaninae). We do not use Circinae, to avoid polyphyly for Accipitrinae and because of its comparatively young age. Similarly, we do not use Milvinae to avoid polyphyly for Buteoninae. We add Aquilinae for the early-diverging, booted eagle group (see Jollie 1977; Amadon and Bull 1988). Species placed as "transitory Accipitrinae" in Fig. 1.2a include *Melierax poliopterus*, *M. canorus*, *Urotriorchis macrourus*, *Mirconisus gabar*, *Accipiter collaris*, *A. superciliosus*, *Kaupifalco monogrammicus*, *Harpagus bidentatus*, and *H. diodon*, and we defer their subfamily consideration until there is a more complete sampling of species in the genus *Accipiter.*

Our most inclusive Accipitriformes phylogeny (Fig. 1.2a) indicates relationships among the primary clades. Relations within various primary groups have already been subject to in-depth phylogenetic surveys with comprehensive species level sampling and several individuals per species and/or multi-locus data, such as those for *Gyps* (Johnson et al. 2006; Arshad et al. 2009), *Buteo/Leucopternis* (Riesing et al. 2003; Kruckenhauser et al. 2004; do Amaral et al. 2006, 2009; Lerner et al. 2008), *Accipiter/Circus* (Lerner and Mindell 2005; Breman et al. 2013; Oatley et al. 2015), *Aquila*, *Hieraaetus*, and relatives (Helbig et al. 2005; Gamauf et al. 2005; Lerner and Mindell 2005; Haring et al. 2007; Griffiths et al. 2007; Lerner et al. 2017). Thus, discussion here focuses on new or alternative relationships indicated by the supermatrix analyses.

Accipiter was found to be non-monophyletic (Fig. 1.2b), consistent with earlier studies based on fewer *Accipiter* species (e.g., Oatley et al. 2015). Non-monophyly stems from placement, among various *Accipiter* species, of eight other Accipitriformes genera (*Circus*, *Megatriorchis*, *Erythrotriorchis*, *Melierax*, *Urotriorchis*, *Micronisus*, *Kaupifalco*, *Harpagus*). Our analyses included 33 of 51 (65%) currently recognized *Accipiter* species (Gill and Donsker 2017). Eighteen *Accipiter* species remain to be analyzed, and additional components of non-monophyly may yet be found. Some of the relationships within this extended *Accipiter* clade are only poorly supported, and those, as well as others, may be due to phylogenetic reconstruction biases. For example, *A. poliogaster* and *M. doriae* are shown as sister taxa, although the two taxa do not share loci in the supermatrix. For these reasons, we refrain from suggesting genus-level taxonomic changes for the *Accipiter* clade. However, some of these relationships are strongly supported (e.g., *Circus* nested within *Accipiter*), and revisions will eventually be needed to retain *Accipiter* monophyly.

Although relationships within the *Accipiter* clade require further analyses, it is clear that at least two *Accipiter* lineages lie outside of the *Accipiter* clade, sensu stricto. One is the crested goshawk (*A. trivirgatus*) which appears as sister to a clade uniting all other Accipitrinae and Buteoninae species (Fig. 1.2b). This is consistent with other studies indicating its distinctiveness among *Accipiter* species (Ong et al. 2011; Breman et al. 2013; Oatley et al. 2015) and the fact that it is the only *Accipiter* species with a crest. Its early diverging position and phylogenetic distinctiveness suggest that *A. trivirgatus*, together with *A. griseiceps* (see Breman et al. 2013), merits genus name revision.

The second *Accipiter* lineage that is not directly related to the larger *Accipiter* clade in our analysis consisted of the semicollared hawk (*A. collaris*) and tiny hawk (*A. superciliosus*) as sister species. Both are from South America, and they comprise the second earliest diverging group (after *A. trivirgatus*) among *Accipiter* species. Their sister relationship to the lizard buzzard (*Kaupifalco monogrammicus*) from Africa suggests long distance dispersal between South America and Africa, but this relationship with *Kaupifalco* is not strongly supported and further data may be needed. These molecular data, coupled with morphological data (Olson 2006), would support the recognition of the genus *Hieraspiza* for the semicollared hawk (*A. collaris)* and tiny hawk (*A. superciliosus)* as noted by Olson (2006).

Aquila was also found to be non-monophyletic, due to a clade with five *Hieraaetus* species being placed among *Aquila* species (Fig. 1.2c). That placement, however, is based on a short internode and not well supported. Thus the current analysis does not provide clear evidence regarding *Aquila* monophyly or polyphyly. We also found *Aviceda* to be non-monophyletic (Fig. 1.2c). However, the relevant support levels are low, and the *Aviceda* taxa do not share loci in our supermatrix. Hence, *Aviceda* non-monophyly may stem from methodological biases, and we do not recommend taxonomic changes before a more comprehensive phylogenetic survey of this genus is made. *Circaetus* (snake-eagles) also appears as non-monophyletic (Fig. 1.2c) due to the inclusion of *Terathopius ecaudatus*, with probability at or above 0.95. Non-monophyly for *Circaetus* is also due, in part, to changing *Dryotriorchis spectabilis* to *Circaetus spectabilis* based on earlier analyses (Lerner and Mindell 2005). Reverting *Circaetus spectabilis* to *Dryotriorchis spectabilis* would effectively maintain monophyly of *Circaetus*, without changing *Terathopius.* Better resolution will require more sampling for this group.

We did not find the Madagascan serpent eagle (*Eutriorchis*) to be nested within the atypical vultures from the genera *Gypohierax*, *Gypaetus*, and *Neophron* as previously found (Lerner and Mindell 2005). Current results (Fig. 1.2c) suggest that *Eutriorchis* is more closely related to various kite genera (*Elanoides*, *Leptodon*, *Chondrohierax*, *Lophoictinia*, *Hamirostra*), honey buzzard (*Pernis*, *Henicopernis*), and cuckoo-hawks (*Aviceda*).

Several diurnal raptor clades that are found in open habitats, including all or parts of *Aquila*, *Buteo*, *Buteogallus*, *Circus*, and *Falco*, began in the periods of increased diversification approximately 5 to 7 million years ago (Fig. 1.2a–d; Fuchs et al. 2015). This period corresponds to the spread of open savannahs dominated by plants adapted to hot-dry environments (C4 plants), a pattern particularly prevalent in Africa (Cerling et al. 1997; Osborne and Beerling 2006; Edwards et al. 2010). The Late Miocene was also a time of major turnover in mammalian communities, including a rapid radiation of Ethiopian rodents (Lecompte et al. 2008). Hence, several independent clades of raptors (various Accipitriformes and Falconiformes) successfully colonized the savannah and grassland from a more forested habitat at a similar time to feed on similar prey species.

Table 1.1 provides the current classification from Gill and Donsker (2017) with some suggested taxonomic revisions. Changes relative to Gill and Donsker are designated by asterisks. Our suggestions provide taxa above the species level that are monophyletic and seek to have genera, families, and subfamilies that are broadly similar in age. We have attempted to be conservative in suggesting changes where sampling and phylogenetic support seems the most reliable. We do not suggest taxonomic revisions for genus *Accipiter* here, due to the need for more complete sampling and weak resolution for some species based on current analyses.

Geography of Species Richness and Phylogenetic Diversity

Phylogenies provide valuable models in assessing the geographic distribution of species and genetic diversity. In order to visualize the geographic distribution of species richness (SR) for all Falconiformes (64 of 64) and 81% of Accipitriformes (206 of 255) species on a global scale, we overlaid their corresponding species range maps from BirdLife International and Handbook of the Birds of the World (2016; http://datazone.birdlife.org) on the World Wildlife Fund (WWF) terrestrial eco-region shapefiles (www.worldwildlife.org/publications/terrestrial-eco-regions-of-the-world; accessed January 2017) using ArcGIS. Only geographic distributions from "Extant" and "Possibly Extant" presence categories and "Native," "Reintroduced," and "Introduced" origin categories were included in the analyses (no Accipitriformes and Falconiformes species entries were classified as "Probably Extant," and no Falconiformes species were classified as "Reintroduced" or "Introduced"). After excluding all entries identified as "Vagrant," "Origin Uncertain," and "Seasonal Occurrence Uncertain," and areas classified as "Rock and Ice" and "Lake," due to their large disjunct distributions, a total of 791 (91%) and 789 (91%) of 867 terrestrial eco-regions possessed at least one Accipitriformes or Falconiformes species, respectively, when considering all seasonal data (Resident, Breeding Season, Non-breeding Season, and Passage).

In addition to species richness, we also explored variation in Falconiformes and Accipitriformes phylogenetic diversity (PD; Faith 1992) among eco-regions. PD is taken as the sum of the total branch lengths accumulated among lineages, derived from the eight loci (ten partitions), concatenated and fossil-calibrated chronogram reported in Fuchs et al. (2015) for Falconiformes, and the final inclusive phylogeny reported here for Accipitriformes (Fig. 1.2a–d). PD can help prioritize biodiversity conservation efforts, by quantifying the evolutionary history for character change (DNA sequence change in our case) within localities. We compare SR and PD measures among eco-regions when all threatened species – based on the IUCN Red List threat status (critically endangered [CR], endangered [EN], vulnerable [VU], and near threatened [NT], including data deficient [DD]) – were excluded, to the situation when only species with Least Concern (LC) status are included. This comparison shows geographic region differences with respect to Falconiformes and Accipitriformes SR and PD measures when assuming all threatened species were extinct and allows a visual comparative measure for assessing eco-region importance in biodiversity conservation. No Falconiformes species are currently listed as data deficient (DD) or critically endangered (CR), while for Accipitriformes, one species is classified as DD (*Erythrotriorchis buergersi*) and 11 species as CR.

The highest concentrations of SR and PD for raptors occur in northern and central South America and in Central and Southern Africa, across seasons and across both threatened and non-threatened IUCN status designations (Figs. 1.3 and 1.4). This is generally consistent with broad patterns for other vertebrates (Schipper et al. 2008; Jenkins et al. 2013; Voskamp et al. 2017; see also www.biodiversitymapping.org). For Accipitriformes, the two eco-regions with the highest SR and PD measures were in Africa (Fig. 1.3), specifically the East Sudanian Savanna (SR = 59, PD = 591) and Victoria Basin forest-savanna (SR = 57, PD = 594) eco-regions, compared to Falconiformes where the two eco-regions with the highest SR and PD measures were in South America (Fig. 1.4), specifically the Eastern Cordillera Real Montane Forest (SR = 18, PD = 180) and Northern Andean páramo (SR = 18, PD = 180) eco-regions. This reflects some basic differences in the evolutionary and biogeographic histories for the two groups.

We also assessed SR and PD based on two seasonal considerations, breeding season only and all seasons combined. By considering species distributions during only the breeding season (785 and 776 eco-regions were occupied by at least 1 Accipitriformes and Falconiformes species, respectively), we can identify areas that may differ in importance depending on species' annual life cycles. However, because of the way the data was compiled, only those geographic areas for a particular species identified as non-breeding season or passage (i.e., migratory) were in areas where it was not also categorized as a resident or breeding season. Therefore, we cannot use these data to quantify overall non-breeding distributions because some species (e.g., peregrine falcon) may include non-breeding or passage distributions that overlap or augment resident or breeding season distributions.

We can, however, identify geographic areas that are important for maintaining both breeding and non-breeding or passage species when occupied at the same time. For example, increased SR is observed in Africa for both Accipitriformes and Falconiformes when including all seasonal distribution data compared to only breeding season, suggesting that African eco-regions maintain a higher number of species when both breeding and non-breeding species occupy the same area (i.e., as with overlapping distributions from Northern Hemisphere wintering and Southern Hemisphere breeding species). Alternatively, these data also suggest some differences between Africa and South America. For example, both Accipitriformes and Falconiformes species in Africa, in general, appear to occupy a wider geographic distribution of eco-regions during the non-breeding season than species in South America. Because the analyses are based on presence/absence data, we do not have the ability to account for changes in abundance within eco-regions.

SR changes more dramatically than PD when currently threatened Falconiformes and Accipitriformes species were considered extinct, and this difference is more pronounced for Falconiformes than Accipitriformes (Figs. 1.3 and 1.4). Eleven of the 15 threatened Falconiformes species are within the relatively young genus

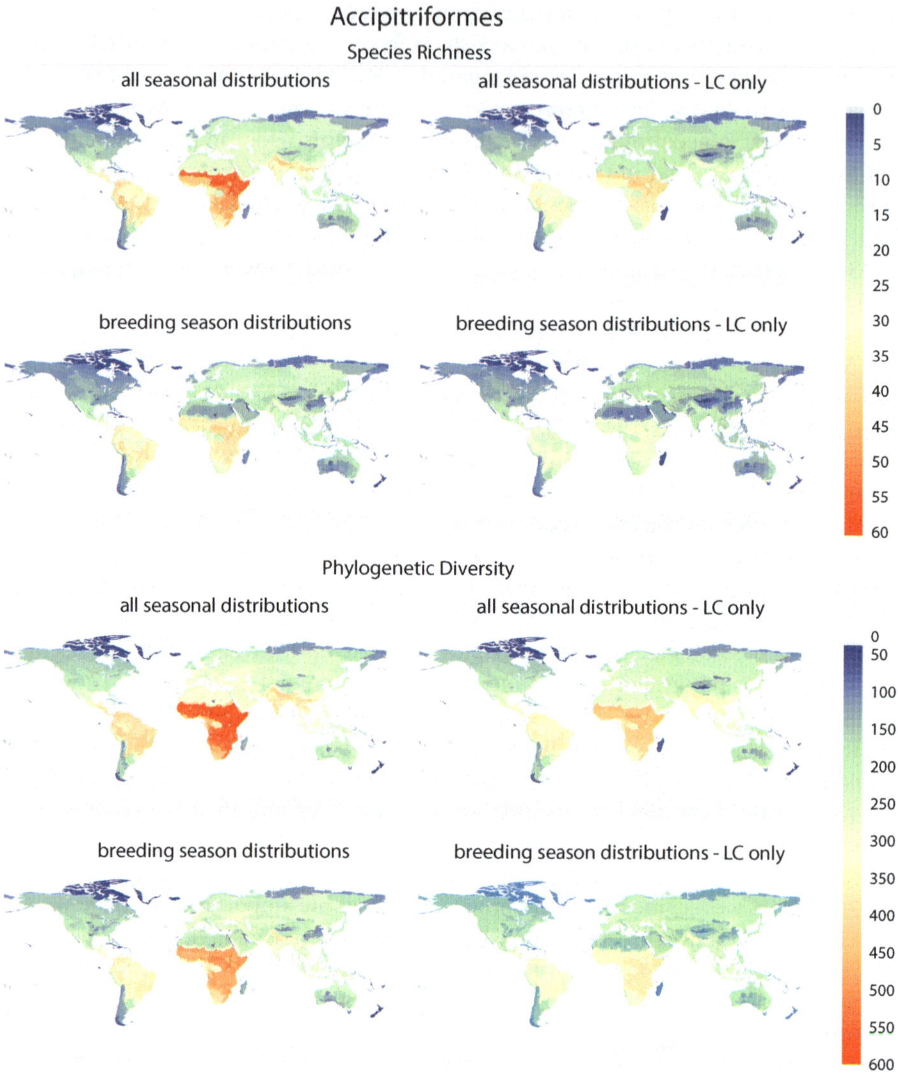

Fig. 1.3 Map showing global distribution of Accipitriformes species richness (SR) and phylogenetic diversity (PD) measures across and within eco-regions. Red denotes higher and blue denotes lower values, respectively. Separate maps show SR and PD distributions for two season designations (all combined and breeding only) and two IUCN status designations (all combined and Least Concern (LC) only). The relative contribution of species having IUCN status as threatened (CR, EN, VU) and near threatened (NT) for SR/PD measures can be assessed by comparing maps on the left with corresponding maps on the right

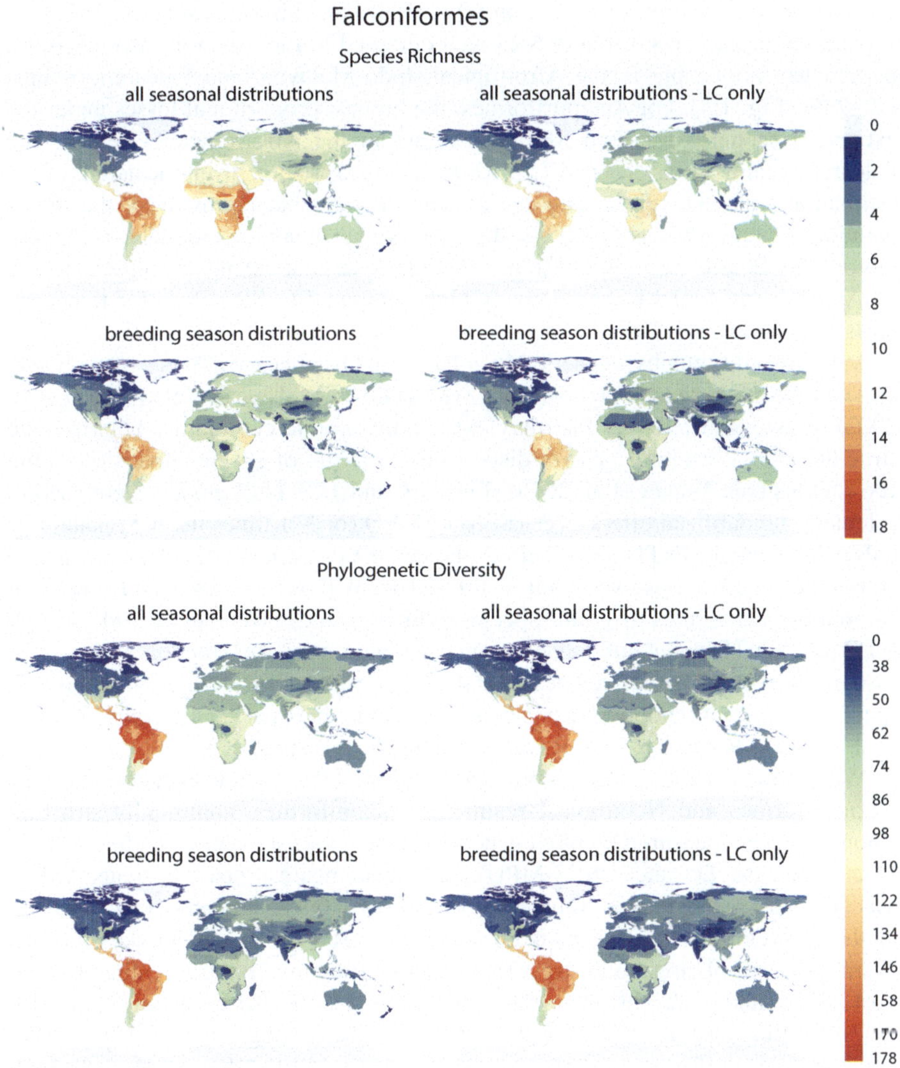

Fig. 1.4 Map showing global distribution of Falconiformes species richness (SR) and phylogenetic diversity (PD) measures across and within eco-regions. Red denotes higher and blue denotes lower values, respectively. Separate maps show SR and PD distributions for two season designations (all combined and breeding only) and two IUCN status designations (all combined and Least Concern (LC) only). The relative contribution of species having IUCN status as threatened (CR, EN, VU) and near threatened (NT) for SR/PD measures can be assessed by comparing maps on the left with corresponding maps on the right

Falco, a species-rich group including 60% of all Falconiformes species. This may explain the higher proportion of SR loss, compared to PD loss, for Falconiformes, particularly noticeable in the Afrotropical, Indo-Malayan, and Palearctic realms (22–24%; Fig. 1.5). For Accipitriformes, the highest proportional losses under the extinction scenario for both SR and PD are in the Afrotropical, Oceanic, and Palearctic realms (Figs. 1.3 and 1.5). Though not surprising, given the nature of PD estimation and relatively young age of many raptor species involved, the trends mentioned suggest SR as a more sensitive proxy in evaluating potential loss of these raptor taxa. Possible exceptions to this include desert and island ecosystems where PD values were found to be relatively high compared to SR values for terrestrial birds (Voskamp et al. 2017).

Measures of both phylogenetic Mean Pairwise Distance (MPD) between all species and the mean nearest taxon distance (MNTD) were also calculated (Webb et al. 2002) to assess differences that may exist in taxonomic community phylogenetic structure (i.e., clustering or over-dispersion) in terms of species diversity within eco-regions (see Tucker et al. 2016) (Figs. 1.6 and 1.7). Null models were used to calculate standardized effect sizes (SES) of MPD (or − 1 times the net relatedness index [NRI]) and MNTD (or − 1 times the nearest taxon index [NTI]) estimates to determine whether observed values differed from those expected under random community assemblies given the species richness values (Webb et al. 2002, 2008). MPD and MNTD mean null models were calculated from 1000 randomly generated assemblages with species richness equal to that observed for the eco-regions and species selected at random ("taxa labels" null model in R package *picante*; Kembel et al. 2010) in R version 3.13 (R Development Core Team 2013).

With the exception of eco-regions within the Afrotropical realm for Accipitriformes and Neotropical realm for Falconiformes, community structure among species, measured by phylogenetic distance with standardized effect size of Mean Pairwise Distance (SES-MPD) and mean nearest taxon distance (SES-MNTD) were not significant (quantiles $p > 0.05$ or < 0.95) indicating that nearly all of the eco-regions shared similar phylogenetic distances to null models, i.e., an overall random phylogenetic structure among eco-regions in those realms. In contrast, 79% and 44% of the 108 eco-regions in the Afrotropical realm for Accipitriformes species, and 33% and 50% of the 172 eco-regions in the Neotropical realm for Falconiformes species, possessed positive SES-MPD values with high quantiles (MPD.obs.$p > 0.95$), indicating significant over-dispersion, or phylogenetic evenness, of species co-occurring within eco-regions than expected by chance, when considering all seasonal distributions and breeding distributions only (Figs. 1.6 and 1.7), respectively. None of the standardized effect size measures of mean nearest taxon distances (SES-MNTD) were significant for Falconiformes. Similar to the SES-MPD results, however, 64% and 58% of the 108 eco-regions in the Afrotropical realm for Accipitriformes possessed positive SES-MNTD values with high quantiles (MNTD.obs.$p > 0.95$) when considering seasonal distributions and breeding distributions, respectively, further indicating significant phylogenetic evenness of Accipitriformes species within eco-regions in Africa.

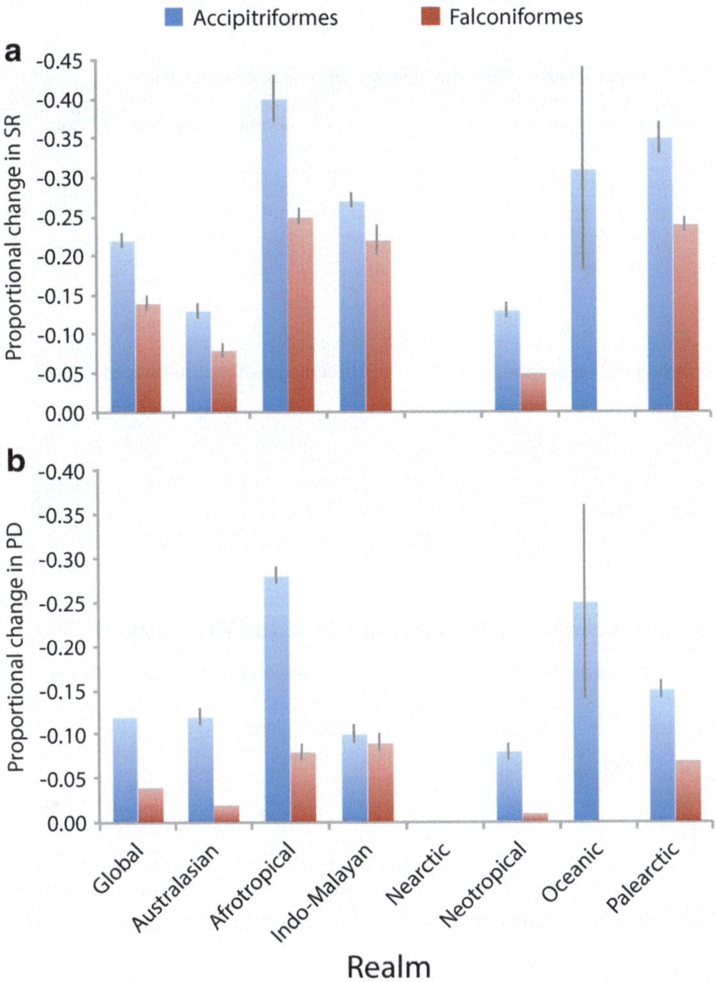

Fig. 1.5 Global and within realm proportional losses (mean ± standard error) of (**a**) species richness and (**b**) phylogenetic diversity within eco-regions for Accipitriformes and Falconiformes. Diversity changes indicate simulated consequences of loss of species currently listed as threatened (CR, EN, VU) and near threatened (NT) by the IUCN. Negative numbers indicate SR/PD loss. All Falconiformes species in the Neotropical and Oceanic realms were identified as least concern (LC)

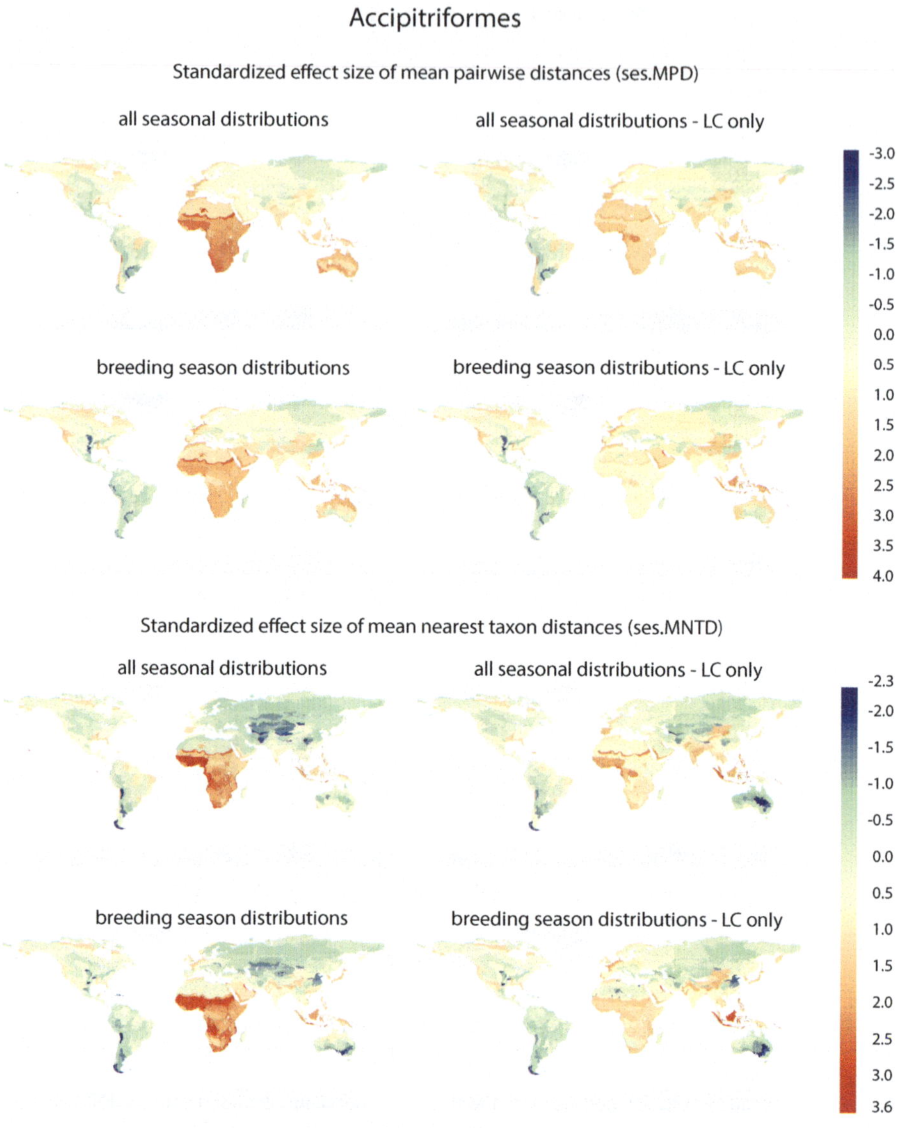

Fig. 1.6 Map showing global distribution of Accipitriformes species mean pairwise distances (MPD) and mean nearest taxon distances (MNTD) measures across and within eco-regions. Red denotes higher and blue denotes lower values, respectively. Separate maps show distributions for two season designations (all combined and breeding only) and two IUCN status designations (all combined and Least Concern (LC) only). The relative contribution of species having IUCN status as threatened (CR, EN, VU) and near threatened (NT) for MPD and MNTD measures can be assessed by comparing maps on the left with corresponding maps on the right

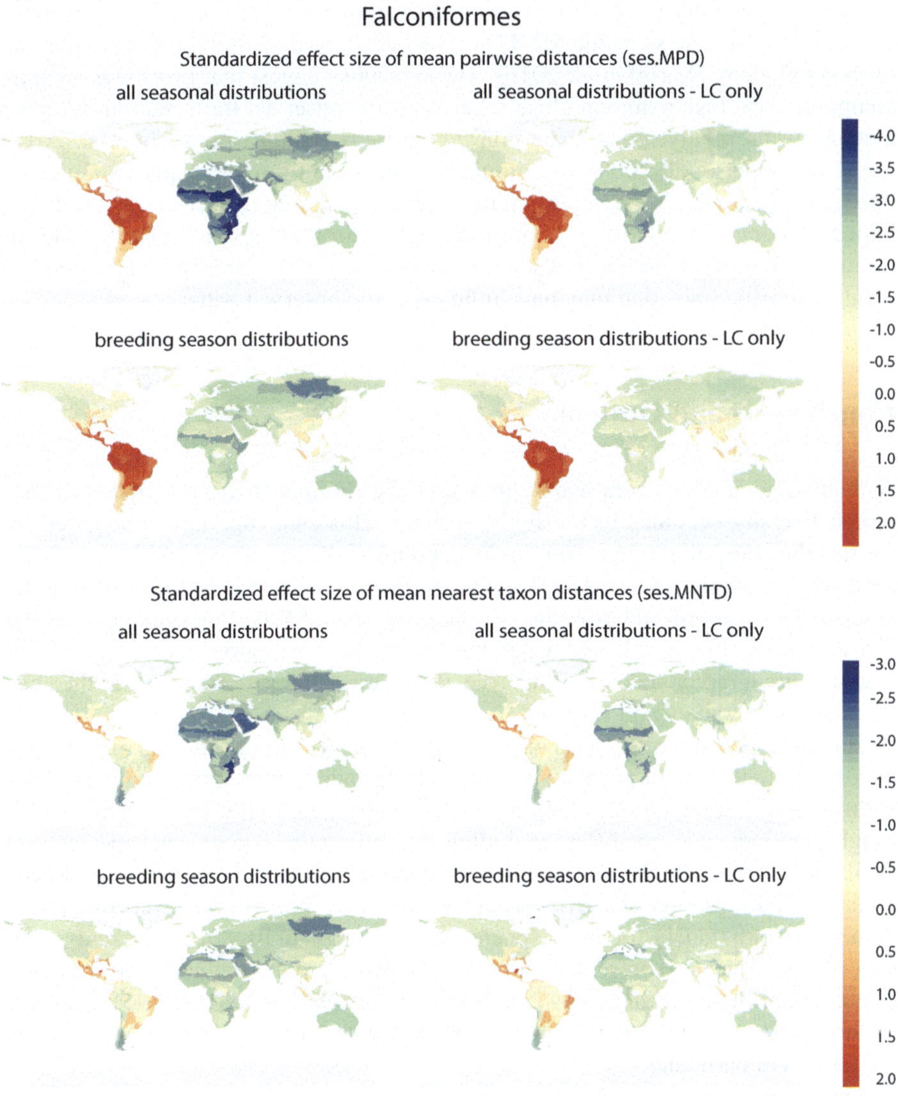

Fig. 1.7 Map showing global distribution of Falconiformes species mean pairwise distances (MPD) and mean nearest taxon distances (MNTD) measures across and within eco-regions. Red denotes higher and blue denotes lower values, respectively. Separate maps show distributions for two season designations (all combined and breeding only) and two IUCN status designations (all combined and Least Concern (LC) only). The relative contribution of species having IUCN status as threatened (CR, EN, VU) and near threatened (NT) for MPD and MNTD measures can be assessed by comparing maps on the left with corresponding maps on the right

MPD is generally more sensitive to deeper branching structure of phylogenetic clustering and evenness, while MNTD is more sensitive to patterns closer to the tips of the phylogeny (Mazel et al. 2016). These results suggest that processes such as competitive exclusion among close relatives with conserved traits, habitat filtering among distant relatives that share convergent traits (HilleRisLambers et al. 2012; see also Cooper et al. 2008), or colonization of distantly related species (Li et al. 2015) may have played an important role organizing Accipitriformes species distributions among eco-regions in Africa and Falconiformes species distributions in Central and South America (but see Gerhold et al. 2015). More work is needed to identify specific traits that may have influenced the observed pattern.

Conclusions and Outlook

As human populations continue to grow and alter Earth's environment, conserving raptor diversity becomes increasingly difficult. This challenge is compounded by the fact that, despite progress, our understanding of the extent and resilience of raptor diversity remains incomplete. How many species of diurnal raptors remain to be described or recognized? How do we diagnose them? What are their geographic distributions? How much and what kind of genetic variation exists within and among diurnal raptor species? What role, if any, does that variation have in the ability of species to adapt and survive?

Having a phylogenetic taxonomy, in which classification reflects evolutionary history, is a step in the right direction – and provides appropriate context – for addressing the research questions mentioned above. And progress is being made. Single phylogenetic analyses including all currently recognized species are available, as working hypotheses, for Falconiformes and Cathartiformes (Fuchs et al. 2015; Johnson et al. 2016). These analyses are limited, however, to a small number of individuals for most species and fewer than 10 genes per study. The supermatrix analyses presented here for Accipitriformes, the most species-rich order, represent an advance; however, there are currently no phylogenetic analyses including all extant species. Additional sampling for *Accipiter* and related genera are needed in particular.

Ideally, biodiversity and ecosystems as a whole is the target of conservation efforts. But, current conservation efforts often focus on species. This gives systematics and the recognition of species particular relevance. It also challenges us to ensure that current species designations adequately represent the existing diversity of unique forms or evolutionary lineages in nature. There are reasons to suspect that the number of bird species, including diurnal raptors, currently recognized is an underestimate (see Barrowclough et al. 2016). One of the reasons is the assumption, inherent in the biological species definition broadly applied to birds, that conspecific populations interbreed or would interbreed if their distributions overlapped. The assumption is rarely tested.

The abundance of avian polytypic species, morphologically distinguishable subspecies and geographically isolated populations, represents opportunities for studies of within species variation to reveal previously unrecognized or cryptic species. At least some recent studies of variation within diurnal raptor species have found levels of genetic differentiation generally characterizing distinct species (e.g., Johnson et al. 2006, 2007a, b). Comprehensive assessment of within species genetic variation has not been done for most diurnal raptor species and doing so would provide a more thorough accounting and better understanding of their diversity. It could also facilitate conservation efforts targeted below the species level, with the opportunity to forestall some species losses. Assessment within species of raptors most at risk should be a high priority.

References

Amadon D, Bull J (1988) Hawks and owls of the world: an annotated list of species. Proc Western Found Vert Zool 3:297–330

Arshad M, Gonzalez J, El-Sayed AA, Osborne T, Wink M (2009) Phylogeny and phylogeography of critically endangered *Gyps* species based on nuclear and mitochondrial markers. J Ornithol 150:419–430

Avise JC, Johns GC (1999) Proposal for a standardized temporal scheme of biological classification for extant species. Proc Natl Acad Sci U S A 96:7358–7363

Avise JC, Liu J-X (2011) On the temporal inconsistencies of Linnean taxonomic ranks. Biol J Linn Soc 102:707–714

Barrowclough GF, Groth JG, Lai JE, Tsang SM (2014) The phylogenetic relationships of the endemic genera of the Australo–Papuan hawks. J Raptor Research 48:36–43

Barrowclough GF, Cracraft J, Klicka J, Zink RM (2016) How many kinds of birds are there and why does it matter? PLoS One 11(11):e0166307

Bigelow RS (1956) Monophyletic classification and evolution. Syst Zool 5:145–146

BirdLife International and Handbook of the Birds of the World (2016) Bird species distribution maps of the world. Version 6.0, http://datazone.birdlife.org/

Boev Z (2012) *Circaetus rhodopensis* sp. n. (Aves: Accipitriformes) from the late Miocene of Hadzhidimovo (SW Bulgaria). Acta Zool Bulg 64:5–12

Breman FC, Jordaens K, Sonet G, Nagy ZT, Van Houdt J, Louette M (2013) DNA barcoding and evolutionary relationships in Accipiter Brisson, 1760 (Aves, Falconiformes: Accipitridae) with a focus on African and Eurasian representatives. J Ornithol 154:265–287

Cerling TE, Harris JM, MacFadden BJ, Leakey MG, Quade J, Eisenmann V, Ehleringer JR (1997) Global vegetation change through the Miocene/Pliocene boundary. Nature 389:153–158

Chesser RT, Burns KJ, Cicero C, Dunn JL, Kratter AW, Lovette IJ, Rasmussen PC, Remsen JV Jr, Rising JD, Stotz DF, Winker K (2016) Fifty-seventh supplement to the American Ornithologists' Union check-list of north American birds. Auk 133(3):544–560

Cooper N, Rodríguez J, Purvis A (2008) A common tendency for phylogenetic overdispersion in mammalian assemblages. Proc R Soc B 275:2031–2037

Cracraft J, Barker FK, Braun MJ, Harshman J, Dyke G, Feinstein J, Stanley S, Cibois A, Schikler P, Beresford P, García-Moreno J, Sorenson MD, Yuri T, Mindell DP (2004) Phylogenetic relationships among modern birds (Neornithes): toward an avian tree of life. In: Cracraft J, Donoghue MJ (eds) Assembling the tree of life. Oxford Univ Press, New York, pp 468–489

do Amaral FSR, Miller MJ, Silveira LF, Bermingham E, Wajntal A (2006) Polyphyly of the hawk genera Leucopternis and Buteogallus (Aves, Accipitridae): multiple habitat

shifts during the Neotropical buteonine diversification. BMC Evol Biol 6:10. https://doi.org/10.1186/1471-2148-6-10
do Amaral FSR, Sheldon FH, Gamauf A, Haring E, Riesing M, Silveira LF, Wajntal A (2009) Patterns and processes of diversification in a widespread and ecologically diverse avian group, the buteonine hawks (Aves, Accipitridae). Mol Phylogenet Evol 53(3):703–715
Drummond AJ, Suchard MA, Xie D, Rambaut A (2012) Bayesian phylogenetics with BEAUti and the BEAST 1.7. Mol Biol Evol 29:1969–1973
Edwards EJ, Osborne CP, CAE S, Smith SA, C4 Grasses Consortium (2010) The origins of C4 grasslands: integrating evolutionary and ecosystem science. Science 328:587–591
Ericson PGP, Anderson CL, Britton T, Elzanowski A, Johansson US, Källersjö M, Ohlson JI, Parsons TJ, Zuccon D, Mayr G (2006) Diversification of Neoaves: integration of molecular sequence data and fossils. Biol Lett 2:543–547
Faith DP (1992) Conservation evaluation and phylogenetic diversity. Biol Conserv 61(1):1–10
Fuchs J, Chen S, Johnson JA, Mindell DP (2011) Pliocene diversification within the south American forest-falcons (Falconidae: Micrastur). Mol Phylogenet Evol 60:398–407
Fuchs J, Johnson JA, Mindell DP (2012) Molecular systematics of the caracaras and allies (Falconidae: Polyborinae) inferred from mitochondrial and nuclear sequence data. Ibis 154:520–532
Fuchs J, Johnson JA, Mindell DP (2015) Rapid diversification of falcons (Aves: Falconidae) due to expansion of open habitats in the late Miocene. Mol Phylogenet Evol 82:166–182
Gamauf A, Haring E (2004) Molecular phylogeny and biogeography of honey-buzzards (genera *Pernis* and *Henicopernis*). J Zool Syst Evol Res 42(2):145–153
Gamauf A, Harring E (2005) Phylogeny of old world Perninae (Accipitridae) based on mitochondrial DNA sequences. Zool Med Leiden 79–3:175–177
Gamauf A, Gjershaug J-O, Røv N, Kvaløy K, Haring E (2005) Molecular phylogeny of the hawk-eagles (genus *Spizaetus*). Zool Med Leiden 79-3:179–180
Gerhold P, Cahill Jr JF, Winter M, Bartish IV, Prinzing A (2015) Phylogenetic patterns are not proxies of community assembly mechanisms (they are far better). Funct Ecol 29:600–614
Gill F, Donsker D (eds) (2017) IOC World Bird List (v 7.1). https://doi.org/10.14344/IOC.ML.7.1
Griffiths CS (1999) Phylogeny of the Falconidae inferred from molecular and morphological data. Auk 116:116–130
Griffiths CS, Barrowclough GF, Groth JG, Mertz L (2004) Phylogeny of the Falconidae (Aves): a comparison of the efficacy of morphological, mitochondrial, and nuclear data. Mol Phylogenet Evol 32(1):101–109
Griffiths CS, Barrowclough GF, Groth JG, Mertz LA (2007) Phylogeny, diversity, and classification of the Accipitridae based on DNA sequences of the RAG-1 exon. J Avian Biol 38(5):587–602
Hackett SJ, Kimball RT, Reddy S, Bowie RC, Braun EL, Braun MJ, Chojnowski JL, Cox WA, Han KL, Harshman J, Huddleston CJ, Marks BD, Miglia KJ, Moore WS, Sheldon FH, Steadman DW, Witt CC, Yuri T (2008) A phylogenomic study of birds reveals their evolutionary history. Science 320(5884):1763–1768
Haring E, Kvaløy K, Gjershaug J-O, Røv N, Gamauf A (2007) Convergent evolution and paraphyly of the hawk-eagles of the genus *Spizaetus* (Aves, Accipitridae) – phylogenetic analysis based on mitochondrial markers. J Zool Syst Evol Res 45:353–365
Helbig AJ, Kocum AJ, Seibold A, Braun MJ (2005) A multi-gene phylogeny of aquiline eagles (Aves: Accipitriformes) reveals extensive paraphyly at the genus level. Mol Phylogenet Evol 35:147–164
Hennig W (1966) Phylogenetic systematics. U. Illinois Press, Urbana
HilleRisLambers J, Adler PB, Harpole WS, Levine JM, Mayfield MM (2012) Rethinking community assembly through the lens of coexistence theory. Annu Rev Ecol Evol Syst 43:227–248
Holt BG, Jønsson KA (2014) Reconciling hierarchical taxonomy with molecular phylogenies. Syst Biol 63(6):1010–1017
Jarvis ED, Mirarab S, Aberer AJ, Li B, Houde P et al (2014) Whole genome analyses resolve the early branches in the tree of life of modern birds. Science 346:1320–1331

Jenkins CN, Pimm SL, Joppa LN (2013) Global patterns of terrestrial vertebrate diversity and conservation. Proc Natl Acad Sci U S A 110:E2602–E2610

Johnson JA, Lerner HR, Rasmussen PC, Mindell DP (2006) Systematics within Gyps vultures: a clade at risk. BMC Evol Biol 6(1):1. https://doi.org/10.1186/1471-2148-6-65

Johnson JA, Thorstrom R, Mindell DP (2007a) Systematics and conservation of the hook-billed kite including the island taxa from Cuba and Grenada. Anim Conserv 10(3):349–359

Johnson JA, Burnham KK, Burnham WA, Mindell DP (2007b) Genetic structure identified among continental and island populations of gyrfalcons. Mol Ecol 16:3145–3160

Johnson JA, Brown JW, Fuchs J, Mindell DP (2016) Multi-locus phylogenetic inference among new world vultures (Aves: Cathartidae). Mol Phylogenet Evol 105:193–199

Jollie MT (1977) A contribution to the morphology and phylogeny of the Falconiformes, parts 2–3. Evol Theory 2:115–300

Jønsson KA, Fabre PH, Kennedy JD, Holt BG, Borregaard MK, Rahbek C, Fjeldså J (2016) A supermatrix phylogeny of corvoid passerine birds (Aves: Corvides). Mol Phylogenet Evol 94:87–94

Kembel SW, Cowan PD, Helmus MR, Cornwell WK, Morlon H et al (2010) Picante: R tools for integrating phylogenies and ecology. Bioinformatics 26:1463–1464

Kruckenhauser L, Haring E, Pinsker W, Riesing MJ, Winkler H, Wink M, Gamauf A (2004) Genetic vs. morphological differentiation of old world buzzards (genus *Buteo*, Accipitridae). Zool Scr 33:197–211

Lecompte E, Aplin K, Denys C, Catzeflis F, Chades M, Chevret P (2008) Phylogeny and biogeography of African Murinae based on mitochondrial and nuclear gene sequences, with a new tribal classification of the subfamily. BMC Evol Biol 8:199

Lerner HR, Mindell DP (2005) Phylogeny of eagles, old world vultures, and other Accipitridae based on nuclear and mitochondrial DNA. Mol Phylogenet Evol 37(2):327–346

Lerner HR, Klaver MC, Mindell DP (2008) Molecular phylogenetics of the Buteonine birds of prey (Accipitridae). Auk 125(2):304–315

Lerner HR, Christidis L, Gamauf A, Griffiths C, Haring E, Huddleston CJ, Kabra S, Kocum A, Krosby M, Kvaløy K, Mindell DP, Rasmussen P, Røv N, Wadleigh R, Wink M, Gjershaug JO (2017) Supermatrix phylogeny and new taxonomy of the booted eagles (Accipitriformes: Aquilinae). Zootaxa 4216(4):301–320

Li S, Cadotte MW, Meiners SJ, Hua Z, Jiang L, Shu W (2015) Species colonization, not competitive exclusion, drives community overdispersion over long-term succession. Ecol Lett 18:964–973

Li Z, Clarke JA, Zhou Z, Deng T (2016) A new old world vulture from the late Miocene of China sheds light on Neogene shifts in the past diversity and distribution of the Gypaetinae. Auk 133:615–625

Mahmood MT, McLenachan PA, Gibb GC, Penny D (2014) Phylogenetic position of avian nocturnal and diurnal raptors. Genome Biol Evol 6(2):326–332

Manegold A, Pavia M, Haarhoff P (2014) A new species of *Aegypius* Vulture (Aegypiinae, Accipitridae) from the early Pliocene of South Africa. J Vertebr Paleontol 34(6):1394–1407

Mazel F, Davies TJ, Gallien L, Renaud J, Groussin M, Münkemüller T, Thuiller W (2016) Influence of tree shape and evolutionary time-scale on phylogenetic diversity metrics. Ecography 39:913–920

McCormack JE, Harvey MG, Faircloth BC, Crawford NG, Glenn TC, Brumfield RT (2013) A phylogeny of birds based on over 1,500 loci collected by target enrichment and high-throughput sequencing. PLoS One 8(1):e54848

Miller MA, Pfeiffer W, Schwartz T (2010) Creating the CIPRES Science Gateway for inference of large phylogenetic trees. Proc Gateway Computing Environments workshop (GCE). New Orleans, 1–8

Milne I, Lindner D, Bayer M, Husmeier D, McGuire G, Marshall DF, Wright F (2009) TOPALi v2: a rich graphical interface for evolutionary analyses of multiple alignments on HPC clusters and multi-core desktops. Bioinformatics 25:126–127

Oatley G, Simmons RE, Fuchs J (2015) A molecular phylogeny of the harriers (*Circus*, Accipitridae) indicate the role of long distance dispersal and migration in diversification. Mol Phylogenet Evol 85:150–160

Olson SL (2006) Reflections on the systematics of *Accipiter* and the genus for *Falco superciliosus* Linnaeus. Bull BOC 126:69–70

Ong PS, Luczon AU, Quilang JP, Sumaya AMT, Ibanez JC, Salvador DJ, Fontanilla IKN (2011) DNA barcodes of Philippine accipitrids. Mol Ecol Resour 11:245–254

Osborne CP, Beerling DJ (2006) Nature's green revolution: the remarkable evolutionary rise of C4 plants. Philos Trans R Soc B 361:173–194

Peters JL (1931) Check-list of the birds of the world, vol 1. Harvard Univ Press, Cambridge

Philippe H, de Vienne DM, Ranwez V, Roure B, Baurain D, Delsuc F (2017) Pitfalls in supermatrix phylogenomics. European J Taxonomy 283:1–25

Prum RO, Berv JS, Dornburg A, Field DJ, Townsend JP, Lemmon EM, Lemmon AR (2015) A comprehensive phylogeny of birds (Aves) using targeted next-generation DNA sequencing. Nature 526:569–573

Rambaut A, Suchard MA, Xie D, Drummond AJ (2014) Tracer v1.6, Available from. http://beast.bio.ed.ac.uk/Tracer

R Development Core Team (2013) R: A Language and Environment for Statistical Computing. R Foundation for Statistical Computing, Vienna, Austria

Riesing MJ, Kruckenhauser L, Gamauf A, Haring E (2003) Molecular phylogeny of the genus Buteo (Aves: Accipitridae) based on mitochondrial marker sequences. Mol Phylogenet Evol 27(2):328–342

Schipper J, Chanson JS, Chiozza F, Cox NA, Hoffmann M, Katariya V et al (2008) The status of the world's land and marine mammals: diversity, threat, and knowledge. Science 322:225–230

Sibley CG, Ahlquist JE (1990) Phylogeny and classification of birds. Yale Univ Press, New Haven, A Study in Molecular Evolution

Swann HK (1922) A synopsis of the Accipitres (diurnal birds of prey): comprising species and subspecies described up to 1920, with their characters and distribution. Wheldon and Wesley, London

Tucker CM, Cadotte MW, Carvalho SB, Davies TJ, Ferrier S, Fritz SA, Grenyer R, Helmus MR, Jin LS, Mooers AO, Pavoine S, Purschke O, Redding DW, Rosauer DF, Winter M, Mazel F (2016) A guide to phylogenetic metrics for conservation, community ecology and macroecology. Biol Rev. https://doi.org/10.1111/brv.12252

Voskamp A, Baker DJ, Stephens PA, Valdes PJ, Willis SG (2017) Global patterns in the divergence between phylogenetic diversity and species richness in terrestrial birds. J Biogeogr 44:709–721

Webb CO, Ackerly DD, McPeek MA, Donoghue MJ (2002) Phylogenies and community ecology. Annu Rev Ecol Evol Syst 33:475–505

Webb CO, Ackerly DD, Kembel SW (2008) Phylocom: software for the analysis of phylogenetic community structure and trait evolution. Bioinformatics 24:2098–2100

Wink M, Sauer-Gürth H (2004) Phylogenetic relationships in diurnal raptors based on nucleotide sequences of mitochondrial and nuclear marker genes. In: Chancellor R, Meyburg B-U (eds) Raptors worldwide. World working group on birds of prey and owls, Berlin, pp 483–498

Chapter 2
Behavioural Ecology of Raptors

Juan José Negro and Ismael Galván

Introduction

There are two distantly related orders of diurnal raptors currently recognized (Jarvis et al. 2014): the Falconiformes, or true falcons and allies on one side, phylogenetically close to Passeriformes and Psittaciformes, and the Accipitriformes, with the eagles and relatives on the other side. They both sum up to more than 300 species with numerous subspecies distributed practically all around the world (Ferguson-Lees and Christie 2001). The most cosmopolitan of all, the peregrine falcon *Falco peregrinus*, with 25 described subspecies, inhabits icy landscapes above the Arctic Circle but also equatorial areas, semi-deserts such as the Namibian inland in the Southern Hemisphere and high mountains up to 4000 m.a.s.l. Apart from their widespread occurrence and being essentially predatorial or necrophagous, raptors tend to present narrow dietary adaptations, and some are almost exclusively bird eaters, such as the true falcons and the sparrowhawks, while others are partial to snakes – short-toed eagles of the genus *Circaetus* and the laughing falcon *Herpetotheres cachinnans* – or even to skeletal parts of ungulates, as with the bone-breaker bearded vulture *Gypaetus barbatus*. There is even a species with a liking for vegetal items, the palm vulture *Gypohierax angolensis*, which "preys", among other things, on dates hanging from palm trees.

There are raptorial birds weighing more than 10 kg with wingspans longer than 3 m – i.e. the Andean condor *Vultur gryphus* – but also dwarfs such as the pigmy falcon *Polihierax semitorquatus* with a body mass just about 50–70 g. Being apex predators in every possible land habitat on Earth except Antarctica imposes numerous constraints, and birds of prey have evolved numerous fascinating adaptations to make their living. Raptors are thus rich in tricks to procure food for themselves, to

J. J. Negro (✉) · I. Galván
Department of Evolutionary Ecology, Doñana Biological Station (EBD-CSIC), Seville, Spain
e-mail: negro@ebd.csic.es

J. H. Sarasola et al. (eds.), *Birds of Prey*,
https://doi.org/10.1007/978-3-319-73745-4_2

find mates or to gain protection against other predators, as many of the smaller species may fell prey to larger diurnal or nocturnal raptors, mammalian carnivores or even snakes.

In this chapter, we describe and discuss adaptations of raptors in the realm of behavioural ecology, which studies the adaptive value of behavioural traits. Even though raptors are chronically scarce and tend to be difficult to study, they have caught the attention of researchers and conservationists from long ago. This has permitted to amass a valuable corpus of behavioural information on the group that we try to summarize here. In many instances, behavioural science and plain "ecology" intersect. This should not be seen as a problem. On the contrary, the study of adaptations pertains to evolutionary ecology, which feeds on both ecological data and behavioural observations. Plus the more we know about raptors in every respect, the better we will preserve them and their habitats. And this is, admittedly, the ultimate goal of every researcher and manager concerned with raptors, which unfortunately plague the red lists of endangered species or species in vulnerable situation in a majority of countries. As we will see in the following pages, both birds of prey and raptor biologists have proven wrong one of the most prominent and respected raptor investigators of all times, Dr. I. Newton, when he said many years ago that raptor studies have contributed little to mainstream ecology (Newton 1979). Raptors, on the contrary, have given a wealth of information on practically every domain of animal science, ecology and genetics. With the exception of those species which inhabit remote places and very often at very low densities, raptors have provided a tremendous amount of data on how land ecosystems work and on how to tackle numerous conservation challenges.

Courtship, Mating and Raising a Brood

Breeding attempts happen once a year, as with most raptorial species in temperate regions, multiple times within a year (as with the *Elanus* species) or every 2 years as in the case of the gigantic condors (both Andean and Californian) and the larger tropical eagles, such as the crowned eagle *Stephanoaetus coronatus* in African jungles or the harpy eagle *Harpia harpyja* in New World forests. But it all happens with a courtship period in which the male, typically the smaller sex because raptors show reversed sexual size dimorphism, often performs aerial acrobatic displays to attract a mate or to secure the one he is already paired to.

Reversed Sexual Size Dimorphism (RSD)

While in a majority of vertebrate species males are larger than females, raptors constitute a flagrant exception. With the exception of some scavengers, female birds of prey are heavier and dominant over their male partners. The direction of the size difference was already noted by Darwin (1871), and the search for an explanation has produced abundant literature and no less than 20 hypotheses that Kruger (2005) grouped into three main categories:

1. Ecological hypotheses, proposing that raptors being mainly territorial lessen intra-pair competition by differing in size and aiming at different target prey.
2. Role differentiation hypotheses, according to which males and females have divided their roles in raising their progeny. A larger female would be more endowed for producing viable eggs and incubating them, whereas a smaller male could be a more efficient hunter.
3. Behavioural hypotheses state that either females gain dominance over their male mates, or that females compete among themselves for access to males, or that smaller males are more agile and more able hunters.

The above-mentioned study by Kruger (2005), one of the last comparative studies published so far, favours the "small male hypothesis", according to which males have become smaller for more efficient hunting or territorial defence. Almost concurrently, MacDonald et al. (2005), who studied RSD in a single large *Falco* species, *Falco berigora*, reported that their findings were consistent with the "intrasexual competition hypothesis", favouring an increase in female size. Therefore, there is no current agreed paradigm on the evolution and function of RSD in raptors, and it remains a challenging question for scientists to come. We can only emphasize that RSD is more extreme in falcons and hawks, hunters of elusive prey such as birds, and practically non-existent in carrion eaters which do not actually hunt or defend a hunting territory.

Courtship Displays

In both Falconiformes and Accipitriformes, reported courtship activities involve aerial manoeuvers, sometimes accompanied by vocalizations, and the feeding of females by males, an activity called courtship-feeding by some authors (e.g. Poole 1985). However, calls alone or displays on the ground or at the nest are rare. Certain species, such as hawks, eagles and falcons, are rather vocal, nonetheless. Vultures, on the contrary, are all practically mute in most circumstances. One exception may be the bearded vulture, which emits chuckling noises during courtship and mating.

Undulating flights at the time of mating and near nesting platforms have been observed in large eagles of the *Aquila* genus, such as the golden eagle *Aquila chrysaetos* (e.g. see Newton 1979). It seems that males initiate these flights but may be followed by their female mates. Harmata (1982) has challenged the view that these flights serve in attracting mates or reinforcing pair bonds. This author, who witnessed numerous undulating flights outside the breeding season, believes that they are addressed towards inter- or intraspecific intruders in the eagles' territory. In other words, it would be a strategy for the eagles to conspicuously advertise territory ownership.

Fig. 2.1 Two male lesser kestrels (*Falco naumanni*) fighting by grabbing talons at the entrance of a nesting cavity in an old church in Extremadura, western Spain, during the courtship period. The winner in these rather violent contests will hold the place and will start attracting a female partner. (Credit @ JJ Negro)

In some colonial species, for instance, the urban-breeding lesser kestrel *Falco naumanni*, the only asset defended by the birds is the nest site and immediate surroundings. In this species, males arrive 1 or 2 weeks earlier than the females to the colonies (Negro 1997) and choose a nest site – a hole or crevice – that they defend vigorously against other males (Fig. 2.1). Once they get uncontested ownership of the nest site, they advertise it to passing-by females by fast flying into it displaying the inside of the wings which are perceived at a distance as white flashing flags (Fig. 2.2). In other colonial species such as the Montagu's harrier *Circus pygargus*, aerial displays are more frequently performed by males, during the pre-laying period, in large colonies and with high abundance of prey (voles) (Arroyo et al. 2013). The authors of the latter study suggested that these displays may be costly to perform and thus may signal individual quality to potential mates. It is difficult to think, however, that aerial displays are especially costly in energetic terms, and thus low-quality individuals would not be prevented from performing displays because of these costs. Instead, aerial displays may be amplifiers of individual quality

Fig. 2.2 Adult male lesser kestrel (*Falco naumanni*) showing the white underwings used to attract females to the nesting site that he has secured and now defends from other intruding males. White is an achromatic colouration but the most visible at a distance. Male kestrels perform highly conspicuous acrobatic flights to show the nest cavity to prospecting females after their arrival to the colonies from Africa. (Credit @ JJ Negro)

(Barber and Folstad 2000), their honesty residing in the fact that high-quality individuals would benefit from displaying while low-quality individuals would enhance the perception of their low quality by conspecifics (Hasson 1997).

Courtship-feeding, now commonly referred as mate-feeding, is widespread in birds of prey (Newton 1979). Males feed their mate females before egg laying starts with the apparent purpose of improving the female's body condition. It has been debated whether mate-feeding precedes or elicits copulation, and it has been seen by some authors as a reward for sex. However, when thorough observations have been conducted, it becomes clear that mate-feeding occurs well after copulations are initiated and ceases as soon as the female may provide for herself. Poole (1985) proposed that mate-feeding in osprey has the added value of "reducing female wandering and to increase a male's confidence of paternity". This author argues, therefore, that mate-feeding in ospreys may function to ensure mate fidelity. As with the remaining raptor females fed by their mates, ospreys may even find it difficult to fly and hunt efficiently at the peak of weight gain, so it would be difficult to disentangle cause and effect. To test Poole's ideas it would be necessary to determine whether female ospreys at the nest receive invitations for extra-pair copulations, and also whether they ever accept them or just expel intruding males.

In the case of the lesser kestrel, a species for which one of us (JJN) made a functional analysis of mate-feeding (Donazar et al. 1992), the results indicated that this activity was not relevant to pair formation, as it started 54 days, on average, after the pair's first copulation. The most parsimonious explanation was to

Fig. 2.3 A male yellow-headed caracara (*Milvago chimachima*) feeding his partner. Food supply by males during courtship or incubation (i.e., courtship feeding) is common in raptors. In other birds, this behaviour is known to compensate for nutritional limitations of females during reproduction, thus the same function is assumed to have driven its evolution in raptors. (Credit @ I Galván)

increase the female's body mass as soon as possible to allow the laying of earlier and larger clutches. Mate-feeding started about 16 days before the laying of the first egg and lasted an average of 24 days, just up to the point when the clutch was completed and the females commenced to forage for food by themselves. It must be mentioned that mate-feeding has been proven to function as a compensatory energetic strategy that affects breeding success in other birds (Galván and Sanz 2011), and thus this should be considered as the most likely function also in raptors (Fig. 2.3).

Aerial Food Transfers

The transfer of food items in the air, or food passes (Simmons 2000), is commonly practised by harriers (*Circus* spp.) and the *Elanus* kites. The harriers are mostly ground nesters, and it seems that they have evolved the passing of food from males to females in the air as a way to minimize the presence of scented remains by the nest that might attract ground predators. Curiously enough, harriers and the *Elanus* kites carry out sky dancing and food passes during the pre-laying periods, and at least in harriers, the two activities correlate (Wiącek 2006). The coordination of the couple on these food passes is remarkable, and successful transfer of the prey approaches 100%.

Food passes in the air have also been recorded from an adult male to juvenile white-tailed kite *Elanus leucurus* and also in species other than harriers and the *Elanus* kites. There are recorded observations, for instance, of adult peregrine falcons transferring prey on the wing.

Copulation Patterns and Sperm Competition

A majority of bird species are considered to be monogamous forming pair bonds that last at least a breeding cycle (Lack 1968). Birds of prey are no exception to this rule, and again most species are socially monogamous. If anything, some species add some complexity to this pattern, and a varying proportion of the individuals engage in reproductive trios or even quartets. This is the case, for instance, of the bearded vulture (Heredia and Donázar 1990; Bertran and Margalida 2002) and its close relative the Egyptian vulture (Tella 1993), where a single female may mate with two or more males simultaneously. The most comprehensive survey of group mating so far was published by Kimball et al. (2003). These authors reported cooperative breeding in 42 diurnal raptors including both Falconidae and Accipitridae (now Falconiformes and Accipitriformes, Jarvis et al. 2014), representing about 12% of all extant species of raptors. Considering that numerous raptor species have been poorly studied, many instances of group living may have been overlooked. But even if we just take the proven cases, the proportion of raptors engaging in cooperative breeding should be considered high, as the proportion for class Aves is only 3% (Arnold and Owens 1998).

Apart from blatant group living, some raptors engage in what is called a *mixed reproductive strategy* (Trivers 1972): they are socially bonded to a single mate but engage in extra-pair copulations. Some males, for instance, may share parental care with their males, but do not decline opportunities to fertilize mated females and thus parasitize the parental investment of other males (Birkhead and Møller 1992). One of the first raptorial species in which males were observed to engage in a mixed reproductive strategy was the lesser kestrel (Negro et al. 1992). Similarly, this species also provided one of the first confirmed cases of realized polygyny using DNA markers (Tella et al. 1996), shortly after it was demonstrated for the cooperative breeding Galapagos hawk *Buteo galapagoensis* (Faaborg et al. 1995).

If individual birds in the population practice a mixed reproductive strategy, *sperm competition* takes place (Parker 1970). And given that females, whether fertilized by one or more males, cannot produce more than a certain number of eggs in the clutch, only males in risk of being cuckolded are expected to evolve adaptations to ensure paternity. Two common strategies are mate guarding (the male stays by the female for as long as he can to deter or confront sneaking competitor males and thus impede extra-pair copulations), or if the male cannot possibly be sure of his mate's fidelity, he may try to dilute his competitors' sperm by copulating frequently (Birkhead et al. 1987).

Given that male raptors typically go away to hunt for their female mates during their fertile period and that the latter tend to remain by the nest site, mate guarding does not seem to be an option (Møller 1987). Frequent copulation, however, is indeed a trademark of raptorial birds. The American kestrel *Falco sparverius*, for instance, copulates an estimate of 690 times per season (Balgooyen 1976), in what seems to be a tremendous waste considering that a single successful copulation is enough to fertilize an entire clutch! The tiny American kestrel may be a record-breaking copulator among birds, but closely follow other raptors (reviewed in Negro and Grande 2001), particularly small falcons and sparrowhawks with hundreds of copulations performed over protracted periods of time, sometimes longer than 2 months and often outside the fertile period of the female.

Copulation Frequency and Seasonal Pattern

The review of raptor copulation behaviour has revealed some commonalities: (a) daily copulation rates (generally expressed as copulations/day or copulation/h) tend to show two peaks, one soon after pair formation or reunion and the other centred at the onset of clutch laying. One exception may be the double-brooded black-winged kite *Elanus caeruleus*, for which three peaks in copulation frequency have been observed (Ferrero et al. 2003). (b) Copulations still occur after egg laying, and (c) the incidence of extra-pair fertilizations is low (less than 5%) compared to other groups such as the passerines, where extra-pair fertilization revealed by DNA fingerprinting may affect up to 59% of broods in a population (Perreault et al. 1997) in an age-dependent pattern in which most yearling males attempting to breed for the first time are cuckolded by older, more experienced males.

Although the strength of sperm competition is generally assumed to shape copulation patterns (Birkhead and Møller 1992), it fails to explain raptorial behaviour. First, the early peak of copulations, which for some species takes place about 2 months before laying (Negro et al. 1992), is almost certainly outside the fertile period of the female and that seems unrelated to ensure fertilization of the clutch. In fact, one of the few species for which the fertile period has been determined is the American kestrel (Bird and Buckland 1976), and it starts 1 week before the onset of laying.

If the pressing needs of reproduction and paternity assurance are not always behind the high copulation rates of raptors, is there an alternative explanation? Some years ago we actually proposed (Negro and Grande 2001) the hypothesis that copulations may act as signals of territory ownership. As strange as it sounds, noisy and visually conspicuous copulations may quickly reveal that a pair of raptors is breeding in the area, a vital piece of information for other raptors and perhaps some other species that may prefer to avoid conflicts with well-armed and potentially aggressive neighbours.

Territorial Signalling: A Reason for Frequent Copulation in Raptors

We arrived to the conclusion that copulations may have information content and become communication signals linking two pieces of information, which are in fact prerequisites for this hypothesis to work: (1) copulations in raptors are not hidden but, on the contrary, are easily perceived at a distance because both members of the couple typically vocalize loudly, and copulations often take place on exposed places. (2) Raptors tend to copulate in the area where intruders are not tolerated. Colonial species copulate almost exclusively at the nest site, whereas territorial species, on the other hand, copulate both on the nest site and on the borderline of the defended territory.

One of the difficulties of the sperm competition hypothesis is precisely to explain the first peak of copulations outside the fertile period of the female. The territory-signalling hypothesis actually predicts this peak soon after pair reunion, as it would immediately warn potential intruders of the establishment of a breeding pair.

How Copulations May Become Signals

At least one copulation is needed to fertilize a clutch, and thus copulations must have initially evolved to serve reproduction needs, in raptors or in any sexually reproducing species. But even if copulations evolved primarily for fertilization, they may still gain signalling content. If breeding raptors reduced intrusion rates of conspecifics by copulating conspicuously and frequently in their defended territories, the signalling component of copulations would have caused their frequency to increase. Curiously enough, many copulations performed by raptors are "fake". Often there is not even cloacal contact – this can be seen through binoculars or a spotting scope – and the individuals involved seems to be more interested in spreading a message ("we are a couple, this is our territory and any intruder will be chased away") than in assuring fertilization success. In fact, a majority of fake copulations occur at the beginning of the copulation period, as expected if the main purpose was to inform neighbours and potential intruders of the presence of a territorial couple. Following the signal theory, copulation rituals (i.e. frequent copulation) may have evolved as attention signals, that is, signals that attract (as opposed to camouflage signals, which divert) the recipients' attention to the presence of individuals rather than to their quality (Hasson 1997).

One may wonder, nonetheless, why this strategy of frequent copulation has evolved in raptors and not in other birds. First, raptors tend to have few or no predators and can expose themselves copulating noisily without much risk. In addition, raptors are heavily armed with sharp talons and a strong curved beak and behave aggressively to conspecific and other intruding animals within the limits of their defended territories. It may well be that the cost of copulations – including the fake ones with no sperm transfer – may be lower than that of chasing intruders away or risking injuries or life in dangerous fights.

The Signalling Content of Raptor Colouration

Some people confess to have a problem identifying raptors at a distance because, they say, all look alike. A superficial analysis may lead to think that raptors are generally dark birds with little variation in feather colouration. It is true that raptors tend to be melanized animals in the achromatic part of the colour spectrum, with a predominance of brown hues and some black, grey or white (this implying absence of pigment and thus being a structural colour, Prum 2006). And it is also true that both eumelanin and pheomelanin make the bulk of the pigments found in raptor feathers, where no bright pigments such as the carotenoids have ever been found, although these pigments are prominently displayed in the cere above the beak and on the exposed integument of the legs. However, and only recently, porphyrins have also been reported in the orange-tinged feathers of the breast of juvenile black-winged kites (Negro et al. 2009). Porphyrins, even though they are widespread pigments in nature, are rarely found in the integument of animals, the plumage of some birds representing an exception (see a dramatic case in bustards and allies; Galván et al. 2016). To find them in black-winged kites came as a surprise, particularly because this pigment was interspersed with pheomelanin in the same feather tracts and also because porphyrins quickly degrade when exposed to sunlight. Although we did not analyse pigments in the remaining three species of *Elanus*, i.e. the white-tailed kite, *E. leucurus*, in the New World and the two Australasian species (i.e. the Australian kite *E. axillaris* and the letter-winged *E. scriptus*), the juvenal plumage in all these species includes rusty coloured feathers in the same plumage tracts. We can safely assume that porphyrins were thus present in the ancestral species from which all extant species originated.

The fact that the reddish feathers are only found in the juvenal plumage in *Elanus* when the birds are still in the nest or shortly after they fledge led us to hypothesize a camouflage role for this ephemeral plumage. After all, the *Elanus* kites are small raptors that breed on open and exposed nests built on treetops, and the protective colouration is shown both on the breast and in the upperparts of the body and head but not in the belly (Fig. 2.4). It may well be that the porphyrin-dependent colouration confers protection to the vulnerable young kites when they most need it, but that they benefit more from acquiring the white colouration typical of the adult as soon as they can fend for themselves. The *Elanus* kites tend to gather on communal roosts at night (Parejo et al. 2001), and the white plumage colouration is possibly used as a visual attention signal to recruit individuals in the gathering places, because white is the colour type more visible at a distance. Indeed, social species of birds have overall whiter plumage than nonsocial species, which may facilitate attracting conspecifics (Beauchamp and Heeb 2001). A young and recently fledged

Fig. 2.4 Black-winged kite (*Elanus caeruleus*) in its juvenal plumage showing remains of its rusty colouration due to the pigment porphyrin. Extremadura, Spain. (Credit @ D. Rivera)

Elanus does not need the extra crypsis provided by the porphyrin and, however, may derive more benefits in terms of social acceptance. Thus, and contrary to any other raptor with "delayed plumage maturation", the young kites do not need to wait for the next moult of the body feathers to change its appearance. The photodegradable property of porphyrins gives young *Elanus* kites the chance to get the best of both worlds, i.e. protection and advertising, in just a matter of weeks.

But the seemingly drab and dull-coloured raptors are not all like that. At close range, individuals in many species reveal they do not renounce to glamorous colouration even it often restricted to the exposed skin in the head and/or legs. Also, many raptorial species have "silver wings" or sheen feathers (Galván et al. 2009), a structural mechanism in feathers unrelated to iridescence that increases the conspicuousness or general brightness of the plumage. These sheen feathers are often found in raptors of the order Accipitriformes (e.g. New World vultures, kites, harriers, Egyptian and bearded vultures), as well as in other avian orders including the pelicans (Pelecaniformes), ducks and geese (Anseriformes), storks and ibises (Ciconiiformes), cranes (Gruiformes) and sandgrouses (Pterocliformes).

Raptors may also exhibit behavioural adaptations to enhance their plumage conspicuousness, as some owls do. In this regard, it has been shown that the brightest areas of the plumage of snowy owls *Bubo scandiacus* (face, throat and breast) are always unspotted in contrast to other body parts, and wintering birds display towards conspecifics by orientating to the sun, preferentially on sunny days (Bortolotti et al. 2011). Snowy owls with the least spotted plumage also display more on the ground, maybe to benefit from the reflectance of the albedo from the snow. This indicates that snowy owls integrate plumage colour, behaviour and their environment to maximize the perception of their plumage by conspecifics, thus probably enhancing territorial signalling (Bortolotti et al. 2011). Eagle owls *Bubo bubo* also take advantage of environmental light conditions (i.e. moon phase) to increase the conspicuousness of their white throat patch during vocal displays (Penteriani et al. 2010). Whether diurnal birds of prey have these or similar adaptations is unknown and should be investigated.

Sheen Feathers in Raptors

Iridescence in feathers, obtained most of the time combining structural properties of the feather and the pigment melanin, is the classical way of increasing conspicuousness of the feathers depending on incident light and thus in a way that the animal may theoretically control. This is the case of hummingbirds, which may display an ultra-bright purple throat patch or just black colouration depending on how they position themselves in relation to the sunlight (Prum 2006). There is not a single bird of prey among the 164 bird species with integumentary structural colouration, both iridescent and non-iridescent, examined by electron microscopy and compiled by R. Prum (2006). However, it has been suggested that some New World vultures may actually have iridescent plumage. For the turkey vulture *Cathartes aura*, purple and green reflections have been observed in the otherwise black feathers of its plumage (Ferguson-Lees and Christie 2001). Another species that should have iridescence on its dark back with purple and green reflections is the swallow-tailed kite *Elanoides forficatus*, from the New World and with no extant close relatives, as it is the only species in its genus.

It is clear that very few raptor species have evolved iridescent plumage. But it is equally intriguing that the phenomenon of sheen feathers is widespread in the phylogenetic tree of the Accipitriformes (but not so in falcon, kestrels or caracaras in the Falconiformes). As stated above, sheen feathers increase visibility of their bearers and are often seen in the adult plumage, but not in the juvenal plumage, which tends to be matte in colour. Good examples are provided by bearded and Egyptian vultures (Fig. 2.5) but also Andean condors. In these species, the dull brown colour of the juveniles contrasts with the much brighter plumages of the adult birds sporting both white patches and sheen feathers. The reasons behind these age-related differences in plumage brightness have never been tested experimentally, but they might be related – and this should be the first prediction to test – to the visibility needs of the adults defending a territory and/or searching for mates (these two possibilities are not mutually exclusive). Again, these conspicuous traits may function as attention signals. As we stated above, raptors are visually oriented, have territories that may span several square kilometres and display territory ownership

Fig. 2.5 Bearded vultures (*Gypaetus barbatus*) have different plumages as juveniles or adults. The definitive plumage, which is much lighter in colour, is acquired after about 5 years of age. The first juvenal plumage is very dark and matte; the definitive plumage, shown in the picture, consists of a metallic grey on the upper parts and a lighter colour ranging from white, in unstained individuals, to orange or rusty in individuals which have bathed in waters rich in iron oxides. The grey on top is very bright and shiny when illuminated by sunlight in birds flying or circling in the sky. It is a very good example of sheen feathers in birds (see text for an explanation). (Credit @ JJ Negro)

by undulating flights up in the air, sky dances and conspicuous copulations. Sheen feathers may help further to reveal a displaying adult at a long distance but would be of no help to a nonterritorial individual, typically a juvenile bird with delayed plumage maturation. One exception to this rule is the golden eagle. In this species, the first plumage includes white patches in remiges and rectrices (Fig. 2.6), absent in the full adult plumage (Ferguson Lees and Christie 2001). Without careful behavioural observations and comparative studies, the function of these age-related plumage differences will remain a mystery.

Advertising Condition at Close Range

Raptor feathers owe their colouration to the presence/absence of melanin and their multiple combinations to create patterns (Galván et al. 2017a, b) and, just for juvenile *Elanus* kites, to the added presence of porphyrins. But all this changes in the bare integument of the head and neck of some species and in the legs. At a close range, many raptors do display bright colours due to the concurrence of carotenoid

Fig. 2.6 Golden eagle (*Aquila chrysaetos*) with juvenal plumage showing the white patches under the wings and in the tail. These conspicuous patches are absent in the definitive all-dark plumage of the adult individuals. (Credit @JA Irastorza)

pigments, vascular red colouration with the pumping of blood to the exposed skin or both, as with the coloured face of the adult crested caracara *Polyborus plancus* (Negro et al. 2009).

The most extended expression of carotenoid integumentary colouration in raptors is restricted to the cere of the beak and the legs (Fig. 2.7). These may be yellow or orange, as with many falcons and kestrels (Bortolotti et al. 1996), but also kites (Blas et al. 2013), as well as eagles and hawks (Ferguson-Lees and Christie 2001). In fewer species, these bare parts are red coloured, as with the aptly named red-footed falcon (*Falco vespertinus*) and its close relative the Amur falcon (*Falco amurensis*), but also in the snail kites (both *Rostrhamus sociabilis* and *R. hamatus*) and the bateleur (*Terathopius ecaudatus*). A species with orange-yellow legs but no coloured cere is, for example, the Mississippi kite *Ictinia mississippiensis*. There are also species with no brightly coloured legs and cere. These include the osprey, with creamy cere and legs, but also several species among the Cathartidae, or New World vultures, as well as many other species including snake-eating eagles of the genus *Circaetus*, the black-collared hawk *Busarellus nigricollis* and a group of small falconets of the genera *Microhierax* and *Polihierax*, native of the Indomalayan region, with black ceres and legs. Yet in some species, there are age and related differences in the colouration of the bare parts. In the Chimango caracara (*Phalcoboenus chimango*), for instance, there are no plumage differences, but the adult male has

Fig. 2.7 White-tailed eagle (*Haliaeetus albicilla*) showing the all-yellow bill and cere due to carotenoid pigments. (Credit @ JJ Negro)

bright yellow cere and legs, whereas the female and the juvenile of both sexes present either pinkish or bluish colouration in those areas (Sarasola et al. 2011).

The reason for these different colouration strategies has not been explored for raptors. What is clear, however, is that carotenoids must be ingested in the diet, as they cannot be synthesized directly by birds or any other animal (Galván et al. 2017b). The few bird species where mate-selection studies have been carried out include no raptors (Hill 2006), but for passerines such as the house finch *Carpodacus mexicanus* or the zebra finch *Taeniopygia guttata*, there seems to be a female preference for the more colourful males (Hill 1990). Therefore, carotenoid-dependent colouration may be part of a system to reveal the best possible companions for breeding. Carotenoids are not only pigments in the integumentary parts, such as ceres and legs; they are also important physiological modulators (Bortolotti et al. 1996). An individual choosing a mate based on carotenoid expression may be assessing foraging abilities for specific carotenoids, good genetic makeup for carotenoid absorption and/or processing and also a good general body condition by which more carotenoid can be diverted to ornamentation instead of being used for physiological functions.

Raptors are not usually sexually dichromatic in their plumage, except in a few species including the harriers (*Circus* spp.), the sparrowhawks (*Accipiter* spp.) and some kestrel species (*Falco* spp.) (e.g. see Ferguson-Lees and Christie 2001). Both dichromatic and non-dichromatic species tend to have coloured ceres and legs due to carotenoid pigments, as we discussed above. In this respect, if carotenoid status plays a role in mate choice as a condition-dependent signal, we may assume that selection is possibly bidirectional, with males assessing females and vice versa (Dale et al. 2015). Contrary to feathers, where the pigment is sequestered until the next moulting period, carotenoids in integumentary parts may increase or decrease rapidly in a matter of days or weeks (Negro et al. 2006). Such a strategy would make sense in a group where social monogamy is the rule and pair bonds may last several years but where "divorce" is also known to occur (Serrano et al. 2001). In addition, both males and females contribute significantly to raising a brood (Newton 1979), and a single individual typically cannot do it alone.

The Yellow Face of the Egyptian Vulture

One intriguing case in which the bright yellow-to-orange carotenoid colouration goes beyond the cere and covers the whole face is that of the adult Egyptian vulture (Fig. 2.8). This species stands out among all the Old World vultures because it is the only one with a brightly ornamented skin in the unfeathered head. The main carotenoid responsible for this colouration was determined to be lutein (Negro et al. 2002a), a yellow carotenoid highly prevalent in all green plants. But Egyptian vultures are not herbivores, and they ingest carotenoids, among other potential sources, from the frequent consumption of cow dung. This coprophagous habit is uncommon in birds, but it has been clearly demonstrated that the yellow-faced Egyptian vulture includes ungulate faeces in its diet to increase its facial colouration. The ingestion of pigments via coprophagy makes the following evolutionary scenario possible for the development of carotenoid-dependent ornaments in this species: surplus carotenoids resulting from the massive ingestion of faeces would end up in the skin, where they are clearly visible, and the resulting bright colour may have become a short-term cue of condition that individuals may assess in mating displays and/or advertising dominant status.

Fig. 2.8 Egyptian vulture (*Neophron percnopterus*) displaying a bright yellow face reflecting the ingestion of carotenoid pigments. Wild birds obtain carotenoids from the fresh faeces of ungulates and possibly also from bird egg yolks. (Credit @ JJ Negro)

Flushing Colouration

In the timescale of colour traits in raptors, the shortest category belongs to facial flushing in raptors (Negro et al. 2006). If coloured feathers typically last 1 year between growth and shedding in the next moult and dermal carotenoid-dependent colouration lasts days or weeks, the red colour of blood pumped through the skin can change the appearance of individuals almost instantly.

Charles Darwin, in a rare misinterpretation for this genius, considered facial flushing as a trait exclusive of humans: "blushing is the most peculiar and the most human of all expressions" (Darwin 1872). He overlooked the fact that many birds, including several raptors, do actually flush blood to their bare heads in certain circumstances.

The species flushing facial skin colour are diverse and include several Old World and New World vultures, as well as the serpent eagle (*Sagittarius serpentarius*), the African harrier-hawk (*Polyboroides typus*) and the caracaras (*Polyborus* spp.). They all have in common to be of large size and to present dark plumage colouration. This morphological pattern is common to birds displaying flushing colouration in other avian orders. One of us (JJN) studied the anatomy of flushing skin in crested caracaras and hooded vultures (*Necrosyrtes monachus*). The bare skin showed a much higher density of blood vessels favouring blood flow (Negro et al. 2006). When skin temperature is high, the skin acquires a red colour. Opposite to that, when the skin is cold, there is vasoconstriction, and the perceived colour is that of the subcutaneous connective tissue (i.e. whitish collagen fibres).

Heat dissipation from internal organs to the skin is one of the recognized functions of the blood. The disproportionally large bill of the toucan (Tattersall et al. 2009) and the bare head of the bald ibis (Galván et al. 2017a) are good examples of such function in birds. But even if bare heads and necks evolved in the first place to avoid overheating in large and dark bird species, it is still possible that individuals use flushing to communicate condition or status to conspecifics. Some individuals flush in contexts such as agonistic encounters or when they become excited. The fact that some traits of physiological importance can secondarily become important for something else is not new in avian biology. This could well be the case of feathers. It is still debated whether they evolved first for insulation, for ornamentation or as an adaptation for flight, as the three benefits are definitely there.

Cosmetics in Vultures

So far, we have described cases of colouration generated endogenously, either by depositing pigments (be it melanins, carotenoids, porphyrins or haemoglobin) or by generating structure, which in raptors is responsible for pure white feathers, as in the practically totally white forms of gyrfalcons (*Falco rusticolus*) and variable goshawks (*Accipiter novaehollandiae*) or in the white feather patches of numerous

Fig. 2.9 Bearded vulture (*Gypaetus barbatus*) with the definitive adult plumage and rusty cosmetic colouration visible in the cap and around the neck. Spanish Pyrenees close to Ordesa National Park. (Credit @ JJ Negro)

species, including booted eagles (*Aquila pennata*) or the adult Andean condor. In this section, we will describe the use of adventitious colouration in two species of raptors, the bearded vulture and the Egyptian vulture. Coincidentally, these two species are their respective closest relatives, both are scavengers, and they both live sympatrically in large parts of their distribution range in the Old World. Most Egyptian vulture populations are migratory, except the ones in islands, whereas bearded vultures are sedentary.

Wild adult bearded vultures have an orange tinge in the plumage of their heads, breast and belly, with large individual variation and the occasional presence of birds with white coloration in those same body areas (Fig. 2.9). It is now firmly established that this colouration is cosmetic and derived from bathing in mud pools rich in iron oxides (Negro et al. 1999, 2002a, b; Negro and Margalida 2000). In fact, captive birds with no access to ferruginous waters grow white feathers when they undergo moulting of body feathers.

Cosmetic colours are not exclusive of bearded vultures, and some other species, such as waterfowl and cranes, are known to dye their plumage with iron oxides (Montgomerie 2006). The bearded vulture, nonetheless, is possibly the species where this behaviour is more elaborate (Negro et al. 1999).

Human use of cosmetics serves to establish hypotheses on the use of adventitious colours in other animals, namely, to increase sexual appeal, to threaten enemies, to protect the skin or to camouflage. What is the case in bearded vultures? It remains an open question. While Arlettaz et al. (2002) proposed that the rusty staining of the plumage served the function of protecting the eggs from infections, other possibilities cannot be entirely ruled out. An alternative hypothesis to the "medicinal" value of the iron oxides can be termed the status signalling hypothesis, first proposed by Sievert Rohwer (1975) to explain colour plumage variation in many bird species. There are asymmetries in the pattern of colouration of bearded vultures: females, the dominant sex, are generally more stained than males, and when polyandrous trios share the same territory, alpha males are more stained than beta males. In addition, adult birds seem to apply more iron oxides than immature birds (although the dull brown colour of the young birds makes it difficult to assess the extent of cosmetics in wild specimens). Bearded vultures often fight over food face to face with erected feathers. In these fights, individuals may assess each other for the intensity of colouration, revealing access to ferruginous sources (which are scarce) and demonstrating quality as foragers because they did have time to spare for bathing.

Curiously enough, in the Canary Islands population of Fuerteventura, a very dry place with no cattle ranching and where Egyptian vultures have dull yellow faces compared to other populations (Negro et al. 2002a), the individuals are known to cosmetically apply red dust to their plumage (van Overveld et al. 2017). The purpose of this painting behaviour is not yet known, but according to the authors who first described bathing behaviour, it could be used in pair formation, to keep the pair bond or to show off in sexual conflicts. At this point, it can only be said that bathing in red dust is a different behaviour compared to their "cousins", the bearded vultures. Red dust does not seem to be as limited as the ferruginous springs are for bearded vultures. And the extent and elaboration of dusting are less for Egyptian vultures in Fuerteventura. Are we perhaps observing the initial stages of a behaviour that may later develop into something more sophisticated? Time will tell.

In any case, the breast and belly plumage of both bearded and Egyptian vultures are white, and staining obviously eliminates the immaculateness of white. Lack of plumage uniformity, produced, for example, as a consequence of aggressive interactions, is easily perceived over white plumage, and thus white plumage patches are known to act as amplifiers of dominance in other birds (Galván and Sanz 2008, 2009). Birds with feather imperfections would therefore benefit from concealing their white, but non-immaculate plumage, with dirt. Thus, birds with the capacity of applying dust to their plumage may benefit from this. If this was the case, the cosmetic colouration of bearded and Egyptian vultures would play the opposite function of amplifiers: these would be attenuators, as low-quality individuals (e.g. subordinate individuals with poorly immaculate plumage) would benefit from staining, while high-quality individuals would pay a social (breeding) cost for hiding their immaculateness (Hasson 1997). It seems clear that a high variability exists between individual bearded and Egyptian vultures regarding the frequency with which they apply dust to their plumage and also in the intensity and extent of cosmetic colouration (Negro et al. 1999). Future studies should consider this behaviour in the context of attenuator signals, whose evolutionary stability has recently been shown from a theoretical perspective (Hackett and Ruxton 2018).

Deceptive Plumage Signals in Birds of Prey

Some species of raptors have eyelike patches in the back of their head, mimicking a face and thus sending a deceptive signal of vigilance. These markings have been observed in numerous species of the Falconiformes. No less than 22 species of the genus *Falco* show these "ocelli" in their napes, at least one pygmy falcon (i.e. *Polihierax semitorquatus*) and one New World falconet (i.e. spot-winged falconet (*Spiziapteryx circumcincta*)) (Negro et al. 2007b). Markings in the nape have also been observed in some *Accipiter* hawks and some small owls, where these markings were actually first described. But what is the reason behind these false eyes? The first proposed explanation was that a false face would be directed towards potential predators. By making eye contact, even if the eye is false, a small falcon or owl would tell a larger predator that it has been detected and that the potential victim is warned and is about to escape or defend itself. In other words, it cannot be taken by surprise and off-guard.

But there is an alternative, not mutually exclusive, to this explanation. Negro et al. (2007b) noted that the raptorial species sporting ocelli typically included a large proportion of passerines in their diets. The false faces, they said, may have evolved as an adaptive mechanism to provoke mobbing by small birds, which is indeed observed and includes the approaching of the raptor from behind and very closely, sometimes only a few centimetres away or even touching the head of the mobbed individual.

The immediate benefit for the raptor would be to have a potential prey at reach, particularly if misdirected to the frontal part by the deceptive signal, and also the possibility to evaluate hunting prospects in a given area or to detect the presence of breeding birds raising a brood of nestlings. One of us (JJN) has recently witnessed a Eurasian kestrel *Falco tinnunculus*, which do not even have clear false eyes in the nape, enduring close mobbing by two adult common swallows *Hirundo rustica* (Fig. 2.10). After a session of mobbing lasting about 4 min, the kestrels made a dive towards the nest of the couple of mobbing swallows to prey on the nestlings. We can infer from this observation that the kestrel stood up the mobbing attacks to try to locate the provenance of the birds, which it finally managed to do.

Full Plumage Mimicry

A visual mimic is an organism belonging to a species that resembles another one, called the model, but the two species have not a close phylogenetic relationship. The best known cases of mimicry concern invertebrates, such as bees or butterflies, and two broad types are recognized depending on which species benefits from the resemblance. If the mimic benefits and the model losses from the relationship, this type of mimicry is called Batesian. If both species benefit, this mimicry is called Müllerian (e.g. Speed 1999).

Fig. 2.10 Eurasian kestrel (*Falco tinnunculus*) mobbed by a barn swallow (*Hirundo rustica*). Passerine birds do mob both diurnal and nocturnal birds of prey when they get close to their nests and may threaten their nestlings. Sometimes they pay the price of being captured and eaten by the mobbed raptor or owl. (Credit @ JJ Negro)

Two independent and previously overlooked cases of mimicry involving African snake-eating eagles have been recently proposed (Negro 2008). One mimic model system is formed by both the West African serpent eagle *Dryotriorchis spectabilis* and Cassin's hawk eagle *Spizaetus africanus* (=*Aquila africana*). The other case is formed by the Madagascar serpent eagle *Eutriorchis astur*, and its look-alike is the Madagascar goshawk *Accipiter henstii*. The similarity in size and plumage patterns is outstanding within every system of species. Plus the species live in sympatry either in the jungles of Central Africa (the first system of the two) or in Madagascar. This would be a *sine qua non* condition for such a mimicry system to evolve.

Other species of raptors are said to be involved in mimicry. Raptor biologist Keith Bildstein (2017) reports several cases in two categories. This author distinguishes among "aggressive mimicry" in which a species takes advantage of looking like a nonpredatorial species to deceptively approach potential prey, and "defensive mimicry" in which a nonaggressive species looks like a more powerful species to avoid attacks or mobbing by other birds. Cases of aggressive mimicry includes the tandem formed by the zone-tailed hawk *Buteo albonotatus* and the turkey vulture, as well as the white colour morph of the grey goshawk and the white cockatoos they often take as prey. A case of defensive mimicry concerning raptors is illustrated by Prum (2014) in his general survey of mimicry in class Aves, and involves the heavily armed ornate hawk-eagle *Spizaetus ornatus* as model and the immature of the grey-bellied goshawk *Accipiter poliogaster* as the mimic.

Nest Decorations

Avian nests, particularly in small and defenceless species, are concealed to minimize predation risk. Raptors, particularly the large ones, have few or no potential predators and often have very massive and very exposed nests. Many species which reuse nest platforms add new green material every year (Newton 1979). It has been suggested that green plants may have a role as bactericides, but no conclusive evidence has ever been provided. It may all be a way to advertise nest occupancy to conspecifics or even to other raptor species. For instance, golden eagles *Aquila chrysaetos* and Bonelli's eagles *A. fasciata* are known to use the same rocky outcrops to build their nests in places where they live in sympatry, and they are aggressive to intruders of the same or different species (Carrete et al. 2002). An additional twist concerning raptor nest advertising is added by species such as the black kite *Milvus migrans*, a raptor known to decorate its stick nest with white artificial materials such as plastics and paper (Sergio et al. 2011). These authors hypothesized that decorated nest signal territory occupancy to trespassing conspecifics. But the pattern of signalling was complex. The nests of younger, inexperienced breeders are decorated minimally. Decorations peaked for birds in prime age -7 to 11 years old-, and declined again in older birds. It was suggested that only birds with full force would dare to advertise the nest conspicuously because they were not afraid of chasing off intruders. Lower-quality individuals, even if they attempted to breed, kept a low profile and tried to pass unnoticed.

An interesting follow-up of the previous study on kites demonstrated that decorations actually increase the conspicuousness of the nests from the air, at a kite eye view (Canal et al. 2016). This investigation utilized drones to generate aerial images of the tree nests built by kites and served to assess detectability from the air.

Communal Roosting in Territorial Raptors

The former statement may sound like an oxymoron, but it is true that numerous species of raptors are strictly solitary and repel conspecifics during the day but change their minds as night approaches and gather at communal roost sites. Species showing this alternating social behaviour are numerous and include both small falcons such as the merlin *Falco columbarius*, the *Elanus* kites *Elanus* spp., black kites, red kites, marsh harriers *Circus aeruginosus* and Egyptian vultures. Many other less-studied species may behave in this way too.

The fact is that fully social raptors also tend to gather at night, and it is not surprising to find roost sites of species that do forage together in daytime and/or are colonial breeders. Classical examples are provided by both Old World vultures such as the griffon *Gyps fulvus* and New World vultures such as the turkey vulture *Cathartes aura* and the black vulture *Coragyps atratus*, but also flocking Swainson's hawks *Buteo swainsoni* in their wintering grounds in Argentina, and the highly colonial lesser kestrel, or the Amur falcon, *F. amurensis*. For the latter, forming large roost sites during migration was revealed to be a death sentence for many birds easily massacred by the hundreds of thousands in a remote area of India for local markets (http://www.conservationindia.org/campaigns/amur-massacre). Fortunately, awareness campaigns have halted the slaughter, and no killings have been reported in the last few years.

In some cases, mixed-species roost sites have been found. For instance, black-winged kites in the Iberian Peninsula are known to roost along with merlins, lesser kestrels and magpies *Pica pica*. Although not always all together, often three species were roosting together, either on planted fruit groves or on natural oak trees in a savannah-like habitat with scattered trees over pastureland locally called dehesa (Balbontín et al. 2008).

Why birds of prey gather at night has been explained both in terms of increased vigilance against predators and because the individual reduces its chances of being the one picked by the predator by a risk dilution effect – they are both regarded as general explanations for animal gatherings outside the breeding season. The above-mentioned case involving relatively small-sized species roosting together would point in this direction, particularly knowing that black-winged kites were seen to suffer predation at sunset and at night (the suspects were peregrine falcons, common buzzards and eagle owls). A more popular explanation has been the information centre hypothesis (Ward and Zahavi 1973), according to which individuals may share information about potential sources of food around the roost site.

Other explanations have been put forward, including improvements in thermoregulation at night, or the "two-strategies hypothesis", proposed by Weatherhead (1983). This hypothesis proposes that different individuals join communal roosts for different reasons. Low-rank individuals would join to increase foraging opportunities, but high-rank individuals would gain access to the best places in the roost sites, surrounded by subordinates in the periphery and thus protected by a shield of individuals that would first suffer from predation if predators attacked the communal roost.

An interesting possibility raised by Blanco and Tella (1999) to explain roosting behaviour in choughs *Pyrrhocorax pyrrhocorax* poses that communal roosts in this species might serve to increase mating opportunities for some individuals that would get to know other unpaired individuals of the opposite sex. Roost sites would be like "bars" or "clubs" where individuals looking for mates, either for the first time or because they just lost a former mate, would find new companions. This intriguing but rather logical idea should also be explored in raptors forming nighttime aggregations.

Nonmonogamous Mating in Raptors

Polygyny

Raptors at large are socially monogamous, although we have already explained that individuals of both sexes are known to practise a mixed-reproductive strategy by engaging in extra-pair copulations. This promiscuity is in fact widespread in class Aves, and raptors, if anything, are not the most promiscuous of all. Apart from these cases of non-strict monogamy, a few species overtly engage in polygamous relationships. Polygyny is frequent, for example, among harriers of the genus *Circus* (Simmons 2000). If polygyny is rare in raptors, with less than 5% species practising this mating strategy, a majority of practisers are in fact harriers. From a minimum of one male and two females, some harems are known to go up to one male and seven females in the case of the hen harrier (*Circus cyaneus*) (Kraan and Strien 1969). In a famous long-term study of the same species in the Orkney Islands, polygyny levels were record-breaking, with nests of polygynous males comprising close to 90% of the sample (Picozzi 1984) and several instances of harems of up to six females. The reasons behind polygyny in harriers have been analysed at length by R. Simmons, but the truth is that its adaptive value is not apparent. Females in harems generally perform poorly compared to those in couples, and it can only be speculated that they follow the best-of-a-bad-job strategy in populations with female-biased sex ratios. These polygynous females might also be looking for future opportunities to increase their status and gain access to better territories or good genes from sexy males. R. Simmons' analysis concluded, however, that a better explanation than skewed sex ratios was the deceptive acquisition of females by polygynous male harriers. Some females in the harem would be fooled by males who started to feed them but later ceased to do so and just provided for alpha females. This hypothesis would be supported by higher rates of polygyny in northern latitudes, with higher prey availability compared to temperate zones and also with even higher polygyny rates at peak vole years (voles are the staple prey of the harriers).

Excepting harriers, polygyny is unusual among raptors. Until quite recently, only the Galapagos hawk *Buteo galapagoensis* was known to regularly engage in polygamous relationships. The Galapagos hawk, an island endemic, was the first raptorial species in which mixed paternity was confirmed by molecular analysis (Faaborg et al. 1995). Newton reported 1% of cases among European sparrowhawks *Accipiter nisus* (Newton 1986), with about 1000 nest surveyed over the years. We have only recorded a few cases in the colonial lesser kestrel in Spain. First, two cases of "aborted" polygyny were observed, with the trios breaking up before egg laying actually occurred (Hiraldo et al. 1991). Later on, a polygynous case was confirmed in lesser kestrels by DNA fingerprinting (Tella et al. 1996). This case involved a bigamous male and two females that laid eggs in the same nest at a kestrel colony in Monegros, northeastern Spain, in a recently established kestrel population which was rapidly expanding at the time the reported event took place. The molecular analysis demonstrated that the attending male fathered all four nestlings of the brood and that the two attending females were the mothers of one and three nestlings, respectively.

Polyandry

Until quite recently, and among raptors, only the Galapagos hawk was known to regularly engage in a breeding system described as cooperative polyandry (Faaborg et al. 1995). A female could mate with as many as eight different males, which copulated with her and provided care and food for the brood. Several males were seen to sire nestlings in the broods of the polyandrous females.

Other polyandrous raptors have been discovered and studied in later years: bearded vultures are known to engage in polyandrous trios formed by a female plus an alpha and a beta male (Bertran and Margalida 2002). And a similar association has been reported for Egyptian vultures (Tella 1993). Kimball et al. (2003) reviewed cooperative breeding in diurnal raptors and found a relatively high frequency compared to other bird taxa. In fact, 42 species of both Accipitriformes and Falconiformes (about 14% of all diurnal raptor species) raised their offspring cooperatively. This included polyandry (a majority of cases with extra adult males) but also some polygamous cases and a few instances of helpers at the nest which were young or immature birds of previous breeding events by the resident couple. The high incidence of non-strict monogamy calls for further study to determine why raptors are so singular among birds. In fact, after Kimball et al.'s (2003) review, scarcely studied species have been added to the list little by little. For instance, the striated caracara *Phalcoboenus australis* in the archipelago of Diego Ramirez in Subantarctic Chile, at 56 degrees of South Latitude, and thus one of the southernmost populations for any raptor (Raimilla et al. 2014). In this particular instance, a single male was seen copulating with two different females, and the trio attended the same nest.

Acquisition of Hunting Skills and Play with Objects

To the layperson it may seem puzzling that raptors have innate hunting abilities that they perfect over time without being taught by their parents or other adults. Even though there are descriptions of adults teaching their young how to hunt, such as Meinertzhagen's (1954) romantic depictions of adult ospreys dropping fish in front of their offspring for them to learn how to make the hunt, it is rather clear that raptors learn to hunt by themselves following a trial and error process (Newton 1979). In fact, the post-fledging dependence period for a majority of species is again surprisingly short. Parents, with some exceptions in very large species, tend to leave their offspring on their own soon after they fledge and are capable of flying around. In many instances, adults actually expel their recently fledged progeny from their territories in an aggressive manner.

The best proof of the innate hunting behaviour of raptors is the success of a majority of reintroduction projects using the hacking method (Negro et al. 2007a, b). This raptor propagation technique very often using captive-bred juveniles has been used for practically every possible "type" of raptor, including small species such as the lesser kestrel (Rodríguez et al. 2013), harrier species of the genus *Circus* (Amar et al. 2000) and red kites (Murn and Hunt 2008) but also large eagles such as the Spanish imperial eagle, the white-tailed eagle *Haliaeetus albicilla*, bald eagles *H. leucocephalus*, ospreys or harpy eagles (Campbell-Thompson et al. 2012). Vulture species have also been hacked, including both Old World species, such as the Egyptian vulture and the bearded vulture, and New World species, such as the California condor *Gymnogyps californianus*.

Therefore, no matter how large the raptors are, or the type of food they hunt or scavenge, it is clear that they innately learn to acquire food and do not need the concourse of the adult parents. Nonetheless, this important observation does not preclude the investigation of the actual development of hunting techniques. In this respect, information is sorely lacking. We know little about the process by which falcons such as the peregrine learn to master skydiving, how goshawks learn to avoid killing themselves when flying at full speed in a dense forest or whether hovering in the case of kestrels and other species such as buzzards, short-toed eagles and *Elanus* kites requires some time to be effective in hunting.

One of the few attempts to get a glimpse into the maturation of hunting skills in raptors was an experiment conducted on captive young American kestrels at the Avian Science and Conservation Centre of McGill University in Canada (Negro et al. 1996a). Different objects were offered to captive fledgling kestrels to test whether they preferred to play with objects resembling prey. Individuals were divided into two groups: half of the birds were offered mouse mimics, along with some inanimate objects which were alternated daily. The remaining half of the birds was offered bottle corks and the same type of alternative objects as in the other group. There were no significant differences in the mean number of play instances of birds in each of the two treatments. However, fledglings in group A played significantly more with the mouse mimics, whereas fledglings in group B played with the different objects in proportion to their availability. These results supported the hypothesis that young raptors devote a relatively fixed amount of time to object manipulation and that they prefer to manipulate objects resembling a natural prey. The same objects were offered to adult kestrels, but they were all ignored. This experiment suggested that object play has a function in the maturation of hunting skills of young raptors.

References

Amar A, Arroyo BE, Bretagnolle V (2000) Post-fledging dependence and dispersal in hacked and wild Montagu's harriers *Circus pygargus*. Ibis 142:21–28

Arlettaz R, Christe P, Surai PF, Møller AP (2002) Deliberate rusty staining of plumage in the bearded vulture: does function precede art? Anim Behav 64:F1–F3

Arnold KE, Owens IPF (1998) Cooperative breeding in birds: a comparative test of the life history hypothesis. Proc R Soc Lond B 265:739–945
Arroyo B, Mougeot F, Bretagnolle V (2013) Characteristics and sexual functions of sky-dancing displays in a semi-colonial raptor, the Montagus Harrier *Circus pygargus*. J Raptor Res 47:185–196
Balbontín J, Negro JJ, Sarasola JH, Ferrero JJ, Rivera D (2008) Land-use changes may explain the recent range expansion of the black-winged kite *Elanus caeruleus* in southern Europe. Ibis 150:707–716
Balgooyen TG (1976) Behavior and ecology of the American kestrel (*Falco sparverius* L.) in the Sierra Nevada of California. Univ Calif Publ Zool 103:1–83
Barber I, Folstad I (2000) Schooling, dusk flight and dance: social organisations as amplifiers of individual quality? Oikos 89:191–194
Beauchamp G, Heeb P (2001) Social foraging and the evolution of white plumage. Evol Ecol Res 3:703–720
Bertran J, Margalida A (2002) Social organization of a trio of bearded vultures *Gypaetus barbatus*: sexual and parental roles. J Raptor Res 36:66–70
Bildstein K (2017) The curious nature of birds of prey. Comstock Publishing/Cornell University Press, Ithaca
Bird DM, Buckland RB (1976) The onset and duration of fertility in the American kestrel. Can J Zool 54:1595–1597
Birkhead TR, Møller AP (1992) Sperm competition in birds: evolutionary causes and consequences. Academic, London
Birkhead TR, Atkin L, Møller AP (1987) Copulation behaviour of birds. Behaviour 101:101–138
Blanco G, Tella JL (1999) Temporal, spatial and social segregation of red-billed choughs between two types of communal roosts: a role for mating and territory acquisition. Anim Behav 57:1219–1227
Blas J, Cabezas S, Figuerola F, López L, Tanferna A, Hiraldo F, Sergio F, Negro JJ (2013) Carotenoids and skin coloration in a social raptor. J Raptor Res 47(2):174–184
Bortolotti G, Negro JJ, Tella JL, Marchant T, Bird D (1996) Sexual dichromatism in birds independent of diet, parasites and androgens. Proc R Soc London B 263:1171–1176
Bortolotti GR, Stoffel M, Galván I (2011) Wintering Snowy Owls *Bubo scandiacus* integrate plumage colour, behaviour and their environment to maximize efficacy of visual displays. Ibis 153(1):134–142
Campbell-Thompson E, Hernán Vargas F, Watson RT, Muela A, Cáceres NC (2012) Effect of sex and age at release on the independence of hacked Harpy Eagles. J Raptor Res 46(2):158–167
Canal D, Mulero-Pázmány M, Negro JJ, Sergio F (2016) Decoration increases the conspicuousness of raptor nests. PLoS One 11(7):e0157440
Carrete M, Sánchez-Zapata JA, Martínez JE, Sánchez MA, Calvo JF (2002) Factors influencing the decline of a Bonelli's eagle *Hieraaetus fasciatus* population in southeastern Spain: demography, habitat or competition? Biodivers Conserv 11:975–985
Dale JM, Dey CJ, Delhey K, Kempenaers B, Valcu M (2015) The effects of life history and sexual selection on male and female plumage colouration. Nature 527:367–370
Darwin CR (1871) The descent of man and selection in relation to sex. John Murray, London
Darwin CR (1872) The expression of the emotions in man and animals. John Murray, London
Donazar JA, Negro JJ, Hiraldo JA (1992) Functional analysis of mate-feeding in the Lesser Kestrel *Falco naumanni*. Ornis Scand 23:190–194
Faaborg J, Parker G, DeLay L, de Vries T, Bednarz JC, Maria Paz S, Naranjo J, Waite T (1995) Confirmation of cooperative polyandry in the Galápagos hawk (*Buteo galapagoensis*). Behav Ecol Sociobiol 36:83–90
Ferguson-Lees J, Christie DA (2001) Raptors of the world. Houghton Miffling Company, Boston/New York
Ferrero JJ, Grande JM, Negro JJ (2003) Copulation behaviour of a potentially double-brooded bird of prey, the black-winged kite (*Elanus caeruleus*). J Raptor Res 37:1–7

Galván I, Sanz JJ (2008) The cheek plumage patch is an amplifier of dominance in great tits. Biol Lett 4:12–15
Galván I, Sanz JJ (2009) Cheek plumage uniformity as a social status signal in great tits. Ann Zool Fenn 46:271–282
Galván I, Sanz JJ (2011) Mate-feeding has evolved as a compensatory energetic strategy that affects breeding success in birds. Behav Ecol 22:1088–1095
Galván I, Negro JJ, Bortolotti GR, Margalida A (2009) On silver wings: a fragile structural mechanism increases plumage conspicuousness. J Avian Biol 40:475–480
Galván I, Camarero PR, Mateo R, Negro JJ (2016) Porphyrins produce uniquely ephemeral animal colouration: a possible signal of virginity. Sci Rep 6:39210
Galván I, Palacios D, Negro JJ (2017a) The bare head of the Northern bald ibis (*Geronticus eremita*) fulfills a thermoregulatory function. Front Zool 14:15. https://doi.org/10.1186/s12983-017-0201-5
Galván I, García-Campa J, Negro JJ (2017b) Complex plumage patterns can be produced only with the contribution of melanins. Physiol Biochem Zool 90:600–604
Hackett S, Ruxton GD (2018) The evolutionary stability of attenuators that mask information about animals that social partners can exploit. J Evol Biol in press. https://doi.org/10.1111/jeb.13253
Harmata AR (1982) What is the function of undulating flight display in golden eagles? J Raptor Res 16:103–109
Hasson O (1997) Towards a general theory of biological signaling. J Theor Biol 185:139–156
Heredia R, Donázar JA (1990) High frequency of polyandrous trios in an endangered population of lammergeiers *Gypaetus barbatus* in Northern Spain. Biol Conserv 53:163–171
Hill GE (1990) Female house finches prefer colourful mates: sexual selection for a condition-dependent trait. Anim Behav 40:563–572
Hill GE (2006) Female mate choice for ornamental coloration. In: Hill G, McGraw K (eds) Bird Coloration. Vol. II: Function and Evolution. Harvard University Press, Cambridge, MA, pp 137–200
Hiraldo F, Negro JJ, Donazar JA (1991) Aborted polygyny in the lesser kestrel *Falco naumanni*. Ethology 89:253–257
Jarvis ED et al (2014) Whole-genome analyses resolve early branches in the tree of life of modern birds. Science 346:1320–1331
Kimball RT, Parker PG, Bednarz JC (2003) Occurrence and evolution of cooperative breeding among the diurnal raptors (Accipitridae and Falconidae). Auk 120:712–729
Kraan CVD, Strien NJV (1969) Polygamie bij de Blauwe Kiekendief (Circus cyaneus). Limosa 42(1):34–35
Kruger O (2005) The evolution of reversed sexual size dimorphism in hawks, falcons and owls: a comparative study. Evol Ecol 19:467–486
Lack D (1968) Population studies of birds. Clarendon Press, London
MacDonald PG, Olsen PD, Cockburn A (2005) Selection on body size in a raptor with pronounced reversed sexual size dimorphism: are bigger females better? Behav Ecol 16:48–56
Meinertzhagen R (1954) The education of young Ospreys. Ibis 96:153–155
Møller AP (1987) Copulation behaviour in the goshawk, *Accipiter gentilis*. Anim Behav 35:755–763
Montgomerie R (2006) Analyzing colors. In: Hill GE, McGraw K (eds) Bird coloration, vol I. Harvard University Press, Cambridge, MA, pp 90–147
Murn C, Hunt S (2008) An assessment of two methods used to release red kites (*Milvus milvus*). Avian Biol Res 1:53–57
Negro JJ (1997) Falco naumanni lesser kestrel. Birds of the Western Palearctic. Update 1:49–56
Negro JJ (2008) Two aberrant serpent-eagles may be visual mimics of bird-eating raptors. Ibis 50:307–314
Negro JJ, Grande JM (2001) Territorial signalling: a new hypothesis to explain frequent copulation in raptorial birds. Anim Behav 62(4):803–809

Negro JJ, Margalida A (2000) How Bearded Vultures (*Gypaetus barbatus*) acquire their orange coloration: a comment on Xirouchakis (1998). J Raptor Res 34:62–63
Negro JJ, Donázar JA, Hiraldo F (1992) Copulatory behaviour in a colony of lesser kestrels: sperm competition and mixed reproductive strategies. Anim Behav 43:921–930
Negro JJ, Bustamante J, Milward J, Bird DM (1996a) Captive fledgling American kestrels prefer to play with objects resembling natural prey. Anim Behav 52:707–714
Negro JJ, Villarroel MR, Tella JL, Kuhnlein U, Donázar JA, Bird DM (1996b) DNA fingerprinting reveals a low incidence of extra-pair fertilizations and intraspecific brood parasitism in the lesser kestrel. Anim Behav 51:935–943
Negro JJ, Margalida A, Hiraldo F, Heredia R (1999) The function of the cosmetic coloration of bearded cultures: when art imitates life. Anim Behav 58:F14–F17
Negro JJ, Grande JM, Tella JL, Garrido J, Hornero D, Donázar JA, Sánchez-Zapata A, Benítez JM, Barcell M (2002a) Coprophagy: an unusual source of essential carotenoids. Nature 416:807–808
Negro JJ, Margalida A, Torres MJ, Grande JM, Hiraldo F, Heredia R (2002b) Iron oxides in the plumage of bearded vultures. Medicine or cosmetics? Anim Behav 64:F5–F7
Negro JJ, Sarasola JH, Fariñas F, Zorrila I (2006) Function and occurrence of facial flushing in birds. Comp Biochem Physiol A 143:78–84
Negro JJ, Sarasola JH, Barclay JH (2007a) Augmenting wild populations and food resources. In: Bird DM, Bildstein KL (eds) Raptor research and management techniques. Hancock House, Blaine, pp 401–410
Negro JJ, Bortolotti GR, Sarasola JH (2007b) Deceptive plumage signals in birds: manipulation of predators or prey? Biol J Linn Soc 90:467–477
Negro JJ, Bortolotti GR, Mateo R, García IM (2009) Porphyrins and pheomelanins contribute to the reddish juvenal plumage of black-winged kites. Comp Biochem Physiol B 153:296–299
Newton I (1979) Population ecology of raptors. T. & A.D. Poyser, Berkamsted
Newton I (1986) The Sparrowhawk. T. & A.D. Poyser, Berkamsted
Parejo D, Avilés JM, Ferrero JJ, Rivera D, Casas JM (2001) Communal roosting and diet of black-winged kite (*Elanus caeruleus*) wintering in southwestern Spain. J Raptor Res 35:162–164
Parker GA (1970) Sperm competition and its evolutionary consequences in the insects. Biol Rev 45:525–567
Penteriani V, Delgado MM, Campioni L, Lourenço R (2010) Moonlight makes owls more chatty. PLoS One 5(1):e8696. https://doi.org/10.1371/journal.pone.0008696
Perreault S, Lemon RE, Kuhnlein U (1997) Pattern and correlates of extrapair paternity in American redstarts (*Setophaga ruticilla*). Behav Ecol 8:612–621
Picozzi N (1984) Breeding biology of polygynous Hen harriers *C. c. cyaneus* in Orkney. Ornis Scand 15(1):10
Poole A (1985) Courtship feeding and osprey reproduction. Auk 102:479–492
Prum RO (2006) Anatomy, physics, and evolution of avian structural colors. In: Hill GE, McGraw K (eds) Bird coloration, vol I. Harvard University Press, Cambridge, MA, pp 295–353
Prum RO (2014) Interspecific social dominance mimicry in birds. Zool J Linnean Soc 172:910–941
Raimilla V, Suazo CG, Robertson G, Rau JR (2014) Observations suggesting cooperative breeding by Striated Caracaras (*Phalcoboenus australis*). J Raptor Res 48(2):189–191
Rodríguez A, Negro JJ, Bustamante J, Antolín J (2013) Establishing a lesser kestrel colony in an urban environment for research purposes. J Raptor Res 47(2):214–218
Rohwer S (1975) The social significance of avian winter plumage variability. Evolution 29:593–610
Sarasola, JH, Negro JJ, Bechard, M (2011) Not as similar as thought: Sexual dichromatism in Chimango Caracaras is expressed in the exposed skin but not in the plumage. J Ornithol 152:473–479
Sergio F, Blas J, Blanco G, Tanferna A, Lopez L, Lemus JA et al (2011) Raptor nest decorations are a reliable threat against conspecifics. Science 331:327–330
Serrano D, Tella JL, Forero MG, Donázar JA (2001) Factors affecting breeding dispersal in the facultatively colonial lesser kestrel: individual experience vs. conspecific cues. J Anim Ecol 70:568–578

Simmons R (2000) Harriers of the world: their behaviour and ecology. Oxford University Press, Oxford, UK
Speed M (1999) Batesian, quasi-Batesian or Mullerian mimicry? Theory and data in mimicry research. Evol Ecol 13:755–776
Tattersall GJ, Andrade DV, Abe AS (2009) Heat exchange from the toucan bill reveals a controllable vascular thermal radiator. Science 325:468–470
Tella JL (1993) Polyandrous trios in a population of Egyptian vultures (*Neophron percnopterus*). J Raptor Res 27:119–120
Tella JL, Negro JJ, Villarroel M, Kuhnlein U, Hiraldo F, Donázar JA, Bird DM (1996) DNA fingerprinting reveals polygyny in the lesser kestrel (*Falco naumanni*). Auk 113:262–265
Trivers RL (1972) Parental investment and sexual selection. In: Campbell B (ed) *Sexual selection and the descent of man*, 1871–1971. Aldine, Chicago, pp 136–179
Van Overveld T, de la Riva MJ, Donázar JA (2017) Cosmetic coloration in Egyptian vultures: mud bathing as a tool for social communication? Ecology 98:2216–2218
Ward P, Zahavi A (1973) The importance of certain assemblages of birds as "information centers" for food finding. Ibis 115:517–534
Weatherhead PJ (1983) Two principal strategies in avian communal roosts. Am Nat 121:237–243
Wiącek J (2006) Food transfer in Montagu's harrier Circus pygargus during the courtship. Acta Ornithol 41:88–91

Chapter 3
Breeding and Nesting Biology in Raptors

Luis Tapia and Iñigo Zuberogoitia

Introduction

Raptors are limited by suitable breeding habitat, and they have specific nest-site requirements. Habitats of high quality presumably have the resources required to sustain relatively high rates of survival and reproduction. High-quality individuals would occupy territories of higher quality and would have greater fitness. Many birds may use their own reproductive success to assess the quality of their territories, and breeding failure would act as a determinant for dispersal, increasing an individual's propensity to move to a better habitat. Food supply, nest-site availability, weather conditions and bird experience seem to act through the body condition of the female and are known to limit raptor populations. Quality of nesting territories and breeding success vary widely with different factors.

In this chapter, we review the scope and objectives of breeding and nesting biology studies in birds of prey updating scientific and conservation knowledge in this field.

Breeding and Nesting Habitat Selection

Habitat selection is a consequence of natural selection that favours individuals that preferentially settle in the best habitats, so that they maximize their biological effectiveness (Fuller 2012). Therefore, the habitat selection process is related to habitat quality and long-term population persistence through reproductive success

L. Tapia (✉)
Department of Zoology Genetics and Physical Anthropology, Universidade de Santiago de Compostela, Santiago de Compostela, Galicia, Spain
e-mail: luis.tapia@usc.es

I. Zuberogoitia
Estudios Medioambientales Icarus S.L. Icarus S.L. C/San Vicente, Bilbao, Bizkaia, Spain

J. H. Sarasola et al. (eds.), *Birds of Prey*,
https://doi.org/10.1007/978-3-319-73745-4_3

(Boulinier et al. 2008). In the case of raptors, breeding habitat is an essential resource because it must guarantee food and protection for a long breeding period (Lõhmus 2004; Tapia et al. 2007a). Thus, breeding-nesting habitat selection reports on the basic requirements of raptors in a critical period of their life cycle (Orians and Wittenberger 1991). Management and improvement of breeding raptors' habitat have broader positive effects on the whole of biodiversity, because raptor populations can act as valuable bioindicators of changes and stresses in ecosystems, as they are sensitive to changes in land use, habitat structure and habitat fragmentation and are highly susceptible to local extinctions (Sergio et al. 2005; Bednarz 2007; Burgas et al. 2014; Donázar et al. 2016).

Raptors are limited by suitable breeding habitat, and they have specific nest-site requirements (Newton 1979; Ferguson-Lees and Christie 2001). Each species has its own preferences for the position, height, orientation, shelter, accessibility and visibility of the nesting-site. The nest-site selections include upper parts of trees, caves, cliffs and on the ground (see Tapia et al. 2007a). A shortage of nest sites may also limit the density of a species below the carrying capacity of the habitat, for instance, in cliff-nesting raptors or mature forest-dwelling species (Newton 1998). Some species present a plasticity in nest-site selection, depending on geographic area, habitat or intra- and interspecific densities. Some nest sites are repeatedly occupied, even though the owners may change. In some cases, decades after becoming extinct, when a species returns to an area, they reoccupy the very same nest sites used in the past (Del Hoyo et al. 1994; Hardey et al. 2009).

Many species use old nests, and this strategy can save a good deal of time and energy at the beginning of the breeding season. Resident breeding raptors often have several nests in their territories and use them for breeding alternatively from one year to another (Millsap et al. 2015; Slater et al. 2017). They may also use alternative nest sites for roosting or preparing prey items. Therefore, they are biologically significant and warrant greater conservation consideration (Del Hoyo et al. 1994; Millsap et al. 2015). Management decisions should be based on alternative nest-use patterns within territory (Slater et al. 2017).

Many ecologically relevant aspects related to the mechanism of habitat selection at different spatial scales are involved in the occupancy process of a high-quality territory (Wiens et al. 1987; Newton 1998; Sánchez-Zapata and Calvo 1999). Study designs must be consistent with the abilities of the subject species to perceive and move among existing habitat patches, and researchers should consider the various scales at which habitat features may have influenced (Litvaitis et al. 1994; Morrison et al. 1998).

Breeding raptor habitat requirements sometimes are linked to the distribution of prey. Because it is difficult to observe predatory behaviour in most raptor species, the influence of prey on habitat use by raptors often is inferred by comparing, at the scale of activity areas, measures of prey abundance and raptor use among categories of vegetation types, structure or land uses (Graham and Redpath 1995; Marzluff et al. 1997; Selas 1997a, b; Bakaloudis et al. 1998; Ontiveros et al. 2005; Marti et al. 2007; Rodríguez et al. 2014). Use of land or vegetation structure by raptors often is positively related to prey abundance (Selas and Steel 1998; Ontiveros et al.

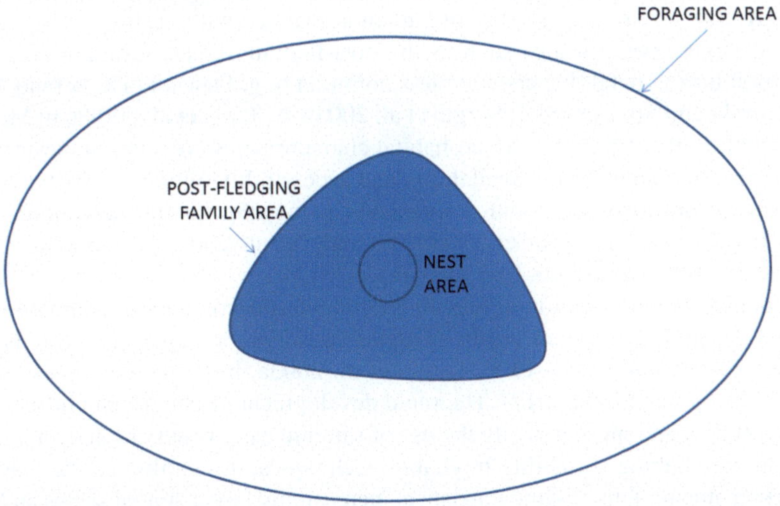

Fig. 3.1 Diagram of spatial scales used by raptors during the breeding season. (From Squires and Kennedy 2006)

2005; Rodríguez et al. 2014), but these relationships can be confounded by density of vegetation. Predation is sometimes more intense in areas where vegetation is sparser and less dense, regardless of prey abundance (Bechard 1982; Thirgood et al. 2003; Ontiveros et al. 2005; Rodríguez et al. 2014; Martínez et al. 2014), illustrating the need to distinguish, when possible, between prey abundance and availability (Tapia et al. 2007a).

Raptors are typically selective with regard to breeding and hunting habitat (Janes 1985). At least three levels of spatial scale used by raptors during the breeding season may be considered – nesting area, post-fledging family area (PFA) and foraging area (see Tapia et al. 2007a). The nesting area is the area immediately surrounding the nesting substrate, often contains alternative nests and may be reused in consecutive years, and these nests should be given the same protection as used nest in land-use planning (Millsap et al. 2015; Slater et al. 2017). The PFA surrounds the nesting area and is defined as the area used by the family group from the time the young fledge until they are no longer dependent on the adults for food. The foraging area is the area used by the provisioning adults and typically encompasses the remainder of the home range during the breeding season (Fig. 3.1). The influence of habitat features at different spatial scales is likely species-specific and can change with body size, mobility and life history requirements.

Raptors typically range over large territories that include heterogeneous habitats and landscapes (e.g. Pedrini and Sergio 2002; Martínez et al. 2003; Sergio et al. 2003b; Kudo et al. 2005; Watson 2010; Bruggeman et al. 2014; Tapia et al. 2017). Raptor populations are limited by the availability of breeding habitat at the microscale level (Bevers and Flater 1999), and different studies have been conducted to elucidate habitat use and preference of nest sites (e.g. Selas 1997a; Martínez and Calvo 2000; Finn et al. 2002; Poirazidis et al. 2004; Squires and

Kennedy 2006). However, studies of habitat associated with nesting activities that occur at larger spatial scales such as the post-fledging areas, foraging areas and areas used during natal dispersal are less common (e.g. Daw and DeStefano 2001; Bosakowski and Speiser 1994; Sergio et al. 2003a, b; Tapia et al. 2008a, b; Mañosa et al. 1998; Balbontín 2005). Macrohabitat characteristics (vegetation cover types, topography, human pressure, availability and accessibility of prey, etc.) of breeding home range are also important components in nesting habitat selection (Janes 1985; Bosakowski and Speiser 1994; McGrady et al. 2002; Sergio et al. 2004; Kudo et al. 2005; Rodríguez-Lado and Tapia 2012).

Modelling the distribution of breeding raptors has become more common in the last two decades, given their utility as raptor conservation management tools (e.g. Sánchez-Zapata and Calvo 1999; Sergio et al. 2003a, b; Bustamante and Seoane 2004; Tapia et al. 2007b, 2017). The rapid development of geographic information systems (GIS), and more recently the use of satellite earth observation data in combination with habitat suitability modelling techniques, has facilitated the handling and management of environmental data at increasingly larger spatial scales and provides repeatable, standardized and verifiable information for long-term monitoring of environmental indicators (Corsi et al. 2000; Sarasola et al. 2008; Franklin and Miller 2009; Pettorelli et al. 2014; Mairota et al. 2015). This type of approach, using raptors as biodiversity surrogates, is particularly suited to highly dynamic landscapes where the spatial interactions between the different ecological drivers strongly determine ecosystem functioning (Sarasola et al. 2008; Rodríguez-Lado and Tapia 2012; Gonçalves et al. 2015; Regos et al. 2017; Tapia et al. 2017).

Studies of breeding raptors' habitat descriptions in management plans should account for temporal changes in habitat preferences. A relatively long time scale would be an examination of the effects of land use and disturbances in dynamic landscapes on the raptors' habitat and protected areas' monitoring and conservation (Regos et al. 2017; Tapia et al. 2017). Some resident raptors use particular habitats during specific periods of the year, and only an assessment of habitat use during the full annual cycle would describe the species habitat preferences (Tapia et al. 2007a, 2008a; Hardey et al. 2009).

Several studies have identified apparent discrepancies between identified habitat preferences and breeding success for different raptor species (see Chalfoun and Schmidt 2012). This ecological effect is often discussed in the context of *Ecological Traps* (suboptimal habitats for breeding or population growth but which attracts individuals). Ecological traps have been observed particularly in environments altered by man that particularly affect birds of prey (Cardador et al. 2015; Torres-Orozco et al. 2016; García-Salgado 2017). Therefore, in order to know the real quality of raptors breeding habitats, it is convenient to study the relationship between habitat preferences and breeding success.

Limitation of breeding density by nest-site shortages is widespread among different species of raptors that use special breeding substrates, whether they are cliffs, burrows or scarce trees in open land (Negro et al. 2007). Breeding raptors often use artificial structures to support their nests or nest in abandoned nests of other species. For such species, nesting platforms can be used to increase nest availability, and this management technique is particularly useful for increasing and maintaining stable raptor populations. Installation of nest boxes can also provide nesting sites for some raptor species. Populations increased faster in areas with nest boxes than in areas without artificial nests and may reduce the effect of rodent pests (Paz et al. 2013).

Territoriality and Habitat Carrying Capacity

One of the basic conditions for the reproduction of raptors is the existence of breeding territories or hunting ranges (Newton 1979). The acceptable criteria for confirming occupation territory vary between raptor species due to the differences in their breeding behaviour and breeding habitat (Hardey et al. 2009). Territorial behaviour is a fundamental spacing system that limits breeding densities of raptors (Newton 1979; Brown 1989). This system informs the neighbours of the occupants of the territory and prevents them from crossing its boundaries. In raptors, with a large ratio of nonbreeders to breeders, it is usual for some individuals to wait several years beyond achieving physiological maturity or adult plumage before they breed (Del Hoyo et al. 1994).

Spatial distribution pattern of territories in suitable habitat is usually more or less regular, except in areas where human persecution is very intense. The natural reasons for deviating from this regularity are the non-uniform distribution of resources, especially food and nest-site availability (Newton 1979). Some raptor species may nest isolated when resources are distributed uniformly and predictably, while they do so in aggregate form when resources are unpredictable or concentrated in certain areas (Solonen 1993; Del Hoyo et al. 1994). This pattern of distribution has also been related to episodes of human disturbs and persecution (Newton 1998; Martínez-Abraín et al. 2010) and by the attraction for co-specifics (Donázar et al. 1993). Interspecific territorial behaviour occurs between sympatric species with morphological similarity and similar trophic requirements and/or nest sites (López-López et al. 2004; Rebollo et al. 2011; Rodríguez et al. 2017). There may be different types of ecological interactions such as mutualism, predation, facilitation, etc. among sympatric breeding raptors, which are essential to understand their spatial relationship (Rebollo et al. 2011).

Territorial behaviour implies a "trade-off" between the energetic cost of searching for and maintaining a territory and its potential benefits (Gordon 1997; Adams 2001). It is commonly assumed, therefore, that the choice of a breeding nest site is produced by natural selection (Martin 1998; Jones 2001), and the frequency of occupancy is generally considered a measure of territory quality (e.g. McIntyre 2002; Sergio and Newton 2003; Rodríguez et al. 2016). In some species human disturbances (e.g. timber harvest) in nest-site areas are associated with lower nest site and territory occupancy. However, if pairs do nest at timber harvest sites, their reproduction appears unaffected by these harvest activities, which would indicate individual differences in tolerance to human disturbances and/or previous breeding experience (Rodríguez et al. 2016).

Site fidelity is generally influenced by the breeding experience of individuals (Serrano et al. 2001; Sergio and Newton 2003; Schmidt 2004). A positive correlation between site fidelity and prior reproductive success (or, rather, between breeding failure and territory change) is a common finding for breeding raptors (Martínez et al. 2006; León-Ortega et al. 2017).

Information on the presence/absence of a raptor species in an area, although potentially an indicator that the area constitutes habitat, tells little about its quality. In contrast, measures of abundance of a given species in an area often are indicative of the relative quality of the area as habitat although it may be misleading in some situations (Van Horne 1983; Pulliam 1988). Perhaps the best indicators for assessing habitat quality for a given species are estimates of productivity and survival or combinations of both (e.g. rate of population change, λ). Unfortunately, these measures are difficult to obtain in short-term studies (see Tapia et al. 2007a). Estimating survival is especially problematic in small populations (Beeisinger and McCullough 2002; Steenhof and Newton 2007) and usually requires monitoring marked birds (e.g. banded, VHF telemetry, GPS satellite telemetry, stable isotopes) over extended periods and large spatial scales (see Bird and Bildstein 2007).

Habitats of high quality presumably have the resources required to sustain relatively high rates of survival and reproduction. Directly measuring the required resources present in an area (e.g. availability of prey or nest sites) is one way to assess habitat quality, but it requires that resources needed by the species in question are known and that resources measured are available for use (Marti et al. 2007). Another approach for assessing habitat quality is based on indicators of population health. Variation in territory or habitat quality "habitat carrying capacity" can act in the same way, reducing the average breeding rate in the population as numbers rise. The concept of carrying capacity encapsulates the notion that, in any area of habitat, resources must ultimately limit the numbers of raptors that can live there. But carrying capacity depends not only on features of the habitat, such as cover, food or nest sites, but also on raptor behaviour. Another concept of carrying capacity, which takes into account resource levels, concerns the "sustainable population size", which is the maximum number of individuals that can use a site over a defined period (see Newton 1998; Beeisinger and McCullough 2002).

While some raptor species live in the same areas and territories year-round, others become widely distributed between the breeding and nonbreeding seasons, and yet others spend the breeding and nonbreeding seasons in geographically separate areas. In addition, many species eat different types of food in the breeding and nonbreeding seasons, which may differ in their availability. For these two reasons, the total carrying capacities of the breeding and nonbreeding habitats of particular populations do not necessarily have to correspond, so that, as raptors return to nesting areas each year, they compete for territories or nest sites (Del Hoyo et al. 1994).

High-quality individuals (e.g. the more experienced) would occupy territories of higher quality (leading to higher occupancy rates for those territories) and would have greater fitness (Calsbeek and Sinervo 2002; Rutz et al. 2006; Martínez et al. 2006). Many birds may use their own reproductive success to assess the quality of their territories, and breeding failure would act as a determinant for dispersal, increasing an individual's propensity to move to a better habitat (Hoover 2003). In fact, for some species, site fidelity is conditioned by variations in habitat quality, and high-quality sites may outweigh the influence of previous breeding experience (Bried and Jouventin 1999; Jiménez-Franco et al. 2013).

Many studies have demonstrated that competitive interactions, both intra- and interspecific, play an important role in the spatial distribution and breeding habitat selection of raptors (e.g. Sergio et al. 2004; Martínez et al. 2008; Rodríguez et al. 2017). Territorial breeding raptors show a negative relationship between density and average productivity (Sergio and Newton 2003), as well as an increase in the proportion of low-quality territories occupied as the population size increases (Ferrer and Donázar 1996; Newton 1998; Krüger and Lindström 2001a).

Natal Dispersal and Floaters: Ecological Importance

Populations of raptors are composed by a sector of territorial breeders and a sector of nonbreeding individuals, "floaters". These individuals may be located far away from the breeding grounds or closely coexisting with territorial holders (e.g. see Prommer et al. 2012; Tanferna et al. 2013; Zuberogoitia et al. 2013b). Survival of a raptor population is strongly dependent on the dynamics of "floaters" and on the number of available settlement areas (Penteriani and Delgado 2009).

A substantial floating population of nonbreeding birds remains within territories. Nowadays it is not possible to determine the status of floater populations of most raptor species because nonbreeding birds are difficult to study due to their cryptic behaviour, differential habitat selection, spatial separation from breeders or potential long-distance dispersal (Tanferna et al. 2013). In some species of raptors, nonbreeding individuals may represent more than 50% of the total population (Kenward 2006). In most raptor species, nonterritorial birds are generally younger than territorial breeders but often show a marked hierarchy, with the more dominant ones acquiring territories before the others.

Floaters normally wait for opportunities to occupy a breeding territory which in turn are conditioned by the health of the breeding populations. The presence of nonbreeders is difficult to detect but repeatedly confirmed by the observations of rapid replacement of lost mates (by human persecution, electrocution, collision fatalities, etc.) in several raptor populations (golden eagle *Aquila chrysaetos*, northern goshawk *Accipiter gentilis*, Eurasian buzzards *Buteo buteo*, etc.) (Newton 1979; Kenward 2006; Watson 2010; Gil-Carrera et al. 2016). But in stable populations with low adult turnover rate, floaters may spend several years waiting for a vacancy. In fact, the time lapse between the loss of a breeder and the establishment of a new individual, the average age at first breeding and the occupancy rate of territories are useful factors to measure the health of the floater population (Zuberogoitia et al. 2009).

Floaters can also negatively interact with those breeding raptor communities that overlap in space (i.e. intraguild predation between dispersing individuals and their intraguild prey). Floating individuals may settle close to breeding pairs; thus, the effects of the intraguild predator on its intraguild prey may be underestimated because researchers often have not monitored the diet of these "invisible" floater individuals (Penteriani and Delgado 2009).

Floaters may prefer to settle in habitats similar to their natal habitats, because (a) this behaviour reduces the costs of assessing suitable new habitats, or (b) experience in natal habitat improves performance if a raptor settles in the same habitat type after dispersing (Stamps 2001). During dispersal, raptors are not territorial and may homogeneously distribute themselves in space; however, dispersal patterns may reveal that the distribution of individuals may be under constraints other than territoriality or location of food resources (Mañosa et al. 1998; Penteriani and Delgado 2009, in this book Chap. 4).

Potential future breeders of populations may spend a large part of their lives in high-risk areas. In fact, stochastic events, such as human persecution or collisions with power lines or vehicles, can seriously increase mortality rates in temporary settlement areas. Because conservation efforts for endangered raptor species populations focus primarily on breeding areas, conservation programs conducted in breeding territories can be ineffective if the genuine problem is in the settlement areas, and dispersing individuals may reveal the locations of crucial areas of conservation interest (Ferrer and Harte 1997; Ferrer et al. 2015).

Quality of Nesting Territories

Food Supply and Breeding Densities

Food supply, nest-site availability, weather conditions and bird experience seem to act through the body condition of the female and are known to limit raptor populations (Newton 1998; Mañosa et al. 1998; Dewey and Kennedy 2001). Clutch size and breeding success increase markedly with an improved food supply, and food shortage may reduce the population size through lowering breeding rates. Such positive relationships involve both numerical and functional responses to population increases in their main prey species (Newton 1998; León-Ortega et al. 2017). This effect may be hard to detect because of time lag between the food shortage and the resultant decline in breeding numbers. In long-lived species as raptors, it may take several years before the effects of poor breeding are reflected in poor recruitment (Watson 2010).

Raptors that have stable food supplies show the higher stability in breeding populations recorded in birds. This level of stability has been recorded in a wide range of species. Moreover, the same species may fluctuate numerically in one region, but not in another, depending on the stability of the local prey supply (Newton 1979, 1998). Raptors need regular trophic resources through the breeding season in order to establish breeding territories. If trophic resources keep at high enough level for feeding adults, they maintain territories throughout the year. This is the normal rule for raptors living at low and medium latitudes. However, in some regions, mainly northern distribution areas, raptors depend on the seasonal presence of resources (migrant birds) or seasonal availability of some key prey species (micromammals, fish), being obligated to migrate or develop a vagrant strategy.

Breeding success varies widely with geographical area, latitude, habitat quality and altitude, but food supply determines a species breeding success at different levels: the proportion of pairs actually breeding, the age of first breeding, clutch size and the quality of the eggs and thus, hatching success and the growth and survival rates of nestlings and fledglings (Del Hoyo et al. 1994). In addition, the availability of the main prey types influences floaters dispersal patterns and movements among the different settlement areas. A reduction in food availability may increase floaters' mortality rates within settlement areas, affecting the stability of breeding populations (Penteriani et al. 2006a, b).

Nest-Site Resources

Nest-site availability has been described as an important limiting factor for some raptor populations (Franco et al. 2005). There are several examples of areas full of favourable nest sites for certain raptor species but lacking regular availability of trophic resources in which these species are absent or keep low densities. On the opposite, other areas plenty of food but with low availability of nest sites show high density of some species but can merge as sink areas due to the impossibility of them to breed (e.g. see Fasciolo et al. 2016).

Most of the Accipitridae species are able to build their own nest with branches, wool of ungulates and even plastics and rubbish (Fig. 3.2). Thus although building a new platform requires energy and time, most of forest raptors, rocky eagles and vultures have several nests shared over their territories (Kochert and Steenhof 2012; Jiménez-Franco et al. 2014). However, most Falconidae species, particularly those of the genera *Falco*, do not build their nest and simply dig a bowl in the sand or gravel of a ledge, pothole or cave of a cliff, building a hole in a tree or reuse platforms of other species (e.g. Del Hoyo et al. 1994, Espie et al. 2004; Olsen 2014). The number of eyries per territory depends on the availability of adequate rocky substrate (cliffs, Zuberogoitia et al. 2015). In certain cases, one eyrie has been continuously used during decades, even centuries (Burnham et al. 2009).

Anthropic Pressure

Both trophic and nest-site availability may be seriously affected by anthropic pressure. Most times this pressure has a negative effect because it produces habitat loss, disturbances or direct persecution (Zuberogoitia et al. 2014; Donázar

Fig. 3.2 Female common buzzard (*Buteo buteo*) feeding her two nestlings. Forest-dwelling raptors build their own platforms, in this case in a pine tree. (Photo credit: Iñigo Zuberogoitia)

et al. 2016; Martínez et al. 2016, in this book Chap. 9), which cause both direct and indirect reductions on breeding raptor populations. Most of the threats are related to urban and suburban areas of all the continents, although new hazards are also affecting pristine areas, pushing some raptor species to critical conservation status (Donald et al. 2013). However, anthropic pressure can reach wider ranges because of the cascade effects of certain activities. For example, bearded vultures and sea eagles were exterminated from most of their European distribution due to direct persecution, although now they are recovering again after expensive reintroduction programs (Helander et al. 2008; Margalida et al. 2008a). Although humans should learn from their errors, similar threat (the use of diclofenac, a veterinary drug) has recently caused the depletion of vulture populations in Asia and Africa and poses a serious threat for European vultures (Ogada et al. 2015; Buechley and Sekerciogly 2016, in this book Chap. 19). Moreover, products of anthropic origin are contaminating most of the ecosystems, being consumed by all the species, being raptors deeply affected due to their top position in the trophic pyramid and bioaccumulation processes (Zuberogoitia et al. 2006, in this book Chaps. 10 and 11).

Also different forms of outdoor recreation have different spatiotemporal activity patterns that may have interactive or cumulative effects on raptor breeding biology and conservation, through human disturbance, physical habitat change or both (Spaul and Heath 2016). However, anthropic actions can turn a habitat previously unsuitable to suitable for a raptor species (e.g. open fields turned into timber plantations may favour forest raptors, Zuberogoitia and Martínez 2011). Novel habitats, as urban areas, also originate opportunities to some adaptable species which find cities plenty of food, although the final consequences for their populations are still unclear (Donázar et al. 2016, in this book Chap. 8).

Intraspecific Interactions. Ideal Free Distribution and Despotic Distribution

The ideal free distribution (IFD) assumes that there are no competitive asymmetries among individuals and that all individuals are equally "free" to occupy any space in the habitat. Thus, in environments where resources have a patchy distribution, relatively high-quality areas are expected to contain more individuals than relatively low-quality areas such that all individuals gain equal access to resources (Calsbeek and Sinervo 2002). However, ideal free distribution is rarely detected in raptors, and dominant individuals occupy the best areas and force less able competitors to unfavourable areas. This behaviour is defined as ideal despotic distribution (IDD). The higher-quality individuals would occupy the best areas, and this, in turn, would result in higher occupancy rates and higher breeding success and would have greater fitness. Under the IDD the best territories should be the most frequently occupied (Sergio and Newton 2003). In addition, because the better territories are occupied first, it is also generally assumed that their owners begin the reproduction sooner, which usually has a positive effect on breeding success (Espie et al. 2004; Pagan et al. 2009; Freund et al. 2017). However, some raptor species do not follow the IDD. Booted eagles (*Hieraaetus pennatus*) in south-eastern Spain followed a random nesting distribution, and territory occupancy rate was not significantly related to reproductive parameters (Pagan et al. 2009). Authors suggest that the lack of strong environmental variability could determine these results.

Individual Quality and Body Condition of Breeders

Under an ideal free distribution, equal competitors select habitats to maximize their individual fitness; however, under ideal despotic distribution or more complex distribution models, individuals are unequal competitors, and resources or territories of highest quality are monopolized by the strongest competitors (Johnson 2007). Zabala and Zuberogoitia (2015) suggested that individuals entering into a breeding population already represent a selection of the best individuals of the floating population, and thus, a first selection against poor quality individuals takes place before they establish as territory holders.

Several studies report higher reproductive success in preferred habitat territories (Newton 1991; Martin 1998; Sergio and Newton 2003), and long-lived raptors have been reported to move from low-quality to high-quality places as they age, and they acquire more experience and dominant status (Newton 1989, 1991). However, this rule is not always followed, and, for example, none of the peregrines monitored by Zabala and Zuberogoitia (2014) in a 17-year study switched territory regardless of some instances of consecutive reproductive failures, similar to the results reported by Krüger and Lindström (2001b) in an 11-year study on common buzzards. Some authors suggest that although site quality is a major determinant of fitness, its effects can be confounded with individual quality, a relationship that has been little studied in large long-lived raptors (Carrete et al. 2008; Cardador et al. 2012). However, individual quality can vary with age (Margalida et al. 2008b; Nussey et al. 2008), and productivity can also be influenced by trophic breadth and other denso-dependent and denso-independent factors (Margalida et al. 2012; Newton 2013).

Zabala and Zuberogoitia (2014) also showed that there was an inter-gender difference in the individual effect, mainly due to the different roles of birds during the reproductive period. For example, in most raptor species, males provide females and offspring with alternate prey items when staple prey is in short availability and experienced males can be better than younger ones in providing alternate prey (Sasvari et al. 2000; Sasvari and Hegyi 2002; Katzner et al. 2005). Higher-quality males seemed to deliver more prey in any circumstances regardless of territory (Zabala and Zuberogoitia 2014; Pérez-Camacho et al. 2015).

Females need to reach a minimum body condition in order to start reproduction and to produce eggs, which demand a high amount of energy. Reserves are accumulated through the weeks or months before laying. Normally, females reach maximum body condition values during the months before laying. Harsh winter conditions can result in poorer body conditions for individuals, less energy available for egg production and a reduced clutch size (Korpimäki 1988; Steenhof et al. 1997; Sasvari and Hehyi 2002).

Interspecific Interactions

The nature of interspecific interactions could be based on both the competition for food resources or nest sites or intraguild predation (IGP) (Holt and Polis 1997). It has been largely proved that the influence of intraspecific interactions may determine the selection of breeding habitat in multispecies assemblages. Katzner et al. (2003) suggested that the coexistence of four large eagles in north-central Kazakhstan was primarily determined by intraspecific nest spacing and that interspecific effects appeared to be secondary. Martínez et al. (2008) showed that interspecific relationships within cliff-nesting raptor community follow a general pattern of dominance related to body mass. In this sense, IGP plays a crucial role in structuring raptor communities (Lourenco et al. 2011). There is a clear separation between super-predator species (apex or top predators) and meso-predator species. When apex predators are removed from a community, other predators may subsequently respond functionally or numerically to this change, a phenomenon known as meso-predator release (Soulé et al. 1988). Conversely, when top predators enter communities from which they had been absent, responses by lower predators may reflect a meso-predator suppression effect. These changes may strongly affect species distribution, density, ecology and behaviour (Rebollo et al. 2011; Buchanan 2012). Changes in IGP relationships may be related to natural processes (e.g. recolonized territories, Hakkarainen et al. 2004) or also are may be masked behind habitat alterations which force raptors to concentrate on favourable habitats. Even scientific activities may affect IGP relationships (Zuberogoitia et al. 2012).

Predation

Although it is not easy to assess the effect of predation on population levels (Newton 1998, 2013), predation largely affects nest-site selection, breeding behaviour and breeding success of raptors (e.g. see Newton 1998; Sergio et al. 2003a; Sergio and Hiraldo 2008). The main predators of raptors (eggs, nestlings and adults) are large owls, carnivores and corvids. Only the largest eagles and vultures are relatively free from predation pressure, although their nests may be occasionally assaulted by carnivores. The nest-site selection of the rest of the species is widely conditioned by predation pressure. Forest-dwelling raptors are more affected by carnivore predation than rocky raptors, mainly due to the climbing ability of forest carnivores such as martens, genets or felids (García-Salgado 2017).

Breeding Cycle Phases

Pairing

In raptors, the division of breeding duties between sexes is more marked than in other birds, and they thus have obligatory biparental care (Korpimäkki and Hakkarainen 2012). In most raptors, one male and one female form a pair and raise a brood together, and they thus typically are monogamous. However, in some cases more than two individuals participate in raising the offspring from a single nest. The involvement of breeding adult birds in parental investment is largely affected by their social mating system: monogamy, polygyny and polyandry may imply different tasks for breeding males and females (Clutton-Brock 1991).

Cooperative breeding is widespread within diurnal raptors, occurring in 22 of 76 genera (29%) and 42 of 304 species (14%, Kimball et al. 2003). The majority of those species consist of groups in which extra birds are primarily adult males (polyandry). Polyandry can again be subdivided into sequential polyandry and cooperative polyandry (Faaborg and Bednarz 1990; Tella 1993; González et al. 2006). Generally, all males participate in copulations with the female; thus, any male in a group may sire offspring and is potentially related to the offspring they assist in rearing. The other pattern, polygyny, occurs regularly in a very small proportion species in which multiple females lay into one nest or lay in separate, widely distributed nests (Kimball et al. 2003). Finally, there are polygynandrous (or communal) groups, composed of multiple females and males, in which all group members may contribute genetically to the offspring produced by the group (Gil et al. 2017).

Courtship

Courtship starts early, even some months before egg laying, and displays consist in mutual soaring, chasing flights by the male on its mate, flapping flights with synchronized movements and diving flights focused on the nesting area with a high level of variations depending on the species (Del Hoyo et al. 1994, in this book Chap. 2). Frequency and duration of mating games increases close to the laying dates, and males start to gift quarries to females. Some species courtship and copulation start 2 months before egg laying (Kenward 2006; Margalida and Bertran 2010; Watson 2010), although the frequency of copulates reach its maximum peak close to the laying date. Females become quite lethargic some days before the deposition of the first eggs, reducing progressively the hunting activity until null values. Simultaneously, males tend to increase feeding rates of females as well as copulation rates (for further discussion on courtship and copulation behaviour in raptors, see in this book Chap. 2).

Nest Selection

Habitat characteristics, of course, also influence the decision for breeding in one nest site, and it largely varies between each species' requirements (e.g. Newton 2013). However, nest-site selection (Fig. 3.3) is also conditioned by social cues, such as past reproductive success of conspecifics (public information) or location of the information producers (location cues: the presence of conspecifics or heterospecifics, Danchin et al. 2004; Mateo-Tomás and Olea 2011; Jiménez-Franco et al. 2014). Moreover, nest-site selection is also conditioned by the knowledge of the territory and individual decisions to prevent future parasitic infestation and to avoid giving cues to predators (Zuberogoitia et al. 2015).

Laying and Hatching Dates

Larger species tend to lay before smaller species, which helps them to complete their long breeding cycles. This intraspecific divergence is also marked at a latitudinal gradient. The southern populations (in the Northern Hemisphere) lay earlier than those located northernmost (Kenward 2006; Zuberogoitia and Martínez 2015).

Fig. 3.3 Male Egyptian vulture (*Neophron percnopterus*) carrying material to the nest while female is waiting for him in order to continue repairing the platform. (Photo credit: Iñigo Zuberogoitia)

The photoperiod and weather are the main factors regulating this gradient. This, in turn, has been adopted by different populations over generations.

Juvenile, first breeding females usually lay later than adults (Biljsma 1993; Kewnward 2006). Therefore, mean laying dates significantly differ between those populations with high proportion of juveniles and those whose age composition is well distributed and tend to older distribution. Laying date is critical since in most avian species, breeding performance decreases over the season with early birds having more success and productivity than late conspecifics (Verhulst and Nilsson 2008) and individuals born earlier in the season being generally more likely to survive and recruit (Wiens et al. 2006; Brommer et al. 2014).

Raptors are among the slower layers in birds. Small species usually lay eggs at the interval of 2 days, medium-sized species at 2–3 days and large ones at 3–6 days (Newton 1979; Margalida et al. 2004). The time lapse between eggs and the beginning of incubation determine the hatching date of every chick and may condition its survival.

Incubation

In most raptors, females perform most or all the incubation, brooding and feeding of the nestlings, while males provide most or all of the food for the family (e.g. Newton 1979, 1986; Cramp and Simmons 1980; Krüger 2005). Incubation involves a transfer of heat between parent and embryo, in order to keep the egg temperature between narrow tolerance limits that are close to the optimal development temperature. In most avian species, this transfer occurs through the brood patch (Lea and Klandorf 2002). A brood patch normally develops in both sexes in raptor species where incubation is biparental (i.e. both sexes share incubation tasks, e.g. vultures, Wolter et al. 2013; Bassi et al. 2016). However, most raptor species are uniparental, where the male does not incubate or contributes for short periods only. In this case the brood patch is absent or poorly developed in males (Newton 1979).

Incubation period (IP) varies between species and depends among other factors on the initial egg mass (IEM), although some species show a divergence from this relationship (e.g. Eurasian sparrowhawk IEM = 23, IP = 35; peregrine falcon IEM = 43, IP = 31; common buzzard IEM = 51, IP = 36; Deeming, 2002). Small falcons (kestrels) have shorter incubation periods (28 days) than do *Accipiter* of the same weight (sparrowhawks); and *Haliaeetus* eagles have shorter incubation periods (35–38 days) than do *Aquila* eagles of similar and lower weights (42–45 days; Newton 1979). Large vultures have the longer incubation periods, from the smaller species (e.g. 31–40 days in *Cathartes* vultures, 39–45 days Egyptian vultures (*Neophron percnopterus*) to the largest vultures (e.g. 53–60 days in both American condors or 54–58 days in Himalayan vultures *Gyps himalayensis*; Campbell 2016).

Clutch Size

The general trend is for larger species to produce the smaller clutches (Del Hoyo et al. 1994). Large vultures only lay one egg, although the Egyptian vulture and the bearded vulture (*Gypaetus barbatus*) are the only Old World vultures whose clutch usually has two eggs (Donázar and Ceballos 1989; Campbell 2016). The maximum number of eggs laid by those species with larger clutches is normally conditioned by food supply. Small falcons that eat rodents tend to lay larger clutches than bird-eating falcons in the same area, and these in turn have larger clutches than insect eaters (Newton 1979). In some exceptional conditions in which some pairs enjoy plenty of food circumstances, clutches with a record number of eggs are detected (Altwegg et al. 2014).

Parental Care and Nestling Development

During the first days after hatching, nestlings are continuously attended by their parents to protect them from weather and predation. In uniparental species, females continue with the brooding process in order to heat offspring, and males relieve them but less often than they usually do during egg incubation, although males of some species do not contribute to this task at all (Dare 2015). The male investment

Fig. 3.4 Female peregrine falcon (*Falco peregrinus*) feeding her four nestlings in an old platform of common raven (*Corvus corax*). (Photo credit: Iñigo Zuberogoitia)

also seems to decrease during post-hatching period in some biparental species (Bassi et al. 2016).

During the first weeks of life, nestlings are fed regularly by their parents, and they show a high growth rate of body mass (muscles and bones). The regular intake of food and the maintenance of stable temperature, thanks to parent care, let nestlings grow adequately (Fig. 3.4). However, when some of these factors fail, the growth rate decreases and delayed development is detected.

The growth rate changes when feathers start to appear in the tail and wings. Some days later, depending on the species, nestlings are able to thermoregulate. The energy intake is redirected to feather growth at the expense of corporal growth. All the energy intake is transformed in plumage and in lesser extent to finish the corporal development. During this period, parental care decreases progressively. Nestlings do not need to be incubated, although females still cover them in adverse weather conditions (excessive sun exposure, low temperatures and heavy precipitations). Females of most species remain close the nest site in order to defend the offspring against intruders or predators, although at this stage female contribution to hunting task increases in order to attend the demand of nestlings.

Fig. 3.5 Peregrines do not build nests; they lay eggs on a depression of the ground or in a platform of other species. In this case, the eyrie is placed on a ledge of a cliff, inside a bush (*Ruscus aculeatus*) that protects the nestlings against rain and low temperatures. The female has just fed the three nestlings and is going to leave the nest in order to finish the quarry herself. (Photo credit: Iñigo Zuberogoitia)

Weather Conditions

Those raptor populations living in environments with relatively stable weather conditions suffer low or scarce losses on breeding productivity (Bosch et al. 2015). However, when weather conditions show higher oscillations, an increasing effect on breeding performance is detected. Persistent or heavy rain and low or high temperatures between the egg-laying period and the first weeks after hatching have adverse effects on the reproduction of most raptors (e.g. see Kostrzewa and kostrzewa 1990; Carrillo and González-Dávila 2010; Zuberogoitia et al. 2011; Mihoub et al. 2012; Anctil et al. 2014; Touati et al. 2017). Rainfall during the first half of chick-rearing period is negatively correlated with breeding success (Fig. 3.5). During this brooding phase, nestlings do not thermoregulate; therefore, rain increases the risk of hypothermia (Elkins 2004). A high amount of rain even if only on one day can be as negative as a number of continuous rainy days (Zuberogoitia et al. 2014). Moreover, rain during the chick-rearing phase prevents adults from foraging, resulting in food shortage for the young (Penteriani 1997; Sergio 2003; Mcdonald et al. 2004; Lehikoinen et al. 2009).

Multiple Brooding

Multiple-brooding occurs infrequently in raptors and is generally restricted to either smaller species with shorter nesting periods in conditions of prolonged food abundance whenever they occur (Newton 1979; Mendelsohn 1981; Curtis et al. 2005). It is also observed in co-operative breeders where additional adults enhance nestling provisioning efficiency, saving both time and energy (Malan et al. 1997). The black sparrowhawk (*Accipiter melanoleucus*) and the northern crested caracara (*Caracara cheriway*) are the only two relatively large, monogamous raptor species in which multiple brooding has been recorded with any frequency (Morrison 1998; Curtis et al. 2005). Nevertheless, there are still many poorly known species, mainly in tropical forests, which may also display this behaviour (e.g. grey goshawk *Accipiter novaehollandiae* (Riddell 2013)).

Sibling Aggression: Cainism

Due to the above-mentioned asynchronous hatching of some raptors, in large eagles and those vulture species that lay two eggs, the last bird to hatch has the biggest disadvantage compared with its older siblings. In favourable years, the antagonist behaviour between siblings remains at low level when parents are able to obtain enough resources for feeding them (Watson 2010). However, some species like the bearded vulture and some eagles (e.g. lesser spotted eagle, tawny eagle and Verreaux's eagles) are considered obligate cainists; in these species, aggression between nest mates always results in a death (Newton 1979; Simmons 1988; Margalida et al. 2004). In medium-sized raptors, sibling aggressions are scarce, only occurring in periods of hunger (e.g. northern goshawks, Kenward 2006; common buzzards, Dare 2015), and not in all species (e.g. peregrine falcon). While in small species, such attacks do not occur, even when the young are starving (Newton 1979). In these cases, adults equally feed all nestlings independently of their age and sex.

Fledging Phase

The fledgling behaviour varies between species, depending on the sex, size, hunting or flying behaviour, habitat requirements and nest-site selection. Small- and medium-sized forest species start to move around the nest some days before their plumages are totally developed. The fledglings explore closest branches during the first days and make short flights to neighbouring trees, scrambling back to the nest whenever a parent arrives with food. Males normally develop earlier than females, and they start to move several days before their female siblings. This behaviour is shared by cliff-nesting species when they breed in cliffs with wide ledges that let fledglings explore the vicinity areas. On the contrary, when the nest site is a narrow ledge, pothole or little cave, fledglings wait until they are able to fly for jumping for the first time. The same occurs in large raptors, whose fledglings practice wing movements in the nest platform in order to tone up the muscles. Most fledglings are able to fly at first attempt, although some of them fail in this first flight and land wherever they can. This causes a non-valued mortality rate, as sometimes they die from starvation in the landing site or are predated by carnivores.

During the post-fledgling stage, most fledglings are noisy when parents approach with food, although hungry broods scream almost continuously. This behaviour makes it easy for predators to locate noisy young (Newton 1986).

Dispersion

Studies of natal dispersal in raptors have focused on four key developmental stages: (1) a post-fledging dependence period ending in emigration from the natal environment; (2) a long transitional phase (often termed "juvenile dispersal", synonymous with transience); (3) provisional settlement in temporary settlement areas, where individuals establish more or less stable home ranges; and (4) settlement at a breeding site (Ferrer 1993; Delgado and Penteriani 2008; Weston et al. 2013). There are huge intra- and interspecies differences among the four phases and the final result of dispersion.

The start of natal dispersion varies between individuals, from some days after fledgling to some months later. Normally, young birds venture out of their natal home ranges (pre-dispersal excursions) prior to their emigration at the start of natal dispersal (Zuberogoitia et al. 2002; Kenward 2006; Weston et al. 2013; Dare 2015). During the juvenile dispersal period, home ranges usually are much larger than those obtained for territorial individuals (Margalida et al. 2016). Some migratory species spent the first years of life on wintering grounds until they mature (Alerstam et al. 2006), although most of them follow the same migratory pattern of adults (Prommer et al. 2012). However, juveniles show more tortuous routes, slower speeds and more stopover days than adults (Mellone et al. 2013). Natal dispersal distances tend to be lower for males than for females; in other words, males tend to come back close to the natal grounds (Philopatric behaviour), while females usually disperse longer (López-López et al. 2013; Newton 2013, for further information on raptor dispersion, see in this book Chap. 4).

Sexual Maturity

Most raptors have one or more immature or subadult plumages before acquiring the definitive adult dress (Newton 1979). Falconidae species and most small- and medium-sized accipitrids moult all the plumage in one year, and therefore immature plumage is changed for adult one during the second calendar year. However, some medium-sized accipitrids need more than one year in order to complete one moult cycle, while large raptors need several years (from 2 to 4 years; Snyder et al. 1987; Zuberogoitia et al. 2013c). In some species the immature plumage is changed first for a subadult plumage and later for an adult dress (Watson 2010; Zuberogoitia et al. 2016).

Bearded vultures reach definitive plumage in their 6th calendar year (Sesé 2011; Zuberogoitia et al. 2016), approximately at the same age at which they start to establish a territory (López-López et al. 2013). This, in turn, reflects the importance of the relationship between the acquisition of definitive plumage, the reach of sexual maturity and the settlement in a breeding territory (Rohwer et al. 2011). However, in the wild, some individuals begin breeding before the acquisition of the fully adult plumage. Birds breed at a younger age than usual when conditions are especially good, either in favourable areas or years, or when depleted population leave territories vacant (Newton 1979). This last case, in fact, has been suggested as reliable warning signals of adult mortality or breeding performance (Balbontin et al. 2003; Ferrer et al. 2003; Zuberogoitia et al. 2009). It is also common to find subadults breeding during the first phases of recently restored populations (e.g. see Cade and Durham 2003; Gil-Carrera et al. 2016).

Immature individuals show a lower breeding performance than adults (Ferrer and Bisson 2003; Margalida et al. 2008a, b), possibly due to a delayed maturation of the gonads. However, immature birds that enter the breeding population in saturated populations are few and might be of higher quality than other individuals of the same age (Zabala and Zuberogoitia 2014).

Management to Increase Reproductive Success

Raptors are long-lived species, and thus adult survival is the demographic parameter that contributes most to breeding success and population growth. Management techniques aimed at increasing productivity:

Clutch manipulation vulnerable eggs can be removed after the start of incubation and replaced with artificial eggs, so that incubation continues. The real eggs are incubated artificially, and the young produced are returned to the nests. The overall production of young obtained should be higher than if the original eggs had been left with the pairs. This method has been used successfully with different raptors (Olsen and Tucker 2003).

Brood manipulation the number of young reared to independence can be augmented by increasing brood size to the normal maximum for a species. This has been done with species that experience death of young nestlings due to fratricide (siblicide or cainism). Brood size is reduced to one by removing nestlings at an age before sibling rivalry develops. These young are hand-reared and then returned to the nest at an age beyond which fratricide is likely.

As for the techniques used to demographic supplementation of wild raptors:

Fig. 3.6 Hacking technique used in a recovery management plan of a threatened golden eagle (*Aquila chrysaëtos*) population. (Photo credit: *GREFA & Estación Biolóxica do Xurés (EBX)*)

Cross-fostering consists of placing young of one species into the nest of another species. Many raptor species have been cross-fostered, either in captivity or in the wild. There always is a risk though that cross-fostered individuals will become imprinted upon the surrogate parental species. This may increase predation risk of one of the species to the other (e.g. cross-fostering of peregrines on goshawk nests) or hybrid production between the two species (e.g. greater spotted eagle and lesser spotted eagle).

Hacking the controlled release of young raptors into the wild, is the most frequently used technique to reintroduce or augment raptor populations. Nestling raptors raised in captivity or in wild nests are translocated alone or in small groups of three to five individuals to the hacking site. The hacking site generally consists of a wooden or metal tower with a large enclosure at the top constructed in such a way as to provide the birds with a view of their surroundings (Fig. 3.6). For some time, individuals are fed in the enclosure, without seeing their handlers. At about the natural fledging time for the species, the front of the enclosure is opened, and the birds inside have the opportunity to fly freely and explore the surroundings. Food continues to be provided in the enclosure for some time after it has been opened, and released individuals often stay in the area for weeks or months before dispersing or migrating. This technique has been successful with different breeding threatened raptor populations (Negro et al. 2007; Gil et al. 2013; Sorenson et al. 2017).

Supplemental feeding may be used to increase raptor breeding success, and it also can be used to increase productivity in poor-quality habitat (Rooney et al. 2015). Scavenging species have been supplemented with food more often than has been done with predatory raptors. Often used to enhance populations of carrion eaters such as vultures, such stations have been referred to as "vulture restaurants" (Negro et al. 2007). This management technique also has been used with territorial endangered raptor species (González and Margalida 2008).

Nest-guarding the goal is to protect the nests of target species from depredation by both wildlife and humans as well as from natural disturbances by actively monitoring individual nests. An alternative management strategy is to establish buffer zones around raptor nests aimed at protecting nests from the effects of recreational activities, forestry management activities, human development, etc. (Negro et al. 2007; Zuberogoitia et al. 2014).

References

Adams ES (2001) Approaches to the study of territory size and shape. Annu Rev Ecol Syst 32:277–303

Alerstam T, Hake M, Kjellén N (2006) Temporal and spatial patterns of repeated migratory journeys by ospreys. Anim Behav 71:555–566

Altwegg R, Jenkins A, Abadi F (2014) Nestboxes and immigration drive the growth of an urban Peregrine falcon *Falco peregrinus* population. Ibis 156:107–115

Anctil A, Franke A, Bêty J (2014) Heavy rainfall increases nestling mortality of an arctic top predator: experimental evidence and long-term trend in peregrine falcons. Oecologia 174:1033–1043

Bakaloudis DE, Vlachos CG, Holloway GJ (1998) Habitat use by short-toed eagles Circaetus gallicus and their reptilian prey during the breeding season in Dadia Forest (North-Eastern Greece). J Appl Ecol 35:821–828

Balbontín J (2005) Identifying suitable habitat for dispersal in Bonelli's eagle: an important issue in halting its decline in Europe. Biol Conserv 126:74–83

Balbontin J, Penteriani V, Ferrer M (2003) Variation in the age of mates as an early warning signal of changes in population trends? The case of Bonelli's eagle in Andalusia. Biol Conserv 109:417–423

Bassi E, Trotti P, Brambilla M, Diana F, Sartirana F, Galli L, Pedrotti L (2016) Parental investment in two large raptors breeding in a high prey density area. J Ornithol. https://doi.org/10.1007/s10336-016-1407-6

Bechard MJ (1982) Effect of vegetative cover on foraging site selection by Swainson's hawk. Condor 84:153–159

Bednarz JC (2007) Study design, data management, analysis, and presentation. In: Bird DM, Bildstein KL (eds) Raptor research and management techniques manual. Hancock House Publishers, Surrey, pp 73–88

Beeisinger AR, Mccullough DR (2002) Population viability analysis. The University of Chicago Press, Chicago

Bevers M, Flather CH (1999) The distribution and abundance of populations limited at multiple spatial scales. J Anim Ecol 68:976–987

Bijlsma RG (1993). Ecologische atlas van de Netherlandse roofrogels. Schuyt & Co

Bird DM, Bildstein KL (Eds) (2007) Raptor research and management techniques. Raptor research foundation. Hancock House Publishers, Surrey, Canada
Bosakowski T, Speiser R (1994) Macrohabitat selection by nesting northern goshawks: implications for managing eastern forests. Stud Avian Biol 16:46–49
Bosch J, Martínez JE, Calvo JF, Zuberogoitia I, Jiménez-Franco MV (2015) Does rainfall affect the productivity of the booted eagle (*Aquila pennata*) during the breeding period in Mediterranean environments? J Ornithol 156:1–8
Boulinier T, Mariette M, Dolige B, Danchin E (2008) Choosing where to breed-breeding habitat choice. In: Danchin E, Giraldeau, LA, Cézzilly F (Eds) Behavioural ecology. Oxford University Press, Oxford, UK
Bried J, Jouventin P (1999) Influence of breeding success on fidelity in long-lived birds: an experimental study. J Avian Biol 20:392–398
Brommer JE, Karell P, Ahola K, Karstinen T (2014) Residual correlations, and not individual properties, determine a nest defense boldness syndrome. Behav Ecol 25:802–812
Brown L (1989) British birds of prey. Collins New Naturalist, London
Bruggeman JE, Andersen DE, Woodford JE (2014) Factors related to northern goshawk landscape use in western Great Lakes region. J Raptor Res 48:228–239
Buchanan JB (2012) Change in merlin hunting behaviour following recovery of peregrine falcon population suggests mesopredator suppression. J Raptor Res 46:349–456
Buechley ER, Sekerciogly CH (2016) The avian scavenger crisis: looming extinctions, trophic cascades, and loss of critical ecosystems functions. Biol Conserv 198:220–228
Burgas D, Byholm P, Parkkima T (2014) Raptors as surrogates of biodiversity along a landscape gradient. J Appl Ecol 51:786–794
Burnham KK, Burnham WA, Newton I (2009) Gyrfalcon *Falco rusticolus* post-glacial colonization and extreme long-term use of nest-sites in Greenland. Ibis 151:514–522
Bustamante J, Seoane J (2004) Predicting the distribution of four species of raptors (Aves: Accipitridae) in southern Spain: statistical models work better than existing maps. J Biogeogr 31:295–306
Cade TJ, Durham W (2003) Return of the Peregrine: a north American saga of tenacity and teamwork. The Peregrine Fund Inc., Boise
Calsbeek R, Sinervo B (2002) An experimental test of the ideal despotic distribution. J Anim Ecol 71:513–523
Campbell MN (2016) Vultures. Their evolution, ecology and conservation. CRC Press, Taylor and Francis Group, Boca Ratón
Cardador L, Carrete M, Mañosa S (2012) Inter-individual variability and conspecific densities: consequences for population regulation and range expansion. PLoS One 7:8
Cardador L, Ll B, Mougeot F, Giralt DM, Bota G, Pomarol M, Arroyo B (2015) Conservation traps and long-term species persistence in human-dominated systems. Conserv Lett 8:456–462
Carrete M, Tella JL, Sanchez-Zapata JA, Moleon M, Gil-Sanchez JM (2008) Current caveats and further directions in the analysis of density-dependent population regulation. Oikos 117:1115–1119
Carrillo J, González-Dávila E (2010) Impact of weather on breeding success of the Eurasian kestrel *Falco tinnunculus* in a semi-arid island habitat. Ardea 98:51–58
Chalfoun AD, Schmidt KA (2012) Adaptive breeding-habitat selection: is it for the birds? Auk 129:589–599
Clutton-Brock TH (1991) The evolution of parental care. Princeton University Press, Princeton
Corsi F, Leeuw J, Skidmore A (2000) Modeling species distribution with GIS. In: Boitani L, Fuller TK (eds) Research techniques in animal ecology. Controversies and consequences. Columbia University Press, New York, pp 389–434
Cramp S, Simmons KEL (1980) Handbook of the birds of Europe the Middle East and North Africa. The birds of the Western Palearctic, Hawks to bustards, vol 2. Oxford University Press, Oxford
Curtis O, Malan G, Jenkins A, Myburgh N (2005) Multiple-brooding in birds of prey: south African black Sparrowhawk *Accipiter melanoleucus* extend the boundaries. Ibis 147:11–16

Danchin E, Giraldeau L, Valone TJ, Wagner RH (2004) Public information: from nosy neighbors to cultural evolution. Science 305:487–491

Dare P (2015) The life of buzzards. Whittles Publishing, Dunbeath

Daw SK, DeStefano S (2001) Forest characteristics of northern goshawk nest stands and post-fledging areas in Oregon. J Wildl Manag 65:59–65

Deeming DC (ed) (2002) Avian incubation. Behaviour, environment, and evolution. Oxford University Press, Oxford

Del Hoyo J, Elliot A, Sargatal J (eds) (1994) Handbook of the birds of the world, New world vultures to guineafowl, vol 2. Lynx Edicions, Barcelona

Delgado MM, Penteriani V (2008) Behavioral states help translate dispersal movements into spatial distribution patterns of floaters. Am Nat 172:475–485

Dewey SR, Kennedy PL (2001) Effects of supplemental food on parental care strategies and juvenile survival of northern goshawks. Auk 118:352–365

Donald PF, Collar NJ, Marsden SJ, Pain DJ (2013) Facing extinction. The world's rarest birds and the race to save them. Christopher Helm, London

Donázar JA, Ceballos O (1989) Growth rates of nestling Egyptian vultures *Neophron percnopterus* in relation to brood size, hatching order and environmental factors. Ardea 77:217–226

Donázar JA, Hiraldo F, Bustamante J (1993) Factors influencing nest sites selection, breeding density and breeding success in the bearded vulture (*Gypaetus barbatus*). J Appl Ecol 30:504–514

Donázar JA, Cortés-Avizanda A, Fargallo JA, Margalida A, Moleón M, Morales-Reyes Z, Moreno-Opo R, Pérez-García JM, Sánchez-Zapata JA, Zuberogoitia I, Serrano D (2016) Roles of raptors in a changing world: from flagships to providers of key ecosystem services. Ardeola 63:181–234

Elkins N (2004) Weather and bird behaviour. T and AD Poyser, London

Espie RHM, James PC, Oliphant LW, Warkentin IJ, Lieske DJ (2004) Influence of nest-site and individual quality on breeding performance in Merlins *Falco columbarius*. Ibis 146:623–631

Faaborg J, Bednarz JC (1990) Galápagos and Harris' hawks: divergent causes of sociality in two raptors. In: Stacey P, Koenig W (eds) Cooperative breeding in birds: long term studies of ecology and behaviour. Cambridge University Press, Cambridge, pp 359–383

Fasciolo A, Delgado MM, Soutullo A, Cortés G, Penteriani V (2016) Limited prospecting behaviour of juvenile eagle owls *Bubo bubo* during natal dispersal: implications for conservation. Bird Study 63:128–135

Ferguson-Lees J, Christie DA (2001) Rapaces del Mundo. Ediciones Omega, Barcelona

Ferrer M (1993) Ontogeny of dispersal distances in young Spanish imperial eagles. Behav Ecol Sociobiol 32:259–263

Ferrer M, Bisson I (2003) Age and territory-quality effects on fecundity in the Spanish imperial eagle (*Aquila adalberti*). Auk 120:180–186

Ferrer M, Donázar JA (1996) Density-dependent fecundity by habitat heterogeneity in an increasing population of Spanish imperial eagles. Ecology 77:69–74

Ferrer M, Harte M (1997) Habitat selection by immature Spanish imperial eagles during the dispersal period. J Appl Ecol 34:1359–1364

Ferrer M, Penteriani V, Balbontín J, Pandolfi M (2003) The proportion of immature breeders as a reliable early warning signal of population decline: evidence from the Spanish imperial eagle in Doñana. Biol Conserv 114:463–466

Ferrer M, Morandini V, Newton I (2015) Floater interference reflects territory quality in the Spanish imperial eagle *Aquila adalberti*: a test of a density-dependent mechanism. Ibis 157:849–859

Finn SP, Varland DE, Marzluff JM (2002) Does northern goshawk breeding occupancy vary with nest-stand characteristics on the Olympic peninsula, Washington. J Raptor Res 36:265–279

Franco AMA, Marques JT, Sutherland WJ (2005) Is nest-site availability limiting lesser kestrel populations? A multiple scale approach. Ibis. https://doi.org/10.1111/j.1474-919x.2005.00437.x

Franklin J, Miller JA (2009) Mapping species distributions: spatial inference and prediction. Cambridge University Press, New York

Freund M, Bahat O, Motro U (2017) Breeding success and its correlation with nest-site characteristics: a study of a griffon vulture colony in Gamla, Israel. J Raptor Res 51:136–144
Fuller RJ (2012) Birds and habitat: relationships in changing landscapes. Cambridge University Press, Cambridge
García-Salgado G (2017) Trophic ecology and nesting habitat selection of the Northern Goshawk (*Accipiter gentilis*) in Eucalyptus plantations in northwestern Spain. PhD. Thesis. Universidad de Alcalá de Henares. España
Gil JA, Chéliz C, Zuberogoitia I, López-López P (2017) First cases of polygyny for the Bearded Vulture *Gypaetus barbatus* in the central Pyrenees. Bird Study 64: 565–568
Gil A, Fernández-Martínez MA, Álvarez E, Galán M, González F, Tapia L (2013) Recovery of golden eagle (*Aquila chrysaëtos*) in the Natural Park Baixa-Limia Serra do Xurés (Galicia, NW Spain). Wingspan 22:11
Gil-Carrera A, Tapia L, Álvarez E, Galán M, Pardavila X, Iglesias JJ, Lamosa A, Izquierdo P (2016) Seguimento, mellora do éxito reproductivo e reforzamento poboacional da Aguia real (*Aquila chrysaëtos*) en Galicia. GREFA. Xunta de Galicia. Fondo Europeo de Desenvolvemento Rexional. Santiago de Compostela, 86pp
Gonçalves J, Alves P, Pôças I, Marcos B, Sousa-Silva R, Lomba Â, Civantos E, Monteiro A, Honrado J (2015) Combining niche models and remote sensing to explore short-term habitat suitability temporal dynamics and improving biodiversity monitoring, 35th EARSeL Symposium–European Remote Sensing: Progress, Challenges and Opportunities. Stockholm, pp. 15–18
González LM, Margalida A (eds) (2008) Biología de la conservación del Águila Imperial Ibérica (*Aquila adalberti*). Organismo Autónomo de Parques Nacionales. Ministerio de Medio Ambiente y Medio Marino y Rural, Madrid
González LM, Margalida A, Sánchez R, Oria J (2006) Cooperative breeding in the Spanish imperial eagle Aquila adalberti: a case of polyandry with male reversed sexual behaviour? Ibis 148:159–163
Gordon DM (1997) The population consequences of territorial behaviour. Trends Ecol Evol 12:63–66
Graham IM, Redpath SM (1995) The diet and breeding density of common buzzards *Buteo buteo* in relation to indices of prey abundance. Bird Study 42:165–173
Hakkarainen H, Mykrä S, Kurki S, Tornberg R, Jungell S (2004) Competitive interactions among raptors in boreal forests. Oecologia 141:420–424
Hardey J, Crick H, Wenham C, Riley H, Etheridge B, Thompson D (2009) Raptors. A field guide for surveys and monitoring. The Stationery Office, Edinburgh
Helander B, Bignert A, Asplund L (2008) Using raptors as environmental sentinels: monitoring the white-tailed sea eagle *Haliaeetus albicilla* in Sweden. Ambio 37:425–431
Holt RD, Polis GA (1997) A theoretical framework for intraguild predation. Am Nat 149:745–764
Hoover JP (2003) Decision rules for site fidelity in a migratory bird, the Prothonotary warbler. Ecology 84:416–430
Janes SW (1985) Habitat selection in raptorial birds. In: Cody ML (ed) Habitat selection in birds. Academic, San Diego, pp 159–188
Jiménez-Franco MV, Martínez JE, Pagán I, Calvo JF (2013) Factors determining territory fidelity in a migratory forest raptor, the booted eagle *Hieraaetus pennatus*. J Ornithol 154:311–318
Jiménez-Franco MV, Martínez JE, Calvo JF (2014) Patterns of nest reuse in forest raptors and their effects on reproductive output. J Zool 292:64–70
Johnson MD (2007) Measuring habitat quality: a review. Condor 109:489–504
Jones J (2001) Habitat selection studies in avian ecology: a critical review. Auk 118:557–562
Katzner TE, Bragin EA, Knick ST, Smith AT (2003) Coexistence in a multispecies assemblage of eagles in Central Asia. Condor 105:538–551
Katzner TE, Bragin EA, Knick ST, Smith AT (2005) Relationship between demographics and diet specificity of imperial eagles *Aquila heliaca* in Kazakhstan. Ibis 147:576–586
Kenward R (2006) The goshawk. T and AD Poyser, London

Kimball RT, Parker PG, Bednarz JC (2003) Occurrence and evolution of cooperative breeding among the diurnal raptors (Accipitridae and Falconidae). Auk 120:717–729

Kochert M, Steenhof K (2012) Frequency of nest use by golden eagles in southwestern Idaho. J Raptor Res 46:239–247

Korpimäki E (1988) Costs of reproduction and success of manipulated broods under varying food conditions in Tengmalms owl. J Anim Ecol 57:1027–1039

Korpimäki E, Hakkarainen H (2012) The boreal owl. Ecology, behaviour and conservation of a Forest-dwelling predator. Cambridge University Press, Cambridge

Kostrzewa A, Kostrzewa R (1990) The relationship of spring and summer weather with density and breeding performance of the buzzard *Buteo buteo*, goshawk *Accipiter gentilis* and kestrel *Falco tinnunculus*. Ibis 132:550–559

Krüger O (2005) The evolution of reversed sexual size dimorphism in hawks, falcons and owls: a comparative study. Evol Ecol 19:467–486

Krüger O, Lindström J (2001a) Habitat heterogeneity affects population growth in northern goshawk *Accipiter gentilis*. J Anim Ecol 70:173–181

Krüger O, Lindström J (2001b) Lifetime reproductive success in common buzzard, *Buteo buteo*: from individual variation to population demography. Oikos 93:260–273

Kudo T, Ozaki K, Takao G, Sakai Yonekawa H, Ikeda K (2005) Landscape analysis of northern goshawk breeding home range in northern Japan. J Wilddl Manag 69:1229–1239

Lea RW, Klandorf H (2002) The brood patch. In: Deeming DC (ed) Avian incubation. Behaviour, environment, and evolution. Oxford University Press, Oxford, pp 100–118

Lehikoinen A, Byholm P, Ranta E, Saurola P, Valkama J, Korpimäki E, Pietiäinen H, Henttonen H (2009) Reproduction of the common buzzard at its northern range margin under climatic change. Oikos 118:829–836

León-Ortega M, Jiménez-Franco MV, Martínez JE, Calvo JF (2017) Factors influencing territorial occupancy and reproductive success in a Eurasian eagle-owl (*Bubo bubo*) population. PLoS One. https://doi.org/10.1371/journal.pone.0175597

Litvaitis JA, Titus K, Anderson EM (1994) Measuring vertebrate use of terrestrial habitats and foods. In: Bookhout TA (ed) Research and management techniques for wildlife and habitats. The Wildlife Society, Bethesda, pp 254–271

Lõhmus A (2004) Raptor Habitat Studies – the state of the art. In: Chancellor RD, Meyburg B-U (eds) Raptors Worldwide. WWGBP/MME, Budapest, pp 279–296

López-López P, García-Ripollés C, García-López F, Aguilar JM, Verdejo J (2004) Patrón de distribución del Águila Real (*Aquila chrysaetos*) y del Águila-azor perdicera (*Hieraaetus fasciatus*) en la provincia de Castellón. Ardeola 51:275–283

López-López P, Zuberogoitia I, Alcantara M, Gil JA (2013) Philopatry, natal dispersal, first settlement and age of first breeding of bearded vultures *Gypaetus barbatus* in central Pyrenees. Bird Study 60:555–560

Lourenco R, Santos SM, Rabaca JE, Penteriani V (2011) Superpredation patterns in four large European raptors. Popul Ecol 53:175–185

Mairota P, Cafarelli B, Didham R, Lovergine FP, Lucas RM, Nagendra H, Rocchini D, Tarantino C (2015) Challenges and opportunities in harnessing satellite remote-sensing for biodiversity monitoring. Eco Inform 30:207–214

Malan G, Crowe TM, Biggs R, Herholdt JJ (1997) The social system of the pale chanting goshawk *Melierax canorus*; monogamy vs polyandry and delayed dispersal. Ibis 139:313–321

Mañosa S, Real J, Codina J (1998) Selection of settlement areas by juvenile Bonelli's eagle in Catalonia. J Raptor Res 32:208–214

Margalida A, Bertran J (2010) Copulatory behavior in the colonial Eurasian griffon vulture *Gyps fulvus*. J Ethol 28:179–182

Margalida A, Bertran J, Boudet J, Heredia R (2004) Hatching asynchrony, sibling aggression and cannibalism in the bearded vulture *Gypaetus barbatus*. Ibis 146:386–393

Margalida A, Heredia R, Razin M, Hernandez M (2008a) Sources of variation in mortality of the bearded vulture *Gypaetus barbatus* in Europe. Bird Conserv Int 18:1–10

Margalida A, Mañosa S, Gonzalez LM, Ortega E, Sanchez R, Oria J (2008b) Breeding of non-adults and effects of age on productivity in the Spanish imperial eagle *Aquila adalberti*. Ardea 96:173–180

Margalida A, Benitez JR, Sanchez-Zapata JA, Avila E, Arenas R, Donázar JA (2012) Long-term relationship between diet breadth and breeding success in a declining population of Egyptian vultures *Neophron percnopterus*. Ibis 154:184–188

Margalida A, Pérez-García JM, Afonso I, Moreno-Opo R (2016) Spatial and temporal movements in Pyrenean bearded vultures (*Gypaetus barbatus*): integrating movement ecology into conservation practice. Sci Rep 6:35746. https://doi.org/10.1038/srep35746

Marti CD, Bechard M, Jaksic FM (2007) Food habits. In: Bird D, Bildstein K (eds) Raptor research and management techniques manual. Raptor research foundation. Hancock House Publishers, Surrey, Canada, pp 129–149. ISBN: 0-88839-639-2

Martin TE (1998) Are microhabitat preferences of coexisting species under selection and adaptive? Ecology 79:656–670

Martínez JE, Calvo JF (2000) Selección de hábitat de nidificación por el Búho real (*Bubo bubo*) en ambientes mediterráneos semiáridos. Ardeola 4:215–220

Martínez JA, Serrano D, Zuberogoitia I (2003) Predictive models of habitat preferences of Eurasian eagle owl *Bubo bubo*: a multiscale approach. Ecography 26:21–28

Martínez JE, Pagán I, Calvo JF (2006) Factors influencing territorial occupancy and reproductive output in the booted eagle. Ibis 148:807–819

Martínez JE, Martínez JA, Zuberogoitia I, Zabala J, Redpath SM, Calvo JF (2008) The effect of intra- and interspecific interactions on the large-scale distribution of cliff-nesting raptors. Ornis Fennica 85:13–21

Martínez JE, Zuberogoitia I, Gómez G, Escarabajal JM, Cerezo E, Jiménez-Franco MV, Calvo JF (2014) Attack success in Bonelli's Eagle *Aquila fasciata*. Ornis Fennica 91:67–78

Martínez JE, Zuberogoitia I, Jiménez-Franco MV, Mañosa S, Calvo JF (2016) Saptio-temporal variations in mortality causes of two migratory forest raptors in Spain. Eur J Wildl Res 62:109–118

Martínez-Abraín A, Oro D, Juan Jiménez J, Stewart G, Pullin A (2010) A systematic review of the effects of recreational activities on nesting birds of prey. Basic Appl Ecol 11:312–319

Marzluff JM, Kimsey BA, Schuek LS, Mcfadzen ME, Vekasy MS, Bednarz JC (1997) The influence of habitat, prey abundance, sex, and breeding success on ranging behavior of prairie falcons. Condor 99:567–584

Mateo-Tomás P, Olea PP (2011) The importance of social information in breeding site selection increases with population size in the Eurasian griffon vulture *Gyps fulvus*. Ibis 153:832–845

Mcdonald PG, Olsen PD, Cockburn A (2004) Weather dictates reproductive success and survival in the Australian brown falcon *Falco berigora*. J Anim Ecol 73:683–692

Mcgrady MJ, Grant JR, Bainbridge IP, Mcleod DRA (2002) A model of golden eagle (*Aquila chrysaetos*) ranging behaviour. J Raptor Res 36(Suppl):62–69

McIntyre CL (2002) Patterns in nesting area occupancy and reproductive success of golden eagles (*Aquila chrysaetos*) in Denali Nacional Park and preserve, Alaska, 1988-99. J Raptor Res 36(Suppl):50–54

Mellone U, López-López P, Limiñana R, Piasevoli G, Urios V (2013) The trans-equatorial loop migration system of Eleonora's falcon: differences in migration patterns between age classes, regions and seasons. J Avian Biol 44:417–426

Mendelsohn JM (1981) A study of the Black-shouldered Kite *Elanus caeruleus*. PhD thesis, University of Natal

Mihoub JB, Mouawad NG, Pilard P, Jiguet F, Low M, Teplitsky C (2012) Impact of temperature on the breeding performance and selection patterns in lesser kestrels *Falco naumanni*. J Avian Biol 43:472–480

Millsap BA, Grubb TG, Murphy RK, Swem T, Watson JW (2015) Conservation significance of alternative nests of golden eagles. Global Ecol Conserv 3:234–241

Morrison JL (1998) Effects of double brooding on productivity of crested caracaras. Auk 115:979–987

Morrison ML, Marcot BG, Mannan RW (1998) Wildlife-habitat relationships. Concepts and applications. University of Wisconsin Press, Madison

Negro JJ, Sarasola JH, Barclay JH (2007) Augmenting wild populations and food resources. In: Bird D, Bildstein K (eds) Raptor research and management techniques manual. Raptor research foundation. Hancock House Publishers, Surrey, Canada, pp 401–441

Newton I (1979) Population ecology of raptors. Poyser, London

Newton I (1986) The Sparrowhawk. T and Ad Poyser, London

Newton I (1989) Lifetime reproduction in birds. Academic, London

Newton I (1991) Habitat variation and population regulation in sparrowhawks. Ibis 133:76–88

Newton I (1998) Population limitation in birds. Academic, London

Newton I (2013) Bird populations. The New Naturalist Library, London

Nussey DH, Coulson T, Festa-Bianchet M, Gaillard JM (2008) Measuring senescence in wild animal populations: towards a longitudinal approach. Funct Ecol 22:393–406

Ogada D, Shaw P, Beyers RL, Buij R, Murn C, Thiollay JM, Beale CM, Holdo RM, Pomeroy D, Baker N, Kruger SC, Botha A, Virani MZ, Monadjem A, Sinclair ARE (2015) Another continental vulture crisis: Africa's vultures collapsing toward extinction. Conserv Lett. https://doi.org/10.1111/conl.12182

Olsen J (2014) Australian high country raptors. CSIRO Publishing, Melbourne

Olsen J, Tucker A (2003) A brood-size manipulation experiment with Peregrine falcons, *Falco peregrinus*, near Camberra. Emu 103:127–132

Ontiveros D, Pleguezuelos JM, Caro J (2005) Prey density, prey detectability and food habits: the case of Bonelli's eagle and the conservation measures. Biol Conserv 123:19–25

Orians GH, Wittenberger JF (1991) Spatial and temporal scales in habitat selection. Am Nat 137:29–49

Pagán I, Martínez JE, Calvo JF (2009) Territorial occupancy and breeding performance in a migratory raptor do not follow ideal despotic distribution patterns. J Zool 279:36–43

Paz A, Jareño D, Arroyo L, Viñuela J, Arroyo B, Mougeot F, Luque-Larena JJ, Fargallo JA (2013) Avian predators as a biological control system of common vole (*Microtus arvalis*) populations in North-Western Spain: experimental set-up and preliminary results. Pest Manag Sci 69:444–450

Pedrini P, Sergio F (2002) Regional conservation priorities for a large predator: golden eagles (Aquila chrysaetos) in the alpine range. Biol Conserv 103:163–172

Penteriani V (1997) Long-term study of a goshawk breeding population on a Mediterranean mountain (Abruzzi Apennines, Central Italy): density, breeding performance and diet. J Raptor Res 31:308–312

Penteriani V, Delgado M (2009) Thoughts on natal dispersal. J Raptor Res 43:90–99

Penteriani V, Fortuna MA, Melián CJ, Otalora F, Ferrer M (2006a) Can prey behaviour induce spatially synchronic aggregation of solitary predators? Oikos 113:497–505

Penteriani V, Otalora F, Ferrer M (2006b) Floater dynamics can explain positive patterns of density-dependence fecundity in animal populations. Am Nat 168:697–703

Pérez-Camacho L, García-Salgado G, Rebollo S, Martínez-Herterkamp S, Fernández-Pereira JM (2015) Higher reproductive success of small males and greater recruitment of large females may explain strong reverse dimorphism (RSD) in the northern goshawk (*Accipiter gentilis*). Oecologia 177:379–387

Pettorelli N, Safi K, Turner W (2014) Satellite remote sensing, biodiversity research and conservation of the future. Philos T Roy Soc B 369:20130190

Poirazidis K, Goutner V, Skartsi T, Stamou G (2004) Modeling nesting habitat as a conservation tool for the Eurasian black vulture (*Aegypius monachus*) in Dadia nature reserve, northeastern Greece. Biol Conserv 111:235–248

Prommer M, Bagyura J, Chavko J, Uhrin M (2012) Migratory movements of central and eastern European Saker falcons (*Falco cherrug*) from juvenile dispersal to adulthood. Aquila 119:111–134

Pulliam HR (1988) Sources, sinks, and population regulation. Am Nat 132:652–661

Rebollo S, Pérez-Camacho L, García-Salgado G, Martínez-Hesterkamp S, Fernández-Pereira JM, Rebollo M, Rebollo P, De la Montaña E (2011) Spatial relationship among northern goshawk, Eurasian sparrowhawk and common buzzard: rivals or partners? In: Zuberogoitia I, Martínez JE (eds) Ecology and conservation of European Forest-dwelling raptors. Diputación Foral de Bizkaia, Bilbao, pp 159–167

Regos A, Tapia L, Gil-Carrera A, Dominguez J (2017) Monitoring protected areas from space: a multi-temporal assessment using raptors as biodiversity surrogates. PLoS One 12(7):e0181769. https://doi.org/10.1371/journal.pone.0181769

Riddell W (2013) Double-brooding and other observations of Grey goshawk in Darwin, northern territory. Aust Field Ornithol 30:00–00

Rodríguez C, Tapia L, Bustamante J, Ferreira E (2014) Crop vegetation structure is more important than crop type in determining where lesser kestrel's forage. Bird Conserv Int 24:438–452. https://doi.org/10.1017/S0959270913000129

Rodríguez SA, Kennedy PL, Parker PH (2016) Timber harvest and tree size near nests explains variation in nest site occupancy but not productivity in northern goshawks (*Accipiter gentilis*). For Ecol Manag 374:220–229

Rodríguez B, Rodríguez A, Siverio F, Siverio M (2017) Factors affecting the spatial distribution and breeding habitat of an insular cliff-nesting raptor community. Curr Zool:1–9. https://doi.org/10.1093/cz/zox005

Rodríguez-Lado L, Tapia L (2012) Suitable breeding habitat for golden eagle (*Aquila chrysaëtos*) in a border of distribution area in northwestern Spain: advantages of using remote sensing information vs land use maps. Vie et Milieu 62:77–85

Rohwer S, Viggiano A, Marzluff JM (2011) Reciprocal tradeoffs between molt and breeding in albatrosses. Condor 113:61–70

Rooney E, Reid N, Montgomery WI (2015) Supplementary feeding increases common buzzard (*Buteo buteo)* productivity but only in poor-quality hábitat. Ibis 157:181–185

Rutz C, Bijlsma RG, Marquiss M, Kenward RE (2006) Population limitation in the northern goshawk in Europe: a review with cases studies. Stud Avian Biol 31:158–197

Sánchez-Zapata JA, Calvo JF (1999) Raptor distribution in relation to landscape composition in semi-arid Mediterranean habitats. J Appl Ecol 36:254–262

Sarasola JH, Bustamante J, Negro JJ, Travaini A (2008) Where do Swainson's hawks winter?. Satellite images to identify potential hábitats. Diver and Distrib 14:742–753

Sasvari L, Hegyi Z (2002) Effects of age composition of pairs and weather condition on the breeding performance of tawny owls, *Strix aluco*. Folia Zool 51:113–120

Sasvari L, Hegyi Z, Csorgo T, Hahn I (2000) Age-dependent diet change, parental care and reproductive cost in tawny owls *Strix aluco*. Acta Oecol 21:267–275

Schmidt KA (2004) Site fidelity in temporally correlated environments enhances population persistence. Ecol Lett 7:176–184

Selas V (1997a) Nest-site selection by four sympatric forest raptors in southern Norway. J Raptor Res 31:16–25

Selas V (1997b) Influence of prey availability on re-establishment of goshawk *Accipiter gentilis* nesting territories. Ornis Fenn 74:113–120

Selas V, Steel C (1998) Large brood sizes of pied flycatcher, sparrowhawk and goshawk in peak microtine years: support for the mast depression hypotesis. Oecologia 116: 449–455

Sergio F (2003) From individual behaviour to population pattern: weather-dependent foraging and breeding performance in black kites. Animal Behaviour 66:1109–1117

Sergio F, Hiraldo F (2008) Intraguild predation in raptor assemblages: a review. Ibis 150(s1):132–145

Sergio F, Newton I (2003) Occupancy as a measure of territory quality. J Anim Ecol 72:857–865

Sergio F, Marchesi L, Pedrini P (2003a) Spatial refugia and the coexistence of a diurnal raptor with its intraguild owl predator. J Anim Ecol 72:232–245

Sergio F, Pedrini P, Marchesi L (2003) Adaptive selection of foraging and nesting habitat by black kites (*Milvus migrans*) and its implications for conservation: a multi-scale approach. Biol Conserv 112:351–362

Sergio F, Marchesi L, Pedrini P (2004) Integrating individual habitat choices and regional distribution of a biodiversity indicator and top predator. J Biogeogr 31:619–628

Sergio F, Newton I, Marchesi L (2005) Top predators and biodiversity. Nature 436:192
Serrano D, Tella JL, Forero MG, Donázar JA (2001) Factors affecting breeding dispersal in the facultatively colonial lesser kestrel: individual experience vs. conspecific cues. J Anim Ecol 70:568–578
Sesé JA (2011) Los plumajes del Quebrantahuesos. In: Lacasa M (ed) El libro de las rapaces. Photodigiscoping, S.C.P., Barcelona, pp 10–25
Simmons R (1988) Offspring quality and the evolution of cainism. Ibis 130:339–357
Slater SJ, Keller KR, Knight RN (2017) Interannual golden eagle (*Aquila chrysaetos*) nest-use patterns in Central Utah: implications for long-term nest protection. J Raptor Res 51:129–135
Snyder NFR, Johnson EV, Clendenen DA (1987) Primary molt of California condors. Condor 89:468–485
Solonen T (1993) Spacing of birds of prey in southern Finland. Ornis Fennica 70:129–143
Sorenson KJ, Burnett LJ, Stake MM (2017) Restoring bald eagle breeding population in Central California and monitoring 25 years of regional population growth. J Raptor Res 51:145–152
Soulé ME, Bolger DT, Alberts AC, Wright J, Sorice M, Hill S (1988) Reconstructed dynamics of rapid extinctions of chaparral-requiring birds in urban habitat islands. Conserv Biol 2:75–92
Spaul RJ, Heath JA (2016) Nonmotorized recreation and motorized recreation in shrub- steppe habitats affects behavior and reproduction of golden eagles (*Aquila chrysaetos*). Ecol Evol. https://doi.org/10.1002/ece3.2540
Squires J, Kennedy PL (2006) Northern goshawk ecology: an assessment of current knowledge and information needs for conservation and management. Stud Avian Biol 31:8–62
Stamps JA (2001) Habitat selection by dispersers: integrating proximate and ultimate approaches. In: Clobert J, Danchin E, Dhondt AA, Nichols JD (eds) Dispersal. Oxford University Press, New York, pp 230–242
Steenhof K, Newton I (2007) Assessing nesting success and productivity. In: Bird D, Bildstein K (eds) Raptor research and management techniques manual. Raptor research foundation. Hancock House Publishers, Surrey, Canada, pp 181–191
Steenhof K, Kochert MN, McDonald TL (1997) Interactive effects of prey and weather on golden eagle reproduction. J Anim Ecol 66:350–362
Tanferna A, López-Jiménez L, Blas J, Horaldo F, Sergio F (2013) Habitat selection by black kite breeders and floaters: implications for conservation management of raptor floaters. Biol Conserv 160:1–9
Tapia L, Kennedy P, Mannan B (2007a) Habitat sampling. In: Bird D, Bildstein K (eds) Raptor research and management techniques manual. Raptor research foundation. Hancock House Publishers, Surrey, Canada, pp 153–169. ISBN: 0-88839-639-2
Tapia L, Domínguez J, Rodríguez L (2007b) Modelling habitat use and distribution of golden eagle (*Aquila chrysaetos*) in a low-density area in NW of the Iberian peninsula. Biodivers Conserv 16:3559–3574
Tapia L, Domínguez J, Rodríguez J (2008a) Hunting habitat preferences of raptors in a mountainous area (northwestern Spain). Pol J Ecol 56:323–333
Tapia L, Domínguez J, Rodríguez J (2008b) Modelling habitat preferences by raptors in two areas of northwestern Spain using different scales and survey techniques. Vie et Milieu 58:257–262
Tapia L, Regos A, Gil-carrera A, Domínguez J (2017) Unravelling the response of diurnal raptors to land-use change in a highly dynamic landscape in northwestern Spain: an approach based on satellite earth observation data. Eur J Wildl Res 63:40. https://doi.org/10.1007/s10344-017-1097-2
Tella JL (1993) Polyandrous trios in a population of Egyptian vultures (*Neophron percnopterus*). J Raptor Res 27:119–120
Thirgood SJ, Redpath SM, Grahan IM (2003) What determines the foraging distribution of raptors on heather moorland? Oikos 100:15–24
Torres-Orozco D, Arroyo B, Pomarol M, Santangeli A (2016) From a conservation trap to a conservation solution: lessons from an intensively managed Montagu's harrier population. Anim Conserv 19:436–443
Touati L, Nedjah R, Samraoui F, Alfarhan A, Gangoso L, Figuerola J, Samraoui B (2017) On the brink: status and breeding ecology of Eleonora's falcon *Falco eleonorae* in Algeria. Bird Conserv Int:1–13. https://doi.org/10.1017/S0959270916000484

Van Horne B (1983) Density as a misleading indicator of habitat quality. J Wildl Manag 47:893–901
Verhulst S, Nilsson J-Å (2008) The timing of birds' breeding seasons: a review of experiments that manipulated timing of breeding. Philos Trans R Soc B: Biol Sci 363:399–410
Watson J (2010) The golden eagle, 2nd edn. T. and A. D. Poyser, London
Weston ED, Whitfield DP, Travis JMJ, Lambin X (2013) When do young birds disperse? Test from studies of golden eagles in Scotland. BMC Ecol 13:42
Wiens JA, Rotenberry JT, Van Horne B (1987) Habitat occupancy patterns of north American shrubsteppe birds: the effects of spatial scale. Oikos 48:132–147
Wiens JD, Noon BR, Reynolds RT (2006) Post-fledging survival of northern Goshawks: the importance of prey abundance, weather and dispersal. Ecol Appl 16:406–418
Wolter K, Mundy P, Dube M (2013) Incubation patch on a male cape griffon. Vulture News 65:37–39
Zabala J, Zuberogoitia I (2014) Individual quality explains variation in reproductive success better than territory quality in a long-lived territorial raptor. PLoS One 9:e90254. https://doi.org/10.1371/journal.pone.0090254
Zabala J, Zuberogoitia I (2015) Breeding performance and survival in the peregrine falcon *Falco peregrinus* support an age-related competence improvement hypothesis mediated via an age threshold. J Avian Biol 45:141–150
Zuberogoitia I, Martínez JE (Eds) (2011) Ecology and conservation of European forest-dwelling raptors. Departamento de agricultura de la Diputación Foral de Bizkaia, Bilbao
Zuberogoitia I, Ruiz Moneo F, Torres JJ (Eds) (2002) El Halcón Peregrino. Servicio Publicaciones de la Diputación Foral de Bizkaia, Bilbao
Zuberogoitia I, Martínez JE (2015) Azor común – *Accipiter gentilis*. En: Salvador A, Morales MB (eds) Enciclopedia Virtual de los Vertebrados Españoles. Museo Nacional de Ciencias Naturales, Madrid. http://www.vertebradosibericos.org/
Zuberogoitia I, Martínez JA, Iraeta A, Azkona A, Zabala J, Jiménez B, Merino R, Gómez G (2006) Short-term effects of the prestige oil spill on the Peregrine falcon (*Falco peregrinus*). Mar Pollut Bull 52:1176–1181
Zuberogoitia I, Martínez JA, Azkona A, Martínez JE, Castillo I, Zabala J (2009) Using recruitment age, territorial fidelity and dispersal as decisive tools in the conservation and management of peregrine falcon (*Falco peregrinus*) populations: the case of a healthy population in northern Spain. J Ornithol 150:95–101
Zuberogoitia I, Castillo I, Zabala J, Iraeta A, Azkona A (2011) Population trends of diurnal forest raptors in Biscay. Pp. 70-80. In: Zuberogoitia I, Martínez JE (eds) Ecology and conservation of European Forest-dwelling raptors. Diputación Foral de Bizkaia, Bilbao
Zuberogoitia I, Arroyo B, O'Donoghue B, Zabala J, Martínez JA, Martínez JE, Murphy SG (2012) Standing out from the crowd: are patagial wing tags a potential predator attraction for harriers (*Circus spp.*)? J Ornithol 153:985–989
Zuberogoitia I, Martínez JE, González-Oreja JA, Calvo JF, Zabala J (2013a) The relationship between brood size and prey selection in a Peregrine falcon population located in a strategic region on the western European flyway. J Ornithol 154:73–82
Zuberogoitia I, González-Oreja JA, Martínez JE, Zabala J, Gómez I, López-López P (2013b) Foraging movements of Eurasian griffon vultures (*Gyps fulvus*): implications for supplementary feeding management. Eur J Wildl Res 59:421–429
Zuberogoitia I, De la Puente J, Elorriaga J, Alonso R, Palomares LE, Martínez JE (2013c) The flight feathers molt of griffon vultures *Gyps fulvus* and associated biological consequences. J Rapt Res 47:292–303
Zuberogoitia I, Zabala J, Martínez JE, González-Oreja JA, López-López P (2014) Effective conservation measures to mitigate the impact of human disturbances on the endangered Egyptian vulture. Anim Conserv 17:410–418
Zuberogoitia I, Zabala J, Martínez JE, Olsen J (2015) Alternative eyrie use in peregrine falcons: is it a female choice? J Zool 296:6–14
Zuberogoitia I, Gil JA, Martínez JE, Erni B, Aniz B, López-López P (2016) The flight feather moult pattern of the bearded vulture (*Gypaetus barbatus*). J Ornithol 157:209–217

Chapter 4
Dispersal in Raptors

David Serrano

Introduction

Dispersal was defined by Howard (1960) as the movement an animal makes from its birth site to the place where it reproduces or would have reproduced if it had survived and found a mate. Iteroparous animals, however, can also move between successive breeding sites during adulthood, so Greenwood (1980) distinguished between *natal dispersal*, the movement from the birth site to the first breeding site, and *breeding dispersal*, the movement between successive breeding sites. *Juvenile dispersal* was later coined as the movements many long-lived species perform during their pre-breeding years, so it refers to a specific part of the natal dispersal process. The opposite of dispersal is *philopatry*, i.e. fidelity to the natal or the previous breeding site. Regarding whether we focus on ecological or evolutionary grounds, two additional terms can be distinguished. Movements of individuals from one place to another irrespective of whether reproduction in the new place is or not successful are defined as *ecological dispersal*, while *effective* or *genetic dispersal* is intended to animals breeding successfully in the place they moved to, thus resulting in gene flow (Greenwood 1980; Johnson and Gaines 1990). Although evolutionary models and population genetics studies often use migration and dispersal as interchangeable terms, the term migration has a very different ecological meaning.

As in any living being, dispersal has deep ecological and evolutionary implications in birds of prey (hereafter raptors, including both diurnal and nocturnal birds of prey), being a key biological process at multiple spatial and temporal scales from genes to ecosystems (Clobert 2001). Dispersal connects fragmented populations and allows organisms to colonize new areas and shift their distribution range, so it is a key process in an increasingly deteriorated world and for understanding species'

D. Serrano (✉)
Department of Conservation Biology, Doñana Biological Station (EBD-CSIC), Seville, Spain
e-mail: serrano@ebd.csic.es

J. H. Sarasola et al. (eds.), *Birds of Prey*,
https://doi.org/10.1007/978-3-319-73745-4_4

responses to climate change. It is therefore of paramount importance from both a basic and an applied perspective.

Raptors form a very diverse group in terms of behaviours, life histories, habitats and distribution ranges, so dispersal is expected to have evolved under very different circumstances. However, many species are long-lived and elusive, occur at low densities in hard-to-reach areas, have a lengthy period of deferred maturity, and are capable of moving long distances in short periods of time, all of which make it difficult to mark and track their lifetime fates (Morrison and Wood 2009). This makes achieving appropriate experimental designs unpractical or directly unfeasible and often constrains seriously adequate sample sizes. Moreover, raptorial birds have been killed by humans for centuries, with the result that many species and populations are scarce and of conservation concern, so understanding dispersal is critical for implementing efficient conservation policies. Fortunately, raptors are so charismatic and fascinating creatures that have encouraged many researchers to get around the inherent problems to their study, devoting huge efforts to the understanding of their biology and ecology. In this review, I begin undertaking an overview of the methods used to study dispersal in raptors with the advantages and biases associated with each one. Then, I describe the most common patterns of dispersal. I continue going through the factors triggering dispersal in raptors, putting current knowledge within the context of the state of the art of the topic in vertebrates. Then, I review the consequences of dispersal at the individual level and its implications for populations and metapopulations. I close by presenting a road map for the future study of raptor dispersal.

Studying Dispersal: Methods, Biases and Patterns

Measuring dispersal is not easy, particularly in highly mobile animals such as raptors. There are a number of direct and indirect techniques to quantifying dispersal (Fig. 4.1) which provide different information, so it is important to be clear about the research objectives and the limitations of each method (Morrison and Wood 2009). If our prime interest is to put dispersal in a context of demographic connectivity and population viability, we should probably rely on capture-mark-recapture methods because large sample sizes can be obtained in a relatively inexpensive way, which will minimize uncertainty in our population estimates. This method has been employed extensively in raptors, either using conventional metal rings (e.g. Morrison and Baird 2016) or plastic rings and tags readable at distance (e.g. Serrano et al. 2001). An interesting alternative to artificial marks for very sensitive, elusive or difficult-to-trap species is the use of natural signatures such as owls' hoots (e.g. Galeotti and Sacchi 2001) or characteristics of the plumage (Bretagnolle et al. 1994). Although recoveries of conventional metal bands may provide interesting information covering wide geographical areas, its main pitfalls are biases due to spatial and temporal variation in reporting probabilities (Korner-Nievergelt et al. 2010) and that the breeding status of the recovered birds must often be presumed (e.g. Millsap et al. 2014). Studies using marks readable at distance allow

Fig. 4.1 Main methods used to study dispersal in raptors. (**a**) Peregrine marked with an alphanumeric band readable at distance. (Photo: Iñigo Zuberogoitia©). (**b**) Attaching a radio-transmitter to an eagle owl fledgling. (Photo: Vincenzo Penteriani©). (**c**) Attaching a GPS device to a griffon vulture. (Photo: David Serrano©). (**d**) Taking a blood sample to an Egyptian vulture fledgling for genetic analysis. (Photo: David Serrano©)

'recapturing' the birds without physical capture but require intensive, large-scale and long-term studies and are not free from other problems either. The most important are that resighting probabilities may decrease with distance across the study area and that birds dispersing beyond its limits are lost from observation (Koenig et al. 1996). A number of authors have proposed methods to circumvent the first problem by correcting for detection probabilities (e.g. Baker et al. 1995, Korner-Nievergelt et al. 2010), but permanent emigration from the study area is much more challenging (Koenig et al. 1996). Moreover, permanent emigration also biases survival probabilities and thereby long-term fitness estimates (Belichon et al. 1996; Zimmerman et al. 2007; Doligez and Pärt 2008). This problem can be minimized by surveying large study areas (Doligez and Pärt 2008; Serrano and Tella 2012) and/or by combining capture-recapture estimates with other methods (see below).

If we are interested in investigating causes of mortality after departure from the natal area, or in knowing the precise itinerary during the dispersal process, we will have to use radio- or satellite-tracking (Cadahía et al. 2010). Modern devices provide detailed information on movement modes and paths, which is invaluable to unravel how raptors travel and orient in different landscapes (Aebischer et al. 2009; Delgado et al. 2010; Poessel et al. 2016). Radio-tracking devices are relatively cheap nowadays but require a high investment in time and effort, and birds can also be lost if they move outside the surveyed area (Wiens et al. 2006b). Satellite telemetry provides nowadays much more accurate data, but it is still prohibitively costly, so sample sizes are often too low, and devices' battery life often precludes tracking the whole dispersal process.

If we are to estimate effective connectivity among populations, we may rely on molecular techniques. There are both indirect and direct genetic methods to estimate contemporary gene flow among populations. Indirect methods rely on the amount and pattern of spatial genetic structure, the basic idea being that there is a simple relationship between population divergence and gene flow. The spatial distribution of genetic variation, though, may be caused by different processes over the past generations, and indirect methods have serious limitations due to a number of model assumptions and problems (Lowe and Allendorf 2010). Since the advent of highly polymorphic nuclear markers such as microsatellites, measurements of current gene flow are possible by assigning individuals probabilistically to candidate populations by their multilocus genotype (Waser and Strobeck 1998). This approach has proven fruitful in measuring gene flow in raptors (Martínez-Cruz et al. 2004; Sonsthagen et al. 2012a) but requires genetic divergence among populations to be effective (Lowe and Allendorf 2010). Measuring dispersal patterns is also possible even in the absence of any genetic structure if all (or nearly all) individuals are genotyped. Studying dispersal through this approach requires in most cases a formidable sampling effort and is only feasible for small populations (Sonsthagen et al. 2017). Molecular techniques have the advantage that non-invasive genetic sampling of moulted feathers permits studying dispersal in species difficult to trap and mark (Booms et al. 2011).

The relative importance of dispersal studies in the scientific raptor literature increased from 1980 to 2005 but decreased somewhat during the last decade (Fig. 4.2a). Capture-recapture has been in general the most frequent approach, only outweighed by radio-tracking in 1991–1995 (Fig. 4.2b). From this period onwards, radio-tracking has been losing importance probably due to the advent of satellite-tracking devices. Finally, an increasing, rather parallel commitment to using molecular techniques and GPS tracking technologies is evident from the early 2000s to the present day (Fig. 4.1b). As we have seen, no method is perfect, so a comprehensive understanding of dispersal minimizing biases and limitations inherent to each approach can be better achieved by means of a combination of different techniques (Korner-Nievergelt et al. 2010; Lowe and Allendorf 2010), which has proven to be fruitful in raptors (Forsman et al. 2002; Martin et al. 2006; Alcaide et al. 2009).

Regarding dispersal patterns, the most basic distinction is between philopatrics and dispersers (dispersal rates). This makes sense under the logic of a change from a familiar to an unfamiliar range (Dobson 2013). This distinction is scale-dependent: individuals may be philopatric to their natal or previous breeding local population, area, territory and/or nest (e.g. Forero et al. 2002; Serrano et al. 2008; Booms et al. 2011). The appropriate scale of study should be biologically meaningful and defined by the research question. When the biological scale of interest is clear and sites or

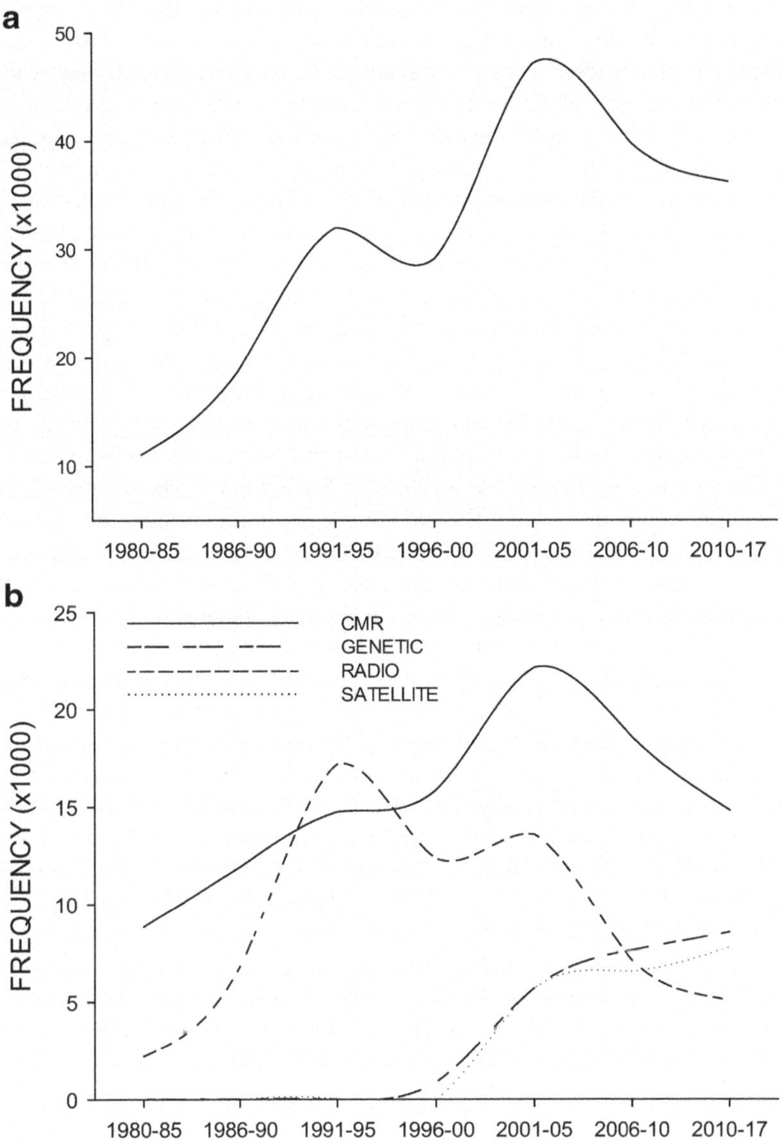

Fig. 4.2 Relative importance of dispersal studies in the scientific raptor literature during the last decades (**a**) and disaggregated by study method (**b**) CMR (capture-mark-recapture), genetic, radio-tracking and satellite-tracking

patches are discrete, as with nest sites and clearly delimited habitat patches, categorizing individuals does not raise serious doubts. In other cases this categorization is necessarily somewhat arbitrary (Limiñana et al. 2011), and randomization procedures may be useful to avoid arbitrary decisions by putting observed patterns into the context of dispersal alternatives (Serrano et al. 2008).

The Euclidean distance between the natal or previous breeding site and the new breeding place, i.e. the realized dispersal distance, is also considered as a basic and biologically relevant descriptor of dispersal. At the population level, the histogram of all Euclidean dispersal distances between the point of origin and destination is usually described by a probability density function called the 'dispersal kernel' ('Dispersal location kernel', see Nathan et al. 2012). In raptors, as in in birds in general, fat-tailed, right-censused dispersal kernels are pervasive (Forsman et al. 2002; Nevoux et al. 2013, Steenhof and Heath 2013; Terraube et al. 2015), with abundant philopatrics or short-distance dispersers at one end and few long-distance dispersers at the other. Therefore, median (and range) dispersal distances are more informative than mean values. Further, dispersal kernels may include or not philopatric individuals (dispersal distances equalling zero), and this must be specified. As dispersal kernels are constrained by the size of the surveyed area, a reliable characterization that does not underestimate long-distance dispersal is important due to its relevant impact on (meta)population persistence, spatial spread and colonization rates (Nathan et al. 2003). For example, Hernández-Matías et al. (2013) showed that the overall fate of the European metapopulation of Bonelli's eagles (Aquila fasciata) was very sensitive to the shape of the dispersal kernel and that the density function which best fitted the empirical natal dispersal data explained past population trajectories worse than a power law function incorporating more long-distance dispersers.

As in virtually all birds species, natal dispersal is of greater magnitude than breeding dispersal in raptors (e.g. Forsman et al. 2002). In most territorial and resident species, recruiting in the natal territory is rarely an available option, so they are often considered as obligate natal dispersers. The overwhelming majority of adults are philopatric to their breeding territory over their entire lives, the very few that disperse mostly moving to nearby territories (e.g. Whitfield et al. 2009). In colonial and semicolonial migratory species, young often settle in their natal colony (Serrano et al. 2003), and adults used to disperse more frequently than territorial residents (Forero et al. 1999; Serrano et al. 2001), although natal dispersal is still much greater than breeding dispersal. As it would be expected, raptors depending heavily on highly fluctuating prey or habitat show the highest rates of breeding dispersal (Korpimäki 1993, Bennetts and Kitchens 2000). These patterns would correspond to a gradient from predictable and evenly distributed to unpredictable and patchily distributed resources in space and time (Wiens 1976, see also below).

Causes of Dispersal in Raptors

In the last decades, a renewed interest in understanding why animals disperse has led to a growing body of empirical and theoretical studies evidencing that movement is influenced simultaneously by social, environmental and phenotypic factors (Clobert et al. 2004), so dispersal is shaped by both external cues (*condition-dependent dispersal*) and the internal state of individuals (*phenotype-dependent dispersal*) (Clobert et al. 2009). Thus, dispersal is seen as a plastic trait varying among individuals and in different ways depending on the ecological context, posing substantial difficulties in identifying consistent patterns and generalities.

Ultimate Causes of Dispersal

Among the ultimate, evolutionary causes that have been selected for dispersal strategies, three have been widely invoked: kin interactions, inbreeding avoidance and habitat variability (Bowler and Benton 2005). Kin selection may favour dispersal strategies if competition among relatives in the natal site is detrimental, while philopatry may be selected and led to kin cooperation if dispersal is limited or delayed (Bowler and Benton 2005). Considerable theoretical attention has been paid to kin selection as a motor of dispersal, but empirical evidence for kin competition promoting dispersal in birds is very limited (e.g. Pasinelli and Walters 2002). In contrast, more effort has been devoted to the role of limited or delayed dispersal on the evolution of cooperative breeding (Koenig and Dickinson 2016). Although cooperative breeding is relatively frequent in raptors, most species form groups of non-relatives (Kimball et al. 2003). On the other side, inbreeding avoidance may select for dispersal as a way of reducing the likelihood of mating with a related individual and avoid inbreeding depression and loss of fitness (Bengtsson 1978), a mechanism also invoked frequently to explain sex-biased dispersal (Pusey 1987). Most studies have reported absence or very low rates of close inbreeding in raptors (Newton and Marquiss 1983; Forero et al. 2002; Forsman et al. 2002; Wiens et al. 2006b), but a few found that these rates are moderately high (James et al. 1987; Millsap and Bear 1990). Evidence suggests that at least for some species mating between close relatives occurs by chance and is more likely between short-distance dispersers (e.g. Warkentin et al. 2013). Inbreeding likelihood has indeed been proved to diminish with dispersal distance (Ortego et al. 2008b). The question that remains is if dispersal has evolved to avoid inbreeding or competition among relatives, as both mechanisms are in practice difficult to separate (Pusey and Wolf 1996). Regarding habitat quality, theoretical models suggest that environmental variation (both in space and time) is one of the most influential causes of the evolution of dispersal (Johnson and Gaines 1990; McPeek and Holt 1992). Perhaps the best evidence in raptors comes from species strongly depending upon fluctuating prey, such as some owls that hardly exhibit breeding site fidelity and track between years areas with high rodent availability (Korpimäki 1987; Therrien et al. 2014; Johnson et al. 2017).

Proximate Condition-Dependent Causes

The proximate causes of dispersal, i.e. those triggering the decisions of individuals over the short term, may often shed light into its evolutionary causes (Bowler and Benton 2005). The environmental causes of dispersal sum up a wide variety of external factors including the physical environment, conspecifics and other living organisms. Direct assessment of habitat quality is the more intuitive way by which animals acquire information and make dispersal decisions. For example, food shortage is known to trigger juvenile (Kennedy and Ward 2003), natal (Byholm et al. 2003; Cattau et al. 2016) and breeding (Korpimäki 1993; Vasko et al. 2011) dispersal. In the same vein, other aspects of habitat quality such as habitat alteration, habitat composition and human disturbance often correlate with dispersal (Kenward et al. 2001; Seamans and Gutiérrez 2007; Di Maggio et al. 2015). Contrarily, however, movements may be higher when food is abundant, because in this situation dispersal is less costly and birds may explore and gain familiarity with potential habitat patches (Bennetts and Kitchens 2000). Weather conditions may also directly influence dispersal in different ways. For example, dispersal distances may be higher in warm periods when thermals favour soaring (Walls et al. 2005) but also under harsh conditions probably because individuals try to move to milder areas (Altwegg et al. 2003). Dominant winds, in turn, have been documented to shape anisotropic dispersal in at least three raptor species (Ferrer 1993c; Walls et al. 2005; Delgado et al. 2010).

A large body of empirical evidence suggests that dispersal depends on population density. Higher rates of dispersal and longer dispersal distances with increasing density have been associated in several raptors with intraspecific competition and some kind of habitat saturation (Negro et al. 1997; Forero et al. 2002; Briggs et al. 2012). However, an opposite pattern where individuals tend to disperse less or recruit where conspecific density is higher has been reported in others (Wiklund 1996; Serrano et al. 2001; Sergio and Penteriani 2005; Hernández-Matías et al. 2010). These contradictory patterns can be explained because the presence and abundance of conspecifics can be used as indirect cues of habitat quality, but beyond a density threshold, recruitment is prevented by intraspecific interactions with residents (Serrano et al. 2003, 2004). Other social factors influencing dispersal were bachelorhood, mate loss and divorce. Single individuals or individuals that lost their mate often disperse from their breeding territories (Forsman et al. 2002; Blakesley et al. 2006; Linkhart and Reynolds 2007; Seamans and Gutiérrez 2007). Divorce has been correlated with probability of dispersal (Forero et al. 1999; Newton 2001; Catlin et al. 2005; Gutiérrez et al. 2011), but it is often difficult to disentangle if birds dispersed after changing mate or changed mate after dispersing (Bai and Severinghaus 2012).

In the case of adults which have already attempted to breed, previous breeding experience may be used as a cue of local habitat quality to take dispersal decisions (Switzer 1997). The use of this kind of private information (sensu Wagner and Danchin 2010) is prevalent in many taxa, individuals following a simple rule of thumb ('win-stay, lose-switch') to decide whether or not to depart from a previous breeding site. Accordingly, breeding dispersal often occurs after breeding failure in raptors (Wiklund 1996; Newton 2001; Serrano et al. 2001; Catlin et al. 2005; Calabuig et al. 2008; Bai and Severinghaus 2012) or even after experimental exposition to predators (Hakkarainen et al. 2001; Krüger 2002). Additional benefits gained from previous residence and experience include other types of familiarity with resources, predators and conspecifics (Tobias 1997; Piper 2011). For example, breeding failure may have a lower impact on breeding dispersal in high- than in low-quality territories (Blakesley et al. 2006). There is also evidence in some species that long-term site quality and years of experience at a given site decreased breeding dispersal probability (Newton 2001; Serrano et al. 2001; Gutiérrez et al. 2011).

Phenotypic Differences

Internal factors such as sex, age or body condition has been widely invoked for explaining dispersal in animals. A general pattern of dispersal in birds is that females are more likely to disperse than males (Greenwood 1980; Clarke et al. 1997), which has been explained because males defend the resources to attract females and would incur in more costs when moving from a familiar area (Greenwood 1980). This explanation has been somewhat challenged in recent times, however, arguing that the role of each sex could be much more complicated than initially realized (Dobson 2013). Another non-exclusive explanation is that sex asymmetries evolved to avoid inbreeding (see *Ultimate causes* above). Although the majority of studies showed female-biased dispersal in raptors (74%, $N = 65$, see Appendix), very few, if any, have succeeded in determining the underlying mechanism. Another intuitive candidate factor correlated with dispersal is individual age. Apart from the differences between natal and breeding dispersers, age has been proven to influence breeding dispersal in several species, older individuals tending to be more philopatric or to disperse shorter distances than younger ones (Forero et al. 1999; Newton 2001; Forsman et al. 2002; Blakesley et al. 2006; Seamans and Gutiérrez 2007; Calabuig et al. 2008; Terraube et al. 2015). This could be explained because many first breeders recruit initially in poor-quality sites but try to disperse to better sites as they grow older (Serrano et al. 2005). Physical condition and body mass (corrected or not for body size) are other factors often examined in the context of dispersal. In some species, individuals in better condition settled at longer distances (Terraube et al. 2015), while in others they settled at shorter distances (Delgado et al. 2010) or did not show any pattern at all (Rosenfield et al. 2016). Other proxies of individual quality reported to affect different aspects of the dispersal process are timing of arrival from the wintering quarters (Serrano et al. 2003; Sergio et al. 2009) and rank within the natal brood (Forero et al. 2002). All these factors may affect differentially males and females or individuals of different age (Byholm et al. 2003; Bai and Severinghaus 2012; Nevoux et al. 2013; Fletcher et al. 2015).

Finally, the genetic control of dispersal in animals is evident in organisms producing specialized dispersal phenotypes such as winged morphs in insects but much less understood in vertebrates where locomotor morphs are rare. Quantitative genetics has provided some evidence for significant additive genetic variation in the individual propensity to disperse in a few bird species (see Charmantier et al. 2011 and references therein). In raptors, evidence is controversial. Some authors examining this issue have not found resemblance in dispersal tactics between relatives (Negro et al. 1997; Lahaye et al. 2001), but others reported similar dispersal patterns between siblings (Newton and Marquiss 1983; Forero et al. 2002; Penteriani and Delgado 2011; Briggs et al. 2012) or between parents and offspring (Steenhof and Heath 2013). However, even in these cases, early parental effects or shared environment often cannot be discarded (Doligez and Pärt 2008). As with other behavioural traits, genome-wide and candidate gene approaches allow advancing in the study of dispersal genetics. The only study in birds showing a relationship between polymorphism in candidate genes and dispersal behaviour was done with a raptor species.

Specifically, significant variation in philopatry, timing of dispersal and dispersal distance was explained by length of alleles in three candidate genes in common buzzards (Buteo buteo) (Chakarov et al. 2013).

During the last two decades, consistent between-individual differences in dispersal have been documented in animals, often associated with a suite of correlated behaviours and morphology, physiology and life history (Ronce and Clobert 2012; Spiegel et al. 2017). Evidence for personality-dependent dispersal syndromes in raptors is however anecdotal. Delgado et al. (2010) studying juvenile eagle owls (Bubo bubo) reported consistent differences between individuals in dispersal behaviour, a similar pattern found in juvenile common buzzards by Kenward et al. (2001). In the facultative colonial lesser kestrel (Falco naumanni), long-distance dispersers tended to settle in empty patches irrespective of whether they moved to the core or the periphery of the population, suggesting that they were behaviourally better skilled to colonizing vacant sites, while philopatrics and short-distance dispersers would be more suited to social environments (Serrano and Tella 2007, 2012).

Interplay Between Internal and External Factors

Mechanistically, dispersal is the result of the interaction between intrinsic and extrinsic factors (Bowler and Benton 2005; Clobert et al. 2009), but most evidence of this interaction in raptors comes from environmental factors affecting both sexes differentially. For example, in situations of high conspecific density and territory saturation, competitive abilities shape dispersal patterns particularly in males due to their prominent role in territory acquisition and defence (Forero et al. 2002; Serrano et al. 2003). Food availability affected dispersal only in female Eurasian kestrels (Falco tinnunculus) (Vasko et al. 2011) but only in male Goshawks (Accipiter gentilis) (Byholm et al. 2003) and at a different spatial scale in male and female Tengmalm's owls (Aegolius funereus) (Korpimäki 1987). Females but not males dispersed more frequently after breeding failure in sparrohawks (Accipiter nisus), black kites (Milvus migrans), flammulated owls (Psiloscops flammeolus) and California spotted owls (Strix occidentalis occidentalis) (Newton and Marquiss 1982; Forero et al. 1999; Blakesley et al. 2006; Linkhart and Reynolds 2007). Conversely, Northern spotted owl (Strix occidentalis caurina) males were more likely to disperse than females when they failed to breed (Forsman et al. 2002).

Dispersal in Raptors Within the Current Research Framework: From Patterns to Processes

Traditionally, behavioural ecologists have neglected to tackle questions regarding animal movement directly, in part because a sound framework was lacking but also because available technologies prevented a detailed knowledge of how animals travel through space. Dispersal is seen nowadays as composed of three sequential stages (Bowler and Benton 2005; Clobert et al. 2009): departure from a natal or breeding site (*departure* or *emigration*), subsequent movement (*transfer, transience, search* or *vagrant*) and establishment in a new breeding area (*settlement* or *immigration*). All these three stages may be governed by different factors, and an understanding of each of them may be important in explaining observed variations in dispersal. On the other hand, like other types of movement, dispersal depends on the interaction between the internal state determining the motivation to move, the motion and navigation capacities of the individual and environmental cues and constraints (Nathan et al. 2008). The motivation to move and the movement capabilities of the individual depend on processes acting at multiple hierarchical levels, from species-specific evolutionary history to the expressed phenotype of a given individual in a certain ecological context at a specific moment in time. Disentangling what factors shape the three stages of dispersal between species, between populations, between individuals and within individuals is a complex but necessary task for a comprehensive understanding of dispersal in raptors (Delgado et al. 2010), as it is in animals in general (Clobert et al. 2009). Importantly, the circumstances experienced at one stage may influence decisions taken at subsequent stages (Clobert et al. 2009), so progressing in dispersal is contingent on understanding the interdependence of the three stages given the (plastic or not) individual phenotype.

As illustrated above, factors explaining realized dispersal patterns comprise the largest body of literature on raptor dispersal, usually interpreted as the determinants of emigration rate. In reality, emigration is a more or less instant behaviour probably involving some innate motivation associated with physiological and locomotor changes (Ritchison et al. 1992). Nonetheless, the internal state of the individual interacting with the external conditions it experienced in the natal or breeding site before dispersal is expected to trigger dispersal decisions (Bonte et al. 2012). For example, timing and age of dispersal initiation may strongly differ between individuals of the same species in response to different factors such as birthplace, sex, laying date, brood size, food supply and weather conditions (Kenward et al. 1993b; Forsman et al. 2002; Balbontín and Ferrer 2005; Walls et al. 2005; Wiens et al. 2006b; Aebischer et al. 2009; Penteriani and Delgado 2011). In many raptors, departure is preceded by excursions or exploratory flights of considerable magnitude, after which individuals return to the natal area (Kenward et al. 1993a; Walls and Kenward 1995; Cadahía et al. 2008; Weston et al. 2013). These movements may be seen as part of the dispersal process itself (Bowler and Benton 2005), influencing, for example, timing and direction of departure. Despite paucity of data, landscape configuration and habitats surrounding the fledgling site may influence post-fledgling movements and orientation, habitat use and initiation of dispersal (Dzialak et al. 2005). Defining when departure takes place (and the transience stage of juvenile dispersal begins) under complex behavioural processes varying between individuals may not be easy, but some remarkable efforts have been made in this direction (Cadahía et al. 2008; Weston et al. 2013).

After departure, animals may change their phenotype and/or carry over effects from the previous stage and collect information on environmental cues and constraints as they move, but transience is the less known stage (Morrison and Wood 2009). These movements may be influenced by pre-dispersal and departure conditions, as exemplified by young Spanish imperial eagles (Aquila adalberti) and burrowing owls (Athene cunicularia) which dispersed at older ages the earlier they hatched, showing also the longest maximum juvenile dispersal distances (Catlin and Rosenberg 2014; Muriel et al. 2015). Movements during this stage may vary with environmental characteristics, as revealed by longer movements of juveniles after habitat degradation or food decrease (Rohner 1996; Kennedy and Ward 2003; Bowling et al. 2012).

Typically, this stage is initiated in juveniles by high rates of movement (Newton and Marquiss 1983; Ferrer 1993a, e.g. Forsman et al. 2002). Tracking devices that allow drawing the precise dispersal pathway have documented dispersal capabilities in a landscape context during these early movements away from the natal area. Alpine eagle owls, for example, followed the principal valleys, but they changed directions abruptly to cross ridges at altitudes of up to 3000 m a.s.l. (Aebischer et al. 2009). These devices offer unique opportunities to disentangle the environmental cues and constraints affecting the mode in which individuals move and adapt their behaviour during dispersal. For instance, snowy owls (Bubo scandiacus) moved rapidly between areas but switch to a searching behaviour for breeding sites when they found low snow cover and depth conditions (Therrien et al. 2015). A general pattern is that juvenile dispersal movements exhibit large differences between individuals (Soutullo et al. 2006; McIntyre et al. 2009; Cadahía et al. 2010; Poessel et al. 2016), so a major challenge is to study how individuals move in a given environmental context depending on their individual phenotype. In eagle owls, path tortuosity strongly depended on landscape properties, so the straightest trajectories were when the owls crossed heterogeneous habitats, this being especially evident in the case of individuals in poor physical condition (Delgado et al. 2010).

Some raptors have among the longest-duration preadult stage of any bird taxa, typically spending several years as floaters before recruiting as breeders (Newton 1979). During these preadult years, they usually engage long-distance wandering movements of different types and use temporary settlement areas (Ferrer 1993a; Balbontín and Ferrer 2009; Cadahía et al. 2010; Nemček et al. 2016; Poessel et al. 2016), presumably to avoid competition and seek for prey-rich areas. A comparable pattern of exploratory movements and settlement has been also described for species breeding at younger ages (Burgess et al. 2009), while others perform wide-ranging wandering movements but do not settle temporary in any area (McIntyre et al. 2009). Some owl species wander to a much lesser extent, and pre-breeding settlement areas are thereby situated very close to the breeding population, even overlapping with home ranges of residents (Rohner 1996; Forsman et al. 2002; Fasciolo et al. 2016). What seems common for most species is that movements and temporary settlement are associated with areas rich in food resources (Ferrer 1993a, b; McIntyre et al. 2009; Fasciolo et al. 2016), and indeed exploratory behaviours may confer familiarity with landscapes important to track resources and survive events of low food availability (Ferrer 1993b; Bennetts and Kitchens 2000; Martin et al. 2006). However, settlement rarely occurs after a unique episode of movement, so this phase may be very complex and may comprise different types of movements whose interpretation is often difficult (Walls and Kenward 1998; Balbontín and Ferrer 2009; McIntyre et al. 2009; Bloom et al. 2011; Poessel et al. 2016). Particularly, the extent to which wandering movements are aimed at exploring and prospecting to gather information on mates and potential recruitment sites is poorly known, although some clues are available. First, mounting evidence shows that juvenile home ranges used to be much larger than in breeders (Belthoff et al. 1993; Rohner 1996; Sternalski et al. 2008; Tanferna et al. 2013; Krüger et al. 2014; Poessel et al. 2016) and in some cases much larger during the breeding than during the non-breeding season (Dwyer et al. 2013). Second, preadults of some species have been reported to return frequently to their areas of origin (Ferrer 1993a; Walls and Kenward 1995; Balbontín and Ferrer 2009; Poessel et al. 2016), where they visit occupied territories, especially those of high quality (Ferrer et al. 2015). Sex-, age- and season-biased exploratory movements and natal returns provide further evidence that birds dedicate at least part of the transience stage to seek for breeding vacancies and mates (Ferrer 1993a; Walls and Kenward 1998; Balbontín and Ferrer 2009; Ferrer et al. 2015).

Finally, settlement at a given site is also a more or less instant behaviour closely connected with the previous transience stage. In eagle owls and Cooper's hawks (Accipiter cooperii), for example, movements during the transience stage seem to determine the area where individuals settle to breed for the first time (Mannan 2010; Fasciolo et al. 2016). Indeed, eagle owls settled in areas similar to those they explored during transience (Delgado et al. 2010). Habitat selection theory predicts that individuals should settle in sites conferring fitness advantages (e.g. Martin 1998), but individuals may find constraints when taking these decisions. First-breeding lesser kestrels were often observed trying to join high-quality colonies from which they were evicted by residents, so most of them settled finally in suboptimal colonies (Serrano and Tella 2007). Not only conditions during transience affect settlement, but even pre-dispersal conditions may carry over to the settlement phase, validating the analyses of realized dispersal distances as functions of pre-dispersal conditions. For example, eagle owl fledgling in the poorest physical condition emigrated later from the natal site and dispersed following straightest trajectories and travelled shorter total distances during transfer but settled further away and after longer times than birds in better conditions (Delgado et al. 2010).

Individuals may also make settlement decisions on the basis of the reproductive performance of conspecifics as a cue integrating all component variables of habitat quality (Danchin et al. 2001). This mechanism has been demonstrated in some raptor species (Sergio and Penteriani 2005; Aparicio et al. 2007). Settlement has been also proposed to be directed towards environments to which the phenotype is more suited to (matching habitat choice, Edelaar and Bolnick 2012). In raptors, several studies support the idea that settlement is not random but biased towards specific habitat features (Hull et al. 2008; Dreiss et al. 2012) or social environments (Serrano and Tella 2007). It has been also suggested that dispersal could be determined by some kind of habitat imprinting during early life, so that individuals could select for settling a similar habitat to that in which they were born (natal habitat preference induction, Davis and Stamps 2004). Evidence in raptors is however controversial: while some raptors were more likely to settle at nest sites and habitats similar to those from which they fledged (Faccio et al. 2013; Fletcher et al. 2015), this does not seem to be the case in others (Delgado et al. 2010; Brown and Collopy 2013).

Consequences of Dispersal

By changing their natal or previous breeding home range, animals experience different environmental and social contexts, and therefore their breeding and/or survival prospects may be affected. Classical habitat selection theory assumes that individuals should choose habitats in a way that maximizes their fitness, but this assumption has been found to not hold in many cases (Chalfoun and Schmidt 2012). Thus, there is an ample room for fitness to vary with dispersal tactics. Estimating the fitness consequences of dispersal has practical difficulties such as not biasing survival estimates due to individuals dispersing permanently beyond the limits of the study area and, as phenotype-dependent dispersal is a common pattern, not confounding the consequences of dispersal with differences between phenotypes (Bonte et al. 2012).

Improvements in habitat or mate quality, and acquisition of mate after mate loss, have been associated with breeding dispersal in birds of prey (Newton and Marquiss 1982; Serrano et al. 2005; Blakesley et al. 2006; Linkhart and Reynolds 2007; Seamans and Gutiérrez 2007). However, it is best to measure the evolutionary consequences of dispersal in terms of fitness payoffs (Bonte et al. 2012). Studying the fitness consequences of natal or breeding dispersal implies by definition that fitness must be estimated after settlement (i.e. conditional of successful dispersal), but fitness can be also estimated during the dispersal process itself (Bonte et al. 2012). This is important because mortality costs during dispersal can be high (Kenward et al. 1999; Wiens et al. 2006a; Bock et al. 2013). Juvenile little owls (Athene noctua), for example, used significantly less sheltered roosting sites than in their natal territory, presumably due to their limited knowledge of the landscape, and this resulted in increased predation risk (Bock et al. 2013). Attempts to estimate mortality rates as function of dispersal distance during this stage are however very limited (Real and Mañosa 2001). In species with deferred maturity, it has been suggested that juvenile survival may depend on the availability of temporary settlement areas and food resources (Penteriani et al. 2005a; Wiens et al. 2006a) or on changes in movement patterns associated with sexual maturity and seeking for a breeding territory (Grande et al. 2009). Even in species in which floaters coexist spatially with breeding residents, home-ranging behaviour and resource use make them more susceptible to human-induced threats (Tavecchia et al. 2012; Tanferna et al. 2013). After settlement, fitness is expected to vary according to the quality of the recruitment site, so a trade-off between recruiting as early as possible and continuing to seek for a good site is expected (Grande et al. 2009; Sergio et al. 2009).

The immediate consequences of natal dispersal has been scarcely investigated, the few studies suggesting that dispersers recruiting in certain habitats suffer higher mortality (Miller et al. 1997; Forsman et al. 2002) or that moving long distances correlated with poor breeding parameters (Newton and Marquiss 1983). By dispersing between breeding sites, however, raptors often improve breeding success (Forero et al. 1999; Newton 2001; Calabuig et al. 2008; Gutiérrez et al. 2011; Martin et al. 2014). In Eurasian kestrels, long-distance dispersing females had better breeding performance, but in males this was conditional on vole abundance (Terraube et al. 2015). As fitness components can be compensated (Belichon et al. 1996), a comprehensive assessment of the consequences of dispersal should ideally include lifetime fitness metrics, although this has only been examined in a handful of species. Studying Swainson's hawks (Buteo swainsoni), Briggs et al. (2012) found no correlation between lifetime reproductive success (LRS) and dispersal distance. In lesser kestrels, however, LRS was negatively related to dispersal distance because long-distance dispersers settled in sites of uncertain quality where immediate survival prospects were lower on average (Serrano and Tella 2012). Sex-specific long-term consequences were reported for black kites and Mauritius kestrels (Falco punctatus). In the former, only males benefited from the fitness advantages of dispersing short distances, although long-distance females mated with more experienced males (Forero et al. 2002). In Mauritius kestrels, females suffered immediate fecundity costs of dispersing long distances but also senescence effects later in life likely as a consequence of increased reproductive effort at middle ages (Nevoux et al. 2013).

Examining the fitness payoffs as functions of dispersal patterns rarely provides information on the underlying mechanisms. These may be a risk cost associated, for example, with probability of mortality due to predation or starvation, but subtler energetic, time or opportunity costs may play a role (see Table 1 in Bonte et al. 2012). Identifying the origin of the differences between dispersal tactics may in theory be facilitated by focusing on the three stages of the dispersal process. Natal habitat-induced phenotypes may differ in their tendency to leave the patch of birth but may also have carry-over effects into later life-history stages which may affect processes during both the transfer and settlement stages and their fitness consequences through propensity to die during dispersal or competitive abilities at settlement (Benard and McCauley 2008). For example, bald eagles (Haliaeetus leucocephalus) born in rural areas survived their first year much better than eagles from suburban areas, probably because rural eagles were more cautious with humans and human-related landscape features (Millsap et al. 2004).

(Meta) population-Level Implications

It is a trivial fact that dispersal has far-reaching consequences at the population level, affecting the size, structure, distribution and dynamics of animal populations. Dispersal is the only natural way by which non-occupied areas may be (re) colonized, and it may be decisive to the persistence of 'sink' populations unable to be sustained in the absence of immigration by rescuing them from extinction (Brown and Kodric-Brown 1977; Pulliam and Danielson 1991). The prevalent role of immigration on local population stability, recovery and growth has been reported for several raptor species (Serrano et al. 2004; Brown and Collopy 2013; Cattau et al. 2016; Lieury et al. 2016). In species with deferred maturity, failing to account for floaters may lead to misleading conclusions about population trajectories and stability because of their important role in buffering breeding population declines (Bart 1995; Penteriani et al. 2005b; Oro et al. 2008). This has important practical consequences, as unperceived declines in the floating population may impact the breeding population some years later, when it could be too late to apply conservation measures (Penteriani et al. 2005b). Moreover, floaters may impact breeders' performance through direct behavioural interference (Bretagnolle et al. 2008). All in all, it is quite clear that floaters and the areas they use must be considered in conservation programmes, although an important hurdle to overcome is that they are often scattered and need cross-border conservation strategies (McIntyre et al. 2009; Cadahía et al. 2010; Penteriani et al. 2011; Nemček et al. 2016). Dispersal may also trigger abrupt population drop and even local extinctions if emigration is not counterbalanced by local recruitment and immigration (Serrano et al. 2004; Le Gouar et al. 2008). Emigration and immigration rates may thereby greatly explain spatio-temporal fluctuations in population size, which will be deeper the tighter their covariation with survival and recruitment (Altwegg et al. 2003).

Since virtually all animal populations live in heterogeneous and fragmented landscapes, all these processes may be critical for the long-term stability and persistence of whole metapopulations (Hanski and Gilpin 1997; Hanski 2001). The scale of movement relative to the scale of fragmentation helps to determine the effect of fragmentation on population trajectories, but the interconnection of habitat demes by dispersal may be complex, and hence natural metapopulations exhibit a variety of structures (Hanski and Gilpin 1997). In raptors, there are a few examples of asymmetrical dispersal among subpopulations (Serrano and Tella 2003; Kauffman et al. 2004), and metapopulation models have pointed out the importance of dispersal from source population for sustaining sink populations located hundreds of kilometres away (Hernández-Matías et al. 2013). In other cases, however, sink populations hardly received immigrants despite a high availability of floaters in source habitats (Kauffman et al. 2004). In this sense, interpatch distance may affect negatively the realized interchange of individuals (Serrano and Tella 2003; Martin et al. 2006), but landscape connectivity may be also influenced by the spatial configuration and permeability of the matrix (Kauffman et al. 2003; Dzialak et al. 2005; Martin et al. 2006). Indeed, spatial barriers are known to inhibit dispersal in some species (Miller et al. 1997; Forsman et al. 2002). Moreover, the spatial scale at which connectivity is important does not have to be reflected by the dispersal potential of the target species, and connectivity across the landscape may be limited even for species with high dispersal potential (Reichert et al. 2016). On the other hand, it is known that some species can perform nearly unrestricted juvenile movements which are often seen as evidence of large-scale connectivity, but this does not necessarily mean that settlement as breeders will be so widespread (Martin et al. 2006, 2007).

Patch size (Martin et al. 2006) and conspecific abundance (Serrano and Tella 2003) may also decrease emigration and increase immigration rates, although conspecifics may exert the opposite effect via competition for nest sites or other resources (Altwegg et al. 2003; Burgess et al. 2008). Conspecifics seem particularly important for reintroduction projects because during the initial phase birds perceive the release area as empty of breeders and may emigrate to other populated areas (Le Gouar et al. 2008). For example, translocated Spanish imperial eagles exhibited longer exploratory movements and wider dispersal ranges than non-translocated conspecifics, presumably due to limited reproductive prospects in the release area. A similar pattern has been found for natal dispersal distances of hacked male peregrines (Falco peregrinus) (Dennhardt and Wakamiya 2013).

On the other hand, effective dispersal involves the movement of genetic material among otherwise separate populations, with pivotal consequences for evolutionary change involving local adaptation, speciation and the evolution of life-history traits. Although gene flow was largely thought to prevent local adaptation (e.g. Lenormand 2002), the opposite may occur if dispersal is non-random (Hull et al. 2008; Edelaar and Bolnick 2012). Contemporary effective dispersal is central for population and conservation genetics, as movement of genes across space determines the relative importance of genetic drift and selection, the extent of inbreeding or outbreeding depression, the level of genetic diversity and the genetic structure of populations. Genetic studies in birds of prey have shown different levels of effective connectivity: extensive connectivity among far apart populations (Huang et al. 2015), isolation by distance (Antoniazza et al. 2010), asymmetric gene flow between mainland and island populations (Ogden et al. 2015) and barriers to dispersal (Sonsthagen et al. 2012b). Molecular approaches have also been successful at identifying the genetic and demographic effects of fragmentation, for example, reduced genetic diversity with isolation (Ortego et al. 2008a), metapopulation dynamics (Sonsthagen et al. 2012a), genetic structure due to genetic drift after population fragmentation (Martínez-Cruz et al. 2007) and recent demographic reduction (Canal et al. 2017).

Conclusions and Future Directions

Through their varied and often unusual lifestyles, raptors open exciting opportunities to understand the complexity of dispersal strategies in animals (Penteriani and Delgado 2009). However, exhaustive research circumventing the many problems and biases inherent to studying dispersal in these highly mobile animals remains scarce. Several conclusions can be extracted from this review.

First, we know the dispersal profile in detail for just a few species. This information is critical to understand between-species variability and the evolution of dispersal at a macroevolutionary level (Paradis et al. 1998) and is the first basic information we need to understand responses to fundamental conservation threats such as climate change and habitat fragmentation.

Second, describing dispersal is not sufficient, and one of the most important challenges is progressing from a phenomenological description of dispersal patterns to a mechanistic understanding of their genesis. Current theoretical and empirical evidence indicates that dispersal is shaped by the interaction between the internal state of dispersers and both the biotic and abiotic environment (Clobert et al. 2009), so many of the phenotype-dependent factors described here are expected to affect dispersal in different ways depending on the environmental context. Although we have some information, many effects are contradictory, based on small sample sizes, or occur in some species or situations but not in others (negative effects have not been shown due to space limitation). Thereby, our knowledge is still rather poor and further effort is necessary. The study of personality-dependent dispersal syndromes is particularly promising because between-individual variation in dispersal propensity may imply covariation with other behavioural, morphological, life-history and ecological traits, with profound consequences ranging from population spread and structure to evolutionary dynamics (Ronce and Clobert 2012). Moreover, dispersal is best understood as an integrated process where emigration, transience and settlement are interconnected (Bowler and Benton 2005). Events and conditions in one stage may affect many processes later in life, including the accumulation of information by individuals. Reconstructing dispersal pathways has already proven fruitful to fill some of these knowledge gaps, but many more data are needed and will emerge as GPS-tagging technologies become smaller, more autonomous and cheaper.

Third, major gaps in our knowledge of the fitness consequences of dispersal persist, and future effort should be directed towards elucidating mortality costs during the transience stage and the lifetime fitness consequences of realized dispersal patterns, particularly in relation to the internal state of individuals and the environment through which dispersers interact.

Fourth, genetic techniques have shed light on the contemporary connectivity of several raptor species at different geographical scales, mostly using neutral markers. Technical developments will allow incorporating routinely genomic approaches for non-model species in the near future, with the potential of correlating gene expression patterns, rather than allelic frequencies, with the dispersal phenotype. Moreover, genome-wide approaches allow sequencing the expression of thousands of genes simultaneously, a critical step for investigating behavioural traits such as dispersal that probably involve many genes of small effect that interact in complex ways.

Fifth, due to the obvious link between dispersal and population dynamics, understanding its causes and consequences is key for managing threatened species and predicting population responses to environmental changes. As most raptor populations live in fragmented landscapes, we also need more information on dispersal and individual interchange in a landscape context. Detailed empirical information on dispersal patterns and underlying factors is necessary for a proper parameterization of spatial models of population viability and target management efforts adequately (Morrison and Wood 2009). Further, failing to account for dispersal in reintroduction and reinforcement programmes may compromise their success and effectiveness.

To conclude, dispersal is a complex multistep behaviour whose understanding remains elusive. It is difficult to characterize dispersal in raptors with broad generalizations that do not take species, population and even individual differences into account, so more carefully designed studies are needed. Emerging advances in tracking technologies and analytical methods offer great promise to deepen on the individual behaviours underlying the different dispersal stages and answer the many unknowns outlined above. Combining different methods is also highly recommended when possible, as they may complement each other offering a more complete and accurate vision of dispersal. Although linking individual decisions and constraints to leave a site and moving across space and settling in another site to the larger-scale dynamics of populations over space and time poses significant conceptual and logistical challenges for raptor ecologists, this is the road we must travel to bring this field forward.

Appendix: Bibliometric Review

I searched in ISI Web of Science for the term 'DISPERSAL' and common names of birds of prey (falcon, falconet, hobby, merlin, kestrel, eagle, hawk, goshawk, sparrowhawk, buzzard, kite, vulture, owl, secretary bird, harrier, owlet, osprey, caracara, condor, baza, lammergeyer, bateleur or gymnogene). I restricted the search to 94 journals of life sciences and some multidisciplinary journals. The search (13 July 2017) rendered 737 papers in English. I filtered the data to exclude search mistakes and papers inferring dispersal from patch occupancy or from past historical diversification and expansion (i.e. phylogeographic studies). The final number of selected papers was 289.

References

Aebischer A, Nyffeler P, Arlettaz R (2009) Wide-range dispersal in juvenile eagle owls (*Bubo bubo*) across the European Alps calls for transnational conservation programmes. J Ornithol 151:1–9

Alcaide M, Serrano D, Tella JL, Negro JJ (2009) Strong philopatry derived from capture-recapture records does not lead to fine-scale genetic differentiation in lesser kestrels. J Anim Ecol 78:468–475

Altwegg R, Roulin A, Kestenholz M, Jenni L (2003) Variation and covariation in survival, dispersal, and population size in barn owls *Tyto alba*. J Anim Ecol 72:391–399

Antoniazza S, Burri R, Fumagalli L, Roulin A, Goudet J (2010) Local adaptation maintains clinal variation in melanin-based coloration of European barn owls (*Tyto alba*). Evolution 64:1944–1954

Aparicio JM, Bonal R, Muñoz A (2007) Experimental test on public information use in the colonial lesser kestrel. Evol Ecol 21:783–800

Bai M-L, Severinghaus LL (2012) Disentangling site and mate fidelity in a monogamous population under strong nest site competition. Anim Behav 84:251–259

Baker M, Nur N, Geupel G (1995) Correcting biased estimates of dispersal and survival due to limited study area: theory and an application using wrentits. Condor 97:663–674
Balbontín J, Ferrer M (2005) Factors affecting the length of the post-fledging period in the Bonelli's eagle *Hieraaetus fasciatus*. Ardea 93:189–198
Balbontín J, Ferrer M (2009) Movements of juvenile Bonelli's eagles *Aquila fasciata* during dispersalMovements of juvenile Bonelli's eagles Aquila fasciata during dispersal. Bird Study 56:86–95
Bart J (1995) Evaluation of population trend estimates calculated using capture-recapture and population projection methods. Ecol Appl 5:662–671
Belichon S, Clobert J, Massot M (1996) Are there differences in fitness components between philopatric and dispersing individuals? Acta Oecol 17:503–517
Belthoff E, Sparks J, Ritchison G (1993) Home ranges of adult and juvenile Eastern screech-owls: size, seasonal variation and extent of overlap. J Raptor Res 27:8–15
Benard MF, McCauley SJ (2008) Integrating across life-history stages: consequences of natal habitat effects on dispersal. Am Nat 171:553–567
Bengtsson BO (1978) Avoiding inbreeding: at what cost? J Theor Biol 73:439–444
Bennetts R, Kitchens W (2000) Factors influencing movement probabilities of a nomadic food specialist: proximate foraging benefits or ultimate gains from exploration? Oikos 91:459–467
Blakesley JA, Anderson DR, Noon BR (2006) Breeding dispersal in the California spotted owl. Condor 108:71
Bloom PH, Michael Scott J, Papp JM, Thomas SE, Kidd JW (2011) Vagrant western red-shouldered hawks: origins, natal dispersal patterns, and survival. Condor 113:538–546
Bock A, Naef-daenzer B, Keil H, Korner-Nievergelt F, Perrig M, Grüebler MU (2013) Roost site selection by little owls *Athene noctua* in relation to environmental conditions and life-history stages. Ibis 155:847–856
Bonte D, Van Dyck H, Bullock JM, Coulon A, Delgado M, Gibbs M, Lehouck V, Matthysen E, Mustin K, Saastamoinen M, Schtickzelle N, Stevens VM, Vandewoestijne S, Baguette M, Barton K, Benton TG, Chaput-Bardy A, Clobert J, Dytham C, Hovestadt T, Meier CM, Palmer SC, Turlure C, Travis JM (2012) Costs of dispersal. Biol Rev Camb Philos Soc 87:290–312
Booms TL, Talbot SL, Sage GK, McCaffery BJ, McCracken KG, Schempf PF (2011) Nest-site fidelity and dispersal of gyrfalcons estimated by noninvasive genetic sampling. Condor 113:768–778
Bowler DE, Benton TG (2005) Causes and consequences of animal dispersal strategies: relating individual behaviour to spatial dynamics. Biol Rev 80:205–225
Bowling AC, Martin J, Kitchens WM (2012) The effect of changes in habitat conditions on the movement of juvenile snail kites *Rostrhamus sociabilis*. Ibis 154:554–565
Bretagnolle V, Mougeot F, Thibault JC (2008) Density dependence in a recovering osprey population: demographic and behavioural processes. J Anim Ecol 77:998–1007
Bretagnolle V, Thibault JC, Dominici JM (1994) Field identification of individual ospreys using head marking pattern. J Wildl Manag 58:175–178
Briggs CW, Collopy MW, Woodbridge B (2012) Correlates and fitness consequences of natal dispersal in Swainson's hawks. Condor 114:764–770
Brown JH, Kodric-Brown A (1977) Turnover rates in insular biogeography: effect of immigration on extinction. Ecology 58:445–449
Brown JL, Collopy MW (2013) Immigration stabilizes a population of threatened cavity-nesting raptors despite possibility of nest box imprinting. J Avian Biol 44:141–148
Burgess M, Black R, Black RA, Jones C, Norris K (2009) The use of agricultural, open and forest habitats by juvenile Mauritius kestrels *Falco punctatus*. Ibis 151:63–76
Burgess MD, Nicoll MA, Jones CG, Norris K (2008) Restricted dispersal reduces the strength of spatial density dependence in a tropical bird population. Proc Biol Sci 275:1209–1216
Byholm P, Saurola P, Lindén H, Wikman M (2003) Causes of dispersal in northern goshawks (*Accipiter gentilis*) in Finland. Auk 120:706
Cadahía L, López-López P, Urios V, Negro JJ (2008) Estimating the onset of dispersal in endangered Bonelli's eagles *Hieraaetus fasciatus* tracked by satellite telemetry: a comparison between methods. Ibis 150:416–420

Cadahía L, López-López P, Urios V, Negro JJ (2010) Satellite telemetry reveals individual variation in juvenile Bonelli's eagle dispersal areas. Eur J Wildl Res 56:923–930

Calabuig G, Ortego J, Cordero PJ, Aparicio JM (2008) Causes, consequences and mechanisms of breeding dispersal in the colonial lesser kestrel, *Falco naumanni*. Anim Behav 76:1989–1996

Canal D, Roques S, Negro JJ, Sarasola JH (2017) Population genetics of the endangered crowned solitary eagle (*Buteogallus coronatus*) in South America. Conserv Genet 18:235–240

Catlin DH, Rosenberg DK (2014) Association of sex, fledging date, and sibling relationships with post-fledging movements of burrowing owls in a nonmigratory population in the imperial valley, California. J Raptor Res 48:106–117

Catlin DH, Rosenberg DK, Haley KL (2005) The effects of nesting success and mate fidelity on breeding dispersal in burrowing owls. Can J Zool 83:1574–1580

Cattau CE, Fletcher RJ Jr, Reichert BE, Kitchens WM (2016) Counteracting effects of a nonnative prey on the demography of a native predator culminate in positive population growth. Ecol Appl 26:1952–1968

Clarke A, Sæther B-E, Røskaft E (1997) Sex biases in avian dispersal: A reappraisal. Oikos 79:429–438

Clobert J (2001) Dispersal. Oxford University Press, Oxford, UK

Clobert J, Ims RA, Rousset F (2004) Causes, mechanisms and consequences of dispersal. In: Hanski I, Gaggiotti OE (eds) Ecology, genetics, and evolution of Metapopulations. Academic, London, pp 307–335

Clobert J, Le Galliard J-F, Cote J, Meylan S, Massot M (2009) Informed dispersal, heterogeneity in animal dispersal syndromes and the dynamics of spatially structured populations. Ecol Lett 12:197–209

Chakarov N, Jonker RM, Boerner M, Hoffman JI, Kruger O (2013) Variation at phenological candidate genes correlates with timing of dispersal and plumage morph in a sedentary bird of prey. Mol Ecol 22:5430–5440

Chalfoun AD, Schmidt KA (2012) Adaptive breeding-habitat selection: is it for the birds? Auk 129:589–599

Charmantier A, Buoro M, Gimenez O, Weimerskirch H (2011) Heritability of short-scale natal dispersal in a large-scale foraging bird, the wandering albatross. J Evol Biol 24:1487–1496

Danchin E, Heg D, Doligez B (2001) Public information and breeding habitat selection. In: Clobert J, Danchin E, Dhondt AA, Nichols JD (eds) Dispersal. Oxford University Press, Oxford, pp 243–258

Davis JM, Stamps JA (2004) The effect of natal experience on habitat preferences. Trends Ecol Evol 19:411–416

Delgado MM, Penteriani V, Revilla E, Nams O (2010) The effect of phenotypic traits and external cues on natal dispersal movements. J Anim Ecol 79:620–632

Dennhardt AJ, Wakamiya SM (2013) Effective dispersal of peregrine falcons (*Falco peregrinus*) in the Midwest, U.S.A. J Raptor Res 47:262–270

Di Maggio R, Mengoni C, Mucci N, Campobello D, Randi E, Sarà M (2015) Do not disturb the family: roles of colony size and human disturbance in the genetic structure of lesser kestrel. J Zool 295:108–115

Dobson FS (2013) The enduring question of sex-biased dispersal: Paul J. Greenwood's (1980) seminal contribution. Anim Behav 85:299–304

Doligez B, Pärt T (2008) Estimating fitness consequences of dispersal: a road to 'knowwhere'? Non-random dispersal and the underestimation of dispersers' fitness. J Anim Ecol 77:1199–1211

Dreiss AN, Antoniazza S, Burri R, Fumagalli L, Sonnay C, Frey C, Goudet J, Roulin A (2012) Local adaptation and matching habitat choice in female barn owls with respect to melanic coloration. J Evol Biol 25:103–114

Dwyer JF, Fraser JD, Morrison JL (2013) Range sizes and habitat use of non-breeding crested caracaras in Florida. J Field Ornithol 84:223–233

Dzialak MR, Lacki MJ, Larkin JL, Carter KM, Vorisek S (2005) Corridors affect dispersal initiation in reintroduced peregrine falcons. Anim Conserv 8:421–430

Edelaar P, Bolnick DI (2012) Non-random gene flow: an underappreciated force in evolution and ecology. Trends Ecol Evol 27:659–665

Faccio SD, Amaral M, Martin CJ, Lloyd JD, French TW, Tur A (2013) Movement patterns, natal dispersal, and survival of Peregrine falcons banded in New England. J Raptor Res 47:246–261

Fasciolo A, Delgado MDM, Cortés G, Soutullo Á, Penteriani V (2016) Limited prospecting behaviour of juvenile eagle owls *Bubo bubo* during natal dispersal: implications for conservation. Bird Study 63:128–135

Ferrer M (1993a) Juvenile dispersal behaviour and natal philopatry of a long-lived raptor, the Spanish imperial eagle Aquila adalberti. Ibis 135:132–138

Ferrer M (1993b) Reduction in hunting success and settlement strategies in young Spanish imperial eagles. Anim Behav 45:406–408

Ferrer M (1993c) Wind-influenced juvenile dispersal of Spanish imperial eagles. Ornis Scand 24:330–333

Ferrer M, Morandini V, Newton I (2015) Floater interference reflects territory quality in the Spanish imperial eagle *Aquila adalberti*: a test of a density-dependent mechanism. Ibis 157:849–859

Fletcher RJ Jr, Robertson EP, Wilcox RC, Reichert BE, Austin JD, Kitchens WM (2015) Affinity for natal environments by dispersers impacts reproduction and explains geographical structure of a highly mobile bird. Proc Biol Sci 282:20151545

Forero M, Donazar J, Hiraldo F (2002) Causes and fitness consequences of Natal dispersal in a population of black kites. Ecology 83:858–872

Forero M, Donázar JA, Blas J, Hiraldo F (1999) Causes and consequences of territory change and breeding dispersal distance in the black kite. Ecology 80:1298–1310

Forsman ED, Anthony RG, Reid JA, Loschl PJ, Sovern SG (2002) Natal and breeding dispersal of northern spotted owls. Wildl Monogr 149:1–35

Galeotti P, Sacchi R (2001) Turnover of territorial Scops owls *Otus scops* as estimated by spectrographic analyses of male hoots. J Avian Biol 32:256–262

Grande JM, Serrano D, Tavecchia G, Carrete M, Ceballos O, Diaz-Delgado R, Tella JL, Donázar JA (2009) Survival in a long-lived territorial migrant: effects of life-history traits and ecological conditions in wintering and breeding areas. Oikos 118:580–590

Greenwood PJ (1980) Mating systems, philopatry and dispersal in birds and mammals. Anim Behav 28:1140–1162

Gutiérrez RJ, Lahaye WS, Zimmerman GS (2011) Breeding dispersal in an isolated population of spotted owls *Strix occidentalis*: evidence for improved reproductive output. Ibis 153:592–600

Hakkarainen H, Ilmonen P, Koivunen V, Korpimäki E (2001) Experimental increase of predation risk induces breeding dispersal of Tengmalm's owl. Oecologia 126:355–359

Hanski I (2001) Population dynamic consequences of dispersal in local populations and in metapopulations. In: Clobert J, Danchin E, Dhondt AA, Nichols JD (eds) Dispersal. Oxford, New York, pp 283–298

Hanski I, Gilpin ME (1997) Metapopulation biology: ecology, genetics, and evolution. Academic, San Diego

Hernández-Matías A, Real J, Moleón M, Palma L, Sánchez-Zapata JA, Pradel R, Carrete M, Gil-Sánchez JM, Beja P, Balbontín J, Vincent-Martin N, Ravayrol A, Benítez JR, Arroyo B, Fernández C, Ferreiro E, García J (2013) From local monitoring to a broad-scale viability assessment: a case study for the Bonelli's eagle in western Europe. Ecol Monogr 83:239–261

Hernández-Matías A, Real J, Pradel R, Ravayrol A, Vincent-Martin N, Bosca F, Cheylan G (2010) Determinants of territorial recruitment in bonelli's eagle (*Aquila fasciata*) populations. Auk 127:173–184

Howard WE (1960) Innate and environmental dispersal of individual vertebrates. Am Midl Nat 63:152–161

Huang AC, Elliott JE, Cheng KM, Ritland K, Ritland CE, Thomsen SK, Hindmarch S, Martin K (2015) Barn owls (*Tyto alba*) in western North America: phylogeographic structure, connectivity, and genetic diversity. Conserv Genet 17:357–367

Hull JM, Hull AC, Sacks BN, Smith JP, Ernest HB (2008) Landscape characteristics influence morphological and genetic differentiation in a widespread raptor (*Buteo jamaicensis*). Mol Ecol 17:810–824

James PC, Oliphant LW, Warkentin IG (1987) Close inbreeding in the merlin (*Falco columbarius*). The Wilson Bulletin 99:718–719

Johnson JA, Booms TL, DeCicco LH, Douglas DC (2017) Seasonal movements of the short-eared owl (*Asio flammeus*) in Western North America as revealed by satellite telemetry. J Raptor Res 51:115–128

Johnson ML, Gaines MS (1990) Evolution of dispersal: theoretical models and empirical tests using birds and mammals. Annu Rev Ecol Syst 21:449–480

Kauffman M, Frick W, Linthicum J (2003) Estimation of habitat-specific demography and population growth for Peregrine falcons in California. Ecol Appl 13:1802–1816

Kauffman MJ, Pollock JF, Walton B (2004) Spatial structure, dispersal, and Management of a Recovering Raptor Population. Am Nat 164:582–597

Kennedy PL, Ward JM (2003) Effects of experimental food supplementation on movements of juvenile northern goshawks (*Accipiter gentilis atricapillus*). Oecologia 134:284–291

Kenward R, Marcström V, Karlbom M (1993a) Post-nestling behaviour in goshawks, *Accipiter gentilis*: I. The causes of dispersal. Anim Behav 46:365–370

Kenward R, Marcström V, Karlbom M (1993b) Post-nestling behaviour in goshawks, *Accipiter gentilis*: II. Sex differences in sociality and nest-switching. Anim Behav 46:371–378

Kenward RE, Marcström V, Karlbom M (1999) Demographic estimates from radio-tagging: models of age-specific survival and breeding in the goshawk. J Anim Ecol 68:1020–1033

Kenward RE, Walls SS, Hodder KH (2001) Life path analysis: scaling indicates priming effects of social and habitat factors on dispersal distances. J Anim Ecol 70:1–13

Kimball RT, Parker PG, Bednarz JC (2003) Occurrence and evolution of cooperative breeding among the diurnal raptors (Accipitridae and Falconidae). Auk 120:717

Koenig W, Van Vuren D, Hooge P (1996) Detectability, philopatry, and the distribution of dispersal distances in vertebrates. Trends Ecol Evol 11:514–517

Koenig WD, Dickinson JL (2016) Cooperative breeding in vertebrates: studies of ecology, evolution, and behavior. Cambridge University Press, Cambridge

Korner-Nievergelt F, Sauter A, Atkinson P, Guelat J, Kania W, Korner-Nievergelt F, Gulat J, Kry M, Kppen U, Robinson R, Schaub M, Thorup K, van Noordwijk A (2010) Improving the analysis of movement data from marked individuals through explicit estimation of observer heterogeneity. J Avian Biol 41:8–17

Korpimäki E (1987) Selection for nest-hole shift and tactics of breeding dispersal in Tengmalm's owl *Aegolius funerus*. J Anim Ecol 56:185–196

Korpimäki E (1993) Does nest-hole quality, poor breeding success or food depletion drive the breeding dispersal of Tengmalm's owls? J Anim Ecol 62:606–613

Krüger O (2002) Interactions between common buzzard *Buteo buteo* and goshawk *Accipiter gentilis*: trade-offs revealed by a field experiment. Oikos 96:441–452

Krüger S, Reid T, Amar A (2014) Differential range use between age classes of southern African bearded vultures *Gypaetus barbatus*. PLoS One 9:e114920

Lahaye W, Gutiérrez RJ, Dunk JR (2001) Natal dispersal of the spotted owl in southern California: dispersal profile of an insular population. Condor 103:691–700

Le Gouar P, Robert A, Choisy J-P, Henriquet S, Lecuyer P, Tessier C, Sarrazin F (2008) Roles of survival and dispersal in reintroduction success of griffon vulture (*Gyps fulvus*). Ecol Appl 18:859–872

Lenormand T (2002) Gene flow and the limits to natural selection. Trends Ecol Evol 17: 183–189

Lieury N, Besnard A, Ponchon C, Ravayrol A, Millon A (2016) Geographically isolated but demographically connected: immigration supports efficient conservation actions in the recovery of a range-margin population of the Bonelli's eagle in France. Biol Conserv 195:272–278

Limiñana R, García JT, González JM, Guerrero Á, Lavedán J, Moreno JD, Román-Muñoz A, Palomares LE, Pinilla A, Ros G, Serrano C, Surroca M, Tena J, Arroyo B (2011) Philopatry and natal dispersal of Montagu's harriers (*Circus pygargus*) breeding in Spain: a review of existing data. Eur J Wildl Res 58:549–555

Linkhart BD, Reynolds RT (2007) Return rate, fidelity, and dispersal in a breeding population of Flammulated owls (*Otus flammeolus*). Auk 124:264

Lowe WH, Allendorf FW (2010) What can genetics tell us about population connectivity? Mol Ecol 19:3038–3051
Mannan RW (2010) Relationship of temporary settling areas to first nesting sites in Cooper's hawks. J Raptor Res 44:143–145
Martin J, Kitchens WM, Hines JE (2007) Natal location influences movement and survival of a spatially structured population of snail kites. Oecologia 153:291–301
Martin J, Nichols JD, Kitchens WM, Hines JE (2006) Multiscale patterns of movement in fragmented landscapes and consequences on demography of the snail kite in Florida. J Anim Ecol 75:527–539
Martin RO, Koeslag A, Curtis O, Amar A (2014) Fidelity at the frontier: divorce and dispersal in a newly colonized raptor population. Anim Behav 93:59–68
Martin TE (1998) Are microhabitat preferences of coexisting species under selection and adaptive? Ecology 79:656–670
Martínez-Cruz B, Godoy JA, Negro JJ (2004) Population genetics after fragmentation: the case of the endangered Spanish imperial eagle (Aquila adalberti). Mol Ecol 13:2243–2255
Martínez-Cruz B, Godoy JA, Negro JJ (2007) Population fragmentation leads to spatial and temporal genetic structure in the endangered Spanish imperial eagle. Mol Ecol 16:477–486
McIntyre CL, Douglas DC, Adams LG (2009) Movements of juvenile gyrfalcons from western and interior Alaska following departure from their natal areas. J Raptor Res 43:99–109
McPeek MA, Holt RD (1992) The evolution of dispersal in spatially and temporally varying environments. Am Midl Nat 140:1010–1027
Miller G, Small R, Meslow EC (1997) Habitat selection by spotted owls during natal dispersal in western Oregon. J Wildl Manag 61:140–150
Millsap B, Breen TIM, McConnell E, Steffer T, Phillips L, Douglass N, Taylor S (2004) Comparative fecundity and survival of bald eagles fledged from suburban and rural natal areas in Florida. J Wildl Manag 68:1018–1031
Millsap B, Harmata A, Stahlecker D, Mikesic D (2014) Natal dispersal distance of bald and golden eagles originating in the coterminous United States as inferred from band encounters. J Raptor Res 48:13–23
Millsap BA, Bear C (1990) Double-brooding by Florida burrowing owls. Willson Bulletin 102:313–317
Morrison J, Baird J (2016) Using banding and encounter data to investigate movements of red-tailed hawks in the northeastern United States. J Raptor Res 50:161–175
Morrison JL, Wood PB (2009) Broadening our approaches to studying dispersal in raptors. J Raptor Res 43:81–89
Muriel R, Morandini V, Ferrer M, Balbontín J (2015) Independence and juvenile dispersal distances in wild and reintroduced Spanish imperial eagles. Biol Conserv 191:300–305
Nathan R, Getz WM, Revilla E, Holyoak M, Kadmon R, Saltz D, Smouse PE (2008) A movement ecology paradigm for unifying organismal movement research. Proc Natl Acad Sci U S A 105:19052–19059
Nathan R, Klein E, Robledo-Arnuncio JJ, Revilla E (2012) Dispersal kernels: review. In: Clobert J, Baguette M, Benton TG, Bullock JM (eds) Dispersal ecology and evolution. Oxford University Press, Oxford, pp 187–210
Nathan R, Perry G, Cronin JT, Strand AE, Cain ML (2003) Methods for estimating long-distance dispersal. Oikos 103:261–273
Negro JJ, Hiraldo F, Donazar JA (1997) Causes of natal dispersal in the lesser kestrel: inbreeding avoidance or resource competition? J Anim Ecol 66:640–648
Nemček V, Uhrin M, Chavko J, Deutschova L, Maderič B, Noga M (2016) Habitat structure of temporary settlement areas of young Saker falcon *Falco cherrug* females during movements in europe. Acta Ornithol 51:93–103
Nevoux M, Arlt D, Nicoll M, Jones C, Norris K (2013) The short- and long-term fitness consequences of natal dispersal in a wild bird population. Ecol Lett 16:438–445
Newton I (1979) Population ecology of raptors. T & AD Poyser, Berkhamsted
Newton I (2001) Causes and consequences of breeding dispersal in the Sparrowhawk *Accipiter nisus*. Ardea 89:143–154

Newton I, Marquiss M (1982) Fidelity to breeding area and mate in Sparrowhawks *Accipiter nisus*. J Anim Ecol 51:327–341

Newton I, Marquiss M (1983) Dispersal of Sparrowhawks between birthplace and breeding place. J Anim Ecol 52:463–477

Ogden R, Heap E, McEwing R, Tingay R, Whitfield DP (2015) Population structure and dispersal patterns in Scottish golden eagles *Aquila chrysaetos* revealed by molecular genetic analysis of territorial birds. Ibis 157:837–848

Oro D, Margalida A, Heredia R, Donázar JA (2008) Testing the goodness of supplementary feeding to enhance population viability in an endangered vulture. PLoS One 3:e4084

Ortego J, Aparicio JM, Cordero PJ, Calabuig G (2008a) Individual genetic diversity correlates with the size and spatial isolation of natal colonies in a bird metapopulation. Proc R Soc B Biol Sci 275:2039–2047

Ortego J, Calabuig G, Aparicio JM, Cordero PJ (2008b) Genetic consequences of natal dispersal in the colonial lesser kestrel. Mol Ecol 17:2051–2059

Paradis E, Baillie SR, Sutherland WJ, Gregory RD (1998) Patterns of natal and breeding dispersal in birds. J Anim Ecol 67:518–536

Pasinelli G, Walters JR (2002) Social and environmental factors affect natal dispersal and philopatry of male red-cockaded woodpeckers. Ecology 83:2229–2239

Penteriani MF, Ferrer M, Penteriani V, Otalora F (2005a) Floater survival affects population persistence. The role of prey availability and environmental stochasticity. Oikos 108:523–534

Penteriani V, Delgado M d M (2009) Thoughts on natal dispersal. J Raptor Res 43:90–98

Penteriani V, Delgado MM (2011) Birthplace-dependent dispersal: are directions of natal dispersal determined a priori? Ecography 34:729–737

Penteriani V, Ferrer M, Delgado MM (2011) Floater strategies and dynamics in birds, and their importance in conservation biology: towards an understanding of nonbreeders in avian populations. Anim Conserv 14:233–241

Penteriani V, Otalora F, Sergio F, Ferrer M (2005b) Environmental stochasticity in dispersal areas can explain the "mysterious" disappearance of breeding populations. Proc Biol Sci 272:1265–1269

Piper WH (2011) Making habitat selection more "familiar": a review. Behav Ecol Sociobiol 65:1329–1351

Poessel SA, Bloom PH, Braham MA, Katzner TE (2016) Age- and season-specific variation in local and long-distance movement behavior of golden eagles. Eur J Wildl Res 62:377–393

Pulliam HR, Danielson BJ (1991) Sources, sinks, and habitat selection: a landscape perspective on population dynamics. Am Nat 137:50–66

Pusey A (1987) Sex-biased dispersal and inbreeding avoidance in birds and mammals. Trends Ecol Evol 2:295–299

Pusey M, Wolf M (1996) Inbreeding avoidance in animals. Trends Ecol Evol 11:201–206

Real J, Mañosa S (2001) Dispersal of juvenile and immature Bonelli's eagles in northeastern Spain. J Raptor Res 35:9–14

Reichert BE, Fletcher RJ Jr, Cattau CE, Kitchens WM (2016) Consistent scaling of population structure across landscapes despite intraspecific variation in movement and connectivity. J Anim Ecol 85:1563–1573

Ritchison G, Belthoff J, Sparks E (1992) Dispersal restlessness: evidence for innate dispersal by juvenile eastern screech-owls? Anim Behav 43:57–65

Rohner C (1996) The numerical response of great horned owls to the snowshoe hare cycle: consequences of non-territorial `floaters' on demography. J Anim Ecol 65:359–370

Ronce O, Clobert J (2012) Dispersal syndromes. In: Clobert J, Baguette M, Benton TG, Bullock JM (eds) Dispersal ecology and evolution. Oxford University Press, Oxford, pp 119–138

Rosenfield R, Bielefeldt J, Haynes T, Hardin M, Glassen F (2016) Body mass of female Cooper's hawks is unrelated to longevity and breeding dispersal: implications for the study of breeding dispersal. J Raptor Res 50:305–312

Seamans ME, Gutiérrez RJ (2007) Habitat selection in a changing environment: the relationship between habitat alteration and spotted owl territory occupancy and breeding dispersal. Condor 109:566

Sergio F, Blas J, Hiraldo F (2009) Predictors of floater status in a long-lived bird: a cross-sectional and longitudinal test of hypotheses. J Anim Ecol 78:109–118
Sergio F, Penteriani V (2005) Public information and territory establishment in a loosely colonial raptor. Ecology 86:340–346
Serrano D, Carrete M, Tella JL (2008) Describing dispersal under habitat constraints: a randomization approach in lesser kestrels. Basic and Applied Ecology 9:771–778
Serrano D, Forero M, Donazar J, Tella J (2004) Dispersal and social attraction affect colony selection and dynamics of lesser kestrels. Ecology 85:3438–3447
Serrano D, Oro D, Ursua E, Tella J (2005) Colony size selection determines adult survival and dispersal preferences: Allee effects in a colonial bird. Am Nat 166:E22–E31
Serrano D, Tella J (2003) Dispersal within a spatially structured population of lesser kestrels: the role of spatial isolation and conspecific attraction. J Anim Ecol 72:400–410
Serrano D, Tella J, Donazar J, Pomarol M (2003) Social and individual features affecting natal dispersal in the colonial lesser kestrel. Ecology 84:3044–3054
Serrano D, Tella J, Forero M, Donazar J (2001) Factors affecting breeding dispersal in the facultatively colonial lesser kestrel: individual experience vs. conspecific cues. J Anim Ecol 70:568–578
Serrano D, Tella JL (2007) The role of despotism and heritability in determining settlement patterns in the colonial lesser kestrel. Am Nat 169:E53–E67
Serrano D, Tella JL (2012) Lifetime fitness correlates of natal dispersal distance in a colonial bird. J Anim Ecol 81:97–107
Sonsthagen S, McClaren E, Doyle F, Titus K, Sage G, Wilson R, Gust J, Talbot S (2012a) Identification of metapopulation dynamics among Northern goshawks of the Alexander archipelago, Alaska, and coastal British Columbia. Conserv Genet 13:1045–1057
Sonsthagen S, Rosenfield R, Bielefeldt J, Murphy R, Stewart A (2012b) Genetic and morphological divergence among Cooper's hawk (*Accipiter cooperii*) populations breeding in north-central and western North America. Auk 129:427–437
Sonsthagen SA, Williams JC, Drew GS, White CM, Sage GK, Talbot SL (2017) Legacy or colonization? Posteruption establishment of peregrine falcons (*Falco peregrinus*) on a volcanically active subarctic island. Ecol Evol 7:107–114
Soutullo A, Urios V, Ferrer M, Peñarrubia SG (2006) Dispersal of golden eagles *Aquila chrysaetos* during their first year of life. Bird Study 53:258–264
Spiegel O, Leu ST, Bull CM, Sih A (2017) What's your move? Movement as a link between personality and spatial dynamics in animal populations. Ecol Lett 20:3–18
Steenhof K, Heath JA (2013) Local recruitment and natal dispersal distances of American kestrels. Condor 115:584–592
Sternalski A, Bavoux C, Burneleau G, Bretagnolle V (2008) Philopatry and natal dispersal in a sedentary population of western marsh harrier. J Zool 274:188–197
Switzer PV (1997) Past reproductive success affects future habitat selection. Behav Ecol Sociobiol 40:307–312
Tanferna A, López-Jiménez L, Blas J, Hiraldo F, Sergio F (2013) Habitat selection by black kite breeders and floaters: implications for conservation management of raptor floaters. Biol Conserv 160:1–9
Tavecchia G, Adrover J, Navarro AM, Pradel R (2012) Modelling mortality causes in longitudinal data in the presence of tag loss: application to raptor poisoning and electrocution. J Appl Ecol 49:297–305
Terraube J, Vasko V, Korpimäki E (2015) Mechanisms and reproductive consequences of breeding dispersal in a specialist predator under temporally varying food conditions. Oikos 124:762–771
Therrien JF, Gauthier G, Pinaud D, Bêty J (2014) Irruptive movements and breeding dispersal of snowy owls: a specialized predator exploiting a pulsed resource. J Avian Biol 45:536–544
Therrien JF, Pinaud D, Gauthier G, Lecomte N, Bildstein KL, Bety J (2015) Is pre-breeding prospecting behaviour affected by snow cover in the irruptive snowy owl? A test using state-space modelling and environmental data annotated via Movebank. Mov Ecol 3:1

Tobias J (1997) Asymmetric territorial contests in the European robin: the role of settlement costs. Anim Behav 54:9–21

Vasko V, Laaksonen T, Valkama J, Korpimäki E (2011) Breeding dispersal of Eurasian kestrels *Falco tinnunculus* under temporally fluctuating food abundance. J Avian Biol 42:552–563

Wagner RH, Danchin É (2010) A taxonomy of biological information. Oikos 119:203–209

Walls S, Kenward R, Holloway G (2005) Weather to disperse? Evidence that climatic conditions influence vertebrate dispersal. J Anim Ecol 74:190–197

Walls SS, Kenward RE (1995) Movements of radio-tagged common buzzards *Buteo buteo* in their first year. Ibis 137:177–182

Walls SS, Kenward RE (1998) Movements of radio-tagged buzzards *Buteo buteo* in early life. Ibis 140:561–568

Warkentin IG, Lieske DJ, Espie RHM (2013) Close inbreeding and dispersal in merlins: further examination. J Raptor Res 47:69–74

Waser PM, Strobeck C (1998) Genetic signatures of interpopulation dispersal. Trends Ecol Evol 13:43–44

Weston ED, Whitfield DP, Travis JM, Lambin X (2013) When do young birds disperse? Tests from studies of golden eagles in Scotland. BMC Ecol 13:42

Whitfield DP, Douse A, Evans RJ, Grant J, Love J, McLeod DRA, Reid R, Wilson JD (2009) Natal and breeding dispersal in a reintroduced population of white-tailed eagles *Haliaeetus albicilla*Natal and breeding dispersal in a reintroduced population of white-tailed eagles Haliaeetus albicilla. Bird Study 56:177–186

Wiens JA (1976) Population responses to patchy environments. Annu Rev Ecol Syst 7:81–120

Wiens JD, Noon B, Reynolds R (2006a) Post-fledging survival of Northern goshawks: the importance of prey abundance, weather, and dispersal. Ecol Appl 16:406–418

Wiens JD, Reynolds RT, Noon BR (2006b) Juvenile movement and natal dispersal of Northern goshawks in Arizona. Condor 108:253–269

Wiklund CG (1996) Determinants of dispersal in breeding merlins (*Falco columbarius*). Ecology 77:1920–1927

Zimmerman GS, Gutiérrez RJ, Lahaye WS (2007) Finite study areas and vital rates: sampling effects on estimates of spotted owl survival and population trends. J Appl Ecol 44:963–971

Chapter 5
Raptor Migration

Keith L. Bildstein

Introduction

The phenomenon of raptor migration has fascinated humanity since antiquity and with good reason. As one of the most behaviorally flexible aspects of raptor biology, migration has allowed this trophic lineage to prosper in ways that few terrestrial top predators have been able to achieve.

Old World accounts of raptor migration date from the Bible. European accounts of New World raptor migration date from within 30 years of European settlement and indigenous accounts predate those (Bildstein 2006). Today premiere raptor migration watchsites in North America, Mesoamerica, Europe, and eastern Asia continue to document the flights of visible migrants along major migration corridors. This current network of watchsites allows us to monitor regional and, in some instances, global populations of many species (Bildstein 1998, 2006). Migration watchsite counts, together with the charismatic nature of the birds themselves, and, importantly, the more recent use of satellite tracking, mean that we now know much about the mechanics and geography of the phenomenon (Bildstein 2017). Indeed, in many ways, the migratory movements of birds of prey are better documented than are those of most avian taxa.

Below I build upon this record, together with knowledge found in the greater ecological, evolutionary biology and raptor-conservation literature, to summarize what we know and what we do not know about raptor migration, particularly as it pertains to both raptor biology and conservation. I begin by discussing the origins and maintenance of raptor migration itself. I then discuss the ecological underpinnings and limitations of this extraordinarily flexible behavioral tool. I finish with an overview of how all of this both explains and constrains the current conservation status of birds of prey.

K. L. Bildstein (✉)
Acopian Center of Conservation Learning, Hawk Mountain Sanctuary, Orwigsburg, PA, USA
e-mail: bildstein@hawkmtn.org

J. H. Sarasola et al. (eds.), *Birds of Prey*,
https://doi.org/10.1007/978-3-319-73745-4_5

Evolutionary Considerations

Evidence suggests that birds of prey have been migrating for a long time. Even so said raptor-like Falconiformes first appear in the fossil record of the Early Eocene 55–50 million years ago (Bildstein 2006), and raptor migration could not have occurred before raptors themselves. The Early Eocene was warmer and less seasonable than the Late Eocene of 42–38 million years ago, which was cooler and more seasonable. That, together with the appearance of somewhat modern raptors in fossil record in the Late Eocene, suggests that raptor migration most likely evolved at this time (Bildstein 2006).

Most migration specialists now recognize that bird migration, including raptor migration, is a genetically based (Berthold 2001), albeit flexible behavioral "tool" that responds to many intrinsic and extrinsic factors, including flight mechanics (Kerlinger 1989), intraspecies competition (Cox 1968), and continental geography, climate, and the seasonal availability of food resources (Bildstein 2006, 2017). The genetic link suggests that birds, including raptors, have genes for both sedentary and migratory behavior; that natural selection can result in sedentary populations, as well as in partially migratory and completely migratory populations; and that reversals among such lifestyles occur (Cox 1968; Bildstein 2017).

The best experimental evidence for the role of genetics in raptor migration comes in the form of displacement experiments conducted in the 1930s when German ornithologist Rudolf Drost "displaced" several hundred migrating adult and juvenile Eurasian sparrowhawks (*Accipiter nisus*) he caught on an island in the North Sea off the coast of Germany (Drost 1938). The birds were aged by plumage and banded. Some individuals were then transported 600 km (410 mi) east-southeast to eastern Poland, where they were released. A control group was released at the capture site. Recaptures of several dozen control birds indicated that these individuals migrated southwest and overwintered in the Netherlands, Belgium, northwestern France, and Portugal. Three dozen "displaced" individuals also were recaptured in winter. Displaced adults apparently migrated west upon being released, and most recoveries were closer to the population's "targeted" wintering grounds. Simply put, this age group corrected for the displacement. Displaced juveniles, however, apparently continued to move southwest, and most were recaptured in central Europe, east-southeast of the Scandinavian population's traditional wintering areas. The results suggest that first-autumn raptors possess genetic instructions regarding directional orientation but lack the navigational skills needed to correct their course if displaced. Experienced adults on the other hand have navigational abilities and are able to make midcourse corrections and reach their intended destinations.

Although similar experiments have not been performed on other raptors, observations of wind-drifted and storm-deflected migrants suggest that natural displacements also affect juvenile more so than adult raptors, indicating that both initial genetics and eventual learning play important roles in the development and maintenance of migration geography (Bildstein 2017). The evidence outlined above suggests that partial migration maintains enough genetic variability in migratory raptors so that as conditions change, natural selection can track the changes, and populations can evolve appropriately. Examples of such changes involving sharp-shinned hawks (*Accipiter striatus*), North American populations of Merlins (*Falco columbarius*), and other raptors support this idea (Bildstein 2006).

In the early twentieth century, for example, Merlins belonging to a pale gray subspecies called the Richardson's Merlin began to expand the northern limits of their wintering range from Colorado and Wyoming to southwestern Canada. Reports of the expansion place overwintering Merlins in Saskatchewan in 1922 and in Alberta in 1948. The expansion, which continued well into the latter half of the twentieth century, was especially apparent in urban areas. By 1970, Richardson's Merlins were both overwintering and breeding in several western Canadian cities (Sodhi et al. 1993). As of early 2017, nonmigratory "city" populations of Merlins now occur in many urban areas across southern Canada and the northern United States. The shift from migratory to nonmigratory behavior in these populations, and their northward range expansion, coincided with the regional expansion of the species' predominant urban prey, the house sparrow (*Passer domesticus*), an exotic introduced into North America in the 1850s that had spread into the American West in the early twentieth century, suggesting that prey availability contributed substantially to the migratory changes. Declining human persecution at the time also may have played a role.

A similar shift in the overwintering behavior in sharp-shinned hawks in the last two decades of the twentieth century also appears to have resulted from so-called migration short stopping, in which naïve first-year migrants are inclined to migrate less on their outbound migrations when food availability increases along the migratory route and individuals stop or slow their migratory movements to take advantage of it. For sharpshins, the food in question was increased numbers of backyard "feeder birds" in northeastern North America brought about by the growing use of backyard bird feeders and a series of especially warm winters beginning in the 1980s and continuing into the 1990s. Again, reduced human persecution appears to have induced the shift (Viverette et al. 1996). Since then other researchers have described migration short stopping in American kestrels (*Falco sparverius*) in the American West, as well as in red kites (*Milvus milvus*), marsh harriers (*Circus aeruginosus*), short-toed snake eagles (*Circaetus gallicus*), and booted eagles (*Hieraaetus pennatus*) in Europe (Bildstein 2017).

Overall, mounting evidence suggests that raptor migration is best viewed as a genetically based but flexible behavioral tool that can be used by populations to adapt to changing conditions, rather than a static trait unamenable to such actions.

Ecological Considerations

Earth's atmosphere provides raptors with three things necessary for successful long-distance migration: oxygen for metabolism, a gaseous medium for generating lift, and vertical updrafts for soaring flight (Bildstein 2006). An additional component, horizontal wind, can help or hinder migrants. Most atmospheric mass is within 5 km (3 mi) of the surface. This layer of dense air is where most "weather" occurs; it also is where temperatures and oxygen pressures are most similar to those at the earth's surface and where most raptor migration occurs (Bildstein 2006).

Observations of visible migrants at raptor migration watchsites indicate that many raptors migrate within several hundred meters of the surrounding landscape. Nevertheless radar studies, studies involving satellite-tracked raptors, and studies in which migrants were followed in fixed-wing aircraft suggest that many other migrants travel at heights that make them impossible for observers to see from the ground (Kerlinger 1989; Bildstein 2006). Turkey vultures (*Cathartes aura*), broad-winged hawks (*Buteo platypterus*), and Swainson's hawks (*B. swainsoni*) followed in fixed-winged aircraft in tropical Panama, for example, migrated at up to 4000 to 5000 m above the ground (Smith 1985). And bald eagles (*Haliaeetus leucocephalus*) migrating across Canada and the western United States regularly reach altitudes of up to 4500 m (Harmata 2002). Radio-tracked griffon vultures (*Gyps fulvus*) are known to soar at up to 10,000 m, and a Rüppell's vulture (*G. rueppellii*) was struck by an airliner over West Africa, traveling at an altitude of 11,300 m at the time of the collision (Laybourne 1974; Bildstein 2006).

Thermals and Other Updrafts

Updrafts, including thermals and other vertical winds, create opportunities for low-cost soaring flight, and as such almost always assist migrants. Thermals are pockets of warm, rising air that form when different surfaces in landscapes receive and absorb different amounts of sunlight. Dry surfaces, including rocky outcrops and parched fields, heat more quickly than do "evaporatively cooled" wet surfaces, including water and living vegetation. In hilly areas, surfaces oriented perpendicular to incoming solar radiation warm more quickly than do shaded and less-perpendicular surfaces.

Land-based thermals typically require sunlight to sustain them. Because of this, terrestrial thermals typically occur on bright, sunny days between midmorning and midday, after the sun is high enough to differentially warm the landscape and create thermals and before strong afternoon horizontal winds begin to tear them apart. Because they are fueled by sunlight outside of the tropics, thermals are stronger in summer than in winter and are weaker overall than they are in the tropics. In addition, because the sun rises more vertically, and, thus, more rapidly, timewise, in the tropics, tropical thermals form earlier in the morning and extend to later in the afternoon than do temperate-zone thermals. Overland, thermals routinely reach heights of more than 2000 m (6600 ft) outside of the tropics and higher within them. The speed at which air rises within the thermal decreases significantly both with height above the surface and with distance from the center of the cell. Not surprisingly, raptors soaring within terrestrial thermals seek out the cores of the vortices and circle within them and usually leave thermals before reaching the top (Bildstein

2017). Although most terrestrial thermals are isolated from one another, thermal streets, or linear arrays of thermals aligned with the prevailing wind or with shorelines, are sometimes created during periods of intense solar radiation, particularly in the tropics. Large and powerful terrestrial thermal updrafts can extend for hundreds of kilometers. When such formations are aligned along the so-called principal axis of migration (sensu Bildstein 2006), migrating raptors are able to soar linearly within them.

Although thermals associated with raptor migration typically form over land, the strongest thermals on earth occur above the Atlantic Ocean's Gulf Stream off the eastern coast of the United States, where warm surface water from the Gulf of Mexico, surrounded by cooler temperate-zone water, creates thermal updrafts that sometimes extend more than 10 km above the Atlantic (Minobe et al. 2008). Migrating raptors also use such thermals to great effect. A satellite-tracked peregrine falcon (*Falco peregrinus*) using such thermals, together with a tailwind created by a passing hurricane, completed a more than 1500 km (9300 mi), 25 h overwater journey from southern New Jersey to south Florida in mid-October 2008 (Bildstein 2017).

In addition to overwater Gulf Stream thermals, oceanic, or "sea thermals," also occur, and they to, like land-based thermals, play critical roles in shaping the raptor migration geography. Sea thermals regularly form in tropical and subtropical oceans and seas in the trade wind zone 5° to 30° north and south of the equator (Augstein 1983). In these bands, northeasterly winds in the Northern Hemisphere and southeasterly winds in the Southern Hemisphere blow relatively cool, subtropical surface air toward the equator over increasingly warmer surface waters of the equatorial zone heat the cooler air, producing bands of sea thermals. Because this temperature differential exists both day and night, sea thermals occur 24 h a day, which means that migrant raptors using them can soar both day *and* night. Sea thermals allow Chinese sparrowhawks (*Accipiter soloensis*), grey-faced buzzards (*Butastur indicus*), and other migratory raptors to soar hundreds of kilometers over water along the East Asian Oceanic Flyway (Fig. 5.1), and these thermals also make it possible for the Amur falcon (*Falco amurensis*) to undertake the longest overwater passage of any migratory raptor, a 4000 km (2500 mi) outbound passage across the Indian Ocean between India and East Africa. Finally, outbound ospreys (*Pandion haliaetus*) use sea thermals to cross the Caribbean from eastern Cuba and western Hispaniola to the South American mainland in soaring flight.

Horizontal Winds

The degree to which horizontal winds are aligned with the regional principal axis of migration determines whether they help or hinder the passage of migrating raptors. Tailwinds, or winds that are aligned with the preferred direction of travel, help to push migrants forward increasing their travel speed. Headwinds, or winds aligned against the preferred direction of travel, do the opposite. Because wind speed

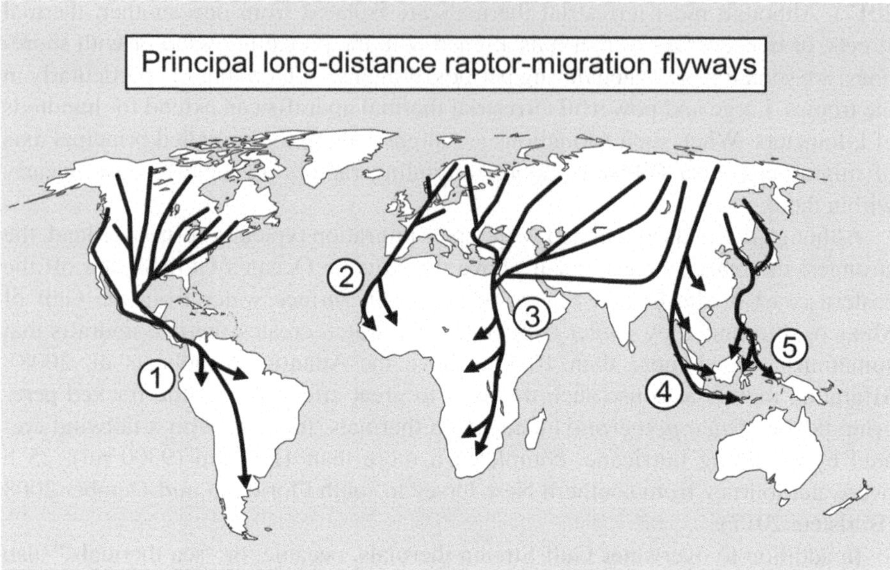

Fig. 5.1 The world's principal raptor migration flyways including (1) the Trans-American Flyway, (2) the Western European-West African Flyway, (3) the Eurasian-East African Flyway, (4) the East-Asian Continental Flyway, and (5) the East-Asian Oceanic Flyway. (After Bildstein (2017))

typically increases with height, raptors migrating into strong headwinds fly lower and those migrating with strong tailwinds fly higher than those migrating in light or variable winds. Indeed, raptors traveling in coastal areas tend to fly lower than those migrating along inland corridors, presumably to reduce the risk of being blown out to sea. Crosswinds, or winds that intersect the preferred direction of travel at perpendicular and near perpendicular angles, can shift migrants from the intended direction of travel via "wind drift" (sensu Alerstam 1990). Evidence suggests that migrants often alter their flight behavior to take advantage of favorable winds, as well as to reduce the costs of unfavorable winds (Kerlinger 1989).

Leading Lines and Diversion Lines

Topography and land-water interfaces often affect the geography of raptor migration. "Leading lines" are said to occur when geographic features such as mountain ranges and rivers attract and channel migrants during migration. Diversion lines occur when geographic features such as high mountain ranges, large unbroken forests, and large bodies of water are avoided by migrants that concentrate along their perimeters.

The northeast-to-southwest-oriented Central Appalachian Mountains of eastern Pennsylvania, USA, create well-known leading lines for outbound migrants in eastern North America. In autumn synoptic weather events called cold fronts, typically pass through the region at an average of 4- to 5-day intervals. As the fronts approach and cross the Appalachian's even-topped, parallel ridges at right angles, their northwest winds are deflected up and over the ridges. The result is a series of parallel leading lines for outbound migrants. The Kittatinny Ridge, the southeastern-most mountain of these ridges, creates the strongest leading line in eastern Pennsylvania, as it offers the last opportunity for southbound migrants to slope-soar there. Ridge adherence along this leading line is considerable, with as many as 95% of all migrants traveling along it continuing to do so after reaching "water gaps" where river valleys bisect the ridge. A series of raptor migration watchsite counts along the Kittatinny suggest that migrants seen along this leading line remain on it for several kilometers to several dozen kilometers.

The strength of the deflection updrafts being produced together with the coincidental occurrence of off-ridge thermals often affect the extent to which mountain ridges such as the Kittatinny function as leading lines. Outbound raptors passing through eastern Pennsylvania, USA, in August and September, for example, thermal-soar more and slope-soar less than do later migrants, apparently because of the availability of strong late-summer thermals then (Maransky et al. 1997). And whereas watchsites in the region typically record their best late-season flights during periods of strong northwest winds, strong early-season flights are most common on days when strong thermals form and winds are light or variable or on days when northeasterly tailwinds dominate (Bildstein 2017).

Flocking

Most migratory raptors flock at least occasionally while migrating. Many such flocks are small single- and mixed-species groups of less than a dozen or so birds that last from a few minutes to less than an hour, and most result when numbers of raptors travel in the same place at the same time. That said several species routinely flock while migrating and on a much larger scale. In these so-called super-flocking species (sensu Bildstein 2017), flocking is all but essential for successful migration. Such migrants actively coalesce into groups of hundreds to tens of thousands of birds that sometimes remain together for days or weeks. Super-flocking occurs mainly in long-distance, transequatorial migrants and is most common in and around the tropics. North America's broad-winged hawk is a well-studied example of this type of obligate flocking (Bildstein 2017).

Most super-flocking migrants, including broad-winged hawks, have acutely timed migrations in which most of each year's flight at individual locations passes within a few days to a week or two, whereas other migrants have more protracted flights. Super-flocking migrants typically begin assembling into flocks soon after they begin to migrate, and most travel in flocks of increasing size long before they reach significant bottlenecks. Taken together this suggests that the birds themselves, and not geographically narrow migration passageways, create the supersized flocks (Bildstein 2006).

In the New World, migrants traveling in enormous mixed-species flocks dominated by turkey vultures, Swainson's hawks, and broad-winged hawks sort themselves by species when flying together in lengthy thermal streets. In such situations, agile and smaller broad-winged hawks stream above the somewhat less buoyant Swainson's hawks, which, in turn, stream above the seemingly less buoyant turkey vultures, whereas decidedly less numerous ospreys, Mississippi Kites (*Ictinia mississippiensis*) and peregrine falcons also pass in well-formed, single-species subgroups, typically at the perimeters of the main flight line.

Downsides of flocking behavior include the potential of enhancing interspecies predation and the spread of contagious diseases, as well as prey robbery within and between species that aerial feed while migrating. Traveling in large flocks also may affect where raptors roost at night and, thus, their habitat needs. In addition, migrating in large flocks can restrict the timing of migration, which, in turn, may result in individuals migrating earlier than their fat levels merit. That said flying together in groups allows many species of birds of prey to achieve the otherwise unachievable goal of migrating long distances between their breeding areas and wintering ground.

Anatomical Considerations

Although the environmental physiologist Schmidt-Nielsen (1972) once quipped that "… flight has the potential of being the most energetically costly form of animal locomotion," evolution has shaped both the flight mechanics and flight behavior of raptors over millions of years, and today birds of prey are astonishingly efficient flying machines.

Even though soaring reduces the energetic costs of long-distance raptor migration substantially (Kerlinger 1989; Bildstein 2006), not all raptors soar extensively while migrating. Indeed although all raptors soar at least occasionally on migration, several species frequently use powered flapping flight en route, and most species often alternate between the two types of flight types. Three phenomena are known to contribute to this. First, many migration pathways lack dependable soaring conditions. Second, the higher metabolic requirements of smaller-bodied migrants necessitate traveling at higher speeds than soaring allows. And third, a raptor's wings are shaped by many selective forces, including nonmigratory foraging and predator-avoidance flight, as well as migratory flight, meaning that not all raptors have wings that are well suited for soaring (Bildstein 2017). Species that depend heavily on thermal soaring, for example, tend to have proportionately larger wings than other migrants, which results in lighter wing loading; this, in turn, permits migrants to circle more tightly in thermals, increasing the efficiency of soaring flight. Many also have so-called "fingered" wing-tips, an anatomical condition in which the tips of the outermost flight feathers are separated both horizontally and vertically when the wings are outstretched, which substantially reduces drag near the wingtips of slow-flying, soaring migrants, increasing their flight efficiency. Both conditions are valuable during low-speed soaring flight (Kerlinger 1989), but both also compromise the efficiencies of other flight types, including high-speed prey-pursuit flight and flapping flight (Bildstein 2017).

Raptor migration specialists recognize five generalized "groupings" of wing shape within the anatomical spectrum of raptors: (1) kites, (2) harriers and osprey, (3) accipiters, (4) buteos and eagles, and (5) falcons. Kites and falcons have relatively long and slender high-aspect-ratio wings that feature pointed tips. Harriers and ospreys also have relatively long and slender high-aspect-ratio wings, but their wing tips are rounder and usually feature wing slotting. Most accipiters have relatively short rounded wings with some wing slotting. Buteos and eagles tend to have large and somewhat oversized, broad wings, which often feature wing slotting (Kerlinger 1989).

Efficient, high-speed linear soaring and gliding, including slope soaring, soaring in large thermal streets, and soaring overwater, all favor relatively high-aspect-ratio wings and high wing loading. Consequently migrants like Chinese sparrowhawks, grey-faced buzzards, and Amur falcons (*Falco amurensis*) that depend on this type of soaring flight to complete their journeys tend to have relatively smaller and more pointed wings than do thermal-soaring migrants.

Raptors that depend on powered flight to complete much of their migratory journeys, such as ospreys, harriers, most accipiters, and most falcons, employ flapping flight while doing so. For example, ospreys and harriers typically intersperse relatively lengthy bouts of slope and thermal soaring with flapping flight. Accipiters often employ undulating flapping flight during which relatively short bursts of flapping flight alternate with similarly brief periods of linear soaring and gliding. Falcons often engage in prolonged bouts of high-speed flapping flight en route, sometimes in weather that grounds most other migrants. Importantly, raptors can reconfigure the shape and attitude of their wings depending on local flight conditions. Peregrine falcons, for example, usually hyperextend their wings to the point of wing slotting when circling in small thermals, and harriers tend to "flatten" their otherwise dihedral wings under similar circumstances. And ospreys, which also circle-soar in thermals on relatively flattened and fully outstretched and slotted wings, quickly reassume their characteristic "M-shaped" profile when flex gliding and linear soaring at higher speeds.

Fueling Migration

Raptors have three options. They can build fat reserves prior to migration and burn fat en route, they can feed regularly on migration, or they can soar. Most species use all three sources of energy, with many depending disproportionately on one or two of these "fuels" (Bildstein 2017).

Most avian migrants fuel their migrations with fat (Blem 1980). Pre-migratory fattening is little studied in raptors, and most of what is known has been inferred almost entirely from examinations of individuals captured while migrating. Migrant fat stores of migrants rarely exceed 20% of their lean body mass. Juveniles have lower fat stores than adults, and females tend to carry more fat on their return migrations than on their outbound flights. Most likely, the latter reflects the fact that females who arrive on their breeding grounds with sufficient fat to produce eggs are advantaged. That said a major problem with studies involving migrants trapped en route is the possibility of sampling bias. Individuals trapped en route are almost always lured with food. This means that food-stressed birds are more likely to be captured than are their less-hungry counterparts. Consequently, the stored-fat values almost certainly represent the lower end of fat stored by most migrants (Bildstein 2017).

Many raptors that migrate alone and that those that travel short distances regularly feed en route. Some feed daily, others do episodically when weather conditions preclude efficient migration, and others do so only when they encounter areas of abundant and available prey en route. Studies of satellite-tracked outbound raptors indicate that many migrate slowly at first and then increase their speeds and slow down again as they approach their destinations. Presumably, the differences reflect the likelihood of feeding en route, with birds feeding more early in migration an attempt to build fat reserves for the journey, and again toward the ends of their migrations, either to avoid metabolizing protein as fat stores are depleted or to sample prey availability on potential wintering areas. Some species time their migrations to take advantage of migratory prey en route. Peregrine falcons, for example, often interrupt migration for a week or more to feed at traditional stopover sites that host concentrations of migratory shorebirds and other prey. Largely bird-eating accipiters, including the sharp-shinned hawk and Eurasian sparrowhawk (*Accipiter nisus*), apparently time and position their migrations to take advantage of the concentrated movements of their songbird prey (cf. Newton 1986).

Many insectivorous raptors routinely hunt swarming insects while migrating, some of which are migrating themselves. Others regularly prey on birds and mammals, either before or after their daily movements or throughout the day during periods of poor weather. Overall, feeding appears to be far more common immediately before and after long-distance travel across inhospitable landscapes, such as deserts and large bodies of water, than at other times (Bildstein 2017). And many long-distance migrants depend on energy-efficient soaring flight to complete their migrations (Bildstein 2017).

Most raptors secure sufficient water when traveling long distances via fat metabolism, which produces metabolic water (Bildstein 2006). That said many raptors drink at migratory stopover sites, especially in Africa and the Middle East. Indeed, some depend so much on water that they use brackish and saltwater sources. Presumably salt poisoning is avoided via salt excretion by super-orbital salt glands (Bildstein 2017).

Age and Gender Differences

It has been known for some time that raptors of different ages and genders frequently differ in their migration schedules. Unfortunately, few universal patterns have emerged concerning these differences. Some researchers argue that adult males, which do most of the hunting on the breeding grounds and, therefore, have a better knowledge of the prey base there than do females, should extend their breeding ground stays later into autumn and return to them earlier in spring than their mates. Others have suggested that males should return to the breeding grounds earlier than females because they are under intense selection pressure to acquire nesting territories before females arrive. In terms of age-related differences, many suggest that inexperienced juveniles should follow rather than precede adults on their first outbound migrations. But although this makes some sense in terms of the relative navigational skills of the two ages classes, and although in many species juveniles do follow adults during the formers' initial outbound movements, in other species, where adults remain on the breeding grounds and molt prior to migration, the reverse can be true. Another possible explanation is that juveniles and adults often follow different routes on outbound migration (Kerlinger 1989; Bildstein 2006).

Another factor complicating age-related differences is that adults often overwinter closer to their breeding grounds than do juveniles, possibly because it allows them to return to these areas earlier in spring. In some species, adult males overwinter farther from their breeding areas than do adult females, whereas in many others the reverse is true. The former is often attributed to behavioral dominance or climate, with smaller males being out-competed by larger females for the closer and presumably preferred wintering areas, whereas the second is believed to result from males attempting to remain as close as possible to the breeding territories so that they are better able to reestablish them in the spring (Kerlinger 1989; Bildstein 2017).

Intratropical Migration

In general tropical raptors are largely sedentary species that time their breeding to take advantage of seasonal fluctuations in prey availability locally. However, some migrate within the tropics, typically among regions that exhibit pronounced wet-dry cycles. Intratropical movements are typically small-scale geographically, and in many cases the species involved appear to be partial rather than complete migrants. Indeed, in many intratropical migrants, the movements produce a shift in the center of species abundance without substantially changing the perimeter of the range. This, and the fact that intratropical migration is little studied, creates an incomplete and, at times, frustrating picture of these movements. That said at least in Africa, most intratropical movements appear to be closely tied to those of the intertropical convergence zone, or ITCZ, a latitudinal band of heavy rains that wavers seasonally north and south across the world's tropical regions. Convective activity within this zone is both intensive and extensive, with moisture-laden clouds forming continually to create daily midday rains, making the so-called doldrums one of wettest places on earth. The ITCZ band of heavy precipitation, which extends farther north and south over land than over water, straddles the tropics year-round. This so-called meteorological equator shifts northward in boreal summer and southward in the winter, creating a series of rainy seasons across more than 20° of tropical latitudes.

"Rains migrants" in the tropics include both the African cuckoo-hawk (*Aviceda cuculoides*), a kite that moves north in July and September on the heels of the retreating ITCZ in West Africa, and the black-winged kite (*Elanus caeruleus*), which migrates to avoid the rains.

Altitudinal Migration

Altitudinal migration consists of relatively short-distance movements between areas of high and low altitudes along mountain slopes. The movements occur because, terrestrially, temperatures decrease of approximately 0.7 °C (1.3 °F) per 100 m up to about 10 km (6 mi) above sea level. Raptors inhabiting exceptionally high-mountain plateaus and mountain ranges, including those in the Himalayas of central Asia, and Andes of South America, are prone to altitudinal migration. Because the distances tend to be short, and the fact that the movements themselves often are coupled with longer latitudinal migrations, altitudinal migration is not well described. Even so at least 45 species of partial and irruptive or irregular migrants are thought to migrate altitudinally (Bildstein 2017).

Satellite Tracking and Migration Studies

Science, including the science of raptor migration, advances most rapidly when new technology is made available to the scientists involved. The advent and use of satellite-tracking technology is a case in point. Developed by the US military at the US Army's Applied Physics Laboratory in Maryland in the 1980s, this new technology now includes the use of 5 g, solar-powered platform transmitter terminals (PTTs) capable of transmitting even a small raptor's position for one or more years. In some cases, larger units attached to larger raptors have been used to track individual birds of prey for a decade or more.

Initially tested on bald eagles (*Haliaeetus leucocephalus*) in the mid-1980s, tracking the movements of raptors by satellite has been used to follow the movements of representatives of at least 40 species of mainly relatively large-bodied raptors. Initially used together with Argos-Tiros satellites, the system, theoretically, can provide locations with an accuracy of ±250 m across the surface of the earth, although with moving targets like raptors locational accuracy typically falls within a range of ±2 km. More recently, PTTs with GPS receivers have increased locational accuracy to ±20 m (Bildstein 2017).

Satellite tracking is expensive. Typically it costs on the order of 1500–4000 US dollars for each tracking device and about 100–800 US dollars per year, per unit for data downloads, with costs depending on the manufacturer and whether the data are sent via Argos-Tiros satellites or GSM cell phone towers. A new system, operating off the International Space Station and expected to lower costs significantly, is projected to be deployed in 2017.

In the United States, a study involving ospreys tracked 117 adults over a total of 164 "osprey years" (Martell et al. 2001) and a study of Alaskan golden eagles tracked more than 70 birds (cf. McIntyre et al. 2008). An ongoing study of North American and South America turkey vultures had tracked the movements of 61 individuals as of early 2017, including an individual that had been tracked for more than a decade (Dodge et al. 2014). Another ongoing study involving Spanish black kites tracked 92 individuals across 364 migration "episodes" as of 2014 (Sergio et al. 2014). These and other investigations have added considerably to our knowledge of raptor migration in ways heretofore unimaginable.

With satellite tracking, the birds "themselves" are now telling *where* and *when* and, to an extent, *how* they migrate, and this, in turn, is allowing us to better understand the phenomenon of raptor migration (Bildstein 2017). No longer are we constrained to generalizing or modeling how raptors migrate. In a very real sense, "Satellite tracking is teaching us more precisely how raptors migrate on individual-bird and individual-year bases, as well as changing the way we think about raptor migration. Until recently, many students of raptor migration have been so enamored with the birds' ability to do what they do that we have tended to characterize the migratory movements of birds of prey as something of a fait accompli, rather than as a flexible and modifiable behavioral work in progress" (Bildstein 2017).

Final Thoughts

It has been said that three forces influence progress in ecological science: *serendipity*, *technology*, and *new paradigms*. Together and alone, these forces shape the questions we ask (Malmer and Enckell 1994). To date, raptor biologists have embraced the first two more than the third. Below I detail the roles – past, present, and future – that each of these forces have had and continue to play in expanding knowledge in the field of raptor migration science.

Serendipity

Beginning in the 1930s, and continuing since, conservation concerns regarding declining population of birds of prey have led to the establishment of raptor migration watchsites at a number of traditional raptor migration "shooting galleries." Such efforts led to long-term raptor-population monitoring along established migration corridors (Broun 1949; Bildstein 2006), which, in turn, helped establish a critical relationship between declining raptor populations and the misuse of DDT and other organochlorine pesticides in mid-twentieth-century North America and elsewhere (Bildstein 2006).

Technology

Modern banding, the initiation of which is credited to Danish schoolteacher Hans Christian Cornelius Mortenson, who pioneered the technique in 1899 when he attached numbered leg bands at first to common starlings (*Sturnus vulgaris*) and shortly thereafter to birds, including roughlegs (*Buteo lagopus*) (Jespersen and Tåning 1950), resulted in the establishment of government-sponsored banding offices in the United States and Canada in the early 1920s. These, in turn, led to more than 4000 raptors being banded by the mid-1930s in the United States alone. Banding offices in Europe soon followed. Since then, scientific banding has blossomed in both North America and Europe with totals of more than a million banded raptors in the United States and Canada and more than 140,000 in Ireland and the United Kingdom as of 2000 (Bildstein 2006). Data resulting from subsequently encountered individuals have led to import new insights regarding the migratory movements of many species of birds of prey, particularly in the United States and Western Europe (Bildstein 2012; Bildstein and Peterjohn 2012), where banding continues to play important roles in promulgating studies in raptor migration biology, most notably in two areas: (1) "assessing demographic statistics, including age at first breeding, survival rates, and mortality factors; and (2) assessing the movements of raptors , including migration, nomadic movements and both natal and breeding dispersal" (Bildstein and Peterjohn 2012). As numbers of banded raptors continue to accumulate, this technique promises to add significant new information regarding the biology of raptor migration.

Importantly, a second new "technology," mentioned above, long-term counts of visible migrants at traditional raptor migration watchsites, which initially began in the United States in the 1930s and shortly thereafter in Europe and, more recently, has occurred in other parts of the world (Zalles and Bildstein 2000; Bildstein 2006), has helped, and continues to help, track the movements and magnitudes of regional and sometimes continental populations of birds of prey. Growing evidence suggests that long-term counts, despite several limitations, usually reflect the long-term status of source populations (Zalles and Bildstein 2000; Bildstein 2006).

Significantly, both of these new techniques (i.e., raptor banding and counts of visible raptor migrants at traditional watchsites) have created critical new information regarding the biology of migratory birds of prey, particularly when carried out by networks of banders and observers engaged in such activities so that the numbers of raptors banded and counted are geographically robust and therefore representative of the regional and continental populations involved (Bildstein et al. 2008; Bildstein and Peterjohn 2012). Notably, both techniques have worked best when individuals within the networks have collaborated on analyzing and publishing their data (Bildstein et al. 2008; Bildstein and Peterjohn 2012).

A third new technology, satellite tracking which allows researchers to follow the migratory movements of individual birds (Bildstein 2017), has yet to fully cross the threshold of networking and collaboration, although mounting evidence suggests that it, too, would benefit substantially by doing so (Dodge et al. 2014). In fact, "the use of satellite tracking, including the implementation of new systems employing GPS-logging, solar-powered mini-tags for use on small, as well as, large raptors (Wikelski et al. 2007), together with other new tracking technology such as cellular-telephone-based tracking units, can play an important role…" in helping raptor migration scientists better understand changes in raptor migration in the face of changing environmental conditions (Bildstein et al. 2008).

New Paradigms

Although only in its infancy, the new paradigm of movement ecology promises overdue new way to look at and understand raptor migration. Movement ecology provides "a unifying framework and common tools" that aims "to promote the development of an integrative theory of organism movement for better understanding the causes, mechanisms, patterns, and consequences of all [raptor] movement…" (Nathan 2008).

These three forces – serendipity, technology, and a new more inclusive biological paradigm – promise to move studies of raptor migration "fast forward" in the future. Indeed, although we have learned much of late and now know quite a bit about raptor migration, we presently stand at threshold of increasing our knowledge of the phenomenon considerably, particularly if we are willing to commit ourselves to collaborative efforts.

Acknowledgments Dating from the early 1980s, my studies of raptor migration would not have been possible without the help and inspiration of many mentors, as well as from dozens of colleagues and students, most of which have made significant contributions to the field on their own. Financial support for my studies has come from many sources including Winthrop University; the Baruch Institute for Marine Biology and Coastal Research, at the University of South Carolina; and more recently from Hawk Mountain Sanctuary, the Acopian family, and the Wallace Research Foundation, among others. I thank them all. This is Hawk Mountain Sanctuary contribution to conservation science number 296.

References

Alerstam T (1990) Bird migration. Cambridge University Press, Cambridge, UK

Augstein E (1983) The atmospheric boundary layer over the tropical oceans. In: Shaw DB (ed) Meteorology over tropical oceans. Royal Meteorological Society, Bracknell, pp 73–104

Berthold P (2001) Bird migration: a general study, 2nd edn. Oxford University Press, Oxford, UK

Bildstein KL (1998) Long-term counts of migrating raptors: a role for volunteers in wildlife research. J Wildl Manag 62:535–445

Bildstein KL (2006) Migrating raptors of the world: their ecology and conservation. Cornell University Press, Ithaca

Bildstein KL (2017) Raptors: the curious natural history of diurnal birds of prey. Cornell University Press, Ithaca

Bildstein KL, Peterjohn BG (2012) The future of banding in raptor science. J Raptor Res 46:3–11

Bildstein KL, Smith JP, Ruelas Inzunza E, Veit RR (eds) (2008) State of North America's birds of prey. Nuttal Ornithological Club and American Ornithologists' Union, Washington, DC
Blem CR (1980) The energetics of migration. In: Gauthreaux SA (ed) Animal migration, orientation, and navigation. Academic Press, New York, pp 175–224
Broun M (1949) Hawks aloft: the story of Hawk Mountain. Cornwall Press, Cornwall
Cox GW (1968) The role of competition in the evolution of migration. Evolution 22:180–192
Dodge S, Bohrer G, Bildstein K, Davidson SC, Weinzierl R, Bechard MJ, Barber D, Kays R, Brandes D, Han J, Wikelski M (2014) Environmental drivers of variability in the movement ecology of Turkey vultures (*Cathartes aura*) in North and South America. Philos Trans R Soc B 369:20130195
Drost R (1938) Über den Einfluss von Verfrachtungen zur Herbstzugzeit auf den Sperber Accipiter nisus (L.). Zugleich ein Beitrag zur Frage nach der Orientierung der Vögel auf dem Zuge ins Winterquartier. Proc Int Ornithol Congr 9:503–521
Harmata AR (2002) Vernal migration of bald eagles from a southern Colorado wintering area. J Raptor Res 36:256–264
Jespersen P, Tåning ÅV (eds) (1950) Studies in bird migration: being the collected papers of H. Chr. C. Mortensen. Munksgaard, Copenhagen
Kerlinger P (1989) Flight strategies of migrating hawks. University of Chicago Press, Chicago
Laybourne RC (1974) Collision between a vulture and an aircraft at an altitude of 37,000 feet. Wilson Bull 86:461–462
Malmer N, Enckell PH (1994) Ecological research at the beginning of the next century. Oikos 71:171–176
Maransky B, Goodrich L, Bildstein K (1997) Seasonal shifts in the effects of weather on the visible migration of red-tailed hawks at Hawk Mountain, Pennsylvania, 1992–1994. Wilson Bull 109:246–252
Martell MS, Henny CJ, Nye PE, Solensky MJ (2001) Fall migration routes, timing and wintering sites of North American Ospreys as determined by satellite telemetry. Condor 103:715–724
McIntyre CL, Douglas DC, Collopy MW (2008) Movements of golden eagles (*Aquila chrysaetos*) from interior Alaska during their first year of independence. Auk 125:214–224
Minobe S, Kuwano-Yoshida A, Komori N, Xie S-P, Small RJ (2008) Influence of the Gulf stream on the troposphere. Nature 452:206–209
Nathan R (2008) An emerging movement ecology paradigm. Proc Natl Acad Sci 105:19050–19051
Newton I (1986) The sparrowhawk. T & AD Poyser, Calton
Schmidt-Nielsen K (1972) Locomotion: energy cost of swimming, running and flying. Science 177:222–228
Sergio F, Tanferna A, De Staphanis R, López Jiménez L, Blas J, Tavecchia G, Preatoni D, Hiraldo F (2014) Individual improvements and selective mortality shape lifelong migratory performance. Nat Mag. https://doi.org/10.1038/nature13696
Smith NG (1985) Thermals, cloud streets, trade winds, and tropical storms: how migrating raptors make the most of atmospheric energy in Central America. In: Harwood M (ed) Proceedings of the Hawk Migration Conference IV. Hawk Migration Associaton of North America, Lynchburg
Sodhi NS, Oliphant LW, James PC, Warkentin IG (1993) Merlin. In: Poole A, Gill F (eds) The birds of North America, no. 44. Birds of North America, Philadelphia
Viverette CB, Struve S, Goodrich LJ, Bildstein KL (1996) Decreases in migrating Sharp-shinned Hawks (*Accipiter striatus*) at traditional raptor-migration watch sites in eastern North America. Auk 113:32–40
Wikelski M, Kays RW, Kasdin NJ, Thorup K, Smith JA, Swenson GW Jr (2007) Going wild: what a global small-animal tracking system could do for experimental biologists. J Exp Biol 210:181–186
Zalles JI, Bildstein KL (2000) Raptor watch: a global directory of raptor migration sites. Birdlife International/Hawk Mountain Sanctuary, Cambridge/Kempton

Chapter 6
Raptors as Seed Dispersers

Néstor Pérez-Méndez and Airam Rodríguez

Introduction

Plants, unlike animals, are organisms with a sessile adult stage, and consequently they need external vectors for moving their propagules (mainly seeds) away from mother plants. The movement of seeds is a central process in the life cycle of plants called seed dispersal (Howe and Smallwood 1982). This process determines the spatial pattern of seed deposition, over which post-dispersal processes such as predation, germination, or seedling survival will act (Nathan and Muller-Landau 2000). Seed dispersal will therefore influence significantly the probability that a single seed becomes an adult individual of the plant population. The advantages of moving away from source plants are multiple. From a demographic point of view, seed dispersal enables to escape seeds to high mortality in the vicinity of maternal plants, where the probability of predation or pathogen infection increases exponentially (Janzen 1970; Connell 1971). In addition, dispersal facilitates the arrival of seeds to suitable microhabitats for germination and survival of seedlings, the colonization of new habitats, and the range expansion of plant populations (Howe and Smallwood 1982). From a genetic point of view, together with pollination, seed dispersal shapes patterns of gene flow within and among plant populations, influencing the characteristics and the cohesiveness of genetic pools at local and regional scales (Loveless and Hamrick 1984). Thus, as whole, seed dispersal will influence the demogenetic features of plant populations, conditioning the regeneration dynamics and the adaptability and resilience of plant populations in changing landscapes.

N. Pérez-Méndez (✉)
Instituto de Investigación en Recursos Naturales, Agroecología y Desarrollo Rural (IRNAD), Universidad Nacional de Río Negro (UNRN), San Carlos de Bariloche, Argentina

A. Rodríguez
Department of Evolutionary Ecology, Doñana Biological Station (EBD-CSIC), Seville, Spain
e-mail: airamrguez@ebd.csic.es

J. H. Sarasola et al. (eds.), *Birds of Prey*,
https://doi.org/10.1007/978-3-319-73745-4_6

The myriads of seed morphologies we can observe in nature give us a picture of how important should be natural selection to ensure dispersal (Ridley 1930). Morphologies range from unadapted seeds without apparent specialized structures to more sophisticated features such as winged or plumed seeds (adapted to wing) or edible coats surrounding seeds, which are adapted to animal dispersal.

Despite that there are no exhaustive studies providing accurate information, the proportion of plants that relies on animals for seed dispersal is quite high, especially in tropical ecosystems (Jordano 2014). Seeds can be dispersed primarily through two different mechanisms: (*i*) by external attach to animal body (epizoochory) through spines, hooks, or sticky coats or (*ii*) through the gut of animals after being swallowed and ultimately defecated or regurgitated (endozoochory). In the latter case, both plants and animals establish a relationship of mutual benefit (mutualisms). Animals provide dispersal services to plants and, in return, obtain a trophic reward from the nutritive fleshy pulp that envelops the seeds. Finally, it is worth to note that the fate of seed is not always determined by primary dispersers, as seeds can be moved secondarily (once or several times) by other animals before arrival to the final location (diplochory).

Frugivorous birds and mammals have been historically proposed as the most important seed dispersers of both endozoochorous (Jordano 2014) and epizoochorous plant species (Sorensen 1986). In fact, literature about seed dispersal assisted by animals is almost monopolized by studies focusing on the ecological and evolutionary relevance of bird- and mammal-mediated dispersal. However, most recent research is beginning to highlight the importance of other overlooked vertebrate groups such as reptiles (Olesen and Valido 2003; Valido and Olesen 2007; Pérez-Méndez et al. 2015; 2016; Pérez-Méndez 2016), amphibians (da Silva et al. 1989; Da Silva and De Britto-Pereira 2006), fishes (Gottsberger 1978; Howe and Smallwood 1982; Costa-Pereira and Galetti 2015), and other non-frugivorous birds (Figuerola 2010; Soons et al. 2016). They have been proved to be quantitatively and qualitatively effective dispersers of many plant species, including those with non-apparent specialized traits for endozoochory or epizoochory (e.g., weeds). For example, Soons et al. (2016) have identified dabbling ducks as a major disperser guild for a broad range of plant species, including fleshy- and dry-fruited plant species. Similarly, Sarasola et al. (2016) stressed the importance of apex predators (e.g., cougars, *Puma concolor*) as secondary seed dispersers after preying on seed predators. They rescue a large proportion of seeds when preying on granivorous animals that otherwise are destroyed within guts. These and other recent studies have documented important ecological functions assisted by animal groups that have been previously overlooked as important seed vectors. Including overlooked groups of animals in the agenda of seed dispersal research is therefore timely and necessary; otherwise we may undervalue unique ecological function of extreme importance for the regeneration dynamics of plant populations.

Raptors are considered apex predators with most of the diet based on vertebrate preys, especially birds and mammals (Ferguson-Lees and Christie 2001). Despite that several examples of raptors consuming fleshy fruits have been documented, their role as primary endozoochorous and epizoochorous seed dispersers remains anecdotal. This is probably the reason why they have been overlooked in the literature of plant dispersal. Yet, the role of raptors as seed dispersers can acquire great relevance if we focus on other mechanisms such as secondary dispersal (López-Darias and Nogales 2016; Hämäläinen et al. 2017).

Here we firstly review the literature on the significance of raptors in the different mechanisms of animal-mediated seed dispersal (primary and secondary dispersal). We then identify the main components of the dispersal process in which raptors are involved, which will determine ultimately their effectiveness as seed dispersers: (*i*) frequency, (*ii*) impact on seed germination, (*iii*) seed arrival to suitable habitats, (*iv*) disruption of seed predation, and (*v*) long-distance seed dispersal. Finally, we emphasize the potential importance of raptor-mediated seed dispersal in influencing demographic performance and genetic characteristics of plants and their ability to cope with landscape and climate change.

Significance of Raptors as Primary Seed Dispersers

Although the "zoochory" term refers to the dispersion of plant propagules (spores, seeds, or vegetative parts) by animals, here we focus on the seed transport by raptors. Depending on the way the seeds are dispersed, several types of zoochory are distinguished (Fig. 6.1). In raptors we have identified the followings:

(a) Endozoochory: dispersion of seeds via ingestion.
(b) Epizoochory: dispersion of seeds via transportation on the surface (plumage or bare parts) of raptors.
(c) Synzoochory: seeds or plant parts containing seeds (e.g., fruits or branches) are deliberately transported, e.g., while storing foods, building nests, or digging burrows, and finally dispersed.

Endozoochory by Raptors

As top predators, raptors are carnivorous birds and usually do not eat plant material. In order to quantify how many raptor species consume fruits and, consequently, disperse seeds, we checked the food section for all the species included in *Raptors of the World* (Ferguson-Lees and Christie 2001) and in the Global Raptor Information Network (The Peregrine Fund 2017). In addition, we searched the ISI Web of Knowledge for other scientific literature. To expand our search, we also checked the references contained in the publication of interest and their citations. Sixteen of the 312 raptors considered by Ferguson-Lees and Christie (2001) consume deliberately fruits at least occasionally (Table 6.1). Given the low number of frugivorous raptor species and the anecdotal information on fruit consumption, it is not surprisingly that raptors are not considered as important primary seed dispersers (Stiles 2000).

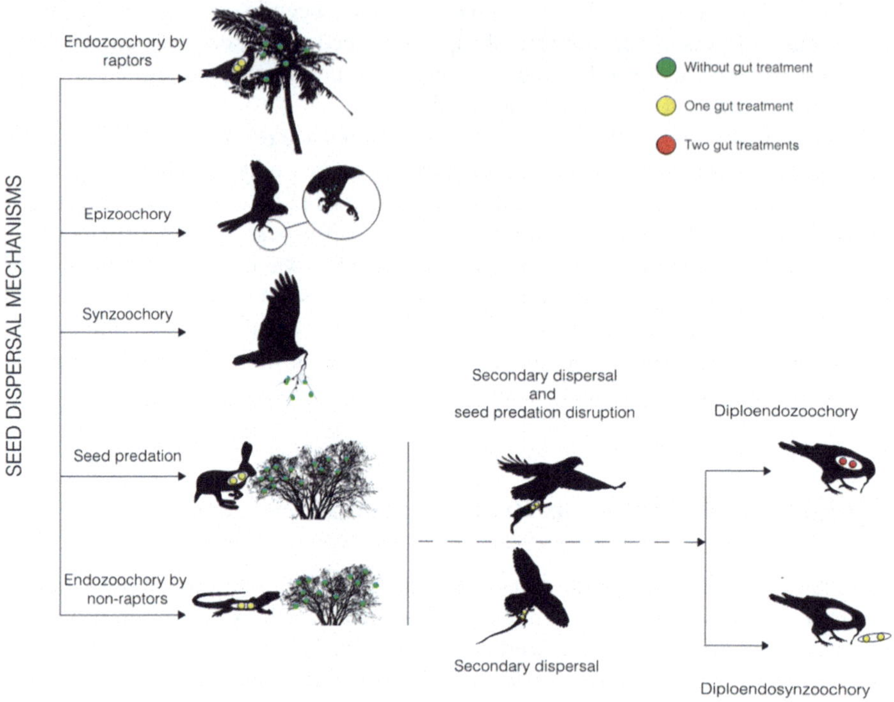

Fig. 6.1 Seed dispersal mechanisms assisted by raptors. Top-down order: endozoochory, fruits are consumed directly by raptors and seeds dispersed within their guts; epizoochory, seeds are attached to the body of raptors and dispersed; synzoochory, seeds or parts of plants containing seeds are deliberately dispersed with a given purpose (e.g., building nests, courting rewards, or other rituals); secondary dispersal after interrupting seed predation by a primary seed predator (e.g., rabbits); secondary dispersal after hunting upon a legitimate primary seed disperser. It is worth to note that seeds being dispersed secondarily may have a double gut treatment if raptors swallow the whole primary consumer (prey). Instead, they may have a single gut treatment if raptors reject guts of primary dispersers before consumption or prey are not consumed

Only two species are considered frugivorous, i.e., fruits constitute a considerable part of their diet, the palm-nut vulture *Gypohierax angolensis* and the black caracara *Daptrius ater,* according to Ferguson-Lees and Christie (2001), although their role as seed dispersers have not been evaluated. The palm-nut vulture eats palm fruits as well as the fruits and grains of other plants. In Guinea-Bissau, oil palm fruits have been estimated at 34.5% (frequency of occurrence of prey remains under nests and roosts) and 45% according to Bayesian mixed models of stable isotope analyses of vulture feathers (Carneiro et al. 2017). In Ivory Coast vegetable items could collectively form more than 50% of its diet (Thiollay 1978). However, Thiollay (1978) estimated the 86% of the diet of the African harrier-hawk *Polyboroides typus* were palm fruits in Ivory Coast. More specific studies on the significance of these two species on the dispersion of seeds of palm trees are needed.

Table 6.1 Evidence of frugivory or seed dispersal by raptors (Accipitriformes, Cathartiformes, and Falconiformes)

Order	Species	Fruits recorded	Seed ingestion	Reference
Cathartiformes	*Cathartes aura*	*Acrocomia sclerocarpa*, *Elaeis guineensis*, *Juniperus*, grapes	?	1, 2, 3
	Coragyps atratus	Bananas, avocados, *Scheelea*, *Elaeis*, *Livistona australis*, and copra (the dried meat of the coconut)	?	1, 3
Accipitriformes	*Buteogallus urubitinga*	*Spondias lutea*	?	3
	Aviceda subcristata	Figs and other fruits	Y	1
	Pernis apivorus	Berries and other fruits	Y	1
	Elanoides forficatus	*Byrsonima*, *Castilla*, *Cupania vernalis*, *Matayba oppositifolia*, *Sapium*, and other fruits	?	1, 2, 3
	Milvus migrans	*Elaeis*	N	1
	Gypohierax angolensis	*Elaeis guineensis*, *Raphia vinifera*, *R. farinifera*, *Phoenix reclinata*, *Antiaris Africana*, other fruits, *Acacia* seeds and cereals	Y	1, 2, 3, 4
	Neophron percnopterus	Rotten pumpkins in garbage dumps	?	1
	Polyboroides typus	*Elaeis guineensis*, *Ficus*	?	1, 2, 4
Falconiformes	*Daptrius ater*	*Desmoncus*, *Mauritia flexuosa*, *Elaeis*, *Ficus*	Y/N	1, 2, 3
	Daptrius americanus	*Poraqueiba*, fruits and soft seeds	Y	1, 2, 3
	Phalcoboenus carunculatus	Unidentified seeds	?	1
	Caracara plancus	*Attalea phalerata*, *Elaeis oleifera*, *Cocos* (and copra), beans, peanuts	N for *A. phalerata*?	1, 2, 3
	Milvago chimachima	*Elaeis*, *Byrsonima*, copra, cork and unidentified seeds	Y	1, 2
	Milvago chimango	rotten apples, cereal seeds	?	1
	Micrastur ruficollis	*Blomia*	?	2
	Falco rufigularis	Small green fruits	Y	2, 3
	Falco ardosiaceus	*Elaeis guineensis*	?	1, 4
	Falco novaeseelandiae	*Leucopogon fraseri*, *Gaultheria depressa*	Y	5

Modified from Galetti and Guimarães (2004)
1 = Ferguson-Lees and Christie (2001), 2 = Galetti and Guimarães (2004), 3 = The Peregrine Fund (2017), 4 = Thiollay (1978), 5 = Young and Bell (2010)

But why raptors consume fruits? The majority of the species recorded, and also the most frequently consumed fruits, are rich in lipids (genera *Elaeis*, *Raphia*, *Cocos*, *Spondias*, etc.). Thus, raptors could consume fruits to satisfy their lipid needs (Galetti and Guimarães 2004). In this regard, an experimental study to determinate the preference of palm-nut vulture between fish and African oil palm fruits (*Elaeis guineensis*) demonstrated that vultures preferred fish (Carneiro et al. 2017). However, experimental conditions could have made that vultures were less inclined to eat fruits as palm fruits were located on the ground in the intertidal area, a non-common place to find fruits (Carneiro et al. 2017).

Epizoochory by Raptors

Epizoochory (also known as exozoochory or ectozoochory) may be conducted by all sorts of animals, mainly mammals and birds (Cousens et al. 2008). Raptors could also contribute to disseminate seeds when their plumages contact the spikes of plants with epizoochorous syndromes of dispersion, i.e., plants with burrs or spines to stick to fur or plumage. This process is typically hard to study as in general, raptors preen their plumage often. In fact, we have not identified any study assessing the relevance of raptors as epizoochorous seed dispersers. However, there are some reports of raptor entanglements in burrs or spikes of plants with typical epizoochorous syndromes which could cause bird death of species like *Elanus* kites, owls, but also other raptors such as kestrels (Mendelsohn 1983; Rodríguez et al. 2009). On Tenerife, Canary Islands, 5.5% of the long-eared owls *Asio otus* and 2.2% of barn owls *Tyto alba* admitted to a wildlife rehabilitation center were due to entanglement with bur bristlegrass *Setaria adhaerens*; however only 2 of the 1200 common kestrels *Falco tinnunculus* admitted were due to this cause (Rodríguez et al. 2009). This leads to think that the velvety plumage seems to be a crucial factor to the adhesion of seeds, and consequently seed dispersal by epizoochory in birds of prey with such soft plumage such as owls, *Elanus* kites or bat hawks *Macheiramphus alcinus* would be highly probable. Although the information came from birds admitted to rehabilitation centers, it is expectable that many raptors escape from the plants dispersing effectively their seeds while preening.

An exceptional type of zoochory is the synzoochory, in which propagules are deliberately transported by animals while storing foods, building nests, or digging burrows (Figs. 6.2 and 6.3). Frugivorous raptors could take fruits, or even kleptoparasite them to other birds, to places where they eat them later (e.g., *Caracara plancus* removing fruits of *Attalea phalerata*; Fig. 1 in Galetti and Guimarães 2004). Although no information is available, the most frugivorous raptor species could also deliver fruits to their offspring at their nests (Carneiro et al. 2017). Excepting most members of the family Falconidae, raptors use plant material to build their nests. Plant material transported to nesting locations may contain fruits and seeds, contributing to dispersal. Although it has never been quantified, we can foresee that this type of seed dispersal is quite limited. However, it could be relevant in some cases, for example, for invasive plant species propagation (Fig. 6.2) or rare events of dispersal.

Fig. 6.2 Osprey nest on Tenerife, Canary Islands. The Crimson fountain grass (*Pennisetum setaceum*) is an invasive plant species in the Canary Islands, where several control programs are conducted to avoid its spread. In the picture, several bunches of Crimson fountain grass are used as nest material (synzoochory). Fortunately, these bunches do not contain seed spikes as they were cut within the control program. (Photo: Beneharo Rodríguez)

Fig. 6.3 A male of Harris's hawk (*Parabuteo unicinctus*) dispersing seeds of *Melia azedarach* by synzoochory. The hawk most likely used the branch as a courting reward for the female (J. Simón Tagtachian, personal communication). (Photo: J. Simón Tagtachian)

Significance of Raptors as Secondary Seed Dispersers

Secondary seed dispersal refers to seeds dispersed in two or more dispersal events by different dispersal agents. In the case of raptors, it normally occurs when a primary seed disperser (e.g., frugivorous animals) or seed predator is hunted and transported from foraging areas to consumption or deposition seed locations (Fig. 6.1). Secondary seed dispersal is the most studied and probably the most relevant seed dispersal mechanism in raptors (Table 6.2), excepting the exceptional cases of the *frugivorous* palm-nut vulture and the black caracara (see primary seed dispersers). The number and frequency of seeds dispersed throughout secondary seed dispersal are probably the highest when compared to other types of dispersal assisted by raptors, and, therefore, it has more important ecological implications. Secondary seed dispersal has been mostly studied in depth in the Canary Islands (Nogales et al. 2002, 2007; Rodríguez et al. 2007; Padilla and Nogales 2009; Padilla et al. 2012; Viana et al. 2015; López-Darias and Nogales 2016). In raptors, the attention has focused on diploendozoochory, that is, seed dispersal involves ingestion of seeds by two or more animal species in sequence. Usually it entails a first consumption by a primary consumer (or seed predator) and subsequently, a second ingestion by the raptor that swallows the primary consumer with the digestive tract containing seeds. However, it has been recently demonstrated that seed dispersal involving the ingestion of the seed by the primary consumer (frugivore or seed predator) but not by its predator could have more important implications for plants (Padilla and Nogales 2009; Viana et al. 2015). We coin this process as "diploendosynzoochory" as it involves a first movement of seeds by the primary consumer (endozoochory) and a secondary intentional movement by the predator (raptor) but without ingesting the seeds (synzoochory) (Fig. 6.1). Unlike diploendozoochory, predators do not ingest the seeds as they reject the prey's digestive tracts before consumption (Padilla and Nogales 2009) or they might forget their prey in larders (Viana et al. 2015); thus seeds only receive the gut treatment of the primary disperser.

Diploendozoochory (Diplochory by Double Endozoochory)

Most of our knowledge comes from the pioneer studies conducted in the Canary Islands. In general, fleshy-fruit plant species, frugivorous primary seed dispersers, and avian predators are involved in the system. In Alegranza Islet, Canary Islands, *Lycium intricatum* seeds were found in the pellets of two predatory birds, the southern grey shrike *Lanius meridionalis* and the common kestrel associated with prey remains of the frugivorous lizard *Gallotia atlantica* (Nogales et al. 1998, 2002). The percentage of viable seeds, measured by the tetrazolium test, varied among treatments, decreasing in the following direction: seeds collected directly from plants (98%), shrikes (88%), lizards (72%), and kestrels (32%) (Nogales et al. 2002). In addition, seeds from fresh fruits and shrike pellets showed higher germination rates than those from lizard droppings and kestrel pellets. Authors discussed that shrikes

Table 6.2 Scientific literature on diplozoochory by diurnal raptors (families Accipitridae, Cathartidae, Falconidae, Pandionidae, and Sagittariidae)

Reference	Species dispersed	Primary consumer	Type of primary consumer	Secondary disperser	Endo or epi
Balgooyen and Moe (1973)	*Calamagrostis canadensis*	?	NA	*Falco sparverius*	Endo
Clarke et al. (2003)	At least 46 taxa	*Alauda arvensis*, *Prunella modularis*, *Carduelis cholris*, *C. cannabina*, *Emberiza citronella*, *E. schoeniclus*	Predators	*Circus cyaneus*	Endo
Darwin (1859) (page 362)	Oat, wheat, millet, canary, hemp, clover, beet	Pigeons; experimentally fish	Predators (pigeons) and probably dispersers (fish)	Hawks and fishing eagles	Endo
Dean and Milton (1988)	?	Granivorous rodents and birds	Predators	*Elanus caeruleus*	Endo
	10 unspecified plant species	?	?	*Falco biarmicus*	Endo
López-Darias and Nogales (2016)	39 species of mainly weeds, few fleshy-fruited plants	*Oryctolagus cuniculus*, *Atlantoxerus getulus* and unidentified birds	Mainly predators	*Buteo buteo*	Endo
	62 species of mainly weeds, few fleshy-fruited plants	*Gallotia atlantica*, *Atlantoxerus getulus*, *Mus domesticus* and unidentified birds	Mainly predators. *Gallotia atlantica* is a seed disperser	*Falco tinnunculus*	Endo
Nogales et al. (2002, 2007)	*Lycium intricatum*, *Rubia fruticosa*, *Asparagus nesiotes*	*Gallotia atlantica*	Disperser	*Falco tinnunculus*	Endo
Padilla and Nogales (2009)	*Rubia fruticosa*	*Gallotia galloti*	Disperser	*Falco tinnunculus*	Endo and epi
Padilla et al. (2012)	76 species	*Gallotia* spp.	Dispersers	*Falco tinnunculus*	Endo and epi
Viana et al. (2015)	*Rhamnus* sp.	*Ficedula hypoleuca*, *Phoenicurus phenicurus*	Disperser	*Falco eleonorae*	Epi
	Rubus sp., *Persicaria* sp.	*Coturnix coturnix*	Predator	*Falco eleonorae*	Exo

Modified from Hämäläinen et al. (2017)

are more effective secondary seed dispersers than kestrels and associated their results to differences in gut passage time and gut enzymatic activity (Nogales et al. 2002). In order to evaluate the generalization of such results, secondary seed dispersal was studied in Lanzarote, Canary Islands, on three fleshy-fruit plant species (*Lycium intricatum*, *Rubia fruticosa*, and *Asparagus nesiotes*) and involving the same primary and secondary dispersers than in the previous study (Nogales et al. 2007). Again seeds from kestrel pellets showed reduced viability and germination than the other treatments (seeds from fresh fruits, consumed by frugivorous lizards, or from shrike pellets) (Nogales et al. 2007). Although no causation has been demonstrated to explain the reduced viability and germination of seeds when they pass through the gut of raptors (i.e., collected in kestrel pellets), other study conducted in the Canaries on another avian predator has provided more evidence in this sense. Unlike the raptors, the grey heron *Ardea cinerea* has no talons and curbed beak to tear away the flesh of their prey. Herons have to swallow their prey whole, and consequently they have evolved a stronger gut enzymatic activity than some raptors such as kestrels. The pass through the gut of the grey heron produces a negative effect on the seeds previously consumed by the lizard *Gallotia galloti* as no viable seeds of at least nine identified plant species were recorded in their pellets (Rodríguez et al. 2007).

To assess the generalization and widespread of secondary seed dispersal in the Canary Islands, Padilla and colleagues visited all the islands and their suitable habitats, finding seeds from 78 plant species inside 2098 shrike pellets and 5304 kestrel pellets. A higher number of plant species were secondarily dispersed by kestrels (76; 97%) than by shrikes (26; 34%) demonstrating the relevance of kestrels as seed dispersers, at least in quantitative terms (Padilla et al. 2012). Other raptor species coexist in the Canaries, and therefore the role of raptors as seed dispersers is not wholly assessed. Raptors not only can predate on frugivorous animals. They also predate on seed predator species which could carry seeds in their guts. On Fuerteventura, López-Darias and Nogales (2016) reported more than 3000 seeds in 300 pellets of common buzzards *Buteo buteo* associated with prey of two introduced mammals, the European rabbit *Oryctolagus cuniculus* and the Barbary ground squirrel *Atlantoxerus getulus*. Many seeds were viable demonstrating that raptors disrupt seed predation and contribute to plant fitness by dispersing seeds that other way would be predated.

Diploendosynzoochory (Diplochory by Endo- and Synzoochory)

Raptors do not necessarily have to disperse secondarily the seeds by prey ingestion. They could reject the body parts where seeds are transported, i.e., guts, feathers, or fur. In the Canary Islands, captive experiments showed that kestrel predation on lizards leads to a secondary seed dispersal with two alternative outcomes. A small proportion of seeds appeared (11%) in the kestrel pellets undergoing double gut treatment (see above diploendozoochory). However, most seeds (89%) were not ingested by the kestrels as they rejected the digestive tracts of lizards when they tore away the flesh (Padilla and Nogales 2009). In fact, germination rates were different for seeds depending on the number of passages through the gut of dispersers. Seeds discarded inside the lizard guts by kestrels have higher germination rates than seeds undergoing double gut treatment (i.e., lizards and kestrels) (Padilla et al. 2012).

Eleonora's falcons *Falco eleonorae* hunt migratory birds over the ocean and store them in larders close to their nests (Viana et al. 2015). Many prey are never consumed by falcons, and thus this process constitutes a case of diploendosynzoochory. Migratory birds could avoid the islands where Eleonora's falcons breed during the migratory journeys, and even if they reach the islands, they may not deposit the seeds they transport on these islands. That could be a relevant dispersal mechanism on those small islands not used as stopover areas, but close to the migratory routes.

Impact on Plant Recruitment

The recruitment outcome of plant populations is largely influenced by the effectiveness of seed dispersal vectors since they remove seeds from source plants until they are deposited to the final location (Schupp 1993; Schupp et al. 2010, 2017). During the process seeds are handled, moved to a certain distance, and finally deposited in microhabitats with variable suitability for germination and seedling survival. Overall, these stages will determine the chances that a single seed survives to post-dispersal processes (e.g., seed predation, seedling establishment, etc.) and ultimately establishes into the population as a new individual plant. More formally, the overall effectiveness of a given vector is defined by the number of seeds it disperses (quantitative component) and the probability that a dispersed seed produces a new adult individual (qualitative component) (Schupp 1993; Schupp et al. 2010, 2017). The qualitative component is determined by (*i*) the effect of handling on the viability and germination of seeds and (*ii*) the suitability of deposition sites. Thus, in order to assess the relevance of raptors as seed dispersers, it is mandatory to understand how they contribute quantitatively and qualitatively to seed dispersal of plants.

Quantitative Component

Few attempts have been done to assess whether frugivory by raptors is really significant in terms of the number of seeds they mobilize in plant communities (Ferguson-Lees and Christie 2001; Galetti and Guimarães 2004; Hämäläinen et al. 2017). Excepting several anecdotal observations on the consumption of fruits by raptors (e.g., Table 6.1), there is no available information about the frequency and intensity of the interaction between predatory birds and fleshy-fruited plant species. Given that trophic ecology of raptors has been broadly studied, it is not easy to think that endozoochory could be a common phenomenon among them. Similarly, information about the quantity of dispersed seeds by epizoochory is very scarce, and, as discussed previously, only indirect evidence proves that epizoochory really occurs (Mendelsohn 1983; Rodríguez et al. 2009).

Only the role of raptors as secondary seed dispersers has been specifically evaluated, although it is still poorly studied across the seed dispersal literature (Nogales et al. 2002, 2007; Padilla and Nogales 2009; Padilla et al. 2012; Viana et al. 2015; López-Darias and Nogales 2016). In fact, the only place where it has been studied consistently is in the Canary Islands, where both kestrels and common buzzards (*Buteo buteo*) act as very important secondary seed dispersers (Nogales et al. 2002, 2007; Padilla and Nogales 2009; Padilla et al. 2012; López-Darias and Nogales 2016). Common kestrels, by preying on frugivorous lizards, mobilize a very large quantity of seeds of a large spectrum of plant species across the entire archipelago (Nogales et al. 2007; Padilla et al. 2012). This is shown by the high proportion of kestrel pellets containing seeds, with an average of 1.1 seeds per pellet (estimated from Padilla et al. (2012); *n pellets* = 5304, *n seeds* = 5546). It is worth to note that these numbers must be much higher as only 11% of dispersed seeds appear in pellets. Most of the times, they reject digestive tracts of lizards before consumption (Padilla and Nogales 2009), and, therefore, a very large fraction of dispersed seeds (89%) is undetectable by using pellets. In turn, common buzzards on Fuerteventura, which prey commonly on herbivorous mammals such as European rabbits and Barbary ground squirrels, disperse seeds frequently as well, as indicated by an occurrence in pellets above 10% for some weed species. An average of 24.3 seeds were found per pellet (estimated from López-Darias and Nogales 2016; *n pellets* = 300, *n seeds* = 7292), although it varied widely among plant species. Even though this phenomenon should be broadly studied from a taxonomical and geographical point of view, results from the Canary Islands suggest that predatory birds may be mobilizing a greater quantity of seeds than previously recognized, leading to potential impacts on the ecosystem functioning.

As shown, importance of raptors as primary seed dispersers has not been specifically evaluated, most likely because it is not a common phenomenon among raptors. Thus, following sections are mostly focused on the relevance of raptors as secondary seed dispersers, although some of the topics addressed may be also applied to primary dispersal.

Qualitative Component

Seed fates will be highly conditioned by the effect of handling by predators and suitability of sites where they are deposited. Thus, in the first term, it will depend on the impact of seed passage through the gut of raptors during the dispersal process. Raptors usually display large gut retention times, have strong enzymatic activity of gastric juices, and ingest a large proportion of solid elements in the diet (e.g., bones). The ability of seeds to deal with these characteristics will define the sign of the interaction with raptors in relation to germination probabilities. This is largely dependent on morphological traits of seeds, specifically hardness of seed coat

(Traveset 1998; Traveset and Verdú 2002). Hard-coated seeds are expected to benefit from gut passage of raptors as they promote the scarification of the endocarp (Traveset 1998; Traveset and Verdú 2002), which usually increases germination speed or breaks seed dormancy. In the other hand, germination of seeds with softer coats may be reduced as they are more sensible to damages caused by acids or by the friction with other elements of the raptors diet (Traveset 1998; Traveset and Verdú 2002). For example, enzymatic action and large retention times of common kestrels in the Canary Islands had a negative effect on seed germination for most plant species, especially for those with soft-coated seeds (Nogales et al. 2007; Padilla et al. 2012). Yet, seeds do not have to be swallowed to be secondarily dispersed by raptors. In fact, common kestrels swallow only a small fraction of seeds, as most of the times they reject the prey's digestive tracts before consumption (Padilla and Nogales 2009). Therefore, great majority of seeds (89%) only receive the gut treatment of the primary disperser; thus they benefit from seed movement but avoid the negative effect of the second gut treatment by raptors (Padilla and Nogales 2009). Feeding behavior of raptors is thus an important issue to be taken into account to define qualitative effectiveness of raptors.

Another element that defines the qualitative effectiveness of frugivores is the ability to deposit seeds on suitable locations for germination. Raptors often use distinct landscape features than other bird groups for perching or nesting. Arrival locations may range from unsuitable sites such as old buildings, caves, or perches along roads, where seed germination is unlikely, to more suitable sites such as nursery trees (Dean and Milton 1988). Raptors can be (*i*) effective dispersers if they deposit seeds at unique locations where alternative seed dispersers are unlikely to arrive and where germination and survival are favored or (*ii*) ineffective dispersers if they disperse seed to unsuitable locations. For example, deposition patterns of seeds generated by common kestrels on Lanzarote (Canary Islands) differ markedly from those produced by other primary (lizards) and secondary dispersers (shrikes) (Nogales et al. 2007). While common kestrels deposit seeds almost exclusively at hills, lizards and shrikes use mainly open grounds and hillocks, respectively, for defecation/regurgitation (Nogales et al. 2007). It would be necessary, however, to determine germination rates across the different microhabitats to assess the qualitative effectiveness of kestrels relative to the alternative seed dispersers. All these components of the dispersal effectiveness have been broadly studied for other plant-animal interactions but remain understudied for systems involving raptor species as seed dispersers.

Disruption of Seed Predation

Recruitment patterns of plants are largely influenced by the outcomes of interactions with mutualistic (e.g., seed dispersers and pollinators) and antagonistic animals (e.g., pre-dispersal seed predators) (Crawley 2000). Granivorous animals, mostly birds and rodents, are expected to reduce plant recruitment as they may "kill" a large proportion of individuals of a new plant cohort. Once granivores remove seeds from source sites, they can keep large amounts of seeds on gizzards (birds) or cheek pouches (rodents) before digesting them. Although some seeds may survive the entire digestive process, most of them are destroyed by the seed predator to get the nutritive tissues. Raptors may disrupt seed predation by hunting on granivores and counteract, at least in part, the negative effect of granivores on plant recruitment. In this manner, they may return a proportion of viable seeds to the seed bank of plant populations that otherwise would be destroyed by granivores. In addition, saved seeds can be benefited by long-distance dispersal resulting from the sequential movements of seed predators and secondary dispersers (see below advantages of long-distance seed dispersal). It is worth to note, however, that this phenomenon is not always positive. The interaction will be positive only if the amount and/or survival chance of viable seeds that are incorporated into the seed bank is higher when raptors intervene than when they do not. As discussed before, it will depend mainly on two factors: (*i*) the effect of seed handling by raptors (e.g., gut passage) on seed viability and (*ii*) the quality of deposition sites.

After the pioneering study of Dean and Milton (1988), several recent studies have suggested that secondary dispersal following the disruption of seed predation may be more frequent and geographically broader than previously considered. This process is not mediated exclusively by raptors but also by other top predators (Sarasola et al. 2016). The high frequency of predator-mediated seed dispersal is not surprising taking into account that predator-prey interactions are ubiquitous key elements of ecosystem functioning worldwide. Considering density of predators, predation rate and quantities of seeds that a single granivore can simultaneously consume, the number of seeds that top predators can mobilize within an ecosystem may be really high. For example, following this criterion Sarasola et al. (2016) estimated that cougars disseminate around 5000 seeds/km^2 annually. Similarly, Dean and Milton (1988) estimated in thousands the number of seeds dispersed by individual raptors during a year. Viana et al. (2015) found that 1.2% of birds captured by Eleonora's falcons and brought to Alegranza Islet, Canary Islands, carried seeds while migrating. Although this percentage could seem circumstantial, migratory birds disperse millions of seeds taking into account that several billions of birds migrate between Europe and Africa every year (Viana et al. 2015). However, as discussed by Hämäläinen et al. (2017), these numbers must be relativized to the overall number of seeds produced within a given area. In addition, it is important to understand not only the numerical but, as discussed before, the qualitative dimensions of the dispersal process. To the best of our knowledge, no attempts have been done to date for evaluating the combined effect of all these components of the process. Only then we will be able to assess the truly relevance of this intriguing mechanism for plant dispersal and demography.

Long-Distance Dispersal

Long-distance seed dispersal (LDD), defined as the movement of seeds outside the stand geographic limits and/or outside the genetic neighborhood area of plants (see Jordano (2017) for a detailed account), plays an important role shaping demographic dynamics and distributions of genetic variation and increases the ability of plants to face environmental changes. It enables the colonization of new habitats or post-disturbance recovery of vegetation (Howe and Smallwood 1982) and mediates the interchange of seeds among often distant plant populations (e.g., in fragmented landscapes) promoting gene flow and maintaining cohesiveness of genetic pools at local and broader spatial scales (Jordano et al. 2007; Pérez-Méndez et al. 2016).

Patterns and frequency of LDD of many plant species (fleshy and non-fleshy) will ultimately depend on foraging patterns, home ranges, and handling of seeds by animal vectors. A large set of seed dispersers, usually ranging from small- to medium-sized species, tend to deposit seeds close to parent plants, being responsible for short dispersal distances and contributing largely to local recruitment and the reinforcement of fine-scale genetic structure (Pérez-Méndez et al. 2016). On the other hand, only a few large-bodied species have the required home ranges and gut retention times to disseminate seeds at much further distances (e.g., Jordano et al. 2007; González-Varo et al.2013; Pérez-Méndez et al. 2016). Among them, apex predators, such as birds of prey, normally far exceed movement distances of small- to medium-sized animals. They may contribute disproportionally, thus, to LDD, promoting the exchange of seeds across large spatial scales. This is especially important when seeds are dispersed secondarily, as it involves the movement of the primary disperser and then, the subsequent movement of raptors.

Despite that LDD is fundamental for plants (Nathan 2006), it is probably the less resilient component of the seed dispersal process (Bueno et al. 2013). This occurs because large-bodied animals are being especially hard-hit by anthropogenic disturbances, and very often, altered ecosystems only retain small- to medium-sized species or individuals (Dirzo et al. 2014; Young et al. 2016; Pérez-Méndez 2016; Pérez-Méndez et al. 2018). This is especially critical in poor-richness ecosystems, such as insular or high mountain assemblages, where functional redundancy among dispersers is quite low (McConkey and Brockelman 2011; McConkey and Drake 2015; González-Castro et al. 2015). Under these scenarios, by predating on remaining small primary consumers, top predators may be the only group of animals that can move seeds at large distances.

Interactions among fleshy-fruited plants, frugivorous *Gallotia* lizards and predatory birds (common kestrels and southern grey shrikes) in the Canary Islands, provide us with an illustrative example of how secondary dispersal mediated by predators contributes disproportionally to LDD on low-redundant functional ecosystems. Frugivorous lizards of the genus *Gallotia* are very important seed dispersers of a wide range of plant species (Valido and Nogales 1994; Valido and Olesen 2007; Rodríguez et al. 2008). Among them, *Gallotia atlantica* in Lanzarote island is one of the main frugivorous species together with a few small-sized passerines. Despite that they disperse a large quantity of seeds, seed dispersal distances are expected to be quite short (a few meters from parent plants), attending to their limited movements (Nogales et al. 2007). In turn, seed dispersal events mediated by small passerines are expected to be similarly short (Jordano et al. 2007). Under these circumstances, only larger predatory birds, which prey on frugivorous lizards, can disperse seeds at much further distances. In fact, mean dispersal distances detected for shrikes and common kestrels were

608 m (range 350–1150) and 775 m (range 275–1350 m), respectively (Nogales et al. 2007). Preservation of raptors in those ecosystems with low richness of seed dispersers is therefore essential to maintain long-distance seed dispersal, a key process for plant dynamics.

Adaptation to Climate Change

Ongoing climate change is forcing plants to move to higher latitudes and elevations worldwide in response to shifts in meteorological conditions to which they are adapted (Chen et al. 2011). Assistance by animals, especially large-bodied species, is essential for fleshy-fruited plant species, as well as some dry-fruit plants, to keep pace with climate change, as very often they are the only vectors providing the latitudinal and altitudinal displacements needed to reach locations within their suitable climatic range (Hampe 2011; Naoe et al. 2016; González-Varo et al. 2017). Recent estimates indicate that geographical range of species in temperate zones has moved to higher latitudes at a rate of 16.9 km per decade in response to climate warming (Chen et al. 2011). Given that raptors may cover several hundred kilometers per day, especially during migration, and have long gut retention times, they potentially provide plants with the necessary dispersal distances to escape climate warming in temperate ecosystems. In the tropical ecosystems, however, temperatures do not change markedly with latitude; thus altitudinal migration is the shortest route to escape climate change. In this sense, plants need to climb an estimated distance of 35.4 m per decade to track shifting temperatures (Chen et al. 2011). Very recently, two studies have highlighted the important role of mammals in helping plants to migrate in altitude by dispersing seeds at much further distances than those required for escaping warming (Naoe et al. 2016; González-Varo et al. 2017). Raptors, which often cover even much longer distances in both altitude and latitude during hunting flights, might be also key players in the fight of plants against climate change.

Invasive Plants

Raptors could facilitate plant invasions by the long distances covered during their movements (e.g., migrations or foraging trips). In this sense, long-distance dispersal accelerates rates of spread, which could be enhanced in human-transformed habitats (Higgins and Richardson 1999). To our knowledge, no studies have assessed the role of raptors on invasive plant dispersal and, consequently, invasion patterns. In this sense, information is quite limited and restricted to anecdotic observations. Padilla et al. (2012) found that at least nine introduced plant species were secondarily dispersed in the pellets of common kestrel in the Canaries. *Lantana camara*, 1 of the 100 worst invasive species (Lowe et al. 2000), was included, although fortunately percentage of germination was zero ($n = 1$). However, other invasive plants, such as *Opuntia dillenii* and *O. maxima*, were more frequent in the pellets and reached higher germination percentages: 13.3% and 5.7%, respectively (Padilla et al. 2012). Plant species, such as the African oil palm, used globally in intensive cultivations could be also dispersed by raptors (CABI 2013). Observations of South American raptors eating fruits of this plant could explain its naturalization in some South American countries (Table 6.1), while naturalization on some islands could be facilitated by secondary seed dispersal events. However, there are no specific studies, and clearly more research is required to quantify the role of raptors in plant invasions.

Conclusions

Overall, raptors are not efficient primary dispersers of seeds, although some species could play an important role on seed dispersal of certain plant species (e.g., the palm-nut vulture or the black caracara). However, raptors could play a significant role as secondary seed dispersers, particularly on island ecosystems, where raptors are the main predators. In this sense, secondary seed dispersal is not uncommon, but more research is needed to understand its overall ecological significance (Hämäläinen et al. 2017). While primary dispersal by raptors may be considered a mutualistic interaction between plants and raptors, secondary dispersal is not. To our best knowledge, raptors do not get any reward from plants by dispersing the seeds. Under these conditions, coevolution between plants and raptors is harder than in the typical case of seed dispersal of fruited plant species mediated by frugivores (e.g., pulp nutrients or seed size). More studies are required on the role of raptors as seed disperser, assessing the deposition sites, germination success, proportion of dispersed seeds in relation to total seed production, and seed dispersal distances in relation to other dispersers.

References

Balgooyen TG, Moe LM (1973) Dispersal of grass fruits-an example of endornithochory. Am Midl Nat 90:545–455

Bueno RS, Guevara R, Ribeiro MC, Culot L, Bufalo FS, Galetti M (2013) Functional redundancy and complementarities of seed dispersal by the last Neotropical megafrugivores. PLoS One 8:e56252

CABI (Centre for Agriculture and Biosciences International) (2013) Invasive Species Compendium: *Elaeis guineensis* (African oil palm). http://www.cabi.org/isc/datasheet/20295. Accessed 15 May 2017

Carneiro C, Henriques M, Barbosa C, Tchantchalam Q, Regalla A, Patrício AR, Catry P (2017) Ecology and behaviour of Palm-nut Vultures (*Gypohierax angolensis*) in the Bijagós Archipelago, Guinea-Bissau. Ostrich 88(2):113–121

Chen IC, Hill JK, Ohlemüller R, Roy D, Thomas CD (2011) Rapid range shifts of species associated with high levels of climate warming. Science 333:1024–1026
Clarke R, Combridge P, Middleton N (2003) Monitoring the diets of farmland winter seed-eaters through raptor pellet analysis. Br Birds 96:360–375
Connell JH (1971) On the role of natural enemies in preventing competitive exclusion in some marine animals and in rain forest trees. In: den Boer PJ, Gradwell GR (eds) Dynamics of populations. Centre for Agricultural Publication and Documentation, Wageningen, pp 298–312
Costa-Pereira R, Galetti M (2015) Frugivore downsizing and the collapse of seed dispersal by fish. Biol Conserv 191:809–811
Cousens RD, Dytham C, Law R (2008) Dispersal in plants. A population perspective. Oxford University Press, Oxford
Crawley MJ (2000) Seed predators and plant population dynamics. In: Fenner M (ed) Seeds: the ecology of regeneration in plant communities. CAB International, Wallingford, pp 167–182
Da Silva HR, De Britto-Pereira MC (2006) How much fruit do fruit-eating frogs eat? An investigation on the diet of Xenohyla truncata (Lissamphibia: Anura: Hylidae). J Zool 270:692–698
Darwin CR (1859) On the origin of species by means of natural selection, or the preservation of favoured races in the struggle for life. Murray, London
Dean WRJ, Milton SJ (1988) Dispersal of seeds by raptors. Afr J Ecol 26:173–176
Dirzo R, Young HS, Galetti M, Ceballos G, Isaac NJB, Collen B (2014) Defaunation in the Anthropocene. Science 345:401–406
Ferguson-Lees J, Christie DA (2001) Raptors of the world. Houghton Mifflin, Boston
Figuerola J (2010) Internal dispersal of seeds by waterfowl: effect of seed size on gut passage time and germination patterns. Naturwissenschaften 97:555
Galetti M, Guimarães PR (2004) Seed dispersal of *Attalea phalerata* (Palmae) by Crested caracaras (*Caracara plancus*) in the Pantanal and a review of frugivory by raptors. Ararajuba 12:133–135
González-Castro A, Calviño-Cancela M, Nogales M (2015) Comparing seed dispersal effectiveness by frugivores at the community level. Ecology 96:808–818
González-Varo JP, López-Bao JV, Guitián J (2013) Functional diversity among seed dispersal kernels generated by carnivorous mammals. J Anim Ecol 82:562–571
González-Varo JP, López-Bao JV, Guitián J (2017) Seed dispersers help plants to escape global warming. Oikos 126:1600–1606
Gottsberger G (1978) Seed dispersal by fish in the inundated regions of Humaita, Amazonia. Biotropica 10:170
Hämäläinen A, Broadley K, Droghini A, Haines JA, Lamb CT, Boutin S, Gilbert S (2017) The ecological significance of secondary seed dispersal by carnivores. Ecosphere 8:e01685
Hampe A (2011) Plants on the move: the role of seed dispersal and initial population establishment for climate-driven range expansions. Acta Oecol 37:666–673
Higgins SI, Richardson DM (1999) Predicting plant migration rates in a changing world: the role of long-distance dispersal. Am Nat 153:464–475
Howe HF, Smallwood J (1982) Ecology of seed dispersal. Annu Rev Ecol Syst 13:201–228
Janzen DH (1970) Herbivores and the number of tree species in tropical forests. Am Nat 104:501–528
Jordano P (2014) Fruits and frugivory. In: Gallagher RS (ed) Seeds: the ecology of regeneration in plant communities. CAB International, Wallingford, pp 18–61
Jordano P (2017) What is long-distance dispersal? And a taxonomy of dispersal events. J Ecol 105:75–84
Jordano P, García C, Godoy JA, García-Castaño JL (2007) Differential contribution of frugivores to complex seed dispersal patterns. Proc Natl Acad Sci U S A 104:3278–3282
López-Darias M, Nogales M (2016) Raptors as legitimate secondary dispersers of weed seeds. Ibis 158:428–432
Loveless MD, Hamrick JL (1984) Ecological determinants of genetic structure in plant populations. Annu Rev Ecol Syst 15:65–95

Lowe S, Browne M, Boudjelas S, De Poorter M (2000) 100 of the world's worst invasive alien species a selection from the global invasive species database. Invasive Species Specialist Group, Auckland
McConkey KR, Brockelman WY (2011) Nonredundancy in the dispersal network of a generalist tropical forest tree. Ecology 92:1492–1502
McConkey KR, Drake DR (2015) Low redundancy in seed dispersal within an island frugivore community. AoB Plants 7:plv088
Mendelsohn J (1983) Causes of mortality in Black-shouldered Kites. Bokmakierie 35:11–15
Naoe S, Tayasu I, Sakai Y, Masaki T, Kobayashi K, Nakajima A, Sato Y, Yamazaki K, Kiyokawa H, Koike S (2016) Mountain-climbing bears protect cherry species from global warming through vertical seed dispersal. Curr Biol 26:R315–R316
Nathan R (2006) Long-distance dispersal of plants. Science 313:786–788
Nathan R, Muller-Landau HC (2000) Spatial patterns of seed dispersal, their determinants and consequences for recruitment. Trends Ecol Evol 15:278–285
Nogales M, Delgado JD, Medina FM (1998) Shrikes, lizards and *Lycium intricatum* (Solanaceae) fruits: a case of indirect seed dispersal on an oceanic island (Alegranza, Canary Islands). J Ecol 86:866–871
Nogales M, Quilis V, Medina FM, Mora JL, Trigo LS (2002) Are predatory birds effective secondary seed dispersers? Biol J Linn Soc 75:345–352
Nogales M, Padilla DP, Nieves C, Illera JC, Traveset A (2007) Secondary seed dispersal systems, frugivorous lizards and predatory birds in insular volcanic badlands. J Ecol 95:1394–1403
Olesen JM, Valido A (2003) Lizards as pollinators and seed dispersers: an island phenomenon. Trends Ecol Evol 18:177–181
Padilla DP, Nogales M (2009) Behavior of kestrels feeding on frugivorous lizards: implications for secondary seed dispersal. Behav Ecol 20:872–877
Padilla DP, González-Castro A, Nogales M (2012) Significance and extent of secondary seed dispersal by predatory birds on oceanic islands: the case of the Canary archipelago. J Ecol 100:416–427
Pérez-Méndez N (2016) Las huellas de la defaunación en el Antropoceno. El colapso de los mutualismos de dispersión de semillas. Ecosistemas 25:134–137
Pérez-Méndez N, Jordano P, Valido A (2015) Downsized mutualisms: consequences of seed dispersers' body-size reduction for early plant recruitment. Perspect Plant Ecol Evol Syst 17:151–159
Pérez-Méndez N, Jordano P, García C, Valido A (2016) The signatures of Anthropocene defaunation: cascading effects of the seed dispersal collapse. Sci Rep 6:24820
Pérez-Méndez N, Rodríguez A, Nogales M (2018) Intra-specific downsizing of frugivores affects seed germination of fleshy-fruited plant species. Acta Oecol 86:38-41
Ridley H (1930) The dispersal of plants throughout the world. Reeve and Ashford, Kent
Rodríguez A, Rodríguez B, Rumeu B, Nogales M (2007) Seasonal diet of the Grey Heron *Ardea cinerea* on an Oceanic Island (Tenerife, Canary Islands): indirect interaction with wild seed plants. Acta Ornithol 42:77–87
Rodríguez A, Nogales M, Rumeu B, Rodríguez B (2008) Temporal and spatial variation in the diet of the endemic lizard *Gallotia galloti* in an insular Mediterranean scrubland. J Herpetol 42:213–222
Rodríguez A, Siverio F, Barone R, Rodríguez B, Negro JJ (2009) An overlooked cost for the velvety plumage of owls: entanglement in adhesive vegetation. Wilson J Ornithol 121:439–441
Sarasola JH, Zanón-Martínez JI, Costán AS, Ripple WJ (2016) Hypercarnivorous apex predator could provide ecosystem services by dispersing seeds. Sci Rep 6:19647
Schupp EW (1993) Quantity, quality and the effectiveness of seed dispersal by animals. Vegetatio 107(108):15–29
Schupp EW, Jordano P, Gómez JM (2010) Seed dispersal effectiveness revisited: a conceptual review. New Phytol 188:333–353

Schupp EW, Jordano P, Gómez JM (2017) A general framework for effectiveness concepts in mutualisms. Ecol Lett 20(5):577–590

da Silva HR, de Britto-Pereira MC, Caramaschi U, de Britto-Pereira MC (1989) Frugivory and seed dispersal by *Hyla truncata*, a Neotropical treefrog. Copeia 1989:781

Soons MB, Brochet A-L, Kleyheeg E, Green AJ (2016) Seed dispersal by dabbling ducks: an overlooked dispersal pathway for a broad spectrum of plant species. J Ecol 104:443–455

Sorensen AE (1986) Seed dispersal by adhesion. Annu Rev Ecol Syst 17:443–463

Stiles EW (2000) Animals as seed dispersers. In: Fenner M (ed) Seeds: the ecology of regeneration in plant communities. CAB International, Wallingford, pp 111–124

The Peregrine Fund (2017) Global raptor information network: species account. http://www.globalraptors.org. Accessed 25 Mar 2017

Thiollay J-M (1978) Les rapaces d'une zone de contact savane-forêt en Côted'Ivoire: Spécialisations alimentaires. Alauda 46:147–170

Traveset A (1998) Effect of seed passage through vertebrate frugivores' guts on germination: a review. Perspect Plant Ecol Evol Syst 1:151

Traveset A, Verdú M (2002) A meta-analysis of gut treatment on seed germination. In: Levey DJ, Silva WR, Galetti M (eds) Seed dispersal and frugivory: ecology, evolution and conservation. CABI, Wallingford, pp 339–350

Valido A, Nogales M (1994) Frugivory and seed dispersal by the lizard *Gallotia galloti* (Lacertidae) in a xeric habitat of the Canary Islands. Oikos 70:403–411

Valido A, Olesen JM (2007) The importance of lizards as frugivores and seed dispersers. In: Dennis A, Green R, Schupp EW, Wescott D (eds) Frugivory and seed dispersal: theory and its application in a changing world. CAB International, Wallingford, pp 124–147

Viana DS, Gangoso L, Bouten W, Figuerola J (2015) Overseas seed dispersal by migratory birds. Proc R Soc B Biol Sci 283:20152406

Young LM, Bell RJH (2010) Frugivory and primary seed dispersal by a New Zealand falcon (*Falco novaeseelandiae*) at Red Tarns, Mt Sebastopol, New Zealand. Notornis 57:94–95

Young HS, McCauley DJ, Galetti M, Dirzo R (2016) Patterns, causes, and consequences of Anthropocene Defaunation. Annu Rev Ecol Evol Syst 47:333–358

Part II
Raptors in Human Landscapes

Chapter 7
Raptors and People: An Ancient Relationship Persisting Today

Juan José Negro

Introduction

Symbolic or religious value has been attributed to raptors in many cultures, being the sustained ritual use of golden eagle feathers among Native Americans perhaps the best known of all cases. This behavior seems to be rooted in more ancient times; bones and claws of white-tailed eagles have been found at many Mesolithic and Neolithic sites in Germany, which indicates these large birds had symbolic meaning for past communities in Central Europe and other geographical areas.

But fossil hominins may have also used raptor feathers as functional, ornamental, and symbolic items. The hypothesis that Neanderthals exploited birds for the use of their feathers or claws as personal items has been posed. Raptors were among the bird taxa found associated with Neanderthals all across the Mediterranean Sea in archaeological sites in Italy, France, Gibraltar, and Croatia. All these discoveries provide evidence of human use for purposes beyond feeding: the overrepresentation of raptor wing bones with clear anthropogenic modifications has been interpreted as related to the procurement of feathers for non-alimentary purposes, in the same way than the cut marks and specific polishes on talons have been interpreted as related to the procurement of claws for jewelry.

Ancient Egyptians also venerated raptors and particularly some falcon species that they embalmed and buried in sacred places. Horus, the falcon God, was often depicted as a man with a falcon head, most likely of a lanner (*Falco biarmicus*). Large eagles of the *Aquila* genus have also been widely used in heraldry for millennia until today. During the Roman Empire, the eagle was synonymous of power and strength. The French Empire also used the eagle prominently as a symbol, and many countries still use it today in coats of arms and emblems all across the world.

J. J. Negro (✉)
Department of Evolutionary Ecology, Doñana Biological Station (EBD-CSIC), Sevilla, Spain
e-mail: negro@ebd.csic.es

J. H. Sarasola et al. (eds.), *Birds of Prey*,
https://doi.org/10.1007/978-3-319-73745-4_7

The bald eagle (*Haliaeetus leucocephalus*) is, for instance, the national emblem of the USA since the eighteenth century, soon after independence.

But perhaps one of the most fascinating relationships among raptors and people is that of falconry. Raptors have been used for hunting for more than 5000 years, and falconry is still popular today in many parts of the world. From the largest eagles to the much smaller falcons, numerous species are trained for the hunt or more recently for scaring off birds at airports, and thus to protect the lives of people reducing the risk of plane crashes.

The scavenging species, and most notably the vultures, provide the essential ecosystem service of quickly eliminating carcasses. The recent and abrupt decline of vulture populations in Asia and Africa has revealed how important this role is for the health and well-being of human populations.

The Oldest Evidence: Neanderthal Use of Raptor Feathers

The Neanderthals (*Homo neanderthalensis*), a hominin species that evolved and lived in temperate regions of western Eurasia predating the arrival of modern humans (i.e., anatomically modern Homo sapiens originating in Africa), may have exploited birds for the use of their feathers or claws as personal ornaments in symbolic behavior (Peresani et al. 2011; Morin and Laroulandie 2012; Finlayson et al. 2012; Radovčić et al. 2015). This discovery stirred a hot debate in academic circles, as well as in the mass media (see, e.g., http://www.bbc.com/news/science-environment-19623929), because it suggested unprecedented cognitive abilities for a species different to ours. Raptors of the orders Accipitriformes and Falconiformes, as well as members of the family Corvidae, the largest species among the order Passeriformes, were the taxa associated with Neanderthal occupation at the archaeological sites of Riparo Fumane (Italy), Combe-Grenal and Les Fieux (France), the Gibraltar cave complex in the south of the Iberian Peninsula, and the Krapina site in Croatia. Among the species found, there were remains of bearded vultures *Gypaetus barbatus*, griffon vultures *Gyps fulvus*, golden eagles *Aquila chrysaetos*, and white-tailed eagles *Haliaeetus albicilla*, but also the tiny (in comparison) kestrels (e.g., *Falco vespertinus*).

Finlayson et al. (2012) have carried out the most comprehensive study of the potential relationship of the Neanderthals and the raptors and corvids to date. These authors asked themselves a number of questions: (a) whether Neanderthals processed and used raptor feathers all across the distribution range of the hominin species, or if this was, on the contrary, a very specific behavior of just a few innovative Neanderthal groups; (b) if this behavior was practiced all along the existence of Neanderthals as a species, or just when they were contacted by *Homo sapiens*, soon before the Neanderthals actually became extinct; (c) whether raptors and corvids were processed in a way compatible with the ornamental use of feathers and not as plain additions to a carnivorous diet; and last, (d) if the Neanderthals were partial to specific types and color of feathers. Finlayson and coworkers compiled a large database of archaeological sites with proven Neanderthal occupation, as well as other sites with no hominin occupation and just containing animal remains (i.e., paleontological sites). These authors even skinned vultures that had died accidentally and extracted remiges and rectrices using lithic tools of the Mousterian type, the one developed by the Neanderthals (https://blogs.scientificamerican.com/observations/caveman-couture-neandertals-rocked-dark-feathers/), to assess the type of cut marks that the Neanderthals may had left when extracting feathers themselves.

Finlayson et al. (2012) found an overrepresentation of raptor and corvid wing bones in Neanderthal sites compared to paleontological sites. Neanderthals were surely using their feathers, as indicated by the high proportion of individual wing bones compared to other skeletal remains, and the fact that there was a statistically significant higher frequency of anthropic marks caused by stone tools on humeri and ulnae – bones that support the large flight feathers – compared to any other bones. Indirect support that the processing by Neanderthals involved feather removal, and not food, comes from the observation that raptors and corvids are not regularly eaten in any extant culture. Neanderthals would use the feathers as the science of ethnography shows us today: tribal humans still widely use raptor feathers (and incidentally feathers of other beautiful species such as parrots or paradise birds) in different parts of the world and always with the same purpose: to decorate themselves (Fig. 7.1), and as status signals (Biebuyk and van den Abbeele 1984). Very often the bird species involved, if not already extinct, are/were chronically scarce or hard to get. The possession of feathers is – or was in historic times – mostly restricted to the elites, as with the colorful bird cloaks of Hawaiian monarchs (Hiroa 1944) or the war headdresses of the Amerindian Chiefs (Mails 1972) made out of golden eagle feathers in juvenal plumage with contrasted light and dark portions in the large flight feathers of both the wings and the tail.

Neanderthals seemed to prefer black-colored raptors, and the corvids associated with their cave dwellings were indeed all black in color. The apparent selection for feathers of a particular coloration would require sophisticated cognitive abilities, and also gives support to Finlayson et al.'s hypothesis of a non-random, and non-utilitarian, use of feathers by the Neanderthals.

The dating of the archaeological sites in Finlayson's database indicates that the extraction of feathers from birds by Neanderthals was a temporally and geographically widespread phenomenon all across their distribution range of the species in temperate Eurasia. Neanderthals become extinct soon after they contacted expanding *Homo sapiens* populations when these started to establish themselves in Europe about 40,000 years ago. Feather use by Neanderthals therefore predated the arrival of modern humans in Europe by several thousand years, and this precludes that this specific skill was copied or acquired from modern humans. Neanderthals, even if they were a different species and not even our direct predecessors in our lineage, had at least some comparable cognitive capacities compared to us, *Homo sapiens*, and that included symbolic expression.

Fig. 7.1 Neanderthal reconstructed figures at the Gibraltar Museum, Gibraltar. The adult figure is ornamented with raptor feathers. (Credit: Juan J. Negro)

The White-Tailed Eagle (*Haliaeetus albicilla*) and the Snowy Owl (*Nyctea scandiaca*) in Prehistoric Europe

Evidence of the use of large raptor feathers after the Neanderthals became extinct, but still in prehistory, comes from Mesolithic and Neolithic sites in Central Europe. Amkreutz and Corbey (2008) discuss the presence of large amounts of white-tailed eagle bone remains in numerous archaeological sites of the Netherlands and Germany more than 6000 years ago, before the advent of agriculture in those areas. The authors argue that even though those past communities may have procured the eagles for subsistence, they favor the alternative idea that their long and attractive feathers were used in ceremonials or as personal ornaments. Citing previous works by fellow archaeologists, Amkreutz and Corbey (2008) dismiss the idea of eagle consumption, noting that the bones more commonly found were those of wings and legs. These authors also remarked that white-tailed eagles should be chronically scarce in the environment, as they are apex predators; and also mention their disgusting taste. In fact, several sites have yielded artifacts made of wing and leg bones of white-tailed eagles. These artifacts included awls, needles, beads, and pendants. Some hollow bones may have been used for "the sucking, sniffing, or blowing of powdered substances or for blow painting."

The importance of sea eagles for ancient European communities is not limited to the Lower Rhine area. White-tailed eagles are very abundant in various Neolithic monuments in the United Kingdom. One of the most renowned is the Neolithic tomb of Isbister, known as Orkney's "Tomb of the Eagles" (http://www.tomboftheeagles.co.uk/). Almost 641 sea eagle bones, the remains of at least 8 different individuals, were found inside a cairn and earned the site its nickname. Although it was initially thought that the eagles were placed in this funerary monument at the time of its foundation more than 5000 years ago, radiocarbon dating on the eagle remains demonstrated they were placed there 1000 years later (still about 4000 years ago). Whether the eagles become growingly important for humans, or more abundant in the environment, is a matter of debate. Perhaps the eagles become a symbol for the tribe or were just left there as an offering to the Gods.

But in prehistoric Europe, and even further back in time, ancient populations were also partial to large owls. At the cave of Bourrouilla, in southern France, the remains of 53 snowy owls (*Nyctea scandiaca*) were found in levels corresponding to Magdalenian occupation, of the Upper Paleolithic more than 12,000 years ago (Eastham 1998).

The owls' bones were subjected to skinning and presented a high level of modification, while the bones of other avian taxa were not systematically used in this way. This fact again points to the idea that the owls were not hunted for meat but for some other nonutilitarian purpose. In addition, there was an overrepresentation of female birds at the cave, which may be related to the sexual dichromatic coloration in this species (males are often pure white, whereas females are dark spotted on the white), but also to the fact that the species may have presented different patterns of habitat occupation if they were hunted outside the breeding season or if they were wintering birds (males tend to be nomadic, whereas females hold territories for themselves along the winter) (Boxall and Ross Lein 1982).

Raptors in the Egypt of the Pharaohs

The everyday life of the ancient Egyptians was dictated by religion and the belief in an afterlife. Animals, both wild and domestic, were very present in their rituals and ceremonies. Thanks to the hieroglyphic writings, other forms of art, or even through the discovery of mummified corpses in animal cemeteries, it is possible to determine the species involved and the importance and meaning they had for the Egyptians. Raptors were revered as sacred animals, and the very Pharaohs were seen as the reincarnation of Horus ("Living Horus on Earth", Fig. 7.2), the divinity represented as a falcon or as a man with the head of a falcon. This cult may have started as early as 2500 years BC. According to Kozloff (2012), the Egyptians knew well the differences among bird species, and Horus was not just any falcon species, but an agile and excellent hunter of birds, the lanner falcon, *Falco biarmicus*. It could be confused with the peregrine falcon, but the yellow orbital ring tends to be more prominent in the lanner, as represented in Egyptian art, and its more brownish color is also diagnostic when Horus was color painted.

Fig. 7.2 Human figure with a falcon head (God Horus). Bronze from Ancient Egypt, dated 1069–664 BC. Displayed at Louvre Museum (Paris). (Credit: Juan J. Negro)

Other raptors are clearly identifiable in Egyptian writing and art: in the Eighteenth Dynasty, the mother figure was a vulture and possibly the griffon (*Gyps fulvus*), later substituted by the larger of the African vultures, the lappet-faced vulture (*Torgus tracheliotus*), also occurring in the Pharaoh's lands. Already in the first half of the nineteenth century, at the onset of Egyptology, several raptors were accurately identified in Egyptian art and writing by, for instance, T.J. Pettigrew (1834), who reported Eurasian hobbies (*Falco subbuteo*), Eurasian kestrels (*F. tinnunculus*), lesser kestrel (*F. naumanni*), the Eurasian sparrowhawk (*Accipiter nisus*), and the osprey (*Pandion haliaetus*).

Of particular interest are the raptor cemeteries found in the Nile Valley. Recent studies suggest that some species were kept in captivity and even bred by dedicated priests at the temples to provide votive animal mummies as offerings to the Gods. These offerings were paid by pilgrims who visited the temples and requested God's blessings in return. According to Morgan and McGovern-Huffman (2008), animal burials are found at all periods throughout Egyptian history, but their use intensified in the late period (ca 664–332 BC) and continued into Roman times. At Saqqara, Buto, Abydos, and Kom Ombo, complexes were dedicated to falcons, representing Horus (Lauer 1976; Wilkinson 2000). Ikram and Iskander (2002) have identified the remains of numerous raptor species among the birds' votive mummies, including falcons and kestrels, but also the short-toed eagle (*Circaetus gallicus*), the white-tailed eagle (*Haliaeetus albicilla*), and the booted eagle (*Aquila pennata*).

The Eagle in Mesopotamia

Overlapping in time with some of the older dynasties of the Egyptians, the Sumerians and later the Assyrians in current Iraq adopted the eagle as a symbol of power (Fig. 7.3). The myth of King Etana, for instance, concerns an eagle: this Babylonian legend says that Etana could not conceive a child, until one day he saved a starving eagle, who in return took him up into the sky to find the plant of birth (Dalley 2000). This led to the birth of his son, Balih. Numerous bas-reliefs in stone depicted human figures with eagle heads and wings. The artists clearly depicted eagles, with powerful and long bills, although a precise identification would be adventurous. The region was possibly home for at least the golden eagle (*Aquila chrysaetos*) and the steppe eagle (*Aquila nipalensis*), although other large eagles, such as the Eastern imperial eagle (*Aquila heliaca*) and the white-tailed eagle, may have inhabited the area in the past or at least be regular visitors on migration.

Very fine examples of eagle-headed figures in alabaster originally placed in the city of Nimrud, in the palace of Ashurnasirpal II, ca. 883–859 B.C., are today exhibited in Berlin at the Pergamon Museum (Fig. 7.4) and in the British Museum in London.

In fact, the double-headed eagles so prevalent in the coats of arms of many royal houses in Europe from early medieval times and persisting today may have originated also in Mesopotamia. It seems that the Hittites already represented power using the figure of a double-headed bird more than 4000 years ago (Charitin 2011), and this symbol was present in Anatolian art in the second millennium BC (Alexander 1989). The eagles were the messengers to the Gods. Soaring high in the sky, becoming practically invisible to the naked eye, the Sumerians thought they transported the souls of the deceased to heaven. Other cultures that followed (Babylonians, Assyrians, and later the Greeks and the Romans) seem to have adopted this thought or independently acquired a similar cultural interpretation. Curiously enough, distant

Fig. 7.3 Colossal figure of a bird of prey excavated at Tell Halaf (Siria) and carved in basalt. It is dated about 3000 years before present. Its significance remains undetermined. On display at the Pergamon Museum (Berlin). (Credit: Juan J. Negro)

Fig. 7.4 Palace reliefs in alabaster depicting an eagle with a human body (left panel) and a bearded man with eagle wings (right panel). The anatomical detail in the different feather tracts is realistic and spectacular. Nimrud (883–859 BC). On display at the Pergamon Museum (Berlin). (Credits: Juan J. Negro)

cultures, such as the Iberians, the population who inhabited the Iberian Peninsula in the westernmost portion of the Mediterranean, may have venerated eagles before the arrival of the Romans, as attested by the art they left behind.

Eagles as Royal Symbols in Europe

Double-headed eagles have been the emblem of empires, nations, and noble houses for millennia in Europe. Examples include the Holy Roman Empire, the Byzantine Empire, the House of Habsburg, and the Romanovs in Russia. A live eagle was said to have been placed inside the funerary pyre of Emperor Augustus. The bird was released when the pyre was ignited to symbolize the flight of the Emperor's soul to heaven (Cocker and Tipling 2013). Eagles with open wings decorated many of the magnificent stone-carved tombs of the Nabateans in the Middle East about 2000 years ago (Fig. 7.5).

Even Napoleon, crowned Emperor at the onset of the nineteenth century, utilized the eagle symbol (one-headed, though). The day after the coronation, Napoleon had an eagle, symbol of military victory for the Romans, placed at the top of the shaft of every flag in the Napoleonic army. In recent times, the same eagle symbolism has been very present in the fascist regimes of the Nazis in Germany, in Mussolini's Italy, or in the Francoist period in Spain. With a different meaning, eagles persist today as a durable symbol connecting to the past, even in the Euro coins (Fig. 7.6).

Fig. 7.5 Entrance to a Nabatean tomb carved in a rocky outcrop. On top there is an eagle figure with spread wings. (Credit: Marco Antonio Bernal)

Fig. 7.6 One-Euro coin from Germany showing eagle with spread wings and talons. (Credit: Juan J. Negro)

Raptors and People in the New World

Native Americans, and particularly those inhabiting North America, have a remarkable culture involving raptors, and particularly large eagles, such as the golden eagle and the bald eagle. The spectacular headdresses of the tribal leaders (Fig. 7.7), for instance, are composed of a fairly large number of flight feathers – both remiges and rectrices – of the eagles, which in the past were often captured alive and by hand using pit traps (Mails 1972). But the warbonnet was not the only feathered artifact: warriors also used the "flag," which was a long pole 2–3 m long decorated with a single row of eagle feathers. Any man with sufficient honors, and not necessarily a Chief, might accumulate feathers for both the flag and the warbonnet. Historically, the warrior who was the first to touch an enemy in battle and escaped unscathed received an eagle feather. For the largest headdresses, several individual eagles (sometimes 6–12) were needed.

All these traditions and cultural and religious beliefs developed before the arrival of the Europeans in the fifteenth century (or before the Vikings established short-term posts in eastern Canada some time before). Therefore, the traditions of the Amerindians involving raptors as deities possibly developed independently in the New World or were brought about by the first peoples arriving there several millennia ago. Perhaps the Clovis people already had these beliefs involving eagles, which were perhaps shared by Siberian Upper Paleolithic populations, and that may also have spread into Asia and Melanesia.

Australian aboriginals in fact venerate the largest true Eagle in the region: the wedge-tail eagle (*Aquila audax*). This eagle, known as Bunjil, is an essential part of many dreaming stories for aboriginal Australians: the eagle is, for instance, a creator deity for the Kulin people of central Victoria. For the Kaurna people of the Adelaide Plains, the eagle's claw is the southern cross constellation up in the sky at night.

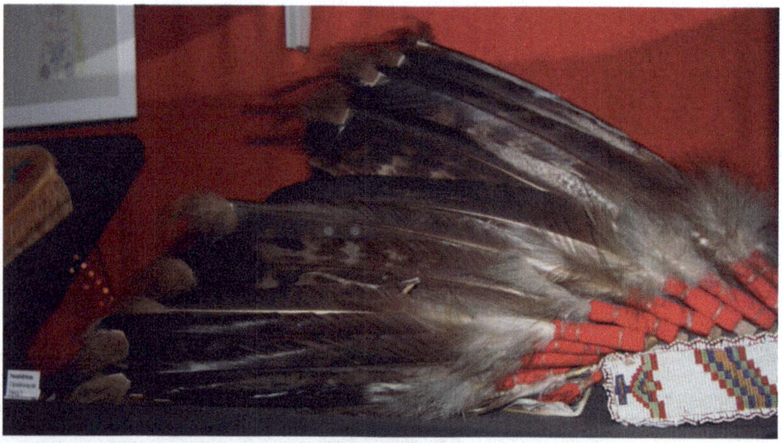

Fig. 7.7 Native-American headdress made out of eagle feathers. Museum of Anthropology. University of British Columbia (Vancouver, Canada). (Credit: Juan J. Negro)

Knowing that the ancestors of the aboriginals arrived in Australia more than 40,000 years ago, it is tempting to think that the myth of the Eagle-God originated somewhere in Eurasia and later spread to Melanesia (first) and after to the New World. Perhaps it was even "copied" first by *Homo sapiens* from *Homo neanderthalensis* living in temperate Europe, as suggested by Finlayson et al. (2012).

The Art of Falconry

Perhaps the most sublime relationship of raptors and people is the cooperative hunting by human and bird through falconry. Falconry is so entrenched in human culture and values that it has been inscribed on the Representative List of the Intangible Cultural Heritage of Humanity by UNESCO in 2010, with new supporting countries added in 2012 and 2016 (https://ich.unesco.org/en/RL/falconry-a-living-human-heritage-01209). With a total of 18 countries, falconry is the largest multinational nomination on the Representative List of the Intangible Cultural Heritage of Humanity. It is now practiced by falconers in many parts of the world, but in some places such as the Asian steppes and the Middle East, it is truly a passion transmitted intergenerationally.

The bond among the birds and the falconers is unique and has little to do with domestication. In fact, until recently falconry was practiced with birds taken from the wild, either as nestlings or as passenger birds. Nowadays the take from nature is highly restricted in most countries for a majority of species due to conservation and ethical reasons. Falconers breed from captive stock and these birds, which have not been artificially selected for tameness as in animal domestication processes, remain perfectly fit for the hunt of wild prey. It is also true, however, that humans have recently "created" new falconry birds by crossing different species and taking advantage of the fact that falcons interbreed easily. In this regard, numerous captive-bred hybrids can be found in the falconry market, including the "perlin" (a crossing among the peregrine falcon and the merlin) or large hybrids, such as gyrfalcon x peregrine or gyrfalcon x saker. Other hybrids are also available, such as peregrine x aplomado falcon or peregrine x lanner. Eagle hybrids have also been created, even with the much smaller Harris's hawk. Other combinations include the "pocket rocket" (gyrfalcon/merlin), Harris's hawk x red-tailed hawk, red-tailed hawk x ferruginous hawk ("ferrutail"), or Harris's hawk x Cooper's hawk. A hybrid's name is a combination of the father first followed by the mother's.

History of Falconry

The International Association for Falconry and Conservation of Birds of Prey (IAF) provides a comprehensive overview of the history of falconry in its official website (http://iaf.org/HistoryFalconry.php), including evidence from art and culture.

The contents derive from a symposium held in Abu Dhabi in 2005 to discuss the origins and future prospects of falconry. The origin of this practice is lost in time, but it may go back millennia. Experts believe the cradle of falconry was in Mesopotamia and/or the Mongolian steppes starting at least 5000 years ago. Some experts go back several millennia up to 10,000 years ago, at the onset of the Neolithic, also in Mesopotamia. There is evidence of falconry before our era (more than 2000 years ago) in Mongolia, China, Korea, India, and the Middle East. Falconry may have reached Japan later, with the first written records in the fourth century of our era. Although falconry may not have been in fashion in classical Greece and in Imperial Roma, there is a beautiful mosaic of Roman inspiration found in Tunisia and dating from the fifth or sixth century of our era. It depicts a man with a goshawk perching on his hand and a hunted partridge hanging from the other hand.

Falconry peaked in the medieval times in Eurasia and Northern Africa, when noblemen irrespective of their beliefs (e.g., Christians, Muslims, Hindus) valued their hawks and falcons more than any other earthly possession. Numerous treatises were written in different countries, first in Latin (tenth century onward) and later, after the thirteenth century, written in vernacular languages such as Spanish, French, Italian, German, English, and Dutch. Expeditions were organized to Greenland to capture white-colored gyrfalcons (Fig. 7.8), the most prized of all falconry birds (Seaver 1996). There are even accounts of falcons used to pay ransom for captured noblemen. And there is of course the tribute of the Maltese Falcon: the Order of St John of Jerusalem had to pay an annual tribute to the Emperor Charles V (Carlos I

Fig. 7.8 Gyrfalcon (*Falco rusticolus*) and falconer in falconry championship in northern Spain. (Credit: Juan J. Negro)

de España) and his mother Queen Joanna of Castile (Reina Juana de Castilla) as monarchs of Sicily, for the granting of Tripoli, Malta, and Gozo. The annual tribute was indeed a live falcon. The order and grand master paid the annual falcon from 1530 to 1798, when the Order of St. John was expelled from the Maltese islands. This story inspired a famous novel written by Dashiell Hammett in 1924 and a legendary film-noir adaptation in 1941, directed by John Huston and starring Humphrey Bogart.

Contemporary Falconry

After the glorious days of medieval falconry, when it was practiced just by the ruling class in their leisure time, falconry has experienced a renaissance in many countries in the second half of the twentieth century, and it has also been exported to places where it was formerly unknown. Falconers now belong to every social class, and they are more limited by available time and personal skill than by monetary or social restrictions. It is now possible to acquire a captive-bred falcon or hawk (untrained) for just some hundred dollars/euros. The problem is one of dedication and perseverance, as a hunting bird demands a lot of attention and skills from the part of the falconer (Fig. 7.9).

There are now falconry clubs and breeders in North America, South America, and South Africa. In the Southern Hemisphere, nonetheless, there are important restrictions in at least two countries: in Australia it is illegal to keep any type of bird of prey in captivity unless the birds are kept for purposes of rehabilitation (for which a license must still be held). In New Zealand, falconry is legal for just one species, the swamp/Australasian harrier (*Circus approximans*).

But falconry is not only a pastime or a hunting technique. For about 50 years, and mainly due to the vision and efforts of the Spanish falconer, conservationist and nature film maker, the late Dr. Felix Rodríguez de la Fuente, falconers and falconry birds are protecting the lives of millions of people at the main airports in the world (Fig. 7.10). These birds are flown to scare away other birds that may collide with aircraft and cause fatal accidents during the risky moments of takeoff and landing. In the USA alone, collisions between wildlife and aircraft, mostly bird strikes, cost civil aviation nearly $500 million a year in direct and associated costs such as aircraft downtime, according to the Federal Aviation Administration. In Spain, with the largest population of vultures and other soaring raptors in western Europe, collisions with aircraft are a big concern. Numerous accidents, particularly involving light aircraft, have caused many fatalities in the twenty-first century.

A major challenge for the integration of falconry in the conservation movement, once the removal of wild animals has proved unnecessary given the success of captive breeding, is the fact that many birds scape back to the wild. As explained above, hybrid birds are in vogue among falconers. These animals should not end up as free-ranging individuals, even if their chances of mating with other birds and raising progeny are slim. In addition, numerous species or subspecies are kept in areas of

Fig. 7.9 Two falconers training their young peregrine falcons (*Falco peregrinus*). Southern Spain. (Credit: Juan J. Negro)

Fig. 7.10 Trained Harris's hawk (*Parabuteo unicinctus*) used to scare off birds in the landing strips of Dorval Airport near Montreal (Canada). (Credit: Juan J. Negro)

the world where they are non-native. Harris hawks (*Parabuteo unicinctus*), for instance, native from the New World, are now commonly bred and used for falconry in Europe. Many have escaped and some have attempted reproduction. In the city of Sevilla (Spain), a pair of escaped Harris hawks built a nest and attempted breeding in recent years. Events like this may pass unnoticed easily in less populated areas, and introduced populations may eventually get established, as it has happened already with invasive parakeet species that cause damage in crops (Senar et al. 2016) and outcompete endangered local species, such as bats and lesser kestrels (Hernández-Brito et al. 2014).

Ecosystem Services Provided by Raptors

Hawks and falcons are apex predators with a potentially prominent role in ecosystems. Whether raptor may or may not help to regulate populations of prey species is a matter of debate, particularly in human-altered environments. But the role of vultures as scavengers is often overlooked. They help to get rid of the carcasses of the larger animals in terrestrial environments in all continents excepting the polar regions and Melanesia. Vultures have helped humans and still do to get rid of rotten corpses around human habitations preventing or reducing the risk posed by some infectious diseases. Vultures have even helped to dispose of human corpses. Some human funerary traditions involve the participation of vultures. The Zoroastrians, for instance, built towers for the excarnation of human bodies by vultures in Iran and India. These places were called Dakhmas or Towers of Silence. The recent vulture population crash in India due to the widespread use of the drug diclofenac to treat domestic animals (Green et al. 2004, in this book Chaps. 10, 19, and 20) has in fact affected these funerary rites as the human bodies took much longer than usual to decompose.

References

Alexander RL (1989) A great queen on the Sphinx Piers at Alaca Hüyük. Anatol Stud 39:151–158

Amkreutz L, Corbey R (2008) An eagle-eyed perspective. *Haliaeetus albicilla* In the Mesolithic and Neolithic of the lower Rhine area. In: Analecta Praehistorica Leidensia 40. Publication of the Faculty of Archaeology Leiden University, Leiden

Biebuyk DP, van den Abbeele N (1984) The power of headdresses: a cross-cultural study of forms and functions. Tendi, Brussels, p 293

Boxall PC, Ross Lein M (1982) Territoriality and habitat selection of female snowy owls (*Nyctea scandiaca*) in winter. Can J Zool 60(10):2344–2350

Charitin J (2011) The Mesopotamian origins of the Hittite double-headed eagle. UW-L J Undergr Res XIV, 1–13

Cocker M, Tipling D (2013) Birds and people. Random House, London

Dalley S (2000) Etana. In: Myths from Mesopotamia. Creation, The Flood, Gilgamesh, and Others. Oxford University Press, Oxford, p 189

Eastham A (1998) Magdalenians and Snowy Owls: bones recovered at the Grotte de Bourrouilla, Arancou (Pyrénées Atlantiques). Paleo 10:95–108

Finlayson C, Brown K, Blasco R, Rosell J, Negro JJ, Bortolotti GR et al (2012) Birds of a feather: Neanderthal exploitation of raptors and Corvids. PLoS One 7(9):e45927. https://doi.org/10.1371/journal.pone.0045927

Green RE, Newton I, Shultz S, Cunningham AA, Gilbert M, Pain DJ, Prakash V (2004) Diclofenac poisoning as a cause of vulture population declines across the Indian subcontinent. J Appl Ecol 41(5):793–800
Hernández-Brito D, Carrete M, Popa-Lisseanu A, Ibáñez C, Tella JL (2014) Crowding in the city: losing and winning competitors of an invasive bird. PLoS One 9(6):e100593
Hiroa TR (1944) The local evolution of Hawaiian feather capes and cloaks. J Polynesian Soc 53(1):1–16
Ikram S, Iskander N (2002) Catalogue general of the Egyptian museum: non- human remains. Supreme Council of Antiquities, Cairo
Kozloff A (2012) Pharaoh was a good egg, but whose egg was he? In: Bailleul-Lesuer R (ed) Between heaven and earth, Oriental institute museum publications, vol 35. Oriental Institute of the University of Chicago, Chicago, pp 59–64
Lauer JP (1976) Saqqara, the royal cemetery of Memphis. Charles Scribner and Sons, Chicago
Mails TE (1972) The mystic warriors of the plains. Marlowe and Company, New York, p 640
Morgan LW, Mcgovern-Huffman S (2008) Noninvasive radiographic analysis of an Egyptian falcon mummy from the late period 1069–332 BC. J Avian Biol 39:1–4
Morin E, Laroulandie V (2012) Presumed symbolic use of diurnal raptors by Neanderthals. PLoS One 7(3):e32856
Peresani M, Fiore I, Gala M, Romandini M, Tagliacozzo A (2011) Late Neanderthals and the intentional removal of feathers as evidenced from bird bone taphonomy at Fumane cave 44 ky B.P., Italy. Proc Natl Acad Sci 108(10):3888–3893
Pettigrew TJ (1834) A history of Egyptian mummies, and an account of the worship and embalming of the sacred animals. Bibliolife, Charleston
Radovčić D, Sršen AO, Radovčić J, Frayer DW (2015) Evidence for Neandertal jewelry: modified white-tailed eagle claws at Krapina. PLoS One 10(3):e0119802
Seaver KA (1996) The frozen Echo. Stanford University Press, Stanford
Senar JC, Domènech J, Arroyo L, Torre I, Gordo O (2016) An evaluation of monk parakeet damage to crops in the metropolitan area of Barcelona. Anim Biodivers Conserv 39:141–145
Wilkinson RH (2000) The complete temples of ancient Egypt. Thames and Hudson Ltd., London

Chapter 8
Costs and Benefits of Urban Living in Raptors

Claudina Solaro

Introduction

Increasing human populations have accelerated urbanization and altered natural habitats. This process began in the eighteenth century with the industrial revolution when workers began moving to cities leaving agricultural jobs for jobs in manufacturing. Global growth in human populations was accompanied by growth of cities, which has increased the demand of goods and services provided by the exploitation of natural ecosystems. Rapid worldwide urbanization has led to a rampant loss of natural habitats and habitat fragmentation, which alarmed to ecologists and conservationists that have focused their researches in last years to understand the response of wildlife to these new scenes. For birds, the number of published studies on urban effects has increased steadily (Marzluff et al. 2001; Marzluff 2017). However, raptors have been poorly studied during much years, mainly due to several limitations imposed by their natural history (i.e., low densities, large home ranges, variable reproductive behaviors, and inaccessible breeding sites) (Donázar et al. 2016) and the high costs necessary for these studies. Nowadays, research of raptors in urbanized habitats has increased considerably. We will focus on a variety of these aspects.

C. Solaro (✉)
Centro para el Estudio y Conservación de las Aves Rapaces en Argentina (CECARA), Universidad Nacional de La Pampa (UNLPam) & Consejo Nacional de Investigaciones Científicas y Técnicas de Argentina (CONICET), Santa Rosa, La Pampa, Argentina
e-mail: csolaro@exactas.unlpam.edu.ar

J. H. Sarasola et al. (eds.), *Birds of Prey*,
https://doi.org/10.1007/978-3-319-73745-4_8

Urban Living in Birds

Birds are probably the most evident and conspicuous wild vertebrates among those living in and adapting to urban environments. Urban parks around the world, for example, hold at least one but usually many bird species, even in those cities most densely populated or located in regions with extreme weather conditions. Birds are also found in a variety of less suitable and more stressful sites inside cities and many of these species breed, forage, and roost on buildings, noisy streets, and avenues.

The study of such avian tolerance to urban environments probably derived from some simple and straightforward questions: How do these birds live and survive in habitats quite different from natural environments? Why are some species found in cities while others are not? What are the consequences and/or profits for these urban-living individuals? Urban living in birds probably started progressively when birds and human began sharing the same physical space, birds because natural habitats were invaded by humans and humans because they chose habitats in which birds lived. This situation was forced by human development and driven to a nonrandom distribution of bird assembles in distinctive ecological habitats that reflect the impacts of urban sprawl on native species. Ecologists studying bird ecology and behavior in urban environments frequently employ the characterization of bird assemblages proposed by Blair (1996), in which three groups are described with regard to their adaptability to urban environments: avoiders, adapters, and exploiters.

Urban avoiders comprise those species particularly sensitive to human-induced changes and which reach their highest densities in less-disturbed natural habitats. These species are the first to disappear in the proximity of humans and the spread of urban development. Urban avoiders include species adapted to live in the interior of old and large forests, migrants, and nesting birds that may be very sensitive to human presence.

Urban adapters include those species able to exploit some additional resources present in moderate levels of urbanized habitats such as suburbs. These species are usually generalist "edge species" that exploit different habitat types and food sources (including human-subsidized foods such as cultivated plants, garbage, and insects attracted to artificial lights), ornamental plantations (for shelter and nesting structure), and open areas. Urban adapters include mostly omnivores and ground foragers, seedeaters, and aerial sweepers such as swifts.

Urban exploiters are species having the skills to exploit urbanized habitats that seem to be very dependent on resources provided by human presence, reaching their highest densities in these developed sites. Urban exploiters experience reduced densities of predators and abundant food resources. These species are usually more common and abundant in cities than in natural habitats. In several cases, urban exploiters are represented by invasive non-native species. In birds, urban exploiters are species originally adapted to cliff and rocky areas and therefore preadapted to exploit shelter and nesting opportunities provided by buildings and other human structures.

Raptors in Urbanized Habitats

The response of raptors to urbanization varies much and depends on species-specific traits, such as home-range requirements, food habits, demography, and behavior, and also on the degree of habitat transformation itself. In general, negative effects of urbanization on the abundance, richness, and diversity of raptor species should be expected (Ferguson 2004; Carrete et al. 2009).

For those species in which urbanization depicts some critical landscape threshold, some minimum degree of urbanization could limit the presence of species with specific habitat requirements that urban landscapes do not provide. Examples include eagles, some large hawks, and falcons (Berry et al. 1998) which have large home-range requirements and could be considered as "urban avoiders" (but see, e.g., McPherson et al. 2016; Kauffman et al. 2004).

There are, however, small- to medium-sized raptors, such as hawks, kestrels, and owls that can be found in urbanized landscapes (Berry et al. 1998; Rullman and Marzluff 2014), and although some species may be more abundant in natural or rural areas, they can reach high abundances in areas with moderate levels of urbanization (Bellocq et al. 2008; Carrete et al. 2009). For these species, preservation of patches of native grasslands or woodlands can enhance the probability of occupation of these areas (Hogg and Nilon 2015). These species which can be considered "urban adapters" can be found usually in areas peripheral to urbanized locations or in suburban environments where their home ranges may extend beyond urban boundaries in order to meet the ecological requirements that urban areas cannot provide.

Breeding in Urban Environments

Since many species of raptors do breed in urbanized habitats, urban areas may not be the suboptimal habitats that they are often assumed to be. Species such as peregrine falcon (*Falco peregrinus*), Cooper's hawk (*Accipiter cooperii*), lesser kestrel (*Falco naumanni*), Eurasian kestrel (*Falco tinnunculus*), Eurasian sparrowhawk (*Accipiter nisus*), and burrowing owl (*Athene cunicularia*), among others, are often seen breeding in urbanized landscapes (Table 8.1). In modern times, these habitats have been described to be free from persecution and, if these habitats or its surrounding area have an adequate food supply, may allow raptors to breed in these areas which might otherwise be considered unsuitable.

Reproductive output in urbanized habitats is highly variable (Chace and Walsh 2006). Burrowing owls, for example, have lower nesting success in natural habitats than they do in human-altered areas on the campus of a university in Las Cruces, New Mexico (Botelho and Arrowood 1996). Mississippi kites (*Ictinia mississippiensis*) also have higher productivity among urban pairs than rural pairs in North America (Parker 1996). Endangered lesser kestrels have larger clutch sizes

Table 8.1 Literature that addresses the variation in reproductive parameters of raptor birds in different gradients of urbanization

Species	Reproductive parameter	↑U, ↓RO	↑U, =RO	↑U, ↑RO	Reference
Bald eagle (*Haliaeetus leucocephalus*)	No differences in reproductive success between areas with different degrees of human activity, except eagles in areas with greater human activity produced more chicks and had larger broods than eagles of forests		X	X	Fraser et al. (1985)
	Number of young fledged per occupied nest site or from successful nests did not differ between suburban and rural nest sites		X		Millsap et al. (2004)
Barn owl (*Tyto alba*)	Unsuccessful nesting sites were associated with more improved grassland, suburban land, and wetlands than successful sites	X			Bond et al. (2005)
Black sparrowhawk (*Accipiter melanoleucus*)	No differences in brood size, reproductive success, and productivity in gradients of urbanization		X		Rose et al. (2017)
Burrowing owl (*Athene cunicularia*)	Human-altered pairs with more nestlings, fledglings, and young than pairs in natural areas			X	Botelho and Arrowood (1996)
	Higher reproductive rate in urban than in more rural areas and lower in areas of very high urban development	X when development is greater than 60%		X when development is less than 60%	Millsap and Bear (2000)
	Higher reproductive success, productivity, and brood size in urban than rural habitats			X	Rebolo-Ifrán et al. (2017)

(continued)

Table 8.1 (continued)

Species	Reproductive parameter	↑U, ↓RO	↑U, =RO	↑U, ↑RO	Reference
Chimango caracara (*Milvago chimango*)	Higher reproductive success and productivity in rural and natural than in suburban habitats	X			Solaro (2015)
Cooper's hawk (*Accipiter cooperii*)	Greater overall proportion of nests that failed in urban than in undeveloped or natural areas	X			Boal and Mannan (1999)
	Similar productivity in urban and rural habitats		X		Rosenfield et al. (1995)
Eastern screech owl (*Otus asio*)	Fledgling per breeding pair and percentage of successful nests increased with urbanization			X	Gehlbach (1996)
Eurasian kestrel (*Falco tinnunculus*)	Lower clutch size, number of hatched offspring, and number of fledged young in urban than in suburban sites	X			Sumasgutner (2013)
	Higher hatching rates and larger fledged brood sizes in vicinity of green backyards. Lower productivity in urban centers than in suburban areas	X			Sumasgutner et al. (2014a)
	Reproductive sites in inner city are associated with lower hatching rates and smaller fledged broods than suburban sites	X			Sumasgutner et al. (2014b)
	Similar reproductive success (clutch size, hatched young, and fledged young) in three areas of gradient of urbanization (city center, mixed zone, and outskirts)		X		Kübler et al. (2005)

(continued)

Table 8.1 (continued)

Species	Reproductive parameter	↑U, ↓RO	↑U, =RO	↑U, ↑RO	Reference
Lesser kestrel (*Falco naumanni*)	Clutch size and fledging success higher in rural than in urban colonies	X			Tella et al. (1996)
Mississippi kite (*Ictinia mississippiensis*)	Higher productivity in urban than in rural colonies			X	Parker (1996)
Red-shouldered hawk (*Buteo lineatus*)	Number of nestlings per active nest did not differ between suburban and rural areas		X		Dykstra et al. (2009)
Red-tailed hawk (*Buteo jamaicensis*)	Nests with high productivity were located in areas of roads and high-density urban land and nests with less productivity were in wetland land			X	Stout et al. (2006)
	Productivity did not differ among urban, suburban, and rural nest sites		X		Stout et al. (1998)

U urbanization, *RO* reproductive output, ↑ increase, ↓ decrease, = the parameter remains the same

and greater fledgling success in rural habitats than in urbanized habitats in Spain, even accounting for reduced food availability in cities and higher losses of nestlings due to starvation in urban colonies (Tella et al. 1996). Although the extent of urbanization surrounding black sparrowhawk (*Accipiter melanoleucus*) nests sites did not seem to have a strong influence on breeding performance, the seasonal declines in productivity differed according to levels of urbanization (i.e., highest productivity had been found in more urbanized habitats earlier in the season, but toward the end of season, breeding performance was better in less urbanized habitats) (Rose et al. 2017). A revision over the response, in reproductive terms, of raptor birds to gradients of urbanization is presented in Table 8.1.

Numerous factors can affect the reproductive success in urbanized landscapes. The way that raptors confront and exploit the resources or characteristics of these habitats will be reflected in their reproductive output. Studies of the breeding biology of raptors in urbanized habitats have cited and described the factors driving their breeding success as being nest site availability, food availability, proximity to green spaces, breeding density, and predation pressure.

Nest Site Availability

For those raptors that do not construct their own nests, the availability of suitable nest sites seems to be a limiting factor (Newton 1979), and in urban settings, buildings provide attractive breeding sites. Although historical centers of cities can offer numerous potential nest sites on buildings and raptors may be present in high densities, these factors do not necessarily mean high-quality habitat; therefore higher rates of nest failure, lower hatching rates, and smaller brood sizes at fledging can be expected for species nesting in urban centers (Sumasgutner et al. 2014a, b).

Diet, Food Abundance, and Prey Availability

Some raptors breeding along urban gradients change their prey preferences as a function of the degree of urbanization (Kübler et al. 2005). For Eurasian kestrels in Vienna, Austria, the diet of individuals breeding in urban centers was more diverse and generalist than the diet of their suburban and rural counterparts (Sumasgutner 2013). The abundance and availability of prey vary along urban gradients, in many cities around the world with many non-native prey species being very abundant in the center (e.g., domestic pigeons *Columba livia*, house sparrows *Passer domesticus*, or common starlings *Sturnus vulgaris* in American cities), and consequently more represented in the diet of individuals that breed in city centers (Haiman 2006; Cava et al. 2012; Sumasgutner et al. 2014a).

Proximity to Open Green Spaces, Parks, or Forest Remnant Inner Cities

Green spaces are very important in cities because these areas have high biodiversity and provide breeding sites to species, especially if these spaces have a high percentage of native plant species (McKinney 2002). These nodes do need to be connected by backyards or other green corridors to ensure dispersal and to serve as valuable foraging sites (Rudd et al. 2002). These habitats can affect the reproductive output, as was demonstrated by Sumasgutner et al. (2014a) who found that individuals breeding close to green backyards had earlier clutch dates, higher hatching rates, and larger fledged brood sizes.

Breeding Density

In social and colonial species, the distance to the nearest neighbor (nearest neighbor distance (NND)) is a measure that depicts the spatial distribution of nests and is widely used to evaluate birds' density (Newton 1998).

Several studies have shown that although individuals of the same species breed more densely (i.e., lower NNDs) in urban core, these sites are associated with higher nest failure, greater predation, increased nest desertion, lower hatching rates, smaller broods at fledging, and greater mortality by diseases (Boal and Mannan 1999; Sumasgutner et al. 2014b). For the social raptor chimango caracara (*Milvago chimango*), smaller NNDs in a suburban colony were associated with smaller clutch and brood sizes (Solaro and Sarasola 2015). In these cases, cities seem to act as ecological traps, habitats that appear favorable for some particular features (i.e., high nest sites availability) but are actually unsuitable or of lower quality (Schlaepfer et al. 2002).

Breeding density can be an important factor of protection from nest competitors too. Black sparrowhawk and Egyptian geese (*Alopochen aegyptiaca*) have recently expanded their South African ranges, and both species breed in urban and suburban habitats. A multiple nest building strategy has been an effective method used by black sparrowhawks to avoid negative interactions (i.e., nest usurpation) with Egyptian geese; thus breeding more densely was the used strategy to maintain productivity in this population of raptor species (Sumasgutner et al. 2016).

Predation Pressure

Human development can displace native predators in urbanized habitats. If prey species are less sensible to human perturbation than their predators, they may be favored by a "predation release" in these habitats (Muhly et al. 2011). For raptors, differential predation risk between urban and rural habitats can affect the breeding habitat selection. Although rural and natural habitats were much more extended, urban habitats were positively selected by burrowing owls in Argentina during the reproductive season due to very low richness and abundance of predators in these habitats, which can be reflected in the high reproductive outputs of urban populations (Rebolo-Ifrán et al. 2017). High productivity has been associated with low predation pressure in urban habitats for Mississippi kite and lesser kestrel too (Parker 1996; Tella et al. 1996). In this sense, cities have been suggested as key conservation hotspots to those species with threatened native habitats but that are able to live in high human-modified habitats (Rebolo-Ifrán et al. 2017).

Fear of Humans and Habitat Selection

Urbanization could be understood also as the process of a species entering urban habitats in which individuals exploit ecological and behavioral features that allow them to adapt to an urban existence. Among these behavioral features, raptors need to accommodate their fear of humans. For practical purposes, this behavioral trait can be quantified by the flight initiation distance (FID), which is the distance between an approaching threat (i.e., human) and a perched bird just as it takes flight (Blumstein 2006). Many studies of fear of humans have highlighted the differences in this behavior when comparing birds living in urban and rural habitats. In general, urban species are less fearful of humans than rural species (Møller 2010) and within species, individuals in urban populations have FIDs consistently lower than individuals in rural populations (Cooke 1980; Gliwicz et al. 1994; Møller 2008; Lin et al. 2012; Díaz et al. 2013).

In birds, inclusion of a species in urban habitats is usually the result of a complex selective process operating on the behavioral variability of populations. Originally, species with more variable FID are more successful in colonizing cities, and then over time, FIDs become reduced and more homogeneous among urban populations. We would expect that only species with a specific subset of behaviors (lower FIDs) are to be represented in urban populations (Møller 2010). Low FIDs in urban birds indicate a reduced perception of predation risk and allow closer approach to potential resources (Sol et al. 2011).

Studies of fear of humans shown by raptors are limited (Carrete and Tella 2010, 2011, 2013; Díaz et al. 2013; Rebolo-Ifrán et al. 2015; Carrete et al. 2016; Cavalli et al. 2016) and are aimed to clarify why and how some species are able to adapt to urbanized environments. In general, these authors have found that urban raptors have significantly shorter mean FIDs than what is found for rural populations in a variety of raptor species around the world.

Studies measuring FID in environments with different degrees of human disturbance have concluded that low mean FIDs in local population of more human disturbed habitats can be attributed to habituation process (Blumstein et al. 2003; Martínez-Abraín et al. 2008; Rodriguez-Prieto et al. 2009; Samia et al. 2015), but these studies have been criticized in that they focused on central tendency measures rather than on individual variation of FID (Carrete and Tella 2010; Lin et al. 2012). Repeatability of individual FID values in burrowing owls has been measured in short-term studies and throughout an adult's life span, and notable individual consistency in this trait has been found (Carrete and Tella 2010, 2013) suggesting that habituation is not occurring in this trait. Individual variability in FID measures and not the population mean value explains the uneven occupation of human-altered territories by individuals with different tolerances to human disturbance, i.e., precluding more frightened individual to colonize more humanized environments. This nonrandom habitat selection hypothesis would be supported by a pattern of more tolerant individuals ("tame" individuals) occupying more disturbed environments rather than by individuals becoming more habituated (Carrete and

Tella 2010, 2011, 2013). Selection acting on this behavioral trait can have important ecological and evolutionary consequences, i.e., if less fearful individuals are positively selected in disturbed environments (Carrete and Tella 2010), then these populations would result in more explorative and aggressive individuals (Evans et al. 2010). This process would change the dynamics among selected individuals and should be considered as an important but unappreciated ecological and evolutionary force (Carrete et al. 2016).

Cavalli et al. (2016) found that FID is significantly lower for urban burrowing owls than for rural owls and that urban individuals show more aggressiveness than rural individuals, indicating that urban environments would select for individuals that are less fearful and more aggressive toward potential predators than individuals in rural environments. Moreover, urban individuals would recognize different predators (human with a dog or human alone) and act appropriately with a differential aggressiveness level toward potential (human) or real (dog) predators, while rural individuals would perceive both human and dogs as real predators and react by flying away earlier (high FID) to avoid aggressive encounters (low aggressiveness) and non-necessary energetic costs.

Heritability of the behavioral trait "fear of humans" has been studied in burrowing owl in order to assess whether FID has a heritable component. Although in their study Carrete et al. (2016) have not found any statistical differences between rural and urban individuals, they have found a high heritability in FID compared with other antipredator behaviors. This is another reason to consider pressures of selection on this behavioral trait as an important evolutionary force in animal populations that are exposed to human disturbance (Fig. 8.1).

In rural or natural habitats, fear of human positively correlates with other behavioral traits as antipredatory and exploratory behaviors, i.e., individuals with larger FID when facing human shows longer latencies in approaching a predator and a new food item. However, in a context as urban habitats in which there is a low predation pressure, antipredatory behavior would be selected against, thus breaking or dismantling the correlation between FID and antipredatory behavior. In this sense, Carrete and Tella (2017) proposed that changes in selection pressures acting in urban areas can change the relation between behavioral traits, maintaining only those that are adaptive in these environments (e.g., FID exploration but not FID-antipredatory behavior).

Behavioral Flexibility

When an animal is exposed to a novel environment, behavioral flexibility can be an advantage to respond more rapidly to these changes. What is behavioral flexibility in urbanized environments? Behavioral flexibility implies learning, cognition, and rapid adjustment to new conditions, which allows animals to exploit a wide variety

Fig. 8.1 Burrowing owls and chimango caracaras are two common raptors usually found in urban and suburban areas in cities of southern South America. (**a**) Brood of burrowing owl in its nest sited few meters from a house. (**b**) Two chimango caracaras, an adult and a fledgling, drink water of an unusual water source, a pool of a house in an urbanized area. (Photo credit: (**a**) J. Hernán Sarasola, (**b**) Claudina Solaro)

Fig. 8.1 (continued)

of resources in different ecological contexts. Then, species that readily exploit a new food resource are preadapted for novel environments in contrast with more specialized species.

Behavioral flexibility has been hypothesized to be advantageous for surviving and reproducing in novel environments (Sol and Lefebvre 2000; Sol et al. 2002; Møller 2009). This flexibility has been associated with the relative size of the brain. Species with a large forebrain (relative to body mass) have behaviors that are more flexible and are more successful for establishing in novel environments. Large brains confer cognitive advantages that allow responses to novel conditions through enhancing their innovation propensity (Sol et al. 2005).

Sol et al. (2011) proposed that common mynas (*Acridotheres tristis*) construct novel behaviors to facilitate colonization of urban habitats in which man-made resources represent much of their diet and individuals need to innovate in order to be able to consume them. They showed that urban individuals had less aversion to approaching novel objects (neophobia), lower risk perception (lower FIDs), and greater exploration (measured as the physic exploration of an apparatus used for the experimental procedure) than rural individuals. These three traits drove urban individual to approach, explore, innovate, and consume the novel resource instead of avoiding it.

Studies on non-captive raptors over their behavioral flexibility in the form of neophobia, exploration, and learning have started in recent years. Beissinger and collaborators (1994) have studied diet specialization in snail kites (*Rostrhamus sociabilis*) in an experiment with wild free-flying birds and found that this species showed risk-averse foraging and neophobia toward eating non-habitual prey and thus maintained their strong specialization in feeding almost exclusively on *Pomacea* snails. This is an example of a species that, considering its specialized diet, would be unsuccessful in urbanized habitats.

In the last decade, six experimental works over behavioral flexibility have been developed in Argentina with wild individuals of the same species: the chimango caracara, a very common raptor in southern South America that lives and breeds in a wide variety of environments including urbanized habitats (Bellocq et al. 2008; Pedrana et al. 2008; Carrete et al. 2009; Solaro and Sarasola 2015, 2017). To clarify the characteristics that enable the ecological success of this species, researchers have hypothesized on neophobia, learning, exploration, and problem-solving capability of urbanized chimango caracaras but have not examined non-urban populations. Juvenile chimango caracaras have been found to have a remarkable ability to obtain food in novel situations and an ability for individual learning (Biondi et al. 2008). Age differentiation in exploration, neophobia, and problem-solving ability were found, which showed juveniles to have a higher explorative tendency, lower neophobia, and greater ability to innovate than adult individuals (Biondi et al. 2010a, 2013, 2015). Chimango caracaras are able to solve a food-related problem by observing the behavior of a conspecific, and this new behavior persists through several days, which shows an adaptive advantage considering the gregarious habits of this species (Biondi et al. 2010b). Guido et al. (2017) studied neophobia and reversal learning in chimango caracara and found that individuals were able to respond to a stimulus-reward association, but reverting to any previously learned association is a harder task. Urban chimango caracaras have been found to cope with novel features of their surroundings (Fig. 8.1). Although similar studies of individuals across a rural-urban gradient have not yet taken place, the studies cited above propose that generalist habits and low neophobia, high exploration, and learning can be critical features to discover and exploit new resource opportunities in modified habitats as developed land or cities.

Threats in Urban Landscapes

Urbanized habitats offer opportunity for those species that can successfully confront novel features and exploit new resources; however, these urban habitats also offer a variety of threats to urban birds that may affect their survival. Increased demands for resources and socioeconomic changes in the modern societies have driven the development of infrastructures for transportation and energy. These infrastructures have shifted mortality factors affecting raptors (Donázar et al. 2016). Current threats to urban birds include collisions with structures and vehicles, intoxication, diseases, and electrocution (Marra et al. 2004; Bradley and Altizer 2007; Hager 2009; Hager and Craig 2014; Cusa et al. 2015). In urbanized environments, birds have died more often from human-related causes than birds in rural environments.

In the United States and Canada, the primary sources of mortality for urban raptors were vehicle collisions (>60% of urban raptors), windows strikes, and electrocutions (Hager 2009). The proportion of raptor casualties from window collisions is similar for both urban and non-urban raptors, suggesting that urban raptors, mainly owls, are not especially vulnerable to this source of mortality. In any case, window collisions affect hawks and falcons more than other raptors (Hager 2009). Collisions have been a major source of mortality for post-fledgling urban raptors, a factor which should be considered in population studies undertaken to estimate survival, mortality, and migration flow.

The major causes of mortality for bald eagles (*Haliaeetus leucocephalus*) in suburban habitats of West Central Florida, USA, were electrocution, vehicle collisions, secondary poisoning from predator control efforts, and disease (Millsap et al. 2004). Poisoning in bald eagle may have been caused by eagles foraging on carcasses of euthanized pets in landfills, a problem resolved by requiring burial of poisoned carcasses. Chlamydial infection in suburban bald eagles has been caused apparently by their interaction with monk parakeets (*Myiopsitta monachus*) because eagle nests are frequently used also by monk parakeets. Monk parakeets have increased only in urban and suburban areas in Florida, and interactions between eagles and parakeets are stronger in suburban than in rural environments. Among urban nesting Cooper's hawks in Arizona, 85% of nestlings tested positive for (and 79.9% of nestlings died due to) trichomoniasis, an avian disease caused by the parasitic protozoan *Trichomonas gallinae* (Boal et al. 1998; Boal and Mannan 1999). In Victoria, British Columbia, *T. gallinae* affected, but did not kill, three nestling Cooper's hawks (Rosenfield et al. 2002). It seems to be a lower prevalence of this parasite in the northern portions of this hawk's breeding range (Rosenfield et al. 2009). The spread of trichomoniasis may be due primarily to the Inca dove (*Columbina inca*), an urban obligate that composes significant part of the diet of urban Cooper's hawks. However, the prevalence of this disease is null or very low among rural nestlings (Boal et al. 1998; Boal and Mannan 1999), which confirms the problem of urban coexistence of parasite hosts (that in general are very abundant) and raptors that can be infected by a variety of diseases. West Nile virus is a mosquito-borne infection with birds being most commonly infected and serving as prime hosts. This virus has been detected among a large number of raptors (Marra et al. 2004; Nemeth et al. 2007; Dusek et al. 2010; Quaglia et al. 2014). Urban and suburban settings provide numerous oviposition sites for mosquitoes. These environments can be important sites to maintain West Nile virus. Several species of urban or suburban raptors positive to West Nile virus have been detected in southeastern Wisconsin, USA (great horned owl *Bubo virginianus*, red-tailed hawk *Buteo jamaicensis*, and Cooper's hawk, Stout et al. 2005) and in Córdoba and Tucumán Argentine (rufous-thighed hawk *Accipiter erythronemius* and American kestrel *Falco sparverius*, Diaz et al. 2011).

Vehicle collision is a major cause of mortality for urban raptors (Donázar et al. 2016). For burrowing owls in an urbanized area of Cape Coral, Florida, vehicle collision was the cause of death for 70% of owl mortality; the same is true for urban bald eagles in Florida (Millsap 2002; Millsap et al. 2004) and Cooper's hawks in Arizona (Boal and Mannan 1999). Nevertheless authors could not envision any viable method in reducing the incidence of collisions with vehicles. While one death by collision was reported, this threat did not seem to be a big problem for Cooper's hawks in an urban environment in Indiana, USA (Roth et al. 2005).

Deaths from collision and electrocution with power lines are an important threat to raptors due to their behavior and size (Rubolini et al. 2005, in this book Chap. 12). Electrocution have been considered as an important source of mortality for raptors breeding in urbanized areas of Arizona, USA (Dwyer 2004). Although collisions are difficult to detect since birds may only be injured and survive to the incident, the potential for collisions with power lines increases with the development of electrical distribution infrastructure. Collision and electrocution risks are higher at specific sites with great development of poles and power lines. In cities, these sites are relatively easy to identify, and retrofitting lethal electric poles can reduce the number of electrocution incidents (Dwyer and William Mannan 2007, in this book Chap. 12). Despite efforts to design mitigation measures against these threats, this problem is not yet resolved (Dwyer 2004).

References

Beissinger S, Donnay T, Walton R (1994) Experimental analysis of diet specialization in the snail kite: the role of behavioral conservatism. Oecologia (100):54–65. https://doi.org/10.1007/BF00317130

Bellocq M, Filloy J, Garaffa P (2008) Influence of agricultural intensity and urbanization on the abundance of the raptor chimango caracara (*Milvago chimango*) in the Pampean region of Argentina. Ann Zool Fenn 45:128–134

Berry M, Bock C, Haire S (1998) Abundance of diurnal raptors on open space grasslands in an urbanized landscape. Condor 100:601–608. Available at: http://hairelab.com/files/Berry-etal1998.pdf [Verified 9 May 2014]

Biondi L, Bó M, Vassallo A (2008) Experimental assessment of problem solving by *Milvago chimango* (Aves: Falconiformes). J Ethol 26:113–118. https://doi.org/10.1007/s10164-007-0035-2

Biondi L, Bó M, Vassallo A (2010a) Inter-individual and age differences in exploration, neophobia and problem-solving ability in a Neotropical raptor (*Milvago chimango*). Anim Cogn 13:701–710. https://doi.org/10.1007/s10071-010-0319-8

Biondi L, García G, Bó M, Vassallo A (2010b) Social learning in the Caracara Chimango, *Milvago chimango* (Aves: Falconiformes): an age comparison. Ethology 116:1–14. https://doi.org/10.1111/j.1439-0310.2010.01794.x

Biondi L, Guido J, Madrid E, Bó M, Vassallo A (2013) The effect of age and sex on object exploration and manipulative behavior in a neotropical raptor, the Chimango Caracara, *Milvago chimango*. Ethology 119:221–232. https://doi.org/10.1111/eth.12056

Biondi L, Guido J, Bó M, Muzio R, Vassallo A (2015) The role of stimulus complexity, age and experience in the expression of exploratory behaviour in the Chimango Caracara, *Milvago chimango*. Anim Cogn 18:139–150. https://doi.org/10.1007/s10071-014-0785-5

Blair RB (1996) Land use and avian species diversity along an urban gradient. Ecol Appl 6:506–519

Blumstein D (2006) Developing an evolutionary ecology of fear: how life history and natural history traits affect disturbance tolerance in birds. Anim Behav 71:389–399. https://doi.org/10.1016/j.anbehav.2005.05.010

Blumstein DT, Anthony LL, Harcourt R, Ross G (2003) Testing a key assumption of wildlife buffer zones: is flight initiation distance a species-specific trait? Biol Conserv 110:97–100. https://doi.org/10.1016/S0006-3207(02)00180-5

Boal CW, Mannan RW (1999) Comparative breeding ecology of Cooper's hawks in urban and exurban areas of southeastern Arizona. J Wildl Manag 63:77. https://doi.org/10.2307/3802488

Boal CW, Mannan RW, Hudelson KS (1998) Trichomoniasis in Cooper's hawks from Arizona. J Wildl Dis 34:590–593. https://doi.org/10.7589/0090-3558-34.3.590

Bond G, Burnside NG, Metcalfe DJ, Scott DM, Blamire J (2005) The effects of land-use and landscape structure on barn owl (*Tyto alba*) breeding success in southern England, U.K. Landsc Ecol 20:555–566. https://doi.org/10.1007/s10980-004-5037-7

Botelho E, Arrowood P (1996) Nesting success of western burrowing owls in natural and human-altered environments. In: Bird CD, Verlan DE, Negro JJ (eds) Raptors in human landscapes: adaptations to built and cultivated environments. Academic Press Limited, New York, pp 61–68

Bradley CA, Altizer S (2007) Urbanization and the ecology of wildlife diseases. Trends Ecol Evol 22:95–102. https://doi.org/10.1016/j.tree.2006.11.001

Carrete M, Tella J (2010) Individual consistency in flight initiation distances in burrowing owls: a new hypothesis on disturbance-induced habitat selection. Biol Lett 6:167–170. https://doi.org/10.1098/rsbl.2009.0739

Carrete M, Tella J (2011) Inter-individual variability in fear of humans and relative brain size of the species are related to contemporary urban invasion in birds. PLoS One 6:e18859. https://doi.org/10.1371/journal.pone.0018859

Carrete M, Tella J (2013) High individual consistency in fear of humans throughout the adult lifespan of rural and urban burrowing owls. Sci Rep 3:3524. https://doi.org/10.1038/srep03524

Carrete M, Tella JL (2017) Behavioral correlations associated with fear of humans differ between rural and urban burrowing owls. Front Ecol Evol 5. https://doi.org/10.3389/fevo.2017.00054

Carrete M, Tella J, Blanco G, Bertellotti M (2009) Effects of habitat degradation on the abundance, richness and diversity of raptors across Neotropical biomes. Biol Conserv 142:2002–2011. https://doi.org/10.1016/j.biocon.2009.02.012

Carrete M, Martínez-Padilla J, Rodríguez-Martínez S, Rebolo-Ifrán N, Palma A, Tella JL (2016) Heritability of fear of humans in urban and rural populations of a bird species. Sci Rep 6:31060. https://doi.org/10.1038/srep31060

Cava JA, Stewart AC, Rosenfield RN (2012) Introduced species dominate the diet of breeding urban Cooper's hawks in British Columbia. Wilson J Ornithol 124:775–782. https://doi.org/10.2307/23324528

Cavalli M, Baladrón AV, Isacch JP, Biondi LM, Bó MS (2016) Differential risk perception of rural and urban burrowing owls exposed to humans and dogs. Behav Process 124:60–65. https://doi.org/10.1016/j.beproc.2015.12.006

Chace J, Walsh J (2006) Urban effects on native avifauna: a review. Landsc Urban Plan 74:46–69. https://doi.org/10.1016/j.landurbplan.2004.08.007

Cooke AS (1980) Observations on how close certain passerine species will tolerate an approaching human in rural and suburban areas. Biol Conserv 18:85–88. https://doi.org/10.1016/0006-3207(80)90072-5

Cusa M, Jackson DA, Mesure M (2015) Window collisions by migratory bird species: urban geographical patterns and habitat associations. Urban Ecosys 18:1427–1446. https://doi.org/10.1007/s11252-015-0459-3

Diaz L, Quaglia A, Flores F, Contigiani M (2011) Virus West Nile en Argentina: un agente infeccioso. Hornero 26:5–28

Díaz M, Møller AP, Flensted-Jensen E, Grim T, Ibáñez-Álamo JD, Jokimäki J, Markó G, Tryjanowski P (2013) The geography of fear: a latitudinal gradient in anti-predator escape distances of birds across Europe. PLoS One 8. https://doi.org/10.1371/journal.pone.0064634

Donázar JA, Cortés-Vizanda A, Fargallo JA, Margalida A, Moleón M, Morales-Reyes Z, Moreno-Opo R, Pérez-Garcia JM, Sanchez-Zapata JA, Zuberogoitia I, Serrano D (2016) Roles of raptors in a changing world: from flagships to providers of key ecosystem services. Ardeola 63:181–234. https://doi.org/10.13157/arla.63.1.2016.rp8

Dusek RJ, Iko WM, Hofmeister EK (2010) Occurrence of West Nile virus infection in raptors at the Salton Sea, California. J Wildl Dis 46:889–895. https://doi.org/10.7589/0090-3558-46.3.889

Dwyer JF (2004) Investigating and mitigating raptor electrocution in an urban environment. MSc Thesis, University of Arizona, Tucson

Dwyer JF, William Mannan R (2007) Preventing raptor electrocutions in an urban environment. J Rap Res 41:259–267. https://doi.org/10.3356/0892-1016(2007)41[259:PREIAU]2.0.CO;2

Dykstra CR, Hays JL, Simon MM (2009) Spatial and temporal variation in reproductive rates of the red-shouldered hawk in suburban and rural Ohio. Condor 111:177–182. https://doi.org/10.1525/cond.2009.080002

Evans J, Boudreau K, Hyman J (2010) Behavioural syndromes in urban and rural populations of song sparrows. Ethology 116:588–595. https://doi.org/10.1111/j.1439-0310.2010.01771.x

Ferguson HL (2004) Winter raptor composition, abundance and distribution around urban Spokane, eastern Washington. In: Shaw WW, Harris LK, VanDruff L (eds) 4th international symposium on urban wildlife conservation. University of Arizona, Tucson, p 368

Fraser JD, Frenzel LD, Mathisen JE (1985) The impact of human activities on breeding bald eagles in north-Central Minnesota. J Mammal 49:585–592

Gehlbach F (1996) Eastern screech owls in suburbia: a model of raptor urbanization. In: Raptors in human landscapes adaptations to built and cultivated environments. Academic Press, London, pp 69–74

Gliwicz J, Goszczynski J, Luniak M (1994) Characteristic features of animal populations under synurbanization - the case of the blackbird and of the striped field mouse. Memorabi Zool 49:237–244

Guido JM, Biondi LM, Vasallo AI, Muzio RN (2017) Neophobia is negatively related to reversal learning ability in females of a generalist bird of prey, the Chimango Caracara, *Milvago chimango*. Anim Cogn. https://doi.org/10.1007/s10071-017-1083-9

Hager SB (2009) Human-related threats to urban raptors. J Rap Res 43:210–226. https://doi.org/10.3356/JRR-08-63.1

Hager SB, Craig ME (2014) Bird-window collisions in the summer breeding season. PeerJ 2:e460. https://doi.org/10.7717/peerj.460

Haiman, A (2006). Prey selection of Cooper's hawks (*Accipiter cooperii*) nesting in urban areas of Berkeley and Albany, California , 1–13

Hogg JR, Nilon CH (2015) Habitat associations of birds of prey in urban business parks. Urban Ecosys 18:267–284. https://doi.org/10.1007/s11252-014-0394-8

Kauffman MJ, Pollock JF, Walton B (2004) Spatial structure, dispersal, and management of a recovering raptor population. Am Nat 164(5):582–597

Kübler S, Kupko S, Zeller U (2005) The kestrel (*Falco tinnunculus* L.) in berlin: investigation of breeding biology and feeding ecology. J Ornithol 146:271–278. https://doi.org/10.1007/s10336-005-0089-2

Lin T, Coppack T, Lin QX, Kulemeyer C, Schmidt A, Behm H, Luo T (2012) Does avian flight initiation distance indicate tolerance towards urban disturbance? Ecol Indic 15:30–35. https://doi.org/10.1016/j.ecolind.2011.09.018

Marra PP, Griffing S, Caffrey C, Kilpatrick MA, McLean R, Brand C, Saito E, Dupuis AP, Kramer L, Novak R (2004) West Nile virus and wildlife. Bioscience 54:393–402. https://doi.org/10.1641/0006-3568(2004)054[0393:WNVAW]2.0.CO;2

Martínez-Abraín A, Oro D, Conesa D, Jiménez J (2008) Compromise between seabird enjoyment and disturbance: the role of observed and observers. Environ Conserv 35:104–108. https://doi.org/10.1017/S0376892908004748

Marzluff JM (2017) A decadal review of urban ornithology and a prospectus for the future. Ibis 159:1–13. https://doi.org/10.1111/ibi.12430

Marzluff JM, Bowman R, Donnelly R (2001) Avian ecology and conservation in an urbanizing world. Kluwer Academic Publishers, Boston

McKinney M (2002) Urbanization, biodiversity, and conservation. Bioscience 52:883–890

McPherson SC, Brown M, Downs CT (2016) Crowned eagle nest sites in an urban landscape: requirements of a large eagle in the Durban metropolitan open space system. Landsc Urban Plan 146:43–50

Millsap BA (2002) Survival of Florida burrowing owls along an urban-development gradient. J Rap Res 36:3–10

Millsap BA, Bear C (2000) Density and reproduction of burrowing owls along an urban development gradient. J Wildl Manag 64:33–41. https://doi.org/10.2307/3802972

Millsap B, Breen T, McConnell E, Steffer T, Phillips L, Douglass N, Taylor S (2004) Comparative fecundity and survival of bald eagles fledged from suburban and rural natal areas in Florida. J Wildl Manag 68:1018–1031. https://doi.org/10.2193/0022-541X(2004)068[1018:CFASOB]2.0.CO;2

Møller AP (2008) Flight distance of urban birds, predation, and selection for urban life. Behav Ecol Sociobiol 63:63–75. https://doi.org/10.1007/s00265-008-0636-y

Møller A (2009) Successful city dwellers: a comparative study of the ecological characteristics of urban birds in the western Palearctic. Oecologia 159:849–858

Møller AP (2010) Interspecific variation in fear responses predicts urbanization in birds. Behav Ecol 21:365–371. https://doi.org/10.1093/beheco/arp199

Muhly TB, Semeniuk C, Massolo A, Hickman L, Musiani M (2011) Human activity helps prey win the predator-prey space race. PLoS One 6:1–8. https://doi.org/10.1371/journal.pone.0017050

Nemeth N, Kratz G, Edwards E, Scherpelz J, Bowen R, Komar N (2007) Surveillance for West Nile virus in clinic-admitted raptors, Colorado. Emerg Infect Dis 13:305–307. https://doi.org/10.3201/eid1302.051626

Newton, I (1979). 'Population ecology of raptors' Ed I. Newton. (T & AD Poyser: London.) Available at: http://books.google.com/books?hl=en&lr=&id=J39XMu7ecjAC&oi=fnd&pg=PP2&dq=Population+ecology+of+raptors&ots=XZ84xVGWX_&sig=eHMuBrUe2lVOPumBJHVN3gZA32Y [Verified 25 July 2014]

Newton I (1998) In: Newton I (ed) Population limitation in birds. Academic Press Limited, San Diego

Parker JW (1996) Urban ecology of the Mississippi kite. In: Bird CD, Verlan DE, Negro JJ (eds) Raptors in human landscapes. Adaptation to built and cultivate environments. New York, Academic Press Limited, pp 45–52

Pedrana J, Isacch J, Bó M (2008) Habitat relationships of diurnal raptors at local and landscape scales in southern temperate grasslands of Argentina. Emu 108:301–310

Quaglia AI, Diaz LA, Argibay H, Contigiani MS, Saggese MD (2014) West Nile and St. Louis encephalitis viruses antibodies surveillance in captive and free-ranging birds of prey from Argentina. EcoHealth 11:603–609. https://doi.org/10.1007/s10393-014-0956-5

Rebolo-Ifrán N, Carrete M, Sanz-Aguilar A, Rodríguez-Martínez S, Cabezas S, Marchant TA, Bortolotti GR, Tella JL (2015) Links between fear of humans, stress and survival support a non-random distribution of birds among urban and rural habitats. Sci Rep 5:13723. https://doi.org/10.1038/srep13723

Rebolo-Ifrán N, Tella JL, Carrete M (2017) Urban conservation hotspots: predation release allows the grassland-specialist burrowing owl to perform better in the city. Sci Rep 7:1–9. https://doi.org/10.1038/s41598-017-03853-z

Rodriguez-Prieto I, Fernández-Juricic E, Martín J, Regis Y (2009) Antipredator behavior in blackbirds: habituation complements risk allocation. Behav Ecol 20:371–377. https://doi.org/10.1093/beheco/arn151

Rose S, Sumasgutner P, Koeslag A, Amar A (2017) Does seasonal decline in breeding performance differ for an African raptor across an urbanization gradient? Front Ecol Evol 5:1–9. https://doi.org/10.3389/fevo.2017.00047

Rosenfield RN, Bielefeldt J, Affeldt JL, Beckmann DJ (1995) Nesting density, nest area reoccupancy, and monitoring implications for Cooper's hawks in Wisconsin. J Rap Res 29:1–4

Rosenfield RN, Bielefeldt J, Rosenfield LJ, Taft SJ, Murphy RK, Stewart AC, Rosenfield RN, Bielefeldt J, Rosenfield LJ, Taft SJ, Murphy RK, Stewart AC (2002) Prevalence of *Trichomonas gallinae* in nestling Cooper' s hawks among three north American populations. Wilson Bull 114:145–147

Rosenfield RN, Taft SJ, Stout WE, Driscoll TG, Evans DL, Bozek M a (2009) Low prevalence of *Trichomonas gallinae* in urban and migratory Cooper's hawks in northcentral North America. Wilson J Ornithol 121:641–644. https://doi.org/10.1676/08-148.1

Roth TC, Lima SL, Vetter WE (2005) Survival and causes of mortality in wintering sharp-shinned hawks and Cooper's hawks. Wilson Bull 117:237–244. https://doi.org/10.1676/04-103.1

Rubolini D, Gustin M, Bogliani G, Garavaglia R (2005) Birds and powerlines in Italy: an assessment. Bird Conserv Int 15:131–145. https://doi.org/10.1017/S0959270905000109

Rudd H, Vala J, Schaefer V (2002) Importance of backyard habitat in a comprehensive biodiversity conservation strategy: a connectivity analysis of urban green spaces. Restor Ecol 10:368–375. https://doi.org/10.1046/j.1526-100X.2002.02041.x

Rullman S, Marzluff JM (2014) Raptor presence along an urban – wildland gradient: influences of prey abundance and land cover. J Rap Res 48:257–272. https://doi.org/10.3356/JRR-13-32.1

Samia DSM, Nakagawa S, Nomura F, Rangel TF, Blumstein DT (2015) Increased tolerance to humans among disturbed wildlife. Nat Commun 6:8877. https://doi.org/10.1038/ncomms9877

Schlaepfer M, Runge M, Sherman P (2002) Ecological and evolutionary traps. Trends Ecol Evol 17:474–480. https://doi.org/10.1016/S0169-5347(02)02580-6

Sol D, Lefebvre L (2000) Behavioural flexibility predicts invasion success in birds introduced to New Zealand. Oikos 90:599–605. Available at: http://onlinelibrary.wiley.com/doi/10.1034/j.1600-0706.2000.900317.x/full [Verified 27 February 2014]

Sol D, Timmermans S, Lefebvre L (2002) Behavioural flexibility and invasion success in birds. Anim Behav 63:495–502. https://doi.org/10.1006/anbe.2001.1953

Sol D, Duncan R, Blackburn T, Cassey P, Lefebvre L (2005) Big brains, enhanced cognition, and response of birds to novel environments. Proc Natl Acad Sci 102:5460–5465. Available at: http://www.pnas.org/content/102/15/5460.short [Verified 27 February 2014]

Sol D, Griffin A, Bartomeus I, Boyce H (2011) Exploring or avoiding novel food resources? The novelty conflict in an invasive bird. PLoS One 6:e19535. https://doi.org/10.1371/journal.pone.0019535

Solaro C (2015) Ecología y comportamiento del chimango (*Milvago chimango*) en ambientes antropizados del centro de Argentina. PhD Thesis, Universidad Nacional del Comahue, Neuquén, Argentina

Solaro C, Sarasola JH (2015) Nest-spacing, not human presence, influences the breeding of Chimango caracaras (*Milvago chimango*) in a peri-urban reserve. Emu 115:72–75. https://doi.org/10.1071/MU14038

Solaro C, Sarasola JH (2017) Natal dispersal and philopatry of Chimango caracaras (*Milvago chimango*) in suburban, rural and natural habitats, determined by band recovery and re-sighting data. Emu 118:158–165. https://doi.org/10.1080/01584197.2017.1321961.

Stout W, Anderson RK, Papp JM (1998) Urban, suburban and rural red-tailed hawk nesting habitat and populations in Southeast Wisconsin. J Rap Res 32:221–228. https://doi.org/10.1080/00958970600794008

Stout WE, Cassini AG, Meece JK, Papp JM, Rosenfield RN, Reed KD (2005) Serologic evidence of West Nile virus infection in three wild raptor populations. Avian Dis 49:371–375. https://doi.org/10.1637/7335-012805R1.1

Stout W, Temple S, Papp J (2006) Landscape correlates of reproductive success for an urban-suburban red-tailed hawk population. J Wildl Manag 70:989–997

Sumasgutner P (2013) Diet specialisation and breeding success along an urban gradient: the kestrel (*Falco tinnunculus*) in Vienna, Austria. Beiträge Zur Jagd Und Wildforschung 38:385–397

Sumasgutner P, Schulze CH, Krenn HW, Gamauf A (2014a) Conservation related conflicts in nest-site selection of the Eurasian kestrel (*Falco tinnunculus*) and the distribution of its avian prey. Landsc Urban Plan 127:94–103. https://doi.org/10.1016/j.landurbplan.2014.03.009

Sumasgutner P, Nemeth E, Tebb G, Krenn H, Gamauf A (2014b) Hard times in the city – attractive nest sites but insufficient food supply lead to low reproduction rates in a bird of prey. Front Zool 11:48. https://doi.org/10.1186/1742-9994-11-48

Sumasgutner P, Millán J, Curtis O, Koelsag A, Amar A (2016) Is multiple nest building an adequate strategy to cope with inter-species nest usurpation? BMC Evol Biol 16:97. https://doi.org/10.1186/s12862-016-0671-7

Tella J, Hiraldo F, Donázar-Sancho J, Negro J (1996) Costs and benefits of urban nesting in the lesser kestrel. In: Bird CD, Verlan DE, Negro JJ (eds) Raptors in human landscapes: adaptations to built and cultivated environments. Academic Press Limited, New York, pp 53–60

Chapter 9
Birds of Prey in Agricultural Landscapes: The Role of Agriculture Expansion and Intensification

Juan Manuel Grande, Paula Maiten Orozco-Valor, María Soledad Liébana, and José Hernán Sarasola

Introduction

The introduction of agricultural activity (including livestock farming) in natural environments involves the replacement of the original system by an assemblage of crops, livestock, associated pests, fences, machinery, several types of installations, and the entire process network that these components determine. Therefore, it affects all the natural processes, from individual behavior and population dynamics to communities' composition and flows of matter and energy and from local to landscape scales (e.g., Ghersa and Martínez-Ghersa 1991; Ojima et al. 1994; McLaughlin and Mineau 1995; Matson et al. 1997; Tella et al. 1998; Marshall et al. 2003; Morelli 2013).

Birds, as any other element of the system, are affected by these changes in different ways depending on particularities of each species such as their residence status, their different trophic requirements, and their degree of specialization in the use of nesting habitats and on the severity of changes in the system (Vickery et al. 1999; Robinson and Sutherland 2002; Fox 2004).

According to social, political, and economic characteristics of a region, changes in land use imposed in modern agroecosystems can imply the total destruction/transformation of the original natural environments (e.g., through deforestation of native forest or through the plowing and cultivation of former natural grasslands or marshlands; Shine and Klemm 1999; Donald 2004; Zak et al. 2008; Graesser et al. 2015), the intensification of productive systems in areas already transformed (Foley et al. 2005; Tilman 1999), or low-intensity changes where part of the original ecosystem is maintained (Brown et al. 2006; Schmiegelow and Mönkönen 2002; Ojima

J. M. Grande (✉) · P. M. Orozco-Valor · M. S. Liébana · J. H. Sarasola
Centro para el Estudio y Conservación de las Aves Rapaces en Argentina (CECARA), Universidad Nacional de La Pampa (UNLPam) & Consejo Nacional de Investigaciones Científicas y Técnicas de Argentina (CONICET), La Pampa, Argentina

J. H. Sarasola et al. (eds.), *Birds of Prey*,
https://doi.org/10.1007/978-3-319-73745-4_9

et al. 1994; Foley et al. 2005). Because of their role as top predators, raptors are particularly sensitive to major ecosystem changes as those imposed by agricultural expansion and intensification and thus could behave as good bioindicators (Newton 1979; Sergio et al. 2005). Here we will summarize examples of negative and positive effects of the expansion and intensification of agriculture and livestock farming on diurnal birds of prey in the late twentieth and early twenty-first centuries.

On the negative side, agriculture and livestock farming expansion and intensification may completely destroy birds of prey habitats or impoverish them by reducing food availability, breeding resources, or intoxicating birds. On the positive side, these changes may provide the species with new habitats, novel food, new breeding resources or reduced competition. Some birds of prey may trigger conflicts with livestock farming through predation or more generally with changes in agricultural practices when they collide with raptors conservation (Fig. 9.1).

Whatever the changes are, birds of prey will respond to them; to understand these responses may allow reducing actual impacts on different species or to design and implement appropriated actions to improve their conservation status in the future.

Agriculture and Livestock Farming as Habitat Destructors

Alteration of natural systems to implement agricultural lands, both for crops and/or livestock farming, is certainly as old as those human activities. However, its global impact has sharply increased in the last three centuries (Meyer and Turner 1992) and keeps growing at accelerated rates (Tilman et al. 2001; Foley et al. 2005). Globally, the main forces behind habitat loss for raptors are deforestation, the clearing and transformation of open forests and scrubland, and the plowing of grassland and marshes.

Deforestation

In the last decades, forest cover has recovered in some areas, e.g., Western and Southern Europe or Eastern United States, but forest loss is still the norm worldwide (FAO 2016). About 3% of forest has been lost worldwide between 1990 and 2015. However, this loss hasn't been homogenous, and, for example, tropical forests have lost up to 10% of their surface in that 25-year period (FAO 2016).

Tropical and subtropical rainforest deforestation occurs through the world tropics, but almost 60% of tropical forests cleared between 2000 and 2005 were in South America with Brazil along clearing 47% of the lost forest. Thirty-seven percent of the lost forest corresponded to Southeastern Asia (with 9.16% of loss only in Indonesia) making this the second world area most affected by tropical forest loss (Hansen et al. 2008).

Fig. 9.1 Schematic main effects of agricultural and livestock farming practices on birds of prey. (**a**) Conflicts between production and birds of prey conservation, during the mid-1990s, the need to protect steppe birds like this lesser kestrel forced the temporal detention of irrigation plans in northeastern Spain, triggering social discomfort among farmers; (**b**) habitat destruction, the expansion of palm oil crops in Southeastern Asia is destroying forest habitats of species like this Wallace's hawk-eagle *Nisetus nanus*; (**c**) reductions of food availability, the abundance of small mammal prey for species like this Eurasian buzzard was reduced along a gradient of agricultural intensification in Western France; (**d**) reduction of breeding habitat, overgrazing and changes of grasslands to crops reduced breeding habitat for ground-nesting raptors in Argentina's Pampas, such as this long-winged harrier; (**e**) use of agrochemicals and veterinary pharmaceutical drugs may affect birds of prey survival, thousands of Swainson's hawks like this died intoxicated with a pesticide in Argentina in the mid-1990s; (**f**) expansion of electricity transmission lines associated with intensification of production systems, thousands of raptors die electrocuted yearly like these two juvenile variable hawk *Geranoaetus polyosoma* in central Argentina; (**g**) construction of water reservoir for livestock in areas with no or scarce superficial water, birds of prey like this variable hawk in central Argentina may drown in high numbers; (**h**) increases in food availability favored by changes in agricultural practices, the expansion of irrigated crops in Northwestern Spain allowed the expansion of common voles, an ideal prey for medium and small raptors such as this Eurasian kestrel; (**i**) creation of new breeding or roosting substrates, implantation of exotic tree stands in ranches and agricultural lands provided new breeding and roosting substrates for chimango caracaras in Argentinean Pampas; (**j**) creation of new habitats for raptors, the building of water channels and ponds for irrigated crops created new breeding habitats for marshland species such as this western marsh-harrier in northeastern Spain. (Photo credits: **a** Arnau Guardia; **b** Kim Chye; **c** and **h** Rafael Benjumea; **d** José H. Sarasola; **e** and **f** Juan M. Grande; **g** Maximiliano Galmes; **i** Claudina Solaro; **j** Manuel de la Riva)

In South America, the main drivers of deforestation are large-scale extensive ranching and slash-and-burn small-scale farming (both particularly in the Brazilian Amazon) and industrial agriculture for the production of soybean and sugarcane (mostly in Argentina, Paraguay and in Brazil south of the Amazon river, Soares-Filho et al. 2006; Hansen et al. 2008; Graesser et al. 2015, Fig. 9.2). Currently, industrial palm oil production seems to be the main driver of deforestation in

Fig. 9.2 Large habitat destruction for turning tropical and subtropical forest into pastures for livestock and crops is one of the main threats for tropical forest raptors. Crops cover nowadays most of the former Atlantic forests in northeastern Argentina and southeastern Brazil (**a**) making almost impossible the persistence of large eagles such as the harpy eagle (**b**). The palm oil industry is one of the main promoters of deforestation in Southeastern Asia (**c**), leading several raptors to critical conservation status. In the past decades, its expansion in the Philippines is threatening the remnants of mature forests where the Critically Endangered Philippine Eagle finds its last strongholds (**d**). (Photo credits: (**a**) Cornell Lab of Ornithology; (**b**) Cornell Lab of Ornithology/Kike Arnal; (**c**) Carlos De Angelo; (**d**) Jorge La Grotteria)

Southeastern Asia (Greenpeace 2013; Stibig et al. 2014, but see Abood et al. 2015, Fig. 9.2). Deforestation of tropical forests in Africa, although important in some areas, is lower, probably reflecting the absence of current agro-industrial-scale clearing there (Hansen et al. 2008; FAO 2016).

Tropical and subtropical forests are vast in South America, and most of the raptor species inhabiting them have wide distribution ranges, e.g., most Neotropical forest eagle and falcon species range from different parts of southern Mexico and Central America to Northern Argentina and Southern Brazil (Ferguson-Lees and Christie

2001; Marks et al. 1999). However, deforestation has been so massive in large areas (e.g., the Amazon basin lost 20% of its surface in Brazil since the 1970s, INPE 2016) that some of those wide-ranging species now have some degree of threat status at local, regional, national, or even international levels (e.g., MMA 2014; Gomes and Sanaiotti 2015; MA y DS and Aves Argentinas 2017).

Species like the harpy eagle (*Harpia harpyja*), the crested eagle (*Morphnus guianensis*), or the orange-breasted falcon (*Falco deiroleucus*), for example, have almost disappeared from the Upper Paranaense Forests in the Atlantic Forest of South America along with the forests itself (less than 8% of the forest remain, Di Bitetti et al. 2003, Fig. 9.2a, b), and are now considered as Near Threatened globally but Extinct or Endangered at national or regional levels (Vargas et al. 2006; MMA 2014; Gomes and Sanaiotti 2015; MA y DS and Aves Argentinas 2017). Besides, genetic studies suggest that habitat loss and fragmentation may cause important loss in genetic diversity, compromising the ability of remaining populations to cope with potential environmental changes in the future (Banhos et al. 2016).

If habitat deforestation has impacted species with large ranges, it is obvious that its impact would be higher in those species with more restricted ranges. In 50% of the 30 Endangered or Critically Endangered diurnal birds of prey for which the IUCN identifies agriculture or aquiculture as a threat, deforestation of tropical and subtropical forests to turn them into agricultural and livestock farming lands was one of the main threats behind their conservation status (Table 9.1).

Thanks to technological advances that allowed to cultivate in semiarid regions, agriculture, and not only cattle ranching, has expanded through deforestation to dry forests like the South American Gran Chaco (Graesser et al. 2015), the Cerrado of Brazil (Morton et al. 2006), or the Chiquitano forest of Bolivia (Graesser et al. 2015). Deforestation for soybean cultivation and extensive livestock farming has already cleared a fifth of the Chaco forests (Vallejos et al. 2015) and an even larger proportion of the espinal (70% of the original forests have been transformed mostly to agriculture, SA y DS 2004); this deforestation has been indicated as a serious threat explaining the retraction of the Endangered crowned solitary eagle (*Buteogallus coronatus*) in Argentina, Brazil, and Paraguay (Collar et al. 1992; Fandiño and Pautasso 2013).

Turning Other Natural Lands into Agriculture and Farming Systems

Outside forest environments, steppes, savannahs, scrublands, and marshlands have also suffered strong transformation pressures.

Steppe environments in Europe and Central Asia have suffered a strong large-scale transformation impact. Steppe habitats turned into agricultural lands showed lower species diversity and a marked scarcity of ground-nesting raptors in Kazakhstan (Sánchez-Zapata et al. 2003), and most steppe raptors are now extinct

Table 9.1 Diurnal bird of prey species Endangered (EN) or Critically Endangered (CR) according to the IUCN for which agriculture and aquaculture (livestock farming and ranching are included) have been identified as a threat (BirdLife International 2017)

Common names	Species	Family	IUCN status	Agricultural/ livestock-related threat
Mauritius kestrel	*Falco punctatus*	FAL	EN	1*,6
Saker falcon	*Falco cherrug*	FAL	EN	3,6,7
White-collared kite	*Leptodon forbesi*	ACC	CR	1*
Cuban kite	*Chondrohierax wilsonii*	ACC	CR	1*,4
Madagascar fish eagle	*Haliaeetus vociferoides*	ACC	CR	1*,2
Egyptian vulture	*Neophron percnopterus*	ACC	EN	4,5,6,7
Hooded vulture	*Necrosyrtes monachus*	ACC	CR	4,5
White-backed vulture	*Gyps africanus*	ACC	CR	1,4,5,6,7
White-rumped vulture	*Gyps bengalensis*	ACC	CR	5,6
Cape vulture	*Gyps coprotheres*	ACC	EN	4,5,6,7
Indian vulture	*Gyps indicus*	ACC	CR	5,6
Rüppell's vulture	*Gyps rueppelli*	ACC	CR	1,5
Slender-billed vulture	*Gyps tenuirostris*	ACC	CR	4,5,6
Lappet-faced vulture	*Torgos tracheliotos*	ACC	EN	4,5,7
White-headed vulture	*Trigonoceps occipitalis*	ACC	CR	4,5
Red-headed vulture	*Sarcogyps calvus*	ACC	CR	5,6
Madagascar serpent eagle	*Eutriorchis astur*	ACC	EN	1*
Madagascar marsh harrier	*Circus macrosceles*	ACC	EN	2,4
Reunion marsh harrier	*Circus maillardi*	ACC	EN	1*,4,6,7
Black harrier	*Circus maurus*	ACC	EN	3
Gundlach's hawk	*Accipiter gundlachi*	ACC	EN	1*
Grey-backed hawk	*Pseudastur occidentalis*	ACC	EN	1*
Crowned solitary eagle	*Buteogallus coronatus*	ACC	EN	1,4,7
Ridgway's hawk	*Buteo ridgwayi*	ACC	CR	1*,4,6,7
Philippine eagle	*Pithecophaga jefferyi*	ACC	CR	1*,4
Steppe eagle	*Aquila nipalensis*	ACC	EN	3,6,7
Javan hawk-eagle	*Nisaetus bartelsi*	ACC	EN	1*
Flores hawk-eagle	*Nisaetus floris*	ACC	CR	1*,4
North Philippine hawk-eagle	*Nisaetus philippensis*	ACC	EN	1*
South Philippine hawk-eagle	*Nisaetus pinskeri*	ACC	EN	1*
Black-and-chestnut eagle	*Spizaetus isidori*	ACC	EN	1*,4

from these transformed habitats in Western Europe (Baumgart 1994; Meyburg and Boesman 2013). Two steppe specialists with large breeding ranges, the saker falcon (*Falco cherrug*) and the steppe eagle (*Aquila nipalensis*), are now considered Endangered by the IUCN, the steppe transformation into agricultural lands being one of the main drivers of such status (Baumgart 1994; Meyburg and Boesman 2013, Table 9.1).

Similar situations occur in different areas of Africa where the expansion of agricultural lands at the expanse of natural habitats is strongly reducing habitat availability for most African resident species (Thiollay 2007, in this book Chap. 18).

In West Africa, typical woodland savannahs are being turned into different crop types (sorghum, millet, maize, beans, groundnut, and cotton) (Brink and Eva 2009). While some birds of prey cope better than others in these agricultural lands (Buij et al. 2013), there is a general trend to a reduction in the number and diversity of raptors in the area following extensive habitat transformation (Thiollay 2007). Large raptors such as vultures and eagles seem to be the most affected victims of these changes and now persist almost exclusively in protected areas (Thiollay 2007; Buij et al. 2013).

In South Africa, the Endangered black harrier (*Circus maurus*) has lost more than 50% of its preferred habitat in the shrubby Fynbos Biome, transformed to agricultural lands (Curtis et al. 2004).

Other environments traditionally considered as unproductive that suffered and still suffer strong pressure are marshlands (Shine and Klemm 1999). Marshlands turning into agricultural lands in different parts of the world have affected several marsh specialist birds of prey. The western marsh-harrier (*Circus aeruginosus*) population in Western Europe was a victim of these transformations. It suffered a large population decline in the 1960–1980 mainly because of marshland desiccation for farming along with other factors linked to agricultural intensification such as food reduction (BirdLife International 2004).

In the last decades, particularly worrying has been the transformation of wetlands into rice fields in Madagascar that according to some authors would now occupy 80% of the former marshlands. These transformations seem to be a leading factor explaining the critical situation of the Endangered Madagascar marsh-harrier (*Circus macrosceles*) and the Critically Endangered Madagascar fish eagle (*Haliaeetus vociferoides*) (De Roland et al. 2009; Razafimanjato et al. 2014, Table 9.1).

Extensive wetland transformation into rice fields (along with other habitat changes and overfishing) may have also impacted negatively piscivorous eagles on Sulawesi, as except the grey-headed fishing eagle (*Ichthyophaga ichthyaetus*), the other fishing eagles (ospreys *Pandion haliaetus*, white bellied sea eagle *Haliaeetus leucogaster*, and lesser fish eagle *Ichthyophaga humilis*) were found to be rare or very local (Thiollay and Rahman 2002).

Agricultural Intensification and Habitat Degradation

Changes in productive practices in already transformed habitats or secondary effects of habitat destruction in nearby natural areas may also reduce the quality of a determined habitat for birds of prey. For example, habitat destruction is usually not homogenous but spatially heterogeneous, with progressively less natural patches in a growing matrix of altered habitats (e.g., Renjifo 1999). This fragmentation process may increase border and isolation effects and increase human pressure on raptors. The presence of houses with chickens or other domestic animals may attract raptors increasing the chance of conflicts and persecution and therefore increasing the impact of deforestation on those raptor species far beyond the mere loss of forest surface (e.g., Vargas et al. 2006; Zuluaga and Echeverry-Galvis 2016).

Overgrazing, the elimination of woodlots, field margins, hedgerows, and isolated trees also reduce the quality of the environment and therefore are expected to reduce food and breeding site availability; the use of veterinary drugs or pesticides may, besides affecting prey availability, directly affect birds' health, while the construction of human infrastructures such as water reservoirs or power lines may reduce individuals' survival (Potter 1997; Tucker and Heath 1994; in this book Chaps. 10, 12, and 19).

Changes in Agricultural Practices

Intensification of farming practices in agriculture includes mechanization, implementation of irrigation systems, and the use of pesticides. These practices allowed expanding agriculture over grazing lands and the introduction of new crops into areas previously barely productive. Within farms, with the aim of maximizing efficiency and production, agricultural lots grew bigger with the aggregation of smaller lots, the plowing of seminatural and marginal habitats, and the clearing of woodlots, field margins, hedgerows, and trees (Tucker and Heath 1994; Potter 1997), thus reducing the availability of refuges, foraging habitats, and sites for resting or nesting for wildlife (Perkins et al. 2002; Bennett et al. 2006; Morelli 2013).

For several harrier species such as Montagu's harrier (*Circus pygargus*), western marsh-harrier, pallid harriers (*Circus macrourus*, globally Near Threatened), and probably also hen harriers (*Circus cyaneus*), changes in crops and farming practices and wetlands' desiccation reduced habitat quality and food resources, while mechanical harvesting operations also increased direct mortality of chicks, being directly related to their declines in many areas of Europe (Arroyo et al. 2003; Amar and Repath 2005; BirdLife International 2003; BirdLife International 2004).

Many small falcon species, such as the American kestrel (*Falco sparverius*), the Eurasian kestrel (*Falco tinnunculus*), or the lesser kestrel (*Falco naumanni*), do relatively well in traditional agroecosystems (White et al. 1994). However, agricultural intensification was directly linked to habitat loss and degradation that ultimately produced a population crash of the last species in Europe (Biber 1994).

As in other parts of Europe, in Spain, irrigation schemes reduced field margins, increased field sizes, and changed crops (Tella et al. 1998; De Frutos et al. 2015). Specific studies showed that in different areas of Spain, lesser kestrels were less abundant and had lower food availability in intensive cropping areas and consequently needed larger home ranges but still had lower breeding success than in traditionally farmed areas (Donázar et al. 1993; Tella et al. 1998; De Frutos et al. 2015). Radio tracking and focal monitoring proved that lesser kestrels indeed selected field margins and traditionally farmed fields and avoided hunting in intensive irrigated crops except when the crop was alfalfa (Tella et al. 1998; Ursúa et al. 2005). In the Guadalquivir Valley in southern Spain, food shortage was so marked in those intensively farmed areas that large fledgling mortalities were detected in some breeding colonies (Hiraldo et al. 1996).

The Eurasian kestrel, along with the Eurasian buzzards (*Buteo buteo*), also decreased across a gradient of agricultural intensification in three study areas in Western France, in parallel with a reduction in their prey (Butet et al. 2010).

The actual decrease in American kestrel numbers in the USA and Canada is still unexplained, but the effects of agricultural intensification are among the candidate factors under study (Smallwood et al. 2009 but see Smith et al. 2017).

Studies with larger raptor species have also shown that in some cases birds negatively selected intensive farmland, e.g., Spanish imperial eagles (*Aquila adalberti*) that usually forage on open Mediterranean scrubland and forests did not use irrigated crops (González et al. 1990; Ferrer 1993), while Swainson's hawks (*Buteo swainsoni*) wintering in Argentina used less frequently crops than pastureland (Canavelli et al. 2003; Filloy and Bellocq 2007) and did not use oleaginous crops such as soybean, the most intensively farmed crops in the area (Sarasola et al. 2008).

The Pampas grasslands of southern South America were devoted mostly to extensive grazing cattle for the last two centuries, while rotation of agriculture and livestock was restricted to areas with particularly rich soils. In the last two decades, industrial agriculture has expanded over traditional agricultural and grazing lands, and cattle farming is now margined to areas with poor soil, flooded or to feedlots (Aizen et al. 2009; Garcia et al. 2013). In Argentina alone, for 2017 the surface of crops covered 30 million hectares of which 20 million were of soybean (Aizen et al. 2009; Bolsa de Cereales de Buenos Aires 2017).

Several studies suggest a negative impact of this process on the raptor community (e.g., Filloy and Bellocq 2007; Carrete et al. 2009). Censuses by car in 2006–2008 found that from the early 1990s, the long-winged harrier (*Circus buffoni*), among other ground breeding species (ground breeding raptors were probably the only birds of prey in the treeless Pampas), disappeared from 10% ($n = 17$) of the Buenos Aires Province's counties, probably mediated by the transformation of seminatural grazing fields to industrial agriculture and the reduction of field margins (Codesido et al. 2011). A long-term monitoring of bird communities during the process showed that even a generalist predator widely distributed in the Pampas agricultural lands such as the southern caracara (*Caracara plancus*) decreased in number along with the intensification process, although another generalist species, the chimango caracara (*Milvago chimango*), seems to be favored by the same process (Gavier-Pizarro et al. 2012). The American kestrel, although traditionally linked to agroecosystems, was found to be more common in grazing lands with livestock rather than in agricultural lands in the area (Filloy and Bellocq 2007) but showed no clear pattern at large scale (Goijman et al. 2015).

Changes in Livestock Farming Practices

In the last decades, livestock production has changed globally. In some areas like South America, increases in grain production and their high market value expanded agriculture over former grazing lands. This led to the movement of extensive livestock production to more marginal areas at the expense of natural habitats such as forests (Graesser et al. 2015). In other areas of the continent as in Europe and some areas of the USA and Canada, there has been a progressive change from extensive to more controlled stabled farming and the use of feedlots, reducing density or directly eliminating extensive livestock production (along with all its ecological interactions) from vast areas (e.g., Donázar et al. 1997; Graesser et al. 2015).

Variations in livestock husbandry and management obviously alter habitat quality for scavengers through the supply of carrion. A general reduction in food availability along with changes in farming practices was alleged historically to be responsible of the drastic reduction of several vulture species in more developed countries (Donázar 1993; Liberatori and Penteriani 2001). Specific effects of these modifications were evident when variations in carcass availability during the bovine spongiform encephalopathy (BSE) crisis in the 2000s drastically diminished food availability for vultures in Spain affecting their demography and behavior (Donázar et al. 2009a, b). A recent study demonstrated that Spanish cinereous (*Aegypius monachus*) and griffon vultures (*Gyps fulvus*) do not cross the border between Spain and Portugal despite habitat is identical in both sides of the border and after considering livestock density effects. This result suggests that the differential removing of dead livestock from the field by differential application of European sanitary legislation across the two countries determines food availability for vultures (and thus their foraging behaviour, Arrondo et al. 2018).

Changes in livestock and carcass management, along with reductions in their natural prey for the expansion of agriculture and overhunting, are responsible for the demise of several vulture species to local and regional extinction in Southeastern Asia and western Africa (Rondeau and Thiollay 2004; Thiollay 2007; Hla et al. 2011). These modifications in management along with the particular cases of diclofenac use in the Indian subcontinent (see below) and some local conflicts originated by wildlife poaching are the main factors responsible of the critical situation of Old World vultures, so that from 16 species, 8 are now Critically Endangered (8 of 14 Critically Endangered raptors are vultures), 3 are Endangered, 2 are Near Threatened, and only 2 are of Least Concern (Table 9.1, in this book Chap. 19).

Besides the effects on vultures, variation in livestock farming practices also affect hunting raptors by affecting their food sources. Increases in livestock density in pasturelands intensify their impact on grasses and soil affecting potential preys of raptors and ultimately birds of prey populations (e.g., Torre et al. 2007). Examples of these effects have been detected in environments as different as Scottish, Spanish, and African grasslands.

Increases in sheep densities in Orkney Island in Scotland strongly reduced unmanaged grasslands, affecting the availability of voles and snipes for hunting male hen harriers and so reducing breeding habitat quality for the species (Amar and Repath 2005). In central Spain, cattle grazing limited grass availability and compacted the soil, producing a reduction in small mammal abundance and richness, thus affecting food availability for raptors (Torre et al. 2007).

In a complete different scenario, in Botswana, structural changes to vegetation caused by overgrazing by livestock outside protected areas seem to have depleted biomass and biodiversity of wildlife, reducing food availability for raptors and thus raptors' abundance (particularly of larger predator birds of prey) in unprotected lands (Herremans and Herremans-Tonnoeyr 2000).

Use of Pesticides and Veterinary Drugs in Modern Agricultural Production

The use of agrochemicals and veterinary drugs in agricultural production are among the most cited degrading factors linked to the agricultural intensification process. Agrochemicals are used to fertilize or control pests in crops (Tucker and Heath 1994; Potter 1997), while veterinary drugs (e.g., synthetic steroid hormones, antibiotics, adrenergic and anti-inflammatory drugs) are used for solving livestock health issues and to increase and optimize livestock growth (USDA National Animal Health Monitoring System 2000; Oaks et al. 2004; Landers et al. 2012; Samuelson et al. 2016).

The use of DDT and other "legacy" pesticides such as OCs had a protagonist role in the demise of many raptor species worldwide (e.g., Newton 1979, in this book Chap. 10). After decades of banning its use, low levels of DDT residuals are still found in Eurasian and North American birds of prey, and some relationship among blood parameters and DDE residuals suggests a pervasive presence of the chemicals that even at low concentrations has a pervasive effect (Ortiz-Santaliestra et al. 2015; Espín et al. 2018).

In Africa, DDT is still legally used to control mosquitoes in areas with chronic malaria and, although illegally, also used in some areas for pest control in agriculture (Wells and Leonard 2006). Recently, residuals of DDE were found in Endangered black harriers in South Africa (García-Heras et al. 2018) and in other birds in Ethiopia (Yohannes et al. 2014). Despite DDE blood concentrations were relatively low, they were correlated with indicators of physiological condition, suggesting that current levels of contaminant exposure in South Africa may still represent a risk for birds of prey (García-Heras et al. 2018), while it certainly is a threat for birds in other areas where DDE levels in birds were high (Yohannes et al. 2014).

In northeastern Asia, some studies have found that DDT metabolites may still pose also a serious threat for raptors there (in this book Chap. 10).

The use of more modern and theoretically safer pesticides reduced large-scale mortalities of raptors, but regularly, some mortality occurs pointing out that wildlife safety is still not properly assessed when approving new agrochemicals or veterinary drugs (see also below, in this book Chap. 10). In Europe, species like the Near Threatened red kite (*Milvus milvus*) along with Eurasian buzzards are regularly poisoned unintentionally (and sometimes intentionally) in fields treated with carbofuran or rodenticides (e.g., Dietrich et al. 1995; Mougeot et al. 2011; Coeurdassier et al. 2014, in this book Chap. 10). The use of powerful rodenticides for protecting crops seems to be or to have been a severe problem across the world, although this situation seems to be improving in some areas with the use of more modern pesticides and specific and careful practices in their use (in this book Chap. 10).

Probably, one of the best known cases of raptor poisoning with pesticides, and upt to now with better ending was the massive mortality of Swainson's hawks wintering in Argentina in the mid-1990s (Goldstein et al. 1999a). A relatively fast interaction among environmental agencies and political authorities from the different countries involved (Argentina, Canada, and USA) led to the banning of the pesticide in Argentina, which stopped hawk's mortalities (Goldstein et al. 1999b; see in this book Chaps. 10 and 16 for further details).

The use of veterinary pharmaceutical products on livestock production has also implications for birds of prey. Although usually there are legal dispositions so no treated meat gets to human consumers (Woodward 2009), if animals die, they are usually discarded in different ways in the field (including vulture restaurants, Cortés-Avizanda et al. 2016), with the risk of incorporating those products or its residuals to the wildlife or the environment (Arnold et al. 2014). A well-known example of this was the use of diclofenac, an anti-inflammatory drug used for treating cattle, in India, Pakistan, and neighboring countries that led several vulture species to the brink of extinction in the subcontinent and globally (Green et al. 2004; Oaks et al. 2004, Fig. 9.3). Inexplicably, this drug has been approved now for veterinary use in Spain, where there are the main stongholds of European vultures (discussed in this book Chap. 19).

Recent studies have also identified various antibiotic residuals in blood samples from three vultures and an eagle species in Spain (Blanco et al. 2016; Casas-Díaz et al. 2016; Blanco et al. 2017a, b). In all cases, the prevalence of antibiotics was very high (44–100% of the sampled birds, Blanco et al. 2016, 2017a, b; Casas-Díaz et al. 2016). The effects of birds' exposure to antibiotics are still unknown. However, a high prevalence and intensity of oral Candida-like lesions were found in sampled cinereous and Egyptian vulture (*Neophron percnopterus*) fledglings which was suggested to be facilitated by their chronic exposure to antibiotics (Blanco et al. 2017a). Regardless of their final effects, all these studies show without doubts that pharmaceutical products used in intensive livestock farming are being transferred to wild ranging birds of prey.

Human-Made Infrastructures

Several infrastructures linked to development in agricultural lands pose a serious risk to raptors.

Fig. 9.3 The use of agrochemicals and veterinary drugs may affect negatively birds of prey. The use of diclofenac for treating livestock decimated several vultures' species across the Indian subcontinent, among them the Egyptian vulture (flying above), now considered globally Endangered. The griffon vultures feeding on carrion in this figure are the only *Gyps* species considered of Least Concern; however, the species disappeared from vast areas of Europe following changes in livestock farming practices in the past, and the recent approval of the use of diclofenac for treating livestock in Spain could rapidly worsen its status. Recent studies have found antibiotic residuals in both species indicating that vultures in Europe are already exposed to veterinary drugs used for treating livestock. (Photo credit Manuel de La Riva)

Livestock farming in areas with scarce or nonexistent superficial water requires the development of infrastructures for the provision of drinking water for livestock. Although the creation of channels and water reservoirs increases the availability of water for wildlife as well as for domestic animals (James et al. 1999; Macchi and Gray 2012), these structures are usually designed to carry or accumulate water and not for wildlife safety.

Raptor mortalities in water tanks have been detected across most arid and semi-arid regions of the world, the USA, South Africa (Anderson et al. 1999), Mongolia (Elllis et al. 2010), Australia (Debus et al. 2006), and Argentina (Sarasola and Maceda 2006).

The quantification of this mortality source for raptors is difficult, but the high number of birds and the species involved in the most exhaustive study (322 of 29 species; Anderson et al. 1999), suggests that the problem may be serious particularly for scarce and endangered species. In fact, among birds most frequently found by Anderson et al. (1999), there were Critically Endangered white-backed vultures (*Gyps africanus*), Endangered Cape vultures (*Gyps coprotheres*) and lappet-faced vultures (*Torgos tracheliotos*), and Vulnerable martial eagles (*Polemaetus bellico-*

sus), species for which drowning in water tanks has been identified as a threat (e.g., Anderson et al. 1999; Van Eeden et al. 2017). The anecdotal data from Mongolia and Argentina also involved Endangered species such as saker falcons and crowned solitary eagles, respectively (Sarasola and Maceda 2006; Ellis et al. 2010). This mortality factor is particularly frustrating since simple correction measures such as the installation of ramps or stable floating structures would certainly reduce or eliminate these mortalities.

Another infrastructure linked to development of agricultural lands may be the increase in power lines around human settlements and along roads in this human-dominated landscape. Power lines are a novel element in the environment that can be used to perch or to nest by birds of prey, but it may also pose a serious threat to them (Bevanger 1998; Ferrer and Janns 1999; Sanchez-Zapata et al. 2003; Galmes et al. in press, in this book Chap. 12).

Agricultural Changes as Providers of New Opportunities

The changes introduced in the environment by the replacement of the original vegetation by an assemblage of crops, the irruption of domestic herbivores, or the intensification of already existing agricultural production systems may increase in some cases the availability of food, breeding, or roosting resources for raptors (Rodríguez-Estrella et al. 1998; Sarasola and Negro 2006; Filloy and Bellocq 2007; Carrete et al. 2009; Cardador et al. 2011; Buij et al. 2013; Ferrer-Sánchez and Rodríguez-Estrella 2015).

New Breeding and Roosting Possibilities

The creation of large open spaces in the past may have provided new breeding and foraging habitats to open space specialists such as some harrier species (*Circus* ssp.), which may have expanded their ranges in Europe and Central Asia, favored by large-scale forest clearings in prehistoric and historic times (Thiollay 1994; García et al. 2011).

Humans usually construct buildings in the field both for living and for keeping grain, animals, tools, or machinery (Tella et al. 1993). Along with houses, barns, and other artificial structures, in many areas ranchers and farmers plant trees to get shadow or as shelter from the wind (Zalba and Villamil 2002). These structures increase structural diversity in the landscape and may be used by birds of prey as nesting, roosting, or feeding sites, allowing many species to increase their breeding ranges into areas formerly unsuitable for them (Tella et al. 1993; Sarasola and Negro 2006).

Fig. 9.4 In some occasions, agricultural activities generate new habitats for birds of prey. Labor houses, and especially abandoned ones (left), were adopted in the agricultural landscapes of Western Europe by lesser kestrels (right) as novel breeding substrates, allowing the species to establish breeding populations in areas far from suitable cliffs. (Photo credits Arnau Guardia)

Across most of their breeding range in Europe, little owls (*Athene noctua*) breed in tree cavities, and Eurasian kestrels breed in cliffs or in other bird's nests (Ferguson-Lees and Christie 2001); however, in Aragon steppes and agricultural lands, they breed in the roofs or walls of abandoned houses (e.g., Tella et al. 1993). The lesser kestrel is in origin a cliff nester (e.g., Ferguson-Lees and Christie 2001). However, nowadays, most breeding pairs of the species in Europe breed in large buildings in towns or, mostly, in farmhouses (usually abandoned ones) in the field (Biber 1994, Fig. 9.4). This adoption of manmade structures may have allowed the species to spread through flat areas with no cliffs (Tella et al. 1993). Similarly, turkey vultures (*Cathartes aura*) in Saskatchewan seem to have benefited by human movement from rural areas to villages and cities in the context of changing agricultural practices. The vulture population has been expanding there since 1982, associated with the use of abandoned barns for breeding in an area naturally lacking the large cavities used by the species to breed (Houston et al. 2007).

Water has limited agriculture options for humans for thousands of years, and humans also for thousands of years have modified water courses and built dams and channels to obtain water for farming (e.g., Herrero and Snyder 1997). Implementation of modern intensified irrigated crops also requires the construction of these structures that may be suitable habitat for birds (e.g., Sebastián-González et al. 2010). western marsh-harriers that suffered other aspects of agricultural intensification are one of the species now favored by these structures and the habitat changes produced by some irrigated crops. In the last two decades, the species has expanded associated with these intensive irrigated systems even into former extensive traditional agricultural areas where they were previously absent (Molina and Martínez 2008; Cardador et al. 2011).

The inclusion of trees in formerly treeless environments in the context of the expansion of the agricultural frontier may have triggered positive effects for some birds of prey. The Argentina's Pampas grasslands were in the past an endless sea of grass with almost no trees or shrubs to the point that settlers had to use dry thistle culms as firewood (see Zalba and Villamil 2002). The expansion of farms and ranches created a myriad of exotic tree stands that completely altered the vegetation structure of the environment to the point that nowadays, those tree stands are a typical sight of the Pampas (Sarasola and Negro 2006). An invasion of grasslands by some native woody species has also occurred from the Espinal ecoregion in the west, promoted by the dispersion of its seeds during cattle movements (Zalba y Villamil 2002). The presence of shrubs and trees, having altered the landscape, has however provided breeding substrate for species that were previously absent in the Pampas or at least must have held extremely low densities such as the American kestrel, the aplomado falcon (*Falco femoralis*), the chimango caracara, the southern caracara, or the roadside hawk (*Buteo magnirostris*) (Narosky and Di Giacomo 1993, authors pers. obs.). It has also provided roosting substrate for communal roosting species such as the chimango caracara, the white-tailed kite (*Elanus leucurus*), or the Swainson's hawk (Sarasola and Negro 2006; Sarasola et al. 2010b, Solaro and Sarasola 2015).

New or more Abundant Food Sources

Changes in agriculture and livestock farming may increase food availability for raptors as well.

The effects on raptors of land abandonment in several marginal areas of Europe have provided direct evidences that human activities to promote cattle grazing and clearing of forests for agriculture, may have increased in the past foraging habitat availability and quality for open space foragers such as the Eurasian buzzards and golden eagles (*Aquila chrysaetos*) (e.g., Pedrini and Sergio 2001; Watson 2010).

The irrigation of semiarid areas increases water availability in environments where water is the limiting factor. This may trigger the expansion and increase in density of some species previously absent or limited by water availability, increasing food availability for raptors (Cardador et al. 2012). Besides providing new habitats for the western marsh-harrier, irrigated crops have higher abundance of small mammals (high-quality prey for harriers) than nonirrigated crops (Molina and Martínez 2008; Cardador et al. 2011), which is reflected in the use of smaller home ranges and more specialized diet in irrigated than in nonirrigated territories in this species (Cardador et al. 2009, 2012).

A clear example of that is the vast expansion of the common vole (*Microtus arvalis*) through Northwestern Spain in the last decades of the twentieth century, after large-scale land use changes turned extensive grasslands and crops (e.g., mostly nonirrigated cereals) to irrigated cereals and alfalfa (Jareño et al. 2015). The high availability of voles, especially in years of outbreaks, provides now ideal food for some medium-sized raptors such as red kites (Mougeot et al. 2011; Paz et al. 2013), although it also triggers conflicts by the use of rodenticides (Olea et al. 2009).

In North America a recent study showed that American kestrels advanced their breeding season through the years to synchronize it with irrigated crops growth and not with the growing of natural scrubland also present in their territories. This suggests that they may track prey in irrigated areas. This may be because prey are of better quality there or appear earlier in the season coinciding with crop growth (Smith et al. 2017).

Several studies in other environments suggest that some crops may provide higher food availability to birds of prey in environments where food was previously limited. Small low-intensity crops within native scrubland increased the abundance and allowed the expansion of some temperate generalist raptors in Baja California (Rodríguez-Estrella et al. 1998). In Africa, rodent specialists such as black-winged kites (*Elanus caeruleus*) or augur buzzard (*Buteo augur*) and some Palearctic migrants (e.g., booted eagle *Aquila pennata*, western marsh-harrier, steppe eagle) may have benefited from rodent-rich cultivation in some areas (Herremans and Herremans-Tonnoeyr 2000; Virani and Harper 2009, Buij et al. 2013, see also in this book Chap. 18 and references therein).

Conflicts with Humans in Agricultural Landscapes

As a result of perceived or real conflicts, humans have long intentionally destroyed nests and shot and poisoned raptors they consider as pests (Newton 1979, 1998; Bildstein 2001).

The main trigger of those conflicts usually lies in the killing or presumed killing of animals valuable for man by raptors (Graham and Mires 2005). The impact of human persecution on raptor populations has been severe and is behind global raptor population declines, behind the local and regional extinction of several species (Newton 1979; Ferguson-Lees and Christie 2001), as well as behind the only global raptor extinction in recent times (White et al. 1994; see below).

The most widespread conflict between raptors and livestock farmers has been the preying of lambs, young goats, or reindeer calves by eagles, mostly by golden eagles in Europe and North America, crowned solitary eagles in Argentina, Verreaux's (*Aquila verreauxii*) and martial eagles in Africa, and wedge-tailed eagles (*Aquila audax*) in Australia, that depending on the areas may end with the killing (usually illegal) of the eagles (Brooker and Ridpath 1980; Davies 1999; Norberg et al. 2006; Sarasola et al. 2010a; Watson 2010, Van Eeden et al. 2017, Fig. 9.5). However, scientific evidence suggests that except locally, losses attributable to eagle predation in lamb-producing ranches are usually small, rarely reaching values above 2–3% (Davies 1999; Watson 2010). Regarding reindeers, golden eagle kills reach similar values, from 0 to 1–2% in most areas (Nybakk et al. 1999; Nieminen 2010) and occasionally up to 3–4% (Norberg et al. 2006). Furthermore, evidences exist in some studies that eagles hunt reindeer calves with lower weights than those surviving (sometimes, 4–6 kg lighter; Nybakk et al. 1999; Norberg et al. 2006). The weight at marking is a good predictor of calf survival (Norberg et al. 2006), and thus golden eagles could be preying more frequently on individuals that have lower survival chances anyway, so the real loss for producers is probably even lower. A study in South Africa found that Verreaux's eagle predation on hyraxes would reduce rodent grazing, increasing pasture availability for lambs so that the benefits to the farmers of having the eagles in their ranches would probably exceed the costs (Davies 1999). In this same direction, in the case of the crowned solitary eagle in central Argentina, the eagle is heavily persecuted accused of killing lambs and young goats, but nest monitoring during 8 years showed that eagles rarely feed on livestock (of 598 prey remains, only one was from a young goat that may have been scavenged), but they prey heavily on poisonous snakes so their presence, again, would be beneficial for producers (Sarasola et al. 2010a).

Despite smaller raptor species may be less capable of hunting livestock, southern and striated caracaras (*Phalcoboenus australis*) are still frequently persecuted accused of killing lambs. Although mostly scavengers, there are evidences that they may occasionally kill small lambs and young goats (Ferguson-Lees and Christie 2001; Catry et al. 2008). The Guadalupe caracara (*Caracara lutosa*) is the only raptor which extinction has been documented in recent times. This raptor, endemic from Guadalupe Island in Mexico, was persecuted to death in the first years of the twentieth century accused of killing young goats and chickens (White et al. 1994). Similar conflicts arise with some forest eagles such as the Philippine (*Pithecophaga jefferyi*), the harpy, or the black-and-chestnut (*Spizaetus isidori*) that may kill chickens or other domestic animals that may be highly valuable for small-scale peasants (Lehman 1959; Bueser et al. 2003; Vargas et al. 2006; Zuluaga and Echeverry-Galvis 2016). This conflict increases as forest clearing and fragmentation allow peasants to get deeper and deeper into the forests (e.g., Vargas et al. 2006).

Most vulture species have been traditionally valued positively by farmers as cleaners of the environment; however, this has not been always the case. The bearded vulture (*Gypaetus barbatus*), despite being a specialist scavenger that feeds mostly on animal bones, was known as lammergeier or "lamb vulture" in the German-speaking Alps in central Europe where it was accused of killing lambs. Intensive persecution for this reason along with hunting for specimens was the cause of its extinction from the Alps around 1920–1930 (Mingozzi and Estève 1997). Fortunately, after a long-lasting and expensive reintroduction program, the species now breeds again in most of its former range in the Alps (Schaub et al. 2009).

The Andean condor (*Vultur gryphus*) and the black vulture (*Coragyps atratus*) in America, and, after the BSE crisis, the griffon vulture in Spain, have been accused of killing sheep, calves, and cows (Avery and Cummings 2004; Margalida et al. 2011; Cailly Arnulphi et al. 2017, Fig. 9.5). Although probably the true impact of vulture predation would be small, scientific research is definitively needed to assess their true impact. Particularly, in the case of the condor, all claims of killing come

Fig. 9.5 Some birds of prey may generate conflicts in agricultural productive systems. Golden eagles (**a**) are blamed as lamb hunters, which they may eventually do. However, in most cases like in the figure, eagles feed on already dead animals (scavenge). In any case, impact for livestock farmers is usually very small or nil (photo credit Manuel De la Riva). (**b**) Andean condors (left) and black vultures (right) are blamed in some areas of South America as livestock (especially calves and lambs) predators. Despite that academia has still to confirm this behavior for the condor and analyze its frequency, retaliations to this species with slow demography could impose a critical toll on its long-term persistence. (Photo credit Manuel De la Riva)

from farmers, but the academia still perceive the species as a scavenger (Cailly Arnulphi et al. 2017). Given the slow demography of this Near Threatened species and its sensibility to mortality, urgent clarification and quantification of impacts of this conflict for both, producers and condors, are needed.

The use of poison baits to kill livestock mammalian predators is a (mostly illegal) practice carried out by farmers globally. However, many scavenging raptors are collaterally killed by this way. Illegal poisoning is in fact one of the main threats for vultures in Europe and Africa and, thus, partially responsible of their critical conservation status (Margalida 2012). In most cases, the poison used is an agricultural pesticide pointing to the need of a further scrutiny on the commercial circuits of these agrochemicals (see in this book Chap. 19 for a further discussion on the use of poison for predator control and its effects on vultures).

When changes in farming practices and raptor's conservation are confronted, further conflicts may be generated. A concrete example was the social irritation produced in northeastern Spain by the temporal interruption of financial support from the European Commission to irrigation schemes in steppes in the late 1990s until no proper IBAs and protected areas were defined to preserve steppe birds, such as the Endangered (at that moment) lesser kestrel (Oñate et al. 1998; Gros Zubiaga 2000). In this same line, the opposition of raptor conservationists to the use of rodenticides that kill raptors as a side effect of rodent control, confronts the urge of farmers to stop rodent damage to their crops (Olea et al. 2009). This kind of conflict extends also to Africa when farmers fight Quelea (*Quelea quelea*) plagues. All the methods used to control Queleas have generated mortalities of raptors, and thus, the solution is complex (McWilliam and Cheke 2004).

Finally, intensively farmed areas may reduce the abundance of some game species like red-legged partridges (*Alectoris rufa*) in intensively irrigated crops (De Frutos et al. 2015), increasing the intensity of traditional conflict between hunters and raptors.

Living in Agroecosystems

Despite the more or less severe transformation of natural habitats, agroecosystems have more or less developed communities of birds of prey (e.g., Tella et al. 1993; Rodríguez-Estrella et al. 1998; Sánchez-Zapata et al. 2003; Carrete et al. 2009; Buij et al. 2013; Ferrer-Sánchez and Rodríguez-Estrella 2015). Furthermore, some studies indicate that even some of the birds of prey presumably more vulnerable to humans because of their size and their large habitat requirements as large eagles, if not persecuted, may have healthy populations in agricultural lands (e.g., Murgatroyd et al. 2016a; Horváth et al. 2011, Fig. 9.6). As in other farmed systems, habitat heterogeneity generated by the contact of farmland and natural habitats may allow Verreaux's and eastern imperial (*Aquila heliaca*) eagles to escape their presumed "food specialization" and successfully occupy agricultural lands (Murgatroyd et al. 2016b; Horváth et al. 2010). While South African Sandveld's Verreaux's eagles were living in a mixed environment (intensive agriculture in the lowlands and natural vegetation in more rugged areas), intensive agriculture covers most of the imperial eagle territories in lowland Eastern Europe (Horváth et al. 2010, 2011). The identification of the factors that allow those eagle species to successfully breed in such altered environments would certainly help to understand what limits birds of prey in agricultural landscapes. In any case, these studies, and similar ones with other species dealing well in intensified environments (Ursúa et al. 2005; Cardador et al. 2011; Buij et al. 2013), indicate that not all intensive productive systems are the same. In general, it seems that to maintain certain habitat heterogeneity, with

Fig. 9.6 Some large eagles are able to occupy and to have high breeding success in intensive agricultural environments. (**a**) Adult Verreaux's eagle with the fledgling and an unhatched egg in its nest in Sandveld agricultural lands in South Africa, where the species breeds at higher rates than in any other population in southern Africa. (Photo credit Megan Murgatroyd). (**b**) Eastern imperial eagle's nest in the middle of a crop field in lowland intensive agricultural lands of Hungary; in the last decades, the species colonized lowland agricultural areas across much of its breeding range in Eastern Europe, breeding there at higher rates than in their traditional natural habitats in the mountains. (Photo credit Marton Horváth). Although adaptations to more diversified diets could play a role in their success, further research is needed to get a deeper understanding of the factors that allow those eagle species to be so successful in agricultural lands

hedgerows and natural field margins, and adequate grazing pressures is positive for most raptors using agroecosystems (e.g., Tella et al. 1998; Ursúa et al. 2005; Amar and Repath 2005; Torre et al. 2007; Butet et al. 2010). However, particular studies should be carried out to assess which factors of productive intensification are leading negative trends on each particular species or community so that specific managing options may be taken on each case.

The role of raptors as ecosystem service providers for humans is still poorly studied. Although more developed in other chapters (in this book Chaps. 18 and 19), a mention here of their role as ecosystem providers in agroecosystems must be done. In agroecosystems vultures and other scavenging birds eliminate animal carcasses from the environment, recycling nutrients, reducing disease spreading, and saving greenhouse emissions and money (see in this book Chap. 19 and references therein). However, the lack of integration of agriculture and livestock production legislation with environmental protection legislation has led to several vulture crises, thus severely affecting those services provided by vultures (e.g., Donázar et al. 2009a, b; Margalida et al. 2010). Although the number of studies is still small, some have already proven that under certain conditions, predatory raptors may indeed play a relevant role in the control of a variety of small mammals, insects, and bird pests for crops (Paz et al. 2013; Shave 2017, in this book Chap. 18) and that this control may be directly translated into economic earnings to producers (Kross et al. 2012).

Conclusions

Although there are evidences of large-scale human landscape transformation several centuries ago, its global impact soared in the last three centuries when cultivated lands increased in the order of 466% (Meyer and Turner 1992). Nowadays cropland and lands for livestock farming represent about 40% of the ice-free land area (Foley et al. 2005), the transformation of natural habitats to modern agricultural and forestry practices occur at unprecedented accelerated rates, and human population growth is far from stopping so this surface will likely increase in coming years (Tilman et al. 2001; Foley et al. 2005).

Agriculture and livestock farming practices, in their multiple ways, are actually a threat for around 83% of the bird of prey species that are Near Threatened or have some degree of global threat according to the IUCN (BirdLife International 2017). But this group of Near Threatened, Vulnerable, Endangered, and Critically Endangered raptors is only 32% of birds of prey species still ranging in the world (Table 9.2). This suggests that the impact of agricultural activities on birds of prey is among the main factors driving species to extinction in this group of birds. In fact, the only raptor species that becomes extinct in modern times was persecuted to death in the frame of a conflict with goat producers (White et al. 1994).

The three chapters of this book dealing with conservation status of birds of prey at regional scales (Chaps. 16, 17, and 18) and the one dealing with a particularly threatened group, the Old World vultures (in this book Chaps. 19), identify agriculture and livestock farming expansion and intensification as one of the main factors

Table 9.2 Number of species of diurnal birds of prey per family, species Near Threatened or with some degree of threat according to IUCN and percentage, and species Near Threatened or with some degree of threat but for which agricultural and livestock farming practices are a threat and percentage they represent of those Near Threatened or with some degree of threat. Extinct species are excluded from the table (BirdLife International 2017)

Birds of prey families	Species	Species NT or higher	% (species)	Species NT or higher by agriculture/livestock	% (of NT or higher)
Cathartidae	7	1	14	0	0
Accipitridae	249	85	34	76	89
Falconidae	64	15	21	8	53
Pandionidae	1	0	0	0	0
Sagittariidae	1	1	100	1	100
Total	322	102	32	85	83

threatening birds of prey. Habitat destruction and degradation, changes in livestock farming practices (including both husbandry and the use of pharmaceutical products), and the use of dangerous pesticides are (probably along with predator-human conflicts) among the leading factors behind the delicate conservation status of most threatened birds of prey.

Generalist or open space users, even in the case of large eagles, may adapt and do well in agricultural environments, particularly if they do not change strongly the physiognomy of the original habitat and if it maintains some degree of habitat heterogeneity. However, it seems clear that this is not the case for forest specialist (Jullien and Thiollay 1996). Particular measures to ensure the conservation of large enough tracts of forest along with corridors that connect them will be needed for this particular group of raptors. The development of sustainable productive systems integrated in the preservation of those forests will thus be critical.

Changes in livestock farming practices, including the use of veterinary pharmaceutical drugs, should imply a wide assessment of their potential impact on raptor communities before being applied. This would prevent drastic population crashes like the one suffered by vultures in Africa and some countries of Southeastern Asia or would limit absurd crisis like the one triggered for European vultures after the irruption of the BSE.

The use of agrochemicals in both crops and livestock clearly deserves further research and a finer scrutiny before their use is approved so events such as the Asian vulture crisis or the mass mortalities of Swainson's hawks do not repeat.

While conflicts among birds of prey and agricultural/farming practices may seem less relevant than other interactions between raptors and those productive activities, conflicts have already caused some local, regional, and even one global raptor extinction. Some actual conflicts already affect long-living species that may be especially sensible to increases in persecution, and therefore, careful attention must be paid to detect these conflicts and to elaborate adequate answers to solve them.

A more integrated production frame including a deep assessment of potential conflicts between agricultural production and conservation of raptors, and a careful research to solve these conflicts, will be needed if healthy birds of prey are to be conserved in agricultural landscapes.

Acknowledgments We want to thank Kike Arnal from the Cornell Lab of Ornithology, Rafa Benjumea from Ecotono Birding, Kim Chye, Carlitos De Angelo, Manuel de la Riva, Maxi Galmes, Arnau Guardia, Marton Horváth, Jorge La Grotteria, Megan Murgatroyd, and Claudina Solaro for kindly providing pictures to illustrate this chapter. We also thank Erin S. Liner and Karen Rodriguez from the Cornell Lab of Ornithology for arranging the authorizations for the use of pictures from the institution, as well as to the Cornell Lab of Ornithology for allowing the use of those pictures. JMG also wishes to thank his family for their patience during the development of this chapter.

References

Abood SA, Lee JSH, Burivalova Z, Garcia-Ulloa J, Koh LP (2015) Relative contributions of the logging, fiber, oil palm, and mining industries to forest loss in Indonesia. Conserv Lett 8:58–67

Aizen MA, Garibaldi LA, Dondo M (2009) Expansión de la soja y diversidad de la agricultura Argentina. Ecol Austral 19:45–54

Amar A, Repath SM (2005) Habitat use by hen harriers *Circus cyaneus* on orkney: implications of land-use change for this declining population. Ibis 147:37–47

Anderson MD, Maritz AW, Oosthuysen E (1999) Raptors drowning in farm reservoirs in South Africa. Ostrich 70:139–144

Arnold KE, Brown AR, Ankley GT, Sumpter JP (2014) Medicating the environment: assessing risks of pharmaceuticals to wildlife and ecosystems. Phil Trans R Soc B 369:20130569

Arrondo E, Moleón M, Cortés-Avizanda A, Jiménez J, Beja P, Sánchez-Zapata JA, Donázar JA (2018) Invisible barriers: differential sanitary regulations constrain vulture movements across country borders. Biol Conserv 219:46–52

Arroyo BE, Bretagnolle V, Garcia JT (2003) Land use, agricultural practices and conservation of Montagu's Harrier. In: Thompson DBA, Redpath SM, Fielding AH, Marquiss M, Galbraith CA (eds) Birds of prey in a changing environment. The Stationery Office, Edinburgh, UK pp 449–463

Avery ML, Cummings JL (2004) Livestock depredations by black vultures and golden eagles. USDA National Wildlife Research Center-Staff Publications, 76. Res Center-Staff Publ. 2004: 76

Banhos A, Hrbek T, Sanaiotti TM, Farias IP (2016) Reduction of genetic diversity of the harpy eagle in Brazilian tropical forests. PLoS One 11:e0148902

Baumgart W (1994) Saker *Falco cherrug*. In: Tucker GM, Heath MF (eds) Birds in Europe: their conservation status. BirdLife International (Conservation Series 3), Cambridge, UK, pp 198–199

Bennett AF, Radford JQ, Haslem A (2006) Properties of land mosaics: implications for nature conservation in agricultural environments. Biol Conserv 133:250–264

Benton TG, Vickery JA, Wilson JD (2003) Farmland biodiversity: is habitat heterogeneity the key? Trends Ecol Evol 18:182–188

Bevanger K (1998) Biological and conservation aspects of bird mortality caused by electricity power lines: a review. Biol Conserv 86:67–76

Biber JP (1994) Lesser Kestrel. In: Tucker GM, Heath MF (eds) Birds in Europe: their conservation status, Birdlife conservation series number 3. Birdlife International, Cambridge, United Kingdom, pp 188–189

Bildstein KL (2001) Raptors as vermin: a history of human attitudes towards Pennsylvania's birds of prey. Endangered species. Update 18:124–128

Birdlife International (2003) International action plan for the Pallid harrier (*Circus macrourus*). Convention on the conservation of European wildlife and natural habitats. Standing committee. 23rd meeting. Strasbourg, 1–4 December 2003

BirdLife International (2004) Birds in Europe: population estimates, trends and conservation status. Cambridge: Bird-life international (BirdLife conservation series No 12)

BirdLife International (2017) IUCN red list for birds. Downloaded from http://www.birdlife.org on 18/12/2017

Blanco G, Junza A, Segarra D, Barbosa J, Barrón D (2016) Wildlife contamination with fluoroquinolones from livestock: widespread occurrence of enrofloxacin and marbofloxacin in vultures. Chemosphere 144:1536–1543

Blanco G, Junza A, Barrón D (2017a) Food safety in scavenger conservation: diet-associated exposure to livestock pharmaceuticals and opportunist mycoses in threatened cinereous and Egyptian vultures. Ecotoxicol Environ Saf 135:292–301

Blanco G, Junza A, Barrón D (2017b) Occurrence of veterinary pharmaceuticals in golden eagle nestlings: unnoticed scavenging on livestock carcasses and other potential exposure routes. Sci Total Environ 586:355–361
Bolsa de Cereales de Buenos Aires (2017) Panorama agrícola semanal. Departamento de Estimaciones Agrícolas. Newsletter, Bolsa de Cereales de Buenos Aires, p 146
Brink AB, Eva HD (2009) Monitoring 25 years of land cover change dynamics in Africa: a sample based remote sensing approach. Appl Geogr 29:501–512
Brooker MG, Ridpath MG (1980) The diet of the wedge-tailed eagle, *Aquila audax*, in western Australia. Wildl Res 7:433–452
Brown A, Martinez Ortiz U, Acerbi M, Corcuera J (eds) (2006) La Situación Ambiental Argentina 2005. Fundación Vida Silvestre Argentina, Buenos Aires
Bueser GLL, Bueser KG, Afan DS, Salvador DI, Grier JW, Kennedy RS, Miranda HC (2003) Distribution and nesting density of the Philippine eagle *Pithecophaga jefferyi* on Mindanao Island, Philippines: what do we know after 100 years? Ibis 145:130–135
Buij R, Croes BM, Gort G, Komdeur J (2013) The role of breeding range, diet, mobility and body size in associations of raptor communities and land-use in a west African savanna. Biol Conserv 166:231–246
Butet A, Michel N, Rantier Y, Comor V, Hubert-Moy L, Nabucet J, Delettre Y (2010) Responses of common buzzard (*Buteo buteo*) and Eurasian kestrel (*Falco tinnunculus*) to land use changes in agricultural landscapes of western France. Agric Ecosyst Environ 138:152–159
Cailly Arnulphi VB, Lambertucci SA, Borghi CE (2017) Education can improve the negative perception of a threatened long-lived scavenging bird, the Andean condor. PLoS One 12:e0185278
Canavelli SB, Bechard MJ, Woodbridge B, Kochert MN, Maceda JJ, Zaccagnini ME (2003) Habitat use by Swainson's hawks on their austral wintering grounds in Argentina. J Raptor Res 37:125–134
Cardador L, Manosa S, Varea A, Bertolero A (2009) Ranging behaviour of marsh-harriers *Circus aeruginosus* in agricultural landscapes. Ibis 151:766–770
Cardador L, Carrete M, Mañosa S (2011) Can intensive agricultural landscapes favour some raptor species? The marsh-harrier in North-Eastern Spain. Anim Conserv 14:382–390
Cardador L, Planas E, Varea A, Mañosa S (2012) Feeding behaviour and diet composition of marsh-harriers *Circus aeruginosus* in agricultural landscapes. Bird Study 59:228–235
Carrete M, Tella JL, Blanco G, Bertellotti M (2009) Effects of habitat degradation on the abundance, richness and diversity of raptors across Neotropical biomes. Biol Conserv 142:2001–2011
Casas-Díaz E, Cristófol C, Cuenca R, Agustí S, Carneiro M, Marco I, Lavín S, Margalida A (2016) Determination of fluoroquinolone antibiotic residues in the plasma of Eurasian griffon vultures (*Gyps fulvus*) in Spain. Sci Total Environ 557:620–626
Catry P, Lecoq M, Strange IJ (2008) Population growth and density, diet and breeding success of striated caracaras *Phalcoboenus australis* on New Island, Falkland Islands. Polar Biol 31:1167–1174
Codesido M, González-Fischer C, Bilenca DN (2011) Distributional changes of landbird species in agroecosystems of Central Argentina. Condor 113:266–273
Coeurdassier M, Riols R, Decors A, Mionnet A, David F, Quintaine T, Truchetet D, Scheifler R, Giradoux P (2014) Unintentional wildlife poisoning and proposals for sustainable management of rodents. Conserv Biol 28:315–321
Collar NJ, Gonzaga LP, Krabbe N, Madroño A, Naraujo LG, Parker TA, Wege DC (1992) Threatened birds of the Americas. Smithsonian Institution Press, Washington, DC
Cortés-Avizanda A, Blanco G, DeVault TL, Markandya A, Virani MZ, Brandt J, Donázar JA (2016) Supplementary feeding and endangered avian scavengers: benefits, caveats, and controversies. Front Ecol Environ 14:191–199
Curtis O, Simmons RE, Jenkins AR (2004) Black harrier *Circus maurus* of the fynbos biome, South Africa: a threatened specialist or an adaptable survivor? Bird Conserv Int 14:233–245
Davies RA (1999) The extent, cost and control of livestock predation. J Raptor Res 33:67–72

De Frutos A, Olea PP, Mateo-Tomás P (2015) Responses of medium-and large-sized bird diversity to irrigation in dry cereal agroecosystems across spatial scales. Agric Ecosyst Environ 207:141–152

De Roland LAR, Thorstrom R, Razafimanjato G, Rakotondratsima MPH, Andriamalala TRA (2009) Surveys, distribution and current status of the Madagascar harrier *Circus macrosceles* in Madagascar. Bird Conserv Int 19:309–322

Debus SJS, Lollback G, Oliver DL, Cairns SC (2006) The birds of Bulgunnia and Mulyungarie stations in the pastoral zone of arid South Australia. South Australian Ornithol 35:27–37

Di Bitetti MS, Placci G, Dietz LA (2003) A biodiversity vision for the Upper Paraná Atlantic forest eco-region: designing a biodiversity conservation landscape and setting priorities for conservation action. World Wildlife Fund, Washington, DC, 116

Dietrich DR, Schmid P, Zweifel U, Schlatter C, Jenni-Eiermann S, Bachmann H, Bühler U, Zbinden N (1995) Mortality of birds of prey following field application of granular carbofuran: a case study. Arch Environ Contam Toxicol 29:140–145

Donald PF (2004) Biodiversity impacts of some agricultural commodity production systems. Conserv Biol 18:17–37

Donázar JA (1993) Los Buitres ibéricos. Biología y Conservación. J.M. Reyero Editor, Madrid

Donázar JA, Negro JJ, Hiraldo F (1993) Foraging habitat selection, land-use changes and population decline in the lesser kestrel *Falco naumanni*. J Appl Ecol 30:515–522

Donázar JA, Naveso MA, Tella JL, Campión D (1997) Extensive grazing and raptors in Spain. In: Pain DJ, Pienkowski MW (eds) Farming and Birds in Europe. Academic Press, London, pp 117–149

Donázar JA, Margalida A, Carrete M, Sánchez-Zapata JA (2009a) Too sanitary for vultures. Science 326:664

Donázar JA, Margalida A, Campión D (Eds.) (2009b) Vultures, feeding stations and sanitary legislation: a conflict and its consequences from the perspective of conservation biology. Munibe 29 (suppl.), Sociedad de Ciencias Aranzadi, Donostia, Spain

van Eeden R, Whitfield DP, Botha A, Amar A (2017) Ranging behaviour and habitat preferences of the martial eagle: implications for the conservation of a declining apex predator. PloS one 12:e0173956

Ellis DH, Roundy TB, Ellis CH (2010) Raptor pit mortality in Mongolia and a call to identify and modify death traps wherever they occur. AMBIO J Hum Environ 39:349–351

Espín S, Terraube J, Arroyo B, Camarero PR, Mateo R, Limiñana R, Vázquez-Pumariño X, Pinilla A, García JT, Mougeot F (2018) Blood concentrations of p, p′-DDE and PCBs in harriers breeding in Spain and Kazakhstan. Sci Total Environ 624:1287–1297

Fandiño B, Pautasso AA (2013) Distribución, historia natural y conservación de *Harpyhaliaetus coronatus* (aves: accipitridae) en el centro–este de Argentina. Natura Neotropicalis 1:41–55

Ferguson–Lees J, Christie DA (2001) Raptors of the world. Houghton Mifflin, Boston

Ferrer M (1993) Juvenile dispersal behaviour and natal philopatry of a long-lived raptor, the Spanish imperial eagle *Aquila adalberti*. Ibis 135:132–138

Ferrer M, Janss GF (1999) Birds and power lines: collision, electrocution and breeding. Quercus, Madrid, Spain, 8

Ferrer-Sánchez Y, Rodríguez-Estrella R (2015) Man-made environments relationships with island raptors: endemics do not cope with habitat changes, the case of the island of Cuba. Biodivers Conserv 24:407

Filloy J, Bellocq MI (2007) Patterns of bird abundance along the agricultural gradient of the Pampean región. Agric Ecosyst Environ 120:291–298

Foley JA, DeFries R, Asner GP, Barford C, Bonan G, Carpenter SR, Chapin FS, Coe MT, Daily GC, Gibbs HK, Helkowski JH, Holloway T, Howard EA, Kucharik CJ, Monfreda C, Patz JA, Prentice IC, Ramankutty N, Zinder PK (2005) Global consequences of land use. Science 309:570–574

Food and Agriculture Organization of the United Nations (2016) Global Forest Resources Assessment 2015. FAO, Rome, Italy

Fox AD (2004) Has Danish agriculture maintained farmland bird populations? J Appl Ecol 41:427–439

Galmes MA, Sarasola JH Grande JM, Vargas FH (2017) Electrocution risk for the endangered crowned solitary eagle and other birds in semiarid landscapes of central Argentina, in press

García JT, Alda F, Terraube J, Mougeot F, Sternalski A, Bretagnolle V, Arroyo B (2011) Demographic history, genetic structure and gene flow in a steppe-associated raptor species. BMC Evol Biol 11:333

García AR, Fleite SN, Vázquez Pugliese D, De Iorio AF (2013) Feedlots and pollution-- a growing threat to water resources of agro-production zone in Argentina. Environ Sci Technol 47:11932–11933

García-Heras MS, Arroyo B, Simmons RE, Camarero PR, Mateo R, Mougeot F (2018) Blood concentrations of PCBs and DDTs in an avian predator endemic to southern Africa: associations with habitat, electrical transformers and diet. Environ Pollut 232:440–449

Gavier-Pizarro GI, Calamari NC, Thompson JJ, Canavelli SB, Solari LM, Decarre J, Goijman AP, Romina P, Suarez RP, Bernardos JN, Zaccagnini ME (2012) Expansion and intensification of row crop agriculture in the pampas and Espinal of Argentina can reduce ecosystem service provision by changing avian density. Agric Ecosyst Environ 154:44–55

Ghersa CM, Martinez-Ghersa MA (1991) Cambios ecológicos en los agroecosistemas de la pampa ondulada. Efectos de la introducción de la soja. Investigación y Ciencia 5:182–188

Goijman AP, Conroy MJ, Bernardos JN, Zaccagnini ME (2015) Multi-season regional analysis of multi-species occupancy: implications for bird conservation in agricultural lands in east-Central Argentina. PLoS One 10:e0130874

Goldstein MI, Lacher TE, Woodbridge B, Bechard MJ, Canavelli SB, Zaccagnini ME, Cobb GP, Scollon EJ, Tribolet R, Hopper MJ (1999a) Monocrotophos-induced mass mortality of Swainson's hawks in Argentina, 1995–96. Ecotoxicology 8:201–214

Goldstein MI, Lacher TE, Zaccagnini ME, Parker ML, Hooper MJ (1999b) Monitoring and assessment of Swainson's hawks in Argentina following restrictions on monocrotophos use, 1996–97. Ecotoxicology 8:215–224

Gomes FB, Sanaiotti TM (2015) A review of the distribution of the crested eagle, *Morphnus guianensis* (Daudin, 1800) (Accipitridae: Harpiinae), including range extensions. Revista Brasileira de Ornitologia-Brazilian. J Ornithol 23:36–63

González LM, Bustamante J, Hiraldo F (1990) Factors influencing the present distribution of the Spanish imperial eagle *Aquila adalberti*. Biol Conserv 51:311–319

Graesser J, Aide TM, Grau HR, Ramankutty N (2015) Cropland/pastureland dynamics and the slowdown of deforestation in Latin America. Environ Res Lett 10:034017

Graham SA, Mires GW (2005) Predation on red tree voles by owls and diurnal raptors. Northwest Nat 86:38–40

Green RE, Newton IAN, Shultz S, Cunningham AA, Gilbert M, Pain DJ, Prakash V (2004) Diclofenac poisoning as a cause of vulture population declines across the Indian subcontinent. J Appl Ecol 41:793–800

Greenpeace (2013) Certifying destruction. Greenpeace International. In: the Netherlands

Gros Zubiaga J (2000) El regadío como necesidad estratégica. El caso de Aragón, REVISTA

Hansen MC, Stehman SV, Potapov PV, Loveland TR, Townshend JR, DeFries RS, Pitman KW, Arunarwati B, Stolle F, Steininger MK, Carroll M, DiMiceli C (2008) Humid tropical forest clearing from 2000 to 2005 quantified by using multitemporal and multiresolution remotely sensed data. Proc Natl Acad Sci 105:9439–9444

Herremans M, Herremans-Tonnoeyr D (2000) Land use and the conservation status of raptors in Botswana. Biol Conserv 94:31–41

Herrero J, Snyder RL (1997) Aridity and irrigation in Aragon, Spain. J Arid Environ 35:535–547

Hiraldo F, Negro JJ, Donázar JA, Gaona P (1996) A demographic model for a population of the endangered lesser kestrel in southern Spain. J Appl Ecol 33:1085–1093

Hla H, Shwe NM, Htun TW, Zaw SM, Mahood S, Eames JC, Pilgrim JD (2011) Historical and current status of vultures in Myanmar. Bird Conser Int 21:376–387

Horváth M, Szitta T, Firmánszky G, Solti B, Kovács A, Moskát C (2010) Spatial variation in prey composition and its possible effect on reproductive success in an expanding eastern imperial eagle (*Aquila heliaca*) population. Acta Zool Acad Sci Hung 56:187–200

Horváth M, Demeter I, Fatér I, Firmánszky G, Kleszó A, Kovács A, Szitta T, Tóth I, Zalai T, Bagyura J (2011) Population dynamics of the eastern imperial eagle (*Aquila heliaca*) in Hungary between 2001 and 2009. Acta Zool Bulg Suppl. 3:61–70

Houston CS, Terry B, Blom M, Stoffel MJ (2007) Turkey vulture nest success in abandoned houses in Saskatchewan. The Wilson J Ornithol 119:742–747

Instituto Nacional de Pesquisas Espaciais. Projeto Prodes—Monitoramento da Floresta Amazônica Brasileira por Satélite (2014) Available: http://www.obt.inpe.br/prodes/index.php. Accessed 11 Dec 2017

James CD, Landsberg J, Morton SR (1999) Provision of watering points in the Australian arid zone: a review of effects on biota. J Arid Environ 41:87–121

Jareño D, Viñuela J, Luque-Larena JJ, Arroyo L, Arroyo B, Mougeot F (2015) Factors associated with the colonization of agricultural areas by common voles *Microtus arvalis* in NW Spain. Biol Invasions 17:2315–2327

Jullien M, Thiollay JM (1996) Effects of rain forest disturbance and fragmentation: comparative changes of the raptor community along natural and human-made gradients in French Guiana. J Biogeogr 23:7–25

Kross SM, Tylianakis JM, Nelson XJ (2012) Effects of introducing threatened falcons into vineyards on abundance of passeriformes and bird damage to grapes. Conserv Biol 26:142–149

Landers TF, Cohen B, Wittum TE, Larson EL (2012) A review of antibiotic use in food animals: perspective, policy, and potential. Public Health Rep 127:4–22

Lehmann FC (1959) Contribuciones al estudio de la fauna de Colombia XIV. Nuevas observaciones sobre *Oroaetus isidori* (Des murs). Novedades Colombianas 1:169–195

Liberatori F, Penteriani V (2001) A long-term analysis of the declining population of the Egyptian vulture in the Italian peninsula: distribution, habitat preference, productivity and conservation implications. Biol Conserv 101:381–389

Macchi L, Grau HR (2012) Piospheres in the dry Chaco. Contrasting effects of livestock *puestos* on forest vegetation and bird communities. J Arid Environ 87:176–187

Margalida A (2012) Baits, budget cuts: a deadly mix. Science 338(6104):192–192

Margalida A, Donázar JA, Carrete M, Sánchez-Zapata JA (2010) Sanitary versus environmental policies: fitting together two pieces of the puzzle of European vulture conservation. J Appl Ecol 47:931–935

Margalida A, Campión D, Donázar JA (2011) European vultures' altered behaviour. Nature 480:457

Margalida A, Campión D, Donázar JA (2014) Vultures vs livestock: conservation relationships in an emerging conflict between humans and wildlife. Oryx 48:172–176

Marks JS, Cannings RJ, Mikkola H (1999) Family Strigidae (typical owls). In: del Hoyo J, Elliott A, Sargatal J (eds) Handbook of the birds of the world. Volume 5.Barn owls to hummingbirds. Lynx Edicions, Barcelona, España, pp 76–242

Marshall EJP, Brown VK, Boatman ND, Lutman PJW, Squire GR, Ward LK (2003) The role of weeds in supporting biological diversity within crop fields. Weed Res 43:77–89

Matson PA, Parton WJ, Power AG, Smith MJ (1997) Agricultural intensification and ecosystem properties. Science 277:504–509

McLaughlin A, Mineau P (1995) The impact of agricultural practices on biodiversity. Agric Ecosyst Environ 55:201–212

Mcwilliam AN, Cheke RA (2004) A review of the impacts of control operations against the red-billed quelea (*Quelea quelea*) on non-target organisms. Environ Conserv 31:130–137

Meyburg BU, Boesman P (2013) Steppe Eagle (*Aquila nipalensis*). In: del Hoyo J, Elliott A, Sargatal J, Christie DA, de Juana E (eds) Handbook of the birds of the world alive. Lynx Edicions, Barcelona

Meyer WB, Turner BL II (1992) Human population growth and global land-use/cover change. Annu Rev Ecol Syst 23:39–61

Mingozzi T, Estève R (1997) Analysis of a historical extirpation of the bearded vulture *Gypaetus barbatus* (L.) in the western alps (France-Italy): former distribution and causes of extirpation. Biol Conserv 79:155–171

Ministerio del Ambiente y Desarrollo Sustentable de la Nación and Aves Argentinas (2017) Categorización de las aves de la Argentina según su estado de conservación. Ministerio de Ambiente y Desarrollo Sustentable de la Nación and Aves Argentinas BirdLife International

Ministério do Meio Ambiente do Brasil (2014) Lista Nacional Oficial das Espécies da Fauna Brasileira Ameaçadas de Extinção. 2014

Molina B, Martínez F (2008) El Aguilucho Lagunero en España. Población en 2006 y métodos de censo. Seo/BirdLife, Madrid

Morelli F (2013) Relative importance of marginal vegetation (shrubs, hedgerows, isolated trees) surrogate of HNV farmland for bird species distribution in Central Italy. Ecol Eng 57:261–266

Morton DC, DeFries R, Shimabukuro YE, Anderson LO, Arai E, Bon Espirito- Santo F, Freitas R, Morisette J (2006) Cropland expansion changes deforestation dynamics in the southern Brazilian Amazon. PNAS 26:14637–14641

Mougeot F, Garcia JT, Viñuela J (2011) Breeding biology, behaviour, diet and conservation of the red kite (*Milvus milvus*), with particular emphasis on Mediterranean populations. Ecology and conservation of European dwelling forest raptors and owls. (Eds. I Zuberogoitia and JE Martínez) 190–204. Editorial Diputación Foral de Vizcaya, Bilbao, Spain

Murgatroyd M, Underhill LG, Rodrigues L, Amar A (2016a) The influence of agricultural transformation on the breeding performance of a top predator: Verreaux's eagles in contrasting land use areas. Condor 118:238–252

Murgatroyd M, Avery G, Underhill LG, Amar A (2016b) Adaptability of a specialist predator: the effects of land use on diet diversification and breeding performance of Verreaux's eagles. J Avian Biol 47:834–845

Narosky T, Di Giacomo A (1993) Las aves de la provincia de Buenos Aires: distribución y estatus (No. 598.2 (821) NAR)

Newton I (1979) Population ecology of raptors. Poyser, Berkhamsted

Newton I (1998) Population limitation in birds. Academic Press, London

Nieminen M (2010) The impact of large carnivores on the mortality of semi-domesticated reindeer (*Rangifer tarandus tarandus* L.) calves in Kainuu, southeastern reindeer-herding region of Finland. Rangifer 30:79–88

Norberg H, Kojola I, Aikio P, Nylund M (2006) Predation by golden eagle *Aquila chrysaetos* on semi-domesticated reindeer *Rangifer tarandus* calves in northeastern Finnish Lapland. Wildl Biol 12:393–402

Nybakk K, Kjelvik O, Kvam T (1999) Golden eagle predation on semidomestic reindeer. Wildl Soc Bull 27:1038–1042

Oaks JL, Gilbert M, Virani MZ, Watson RT, Meteyer CU, Rideout BA, Shivaprasad HL, Ahmed S, Chaudhry MJI, Arshad M, Mahmood S, Ali A, Khan AA (2004) Diclofenac residues as the cause of vulture population decline in Pakistan. Nature 427:630–633

Ojima DS, Galvin KA, Turner BL (1994) The global impact of land-use change. Bioscience 44:300–304

Olea PP, Sanchez-Barbudo IS, Viñuela J, Barja I, Mateo-Tomás P, Piñeiro A, Mateo R, Purroy FJ (2009) Lack of scientific evidence and precautionary principle in massive release of rodenticides threatens biodiversity: old lessons need new reflections. Environ Conserv 36:1–4

Oñate JJ, Malo JE, Suárez F, Peco B (1998) Regional and environmental aspects in the implementation of Spanish Agri-environmental schemes. J Environ Manag 52:227–240

Ortiz-Santaliestra ME, Resano-Mayor J, Hernández-Matías A, Rodríguez-Estival J, Camarero PR, Moleón M, Real J, Mateo R (2015) Pollutant accumulation patterns in nestlings of an avian top predator: biochemical and metabolic effects. Sci Total Environ 538:692–702

Paz A, Jareño D, Arroyo L, Viñuela J, Arroyo B, Mougeot F, Luque-Larena JJ, Fargallo JA (2013) Avian predators as a biological control system of common vole (*Microtus arvalis*) populations in North-Western Spain: experimental set-up and preliminary results. Pest Manag Sci 69:444–450

Pedrini P, Sergio F (2001) Golden eagle *Aquila chrysaetos* density and productivity in relation to land abandonment and forest expansion in the alps. Bird Study 48:194–199
Perkins AJ, Whittingham MJ, Morris AJ, Bradbury RB (2002) Use of field margins by foraging yellowhammers *Emberiza citrinella*. Agric Ecosyst Environ 93:413–420
Potter C (1997) Europe's changing farmed landscape. In: Pain D, Pienkowski MW (eds) Farming and birds in Europe: the common agricultural policy and its implications for bird conservation. Academic Press, San Diego, pp 25–42
Razafimanjato G, Seing Sam T, Rakotondratsima M, Réné de Roland LA, Thorstrom R (2014) Population status of the Madagascar fish eagle *Haliaeetus vociferoides* in 2005–2006. Bird Conserv Int 24:88–99
Renjifo LM (1999) Composition changes in a subandean avifauna after long term forest fragmentation. Conserv Biol 13:1124–1139
Robinson R, Sutherland WJ (2002) Post-war changes in arable farming and biodiversity in great Britain. J Appl Ecol 39:57–176
Rodríguez-Estrella R, Donázar JA, Hiraldo F (1998) Raptors as indicators of environmental change in the scrub habitat of Baja California Sur, Mexico. Conserv Biol 12:921–925
Rondeau G, Thiollay JM (2004) West African vulture decline. Vulture news 51:13–33
Samuelson KL, Hubbert ME, Galyean ML, Löest CA (2016) Nutritional recommendations of feedlot consulting nutritionists: the 2015 New Mexico state and Texas Tech University survey. J Anim Sci 94:2648–2663
Sanchez-Zapata JA, Carrete M, Gravilov A, Sklyarenko S, Ceballos O, Donazar JA, Hiraldo F (2003) Land use changes and raptor conservation in steppe habitats of eastern Kazakhstan. Biol Conserv 111:71–77
Sarasola JH, Maceda JJ (2006) Past and current evidence of persecution of the endangered crowned eagle *Harpyhaliaetus coronatus* in Argentina. Oryx 40:347–350
Sarasola JH, Negro JJ (2006) Role of exotic tree stands on the current distribution and social behaviour of Swainson's hawk, *Buteo swainsoni* in the argentine pampas. J Biogeogr 33:1096–1101
Sarasola JH, Negro JJ, Hobson KA, Bortolotti GR, Bildstein KL (2008) Can a 'wintering area effect' explain population status of Swainson's hawks? A stable isotope approach. Divers Distrib 14:686–691
Sarasola JH, Santillán MÁ, Galmes MA (2010a) Crowned eagles rarely prey on livestock in Central Argentina: persecution is not justified. Endanger Species Res 11:207–213
Sarasola JH, Solaro C, Santillán MÁ, Galmes MA (2010b) Communal roosting behavior and winter diet of the white-tailed kite (*Elanus leucurus*) in an agricultural habitat on the argentine pampas. J Raptor Res 44:202–207
Schaub M, Zink R, Beissmann H, Sarrazin F, Arlettaz R (2009) When to end releases in reintroduction programmes: demographic rates and population viability analysis of bearded vultures in the alps. J Appl Ecol 46:92–100
Schmiegelow FK, Mönkkönen M (2002) Habitat loss and fragmentation in dynamic landscapes: avian perspectives from the boreal forest. Ecol Appl 12:375–389
Sebastián-González E, Sánchez-Zapata JA, Botella F (2010) Agricultural ponds as alternative habitat for waterbirds: spatial and temporal patterns of abundance and management strategies. Eur J Wildl Res 56:11–20
Secretaría de Ambiente y Desarrollo Sustentable (2004) Primer Inventario Nacional de Bosques Nativos. Informe Regional Espinal. SA y DS, Ministerio de Salud y Ambiente de la Nación. Argentina
Sergio F, Newton I, Marchesi L (2005) Conservation: top predators and biodiversity. Nature 436:192–192
Shave ME (2017) Evaluating the Conservation and Agricultural Applications of Orchard Nest Boxes for a Declining Raptor. Michigan State University, Doctoral dissertation
Shine C, Klemm C (1999) Wetlands, water and the law: using law to advance wetland conservation and wise use. UICN, Gland/Switzerland/Cambridge, UK and Bonn/Germany
Smallwood JA, Causey MF, Mossop DH, Klucsarits JR, Robertson B, Robertson S, Mason J, Maurer JM, Melvin RJ, Dawson RD, Bortolotti GR, Parrish JW Jr, Breen TF, Boyd K (2009)

Why are American kestrel (*Falco sparverius*) populations declining in North America? Evidence from nest-box programs. J Raptor Res 43:274–282

Smith SH, Steenhof K, McClure CJ, Heath JA (2017) Earlier nesting by generalist predatory bird is associated with human responses to climate change. J Anim Ecol 86:98–107

Snow AA, Andow DA, Epts PG, Hallerman EM, Power A, Tiedje JM, Olfenbarger LL (2005) Genetically engineered organisms and the environment: current status and recommendations. Ecol Appl 2:377–404

Soares-Filho BS, Nepstad DC, Curran LM, Cerqueira GC, Garcia RA, Ramos CA, Voll E, McDonald A, Lefebvre P, Schlesinger P (2006) Modelling conservation in the Amazon basin. Nature 440:520–523

Solaro C, Sarasola JH (2015) Nest-spacing, not human presence, influences the breeding of Chimango caracaras (*Milvago chimango*) in a peri-urban reserve. Emu 115:72–75

Stibig HJ, Achard F, Carboni S, Rasi R, Miettinen J (2014) Change in tropical forest cover of Southeast Asia from 1990 to 2010. Biogeosciences 11:247

Tella JL, Pomarol M, Muñoz E, López R (1993) Importancia de la conservación de los mases para las aves en los Monegros. Alytes 6:335–350

Tella JL, Forero MG, Hiraldo F, Donázar JA (1998) Conflicts between lesser kestrel conservation and European agricultural policies as identified by habitat use analyses. Conserv Biol 12:593–604

Thiollay JM (1994) Family Accipitridae (hawks and eagles). In: del Hoyo J, Elliott A, Sargatal J (eds) Handbook of the birds of the world, volume 2.New world vultures to guineafowl. Lynx Editions, Barcelona, Spain, pp 52–105

Thiollay JM (2007) Raptor population decline in West Africa. Ostrich-J Afr Ornithol 78:405–413

Thiollay JM, Rahman Z (2002) The raptor community of Central Sulawesi: habitat selection and conservation status. Biol Conserv 107:111–122

Tilman D (1999) Global environmental impacts of agricultural expansion: the need for sustainable and efficient practices. Proc Natl Acad Sci 96:5995–6000

Tilman D, Fargione J, Wolff B, D'antonio C, Dobson A, Howarth R, Schindler D, Schlesinger WH, Simberloff D, Swackhamer D (2001) Forecasting agriculturally driven global environmental change. Science 292:281–284

Torre I, Díaz M, Martínez-Padilla J, Bonal R, Vinuela J, Fargallo JA (2007) Cattle grazing, raptor abundance and small mammal communities in Mediterranean grasslands. Basic Appl Ecol 8:565–575

Tucker GM, Heath MF (1994) Birds in Europe. Their conservation status. BirdLife International, Cambridge, UK

Ursúa E, Serrano D, Tella JL (2005) Does land irrigation actually reduce foraging habitat for breeding lesser kestrels? The role of crop types. Biol Conserv 122:643–648

USDA National Animal Health Monitoring System (2000) Part I: Baseline reference of feedlot management practices, 1999

Valkama J, Korpimäki E, Arroyo B, Beja P, Bretagnolle V, Bro E, Kenward R, Mañosa S, Redpath SM, Thirdgood S, Viñuela J (2005) Birds of prey as limiting factors of gamebird populations in Europe: a review. Biol Rev 80:171–203

Vallejos M, Volante JN, Mosciaro MJ, Vale LM, Bustamante ML, Paruelo JM (2015) Transformation dynamics of the natural cover in the dry Chaco ecoregion: a plot level geo-database from 1976 to 2012. J Arid Environ 123:3–11

Vargas JDJ, Whitacre D, Mosquera R, Albuquerque J, Piana R, Thiollay JM, Márquez C, Sánchez JE, Lezama-López M, Midence S, Matola S, Aguilar S, Rettig N, Sanaiotti T (2006) Estado y distribucion actual del aguila arpia (*Harpia harpyja*) en Centro y Sur America. Ornitologia Neotropical 17:39–55

Vickery PD, Tubaro PL, Da Silva MCJ, Peterjohn BG, Herkert JR, Cavalcanti RB (1999) Conservation of grassland birds in the western hemisphere. Stud Avian Biol 19:2–26

Virani MZ, Harper DM (2009) Factors influencing the breeding performance of the augur buzzard *Buteo augur* in southern Lake Naivasha, Rift Valley, Kenya. Ostrich 80:9–17

Watson J (2010) The golden eagle, 2nd edn. T. and A.D. Poyser, London

Wells M, Leonard L (2006) DDT contamination in South Africa. The international POPs elimination project – IPEP, report. www.ipen.org
White CM, Olsen PF, Kiff LF (1994) Family Falconidae (Falcons and Caracaras). In: Del Hoyo J, Elliott A, Sargatall J (eds) Handbook of the birds of the world, volume 2 new world vultures to guineafowl. Lynx Edicions, Barcelona, Spain
Woodward KN (ed) (2009) Veterinary pharmacovigilance. Adverse environmental effects and veterinary medicinal products. Blackwell Publishing Ltd, Chichester
Yohannes YB, Ikenaka Y, Nakayama SM, Ishizuka M (2014) Organochlorine pesticides in bird species and their prey (fish) from the Ethiopian Rift Valley region, Ethiopia. Environ Pollut 192:121–128
Zak MR, Cabido M, Cáceres D, Díaz S (2008) What drives accelerated land cover change in Central Argentina? Synergistic consequences of climatic, socioeconomic, and technological factors. Environ Manag 42:181–189
Zalba S, Villamil CB (2002) Woody plant invasion in relictual grasslands. Biol Invasions 4:55–72
Zuluaga S, Echeverry-Galvis MÁ (2016) Domestic fowl in the diet of the black-and-chestnut eagle (*Spizaetus isidori*) in the eastern Andes of Colombia: a potential conflict with humans? Ornitología Neotropical 27:113–120

Chapter 10
Toxicology of Birds of Prey

Judit Smits and Vinny Naidoo

Introduction

Toxicity in birds of prey tends to have unique patterns on a global scale, with specific causes responsible for primary and secondary poisonings. The forces governing this may be culturally, socially or economically based, and are often driven by different proximate causes. Considering causation in biology, proximate causation is due to immediate factors, for example, physiological crisies, severe weather events, or in this case exposure to poisons. Ultimate causation is linked with underlying factors for something happening (Mayr 1982), for example, undervaluing natural resources, which is at the root of many of the problems presented in this chapter.

Apart from habitat fragmentation and degradation from anthropogenic activities, the major threat to raptor populations worldwide is exposure to contaminants and toxicants. Both malicious and unintended poisonings plague these top predatory and scavenging birds. Because of their high trophic position, any contaminants that persist and bioaccumulate through the food web will be highest in these animals. The historic use of so-called 'legacy' pollutants, PCBs, DDT and other OC pesticides, have been employed in the sub/tropics for controlling malaria and/or trypanosomiasis, whereas in the temperate zones, they were mainly used to control insects on food crops. In the 1950s, OCs were considered much safer than the arsenicals which were being used and which commonly caused animal toxicity. In the 1970s, they were recognized to compromise reproduction in raptors, which lead to total

J. Smits (✉)
Department of Ecosystem & Public Health, Faculty of Veterinary Medicine,
University of Calgary, Calgary, AB, Canada
e-mail: jegsmits@ucalgary.ca

V. Naidoo
Biomedical Research Centre, Faculty of Veterinary Science, University of Pretoria,
Pretoria, South Africa

J. H. Sarasola et al. (eds.), *Birds of Prey*,
https://doi.org/10.1007/978-3-319-73745-4_10

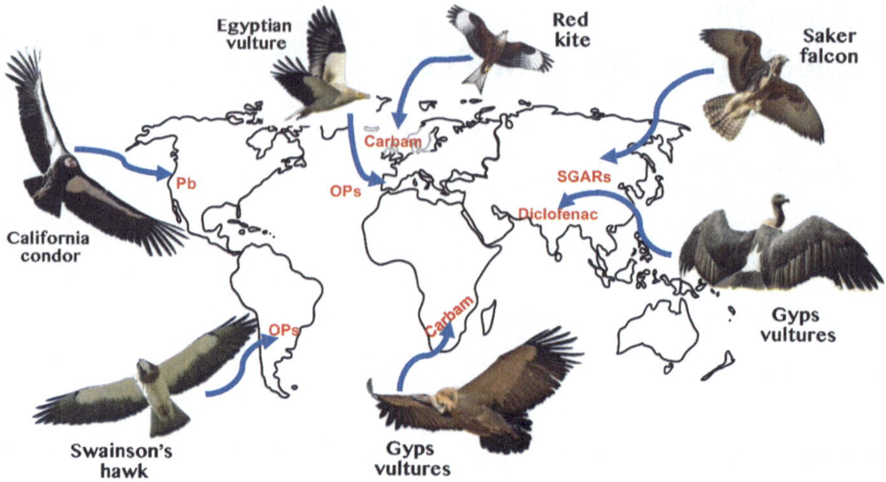

Fig. 10.1 Major poisoning of raptors in various regions of the world is represented by the raptorial species which were most severely affected. The main compounds identified with these toxicities are carbamates (Carbam), the anti-inflammatory drug diclofenac, organophosphate insecticides (OPs), lead (Pb) and second-generation anticoagulant rodenticides (SGARs), although in most areas, multiple compounds have a role in the intentional or unintentional (secondary) toxicity of raptors

bans or strict controls. With rare exception, these pollutants are of minor concern these days. Now, the main toxicants that interfere with the well-being and survival of raptors include insecticides, rodenticides, heavy metals, malicious and intentional poisoning and, most recently, pharmaceuticals (Fig. 10.1). Apart from the more obvious lethal poisonings, the cost to successful survival and fitness from nonlethal, subclinical toxicity from pollutants and poisons is much less understood.

There are estimates that thousands of animals of many species are killed yearly around the world by the legal and illegal use of poisons to control pests (Mateo-Tomás et al. 2012 and references therein). Legal use of pesticides causes secondary, unintended poisonings of scavengers and predators. However, compounds, such as strychnine and endrin banned in Europe and North America, are still used for illegal killing of wildlife (Martínez-Haro et al. 2008).

Because of the numerous challenges inherent in working with wildlife, i.e. dead wildlife are very difficult to detect and when carcasses are found, there is insufficient sample for useful diagnosis (Wobeser et al. 2004; Berny 2007); the true magnitude of any problem is much larger than that represented by the data. Berny (2007) estimates that 18 to 68% of all suspected poisoning cases are detected and recorded, depending on the country, the species and the conditions leading to poisoning.

Raptors, being at the top of the food web and being readily identified and counted, offer a means of understanding forces and changes lower in that food web. This group of predatory and scavenging birds are monitored because of human's

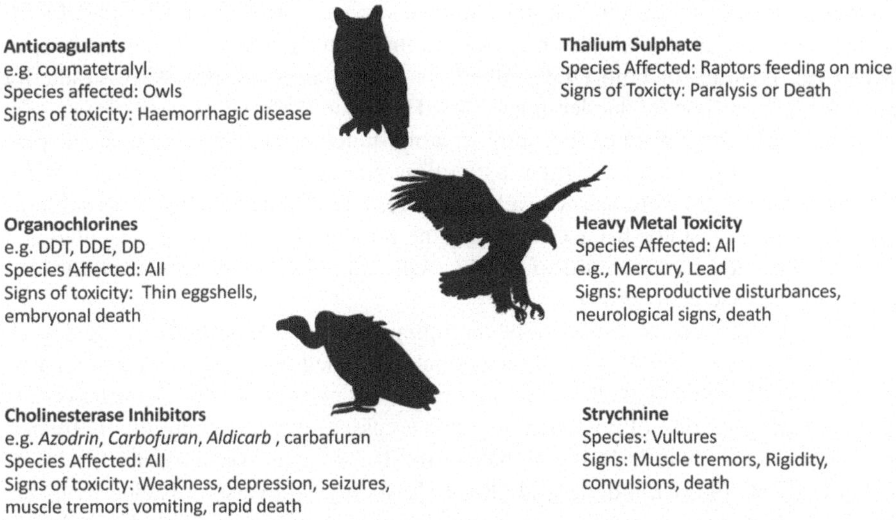

Fig. 10.2 Examples of common toxicants affecting raptors and associated clinical signs following exposure

fascination in them specifically, and to learn through them, what is occurring in the ecosystem in which they live and feed.

We discuss toxicity in raptors by regions of the world, since the toxicity problems and the information available varies greatly from region to region (Fig. 10.2).

Europe and the UK

In Northern and Western Europe, a monitoring system for terrestrial ecosystems was created to track long-range air pollution and other changes driven by anthropogenic forces. Since the 1970s, Norway's contaminant tracking program focused on the white-tailed sea eagle (*Haliaeetus albicilla*) and, more recently, the golden eagle (*Aquila chrysaetos*) and gyrfalcon (*Falco rusticolus*). The peregrine falcon (*Falco peregrinus*), osprey (*Pandion haliaetus*) and merlin (*Falco columbarius*) have been studied more sporadically. Legacy OC contaminants (DDE, dieldrin, hexachlorobenzene and PCBs) have been monitored primarily through egg residues, with brominated and fluorinated contaminants being added most recently (Gjershaug et al. 2008). Egg shell thinning from DDT exposure has improved, but one coastal subpopulation of golden eagles continues to have thinner shells from persistent OC exposure through their diet of coastal birds (Nygard and Gjershaug 2001.) With the ban on DDT and restricted use of mercury, peregrine falcons have returned from near extinction in the mid-1970s to stable populations now.

Sweden has a monitoring program similar to that in Norway, using the white-tailed sea eagle as a sentinel, and measuring persistent pollutants in dead eggs.

Helander et al. (2008) show that reproductive success is still inversely related with residues of DDE and sumPCB, in eggs from the Baltic and Swedish Lapland coast. Mean egg residues of both DDE and PCBs were highest in the 1960s, gradually decreasing over time (Helander et al. 2008; Helander 1985).

A report from Finland of mortality in white-tailed sea eagles over a seven-year period focused on endoparasitism as well as contaminants (Krone et al. 2006). The OCs and PCBs were below levels of concern, but high lead (Pb) was problematic. In a similar study from Greenland, intentional shooting of white-tailed sea eagles accounted for one third of deaths, with some birds having secondary lead poisoning (Krone et al. 2004).

In a six-year retrospective study including birds and mammals in the Czech Republic, raptors made up 39% of the animals poisoned. The carbamate insecticide, carbofuran (Novotný et al. 2011), was found in these 89 cases, with exposure occurring through ingestion of poisoned insects, carcasses or grains intentionally treated with pesticides as bait. In Belgium, half of the 162 buzzards (*Buteo buteo*), submitted to the Toxicology Laboratory at Ghent University, were also poisoned with carbofuran, with the balance of intentional poisonings from aldicarb, carbaryl, strychnine and second-generation anticoagulant rodenticides (SGARs). Other poisoned raptors included Eurasian kestrels (*Falco tinnunculus*), Eurasian sparrowhawks (*Accipiter nisus*) and red kites (*Milvus milvus*) (Vandenbroucke et al. 2010). Similarly, on the Croatian archipelago, carbofuran killed several Euasian buzzards and 17 endangered Eurasian griffons (*Gyps fulvus*) (Muzinic 2007).

The island of Crete is historically an important breeding site for 43 species of raptors, some with international importance such as the bearded vulture (*Gypaetus barbatus*), Golden and Bonelli's (*Hieraaetus fasciatus*) eagles, lanner falcon (*Falco biarmicus*), and Eleanora's falcon (*Falco eleonorae*) among others. A ten-year study here showed that persecution via shooting and poisoning accounted for more than 60% of the mortalities (Xirouchakis 2004), with 20% being due to intentional or secondary poisoning. The anticholinesterases, organophosphate (OP) and carbamate insecticides were the most frequently detected. Intentional killing of peregrine falcons in Switzerland using the carbamate insecticide, methomyl, confirms an ongoing problem with pigeon fanciers being implicated in some cases (Vogler et al. 2015).

A 50-year study of mortality in bearded vultures concentrated in the Pyrenees of France and Spain, confirmed 38% of deaths were due to toxicity from strychnine, OP and carbamates, with most being intentional poisonings (Margalida et al. 2008). In Spain, contaminant-related problems in raptors became evident in the 1980s (Garcia-Fernandez et al. 2008). The biologist community has worked for the decades since then, to track environmental contaminants. Samples such as feathers, unhatched eggs and blood have been used to monitor patterns of metals and persistent organic pollutants (Martinez-Lopez et al. 2009; Blanco et al. 2003). Metals and OC pesticides were the major contaminants considered in the 2008 review (Garcia-Fernandez et al. 2008) which concludes that apart from Pb, other metals are below toxic concentrations. However, it is much more difficult to study and prove subclinical effects from low-level metal toxicity. Changes in behaviour, immunological and thyroid function are only some of the important biological effects that may go unrecognized in free-living birds (Blanco et al. 2004; Nain and Smits 2011; Snoeijs et al. 2005).

The Prestige oil tanker accident that occurred off the northwest coast of Spain in 2002 resulted in egg mortality and nest failure from direct oiling of the egg shells or from parental abandonment of the nests. Poisoned raptors inland along rivers indicated that birds preying on contaminated shorebirds were at highest risk (Zuberogoitia et al. 2006).

Eurasian kestrels in central Italy, undertake minor or no migration making them valuable bioindicators of local pollution. In their study of OC and PCB contaminants in unhatched kestrel eggs, Dell'Omo et al. (2008) found ongoing local sources of pollution around the Aniene, described as a severely polluted branch of the Tiber river. Movalli et al. (2008) also studied the legacy PCB and OC contaminants in Italy in the lanner falcon, an endangered, CITES-listed species in Europe. They examined eggs, blood of nestlings, tissues from dead adults and magpies (*Pica pica*) and doves (*Columba livia*), the falcons' main prey species. Generally, neither OCs nor PCBs were at threatening levels to these falcons. However, subclinical, long-term effects were not studied. From a two-generational study into the effects of PCBs on American kestrels (*Falco sparverius*), there was evidence of disrupted thyroid hormones (Smits et al. 2002), abnormal immune function (Smits and Bortolotti 2001; Smits et al. 2002), behavioural changes (Fisher et al. 2001) and changes in iris colour in juvenile birds (Bortolotti et al. 2003). These subtle changes, which are very difficult to detect, would have long-term consequences on population fitness.

Besides legally applied pesticides, intoxication of raptors and scavenging birds is from poisoned baits illegally used against wolves, bears and other perceived threats to domestic animals (Mateo-Tomás et al. 2012, Sánchez-Barbudo et al. 2012). In a study in Spain of over 1100 cases of suspected poisoning in wild and domestic animals, raptors were the most severely affected group representing 44% of the confirmed poisonings, followed by mammalian carnivores as victims. Organophosphate and carbamate insecticides caused the majority of deaths, followed by rodenticides, strychnine and arsenic. Anticholinesterases, strychnine and arsenic were the pesticides of choice for intentional poisonings (e.g., Fig. 10.3), with clear regional choices of poisons selected. A study, using blood anticholinesterase activity in Eurasian kestrels as a proxy of OP and carbamate intoxication, concluded that these pesticides were not a threat to conservation of these raptors (Vergara et al. 2008). However, in the original study by Fildes et al. (2006), because blood anticholinesterase measurements are effective only within a few days of exposure to the pesticides, those conclusions should be interpreted cautiously. A seven-year study in the French Pyrenees confirmed that 25% of deaths in raptors were poisonings, due mostly to carbamates and OPs plus Pb (Berny et al. 2015). Tavecchia et al. (2012) determined that more than half the deaths in red kites in Mallorca were from poisoning.

Testing rodenticides and other pesticides in Scotland and England falls under the auspices of Wildlife Investigation Scheme (WIIS). When pesticides are suspected in cases of death in wildlife, WIIS investigations are carried out which may trigger a review of the detected pesticide by the UK regulatory body.

Fig. 10.3 An illegal poisoning event with the carbamate insecticide, bendiocarb, in the province of Ciudad Real, Spain, involved 1 red fox (*Vulpes vulpes*), 12 griffon vultures (*Gyps fulvus*) and 1 cinereous vulture (*Aegypius monachus*) (Sánchez-Barbudo et al. 2012). (Photo credit; Rafael Mateo)

The predatory bird monitoring scheme (PBMS), the UK-wide umbrella project responsible for long-term monitoring in avian predators (Walker et al. 2008), tracks specific chemical risks in sentinels such as the barn owl (*Tyto alba*), Eurasian kestrel, red kite and the Eurasian sparrowhawk, plus the piscivorous grey heron (*Ardea cinerea*). Mercury, OC insecticides and industrial PCBs have been tracked since the 1960s. Since the 1980s, second-generation anticoagulant rodenticides (SGARs) were added because of threats to top predatory bids. Ultimately, the role of PBMS is to determine the risk to raptors for these persistent or bioaccumulated contaminants (Walker and Newton 1999). As in most developed countries, declining populations of raptors began to recover in the mid-1980s after OC insecticides were banned (Walker et al. 2008.)

Second-Generation Anticoagulant Rodenticides (SGARs)

Because of their extensive use and food chain transfer to nontarget species, the SGARs are well studied in European raptors. In Denmark, 84% to 100% of owls and other raptors examined had SGARs in their tissues (Christensen et al. 2012).

Of these, up to 37% had potentially lethal levels. The species most severely affected were those on agricultural landscapes: eagle owls (*Bubo bubo*), red kites, Eurasian kestrels, Eurasian buzzards and barn owls. Raptors on more natural habitats, marsh harriers (*Circus aeruginosus*), long-eared owls (*Asio otus*) and short-eared owls (*Asio flammeus*), had much lower exposures. In Danish studies of the mammalian predators, stoats (*Mustela erminea*) and weasels (*Mustela nivalis*), again there was very high prevalence of SGAR exposure with a strong association between their tissue rodenticide concentrations and poor body condition, as well as with death from unexplained causes (i.e., not from trauma). In dead golden eagles and eagle owls in Norway, 30% of the 24 animals examined had SGAR levels above the suspected lethal level of 100 ng/g (Langford et al. 2013). In the same regions gyrfalcons, peregrine falcons and osprey, species which have a different prey base than the affected eagles, had no detectable levels, offering an obvious link with a rodent diet and agricultural land management. In Spain, Sánchez-Barbudo et al. (2012) found equally high levels of lethal poisonings (35%) directly linked with SGARs, owls being particularly highly represented.

After France adopted measures to decrease the poisoned bait used by farmers for rodent control around 2000, there was a dramatic reduction in poisoned wildlife. But a population explosion in voles (*Arvicola terrestris*) in 2011 lead to increased bromadiolone use and related high mortality in common (Eurasian) buzzards and red kites, the latter of which were thought to have experienced compromised reproduction in that same year (Coeurdassier et al. 2014). To address both agricultural and conservation priorities, new means of vole control, such as trapping, physical destruction of tunnels, landscape management and fostering certain predators, are being implemented to avoid these secondary toxicities (Fig. 10.4).

Reintroduced red kites in the UK, which are predators and scavengers, are frequently exposed to the SGARs, bromadiolone and difenacoum, at levels similar to those found in eurasian kestrels but higher than in most other British predators

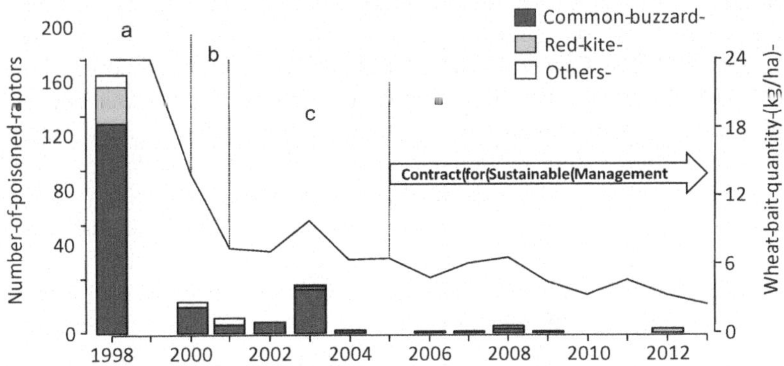

Fig. 10.4 Dramatic temporal decrease in secondary toxicity of raptors, before regulation of rodenticide use (**a**), after legal restrictions were implemented, but with permitted relaxation of the law during vole outbreaks (**b**), followed by no exceptions to tightly restricted applications of the SGAR bromadiolone (**c**), in Eastern France. (Figure adapted from Coeurdassier et al. 2014)

(Shore et al. 2006). Although there is intentional killing of golden eagles and hen harriers (*Circus cyaneus*) with pesticide baits in support of grouse hunting and moor management (Whitfield et al. 2003), red kites being largely scavengers are most likely unintentionally poisoned by the baits set out to target foxes and the predatory raptors. In a 15-year study in the UK, SGARs were found in all of the barn owls confirmed to have died of poisoning (Newton et al. 1990 and Wyllie 2002).

Lead (Pb)

In a 20-year survey of lead exposure in bearded vultures, blood samples from live animals and tissues from dead ones (total $n = 130$) indicated very low levels of Pb (Hernández and Margalida 2009). Similarly, Kenntner et al. (2007) examined heavy metal exposure in golden eagles from the European Alps of Germany, Switzerland and Austria. Apart from two birds with confirmed secondary Pb poisoning, Pb, Cd and Hg were found at low background concentrations. Although preying on waterfowl on heavily hunted wetlands can be a source of Pb exposure in eagles, kites and other raptors (Mateo et al. 2001), and Pb is recognized as the most common metal causing toxicity in raptors in rehabilitation centres and veterinary facilities (Rodriguez et al. 2013), it may be of less importance to the stability of European raptor populations, compared with pesticides (for further details in Pb effects on raptors, see also in this book Chap. 11).

North and South America

In contrast to the situation in Europe and the UK, in North America the majority of poisoning in raptors appears to be secondary or unintended toxicity. A broad study examined more than 500 raptor mortalities over ten years, including cases from Canada, the USA and Europe (Mineau et al. 1999). In North America, poisoning occurred with almost equal frequency from the labelled use of pesticides, as it did from illegal cases of intentional abuse, although not aimed at raptors. These unintended toxicities were from raptors consuming baits set out to kill coyote (*Canis latrans*), red fox (*Vulpes vulpes*), striped skunk (*Mephitis mephitis*) and other perceived nuisance wildlife (Allen et al. 1996; Wobeser et al. 2004). Organophosphates, used both as avicides and as antiparasitic treatments on livestock, have resulted in raptor intoxications. Secondary poisonings occur from consumption of insects from crops recently sprayed with OPs or carbamates, or earthworms and other invertebrates contaminated after soil application of granular pesticides and seed treatments (Mineau et al. 1999). Carbofuran and the OPs applied in low-pH wetlands caused secondary poisoning of raptors months after application. This route of exposure was through poisoned waterfowl consuming active pesticide granules months after they had been applied (Elliott et al. 1996). These pesticide formulations were discontinued largely due to their environmental persistence of up to three years. Sublethal OP exposure, likely more widespread than reported, has been confirmed in migrating snowy owls (*Bubo scandiacus*) (Wildlife Disease Association newsletter, Apr 2016).

Apart from consuming invertebrates, in North America, the major risk factors to birds of prey involve feeding on agricultural areas and opportunistically consuming debilitated prey, whereas in the UK and Europe, being perceived as a pest species poses the greatest threat to raptors. Based upon necropsy reports since approximately 1990, organophosphate and carbamate pesticides appear to have superseded strychnine as major, problematic compounds for illegal killing of vertebrates (Wobeser et al. 2004).

The SGARs are of concern in North America as well as overseas. The difficulty lies in determining when SGAR residues are directly or indirectly responsible for toxicity in raptors. Based on tissue samples from almost 200 great horned owls (*Bubo virginianus*) and red-tailed hawks (*Buteo jamaicensis*), 11% of these has SGAR levels consistent with overt poisoning (Thomas et al. 2011).

In what is likely an under-reported source of poisoning in scavenging raptors is the consumption of domestic animals that have been euthanized with barbiturates, which can produce lethal poisoning unless there is medical intervention (Langelier 1993; Berny et al. 2015). Pharmaceuticals, as seen with the anti-inflammatory drug diclofenac, which almost extirpated Gyps vultures from the Indian subcontinent, maintain potency in treated carcasses since drugs are not metabolized once the animal has died, which results in secondary toxicity of scavengers.

A massive poisoning of raptors occurred on the Argentinian pampas in the mid-1990s, subsequent to a severe outbreak of grasshoppers and caterpillars. The outbreak was controlled with monocrotophos, the OP insecticide eventually proven to be responsible for this kill-off of Swainson's hawks (*Buteo swainsoni)* (Goldstein et al. 1999). Direct ingestion of contaminated invertebrate prey was the likely cause of death in most birds. While on their wintering grounds, these hawks forage and roost together in large congregations which inevitably feed of the same sources of prey. The 5000 plus mortalities were counted or estimated from searches which occurred over just one small section of the known Swainson's hawk wintering range. This count represented approximately 1% of the total population of 513,000 hawks that overwinter in large groups in Argentina, making the actual numbers poisoned, much greater. This die-off resulted in restrictions or banning for the use of this OP (see also in this book Chaps. 9 and 16).

Lead ingested through hunted waterfowl and other game animals, such as deer shot with lead bullets, has provided a constant source of exposure for scavenging raptors (Fisher et al. 2006). Although lead shot has been largely banned for waterfowl hunting, poisoning of raptors continues from Pb fragments consumed with offal from hunted mammals that have been field dressed (Bedrosian et al. 2012). There are efforts to ban Pb ammunition to protect consumers of hunted animals, humans as well as predators and scavengers. As of late January 2017, the Obama administration in the USA banned the use of ammunition and fishing tackle containing Pb, on all public lands in the USA https://www.fws.gov/policy/do219.html. However, the Trump administration reversed this policy.

Two surveys in South America evaluated the status of Pb in raptors including the endangered Andean condor (*Vultur gryphus*) showing levels to be generally low (Saggese et al. 2009; Lambertucci et al. 2011). However, clinical signs of Pb poisoning in the endangered crowned eagle (*Harpyhaliaetus coronatus*) have raised the concern for vigilant monitoring (see also in this book Chap. 16).

Australia and New Zealand

In New Zealand, recorded mortalities of the Australasian harrier (*Circus approximans*) and morepork (*Ninox novaeseelandiae*), also known as the Tasmanian spotted owl, coincided with their brodifacoum-based rat and rabbit eradication programs (Eason and Spurr 1995). Improved pest management strategies entailing controlled and pulse applications, plus secure bait stations for SGARs, successfully decreased secondary toxicity to nontarget species (Eason et al. 2002).

In Australia, very poor reproductive success has been documented in white-bellied sea eagles (*Haliaeetus leucogaster)* over the past 1 ½ decades. Although necropsies with contaminant analyses have ruled out death due to environmental toxicants, Manning et al. (2008) recommended surveying raptors around Australia to establish baseline data for contaminants, to inform management and regulatory decisions supporting long-term viability of native raptor populations. Other raptor researchers in Australia hypothesize that pindone, a warfarin-like rodenticide, may be causally linked with the notable 20-year continuous decline of little eagles (*Hieraaetus morphnoides*) in southern Australia in spite of the fact that their main prey, rabbits, have thriving populations in the regions where the eagle numbers are decreasing (Olsen et al. 2013). Little eagles, wedge-tailed eagles and whistling kites (*Haliastur sphenurus*) inexplicably dead around their failed nests were never analysed for the suspect compound, pindone. Olsen and Osgoode (2006) earlier raised concerns, unanswered to date, about the role of insecticides in the decreasing little eagle populations.

In more than 100 raptors submitted to a veterinary and rehabilitation facility from an agricultural area in Western Australia (Punch 2001), there were no diagnoses of toxicity. This may indicate that the Pb and pesticide poisoning seen in the rest of the world are rare in Australian raptors, or it may reflect a bias in the diagnostic efforts in veterinary clinics, since wounds, fractures and infectious diseases are more commonly encountered and treated. There is brief mention of the possible threat of toxicity from rodenticides and insecticides used on cotton crops, but no cases are discussed (Mooney 1998).

East and South Asia

The large, fish-eating Steller's sea eagles, *Haliaeetus pelagicus*, have breeding grounds limited to coastal eastern Russia. The smaller white-tailed eagle is more widely distributed, ranging from Europe through eastern Asia. A study of dead sea eagles in Hokkaido, Japan, examined both species for causes of mortality. The birds had high tissue residues of PCBs and DDT metabolites at levels which cause egg shell thinning (Fujita et al. 2001). These OC contaminants likely originate in eastern Siberia, based on the atmospheric concentrations of these pollutants in Siberian cities compared with those in Hokkaido.

Secondary Pb poisoning was diagnosed in a subset of these Steller's and white-tailed sea eagles, because of ingestion of bullet fragments while scavenging on deer carcass remains left by hunters (Iwata et al. 2000). Around the same time, veterinarians from the Hokkaido Veterinary Medical Association diagnosed Pb poisoning in eagles from consuming rifle bullet fragments in hunted Sika deer *Cervus nippon* (Kurosawa 2000). Further studies in Japan over the next ten years confirmed high number of deaths from Pb (up to 89 ppm Pb), with fragments of Pb bullets and shotgun pellets in the eagles' gastrointestinal tracts (Saito 2009). Strong desire to protect these birds was demonstrated when these findings, plus support from the Veterinary Medical Associations, resulted in the Environment Ministry in Hokkaido banning all Pb from ammunition from 2001. Regrettably, Pb ammunition continues to poison these raptors because of inconsistent hunter compliance and because the other Japanese provinces have not adopted this regulation.

In the past, population control of small mammals over large areas of Russia and neighbouring countries have been carried out because of their perceived role in epizootics of plague, or because of competition for crops (Shilova 2015 and references therein). From the 1960s to the 1980s, extermination of native rodents in dry steppe, semi-desert and even in the high alpine ecosystems of Russia were accomplished using highly toxic agents such as zinc phosphide and anticoagulants (Boreyko and Parnikoza 2012; Shilova 2015). Ground squirrels and pikas were largely exterminated, killing birds of prey through secondary rodenticide poisoning or starvation. Mongolian pikas account for 64 to 92% of the diet of birds of prey in the alpine steppes. Catastrophic population declines in the eastern imperial eagle (*Aquila heliaca*) and steppe eagle (*Aquila nipalensis*) in the Tuva region of southern Russia were directly linked with the birds' migration routes passing through the rodenticide-treated territories of Mongolia (Karyakin 2010).

On a hopeful note, after 20 years of almost complete absence, and after the large scale application of these rodenticides in Russia was stopped in the 1980s, largely because of the economic collapse (Shilova 2015), some ecosystem recovery has occurred. The population of little bustards (*Tetrax tetrax*) on the northern Trans-Ural steppe of Russia has increased (Korovin 2014), almost reaching levels from before their disappearance. The bustards prefer fallow land but also hunt on grass crops and harvested grain fields. This reversal is associated with the steady reduction in agricultural production in the last two decades.

China

Since 1988, China has had a National Animal Protection Act which is meant to protect all birds of prey, owls included. For example, the saker falcon (*Falco cherrug*) is on Appendix II of the Chinese Red list of threatened species (Xiaodi et al.

2001). However, at the same time, local government policies encourage intense rodent eradication programs over much of the steppe areas, which (i) directly decreases the prey base causing major decreases in raptor numbers (Gombobaatar et al. 2004), and (ii) increases secondary toxicity from rodenticide poisoning. In a newsletter publication, researchers raised the alarm about the very low numbers and poor recruitment in saker falcon populations in central Mongolia (Xiaodi and Fox 2003). The most important reason, apart from persecution and illegal smuggling of birds, has been steppe transformation to agricultural production and therefore widespread use of insecticides, rodenticides and treated seeds (Xiaodi et al. 2001). In five years, the golden eagle population has decreased by 66% (Ma 2013). In central Mongolia, a study of saker falcons determined that anthropogenic threats were responsible for 25% of the known deaths of these raptors, with the majority being from intentional nest destruction to clear power pylons and utility poles (Gombobaatar et al. 2004). Ten percent of total mortality was from poisoning by the SGAR bromadiolone. In direct conflict with the National Animal Protection Act, local governments particularly in Western China encourage falconry through subsidies to families for capturing and taming birds. Although it has been banned for 60 yrs, falconry is becoming popular again, influenced by neighbouring Kyrgyzstan, Mongolia and Kazakhstan which are active in this sport (Ma 2013 and references therein).

Mining developments and rodenticides are also important threats to raptors in China (Ma 2013), but published data are very scarce. Ma's publication was based on information from raptor researchers collected over 10 yrs. With international input, this raptor survey examined the distribution and ecology of falcons, eagles, vultures and buzzards in Xinjiang and Tibet. Intensification of mining activities was tightly correlated with the disappearance of golden eagles from their traditional breeding sites. For example, in the Karamay Mountain Nature Reserve in China (44°50′N, 89°00′E), the reserve boundary was changed three times between 2006 and 2010 to accommodate mining activity, wire fences, high-voltage power installations, asphalt roads and a new railway, dramatically shrinking any protected areas.

Raptor conservation in China is greatly complicated by cultural, economic, political and social factors (MaMing et al. 2014). As poignant examples, raptors are killed to use in Chinese medicine, while humerus bones of vultures and eagles are favoured to make flutes in Tibet and Tajikistan. As well, in the western steppes, raptors are intentionally killed 'en masse' through poisoned mouse baits, so they can be sold to museums, schools, research institutes and other markets.

Middle East

There is a conspicuous and utter lack of published or accessible information on health, population status or causes of mortality in wild raptors and owls in the Middle East including Saudi Arabia. Ostrowski and Shobrak (2001) confirmed

OP or carbamate poisoning in one lappet-faced vulture in Saudi Arabia. The authors reported that broad-scale insecticide spraying occurs regularly to control insects, posing a major threat to survival of scavenging raptors which feed on contaminated ruminant carcasses after spraying events. In contrast, another report from the same author states that nest site disturbance is probably the major threat to these raptors, with insecticide poisoning only having 'some' effect (Shobrak 2003).

In a veterinary facility in Saudi Arabia Naldo and Samour (2004) report that in captive falcons, Pb from bullet fragments consumed in prey shot by the falconers was the main cause of toxicity. A bizarre toxicosis from ammonium chloride was reported in 35 falcons admitted to this veterinary hospital. This unfortunate dosing of falcons with ammonium chloride, a compound known to be highly toxic and painful to the birds, is still a widely practiced tradition meant to make the birds hungry and therefore better hunters.

In a laudable action, Iran announced that it banned the sale and use of diclofenac for veterinary species during a congress on the Conservation of Migratory Birds of Prey in Africa and Eurasia (Raptors MOU), 5–8 October 2015 in Trondheim, Norway. Hopefully, mounting pressure will lead the European Union to follow Iran's example, and also ban veterinary use of diclofenac which is still available in Italy and Spain, although there are safer alternatives for use in food-producing animals (http://www.savevultures.org/save_latestnews.html#iran).

For raptors migrating between Africa and Eurasia, Israel forms the main land bridge where up to 1.2 million raptors have been recorded in a single spring migration (Yosef 1996). The greatest problems of poisoning here as elsewhere around the world have been from the cholinesterase-inhibiting insecticides and the SGAR rodenticides (Bahat 2001; Shlosberg 2011). In Israel, OPs used to control the pest vole (*Microtus guentheri*) on alfalfa crops, killed at least 145 raptors from 10 species (Mendelssohn and Paz 1977). Many species of poisoned raptors have never recovered pre-poisoning population levels (Yom-Tov et al. 2012). Heavy metals, OCs, brominated and fluorinated compounds in griffon vultures in Israel are detected at background levels (Sholsberg et al. 2012), but illegal use of carbamates and OPs against wolves and feral dogs is the major persistent threat to raptors here (Hatzofe, personal commun, Israel Nature and Parks Authority, 2017). Thallium sulphate, a highly toxic, metal-based rodenticide, came into use circa 1950 in Israel as a wheat seed treatment to control field mouse populations. Of 36 species of raptors reported in the area, 12 fed predominantly on field mice. Shortly after the baiting began, thallium poisoning was confirmed in dead birds. Since then, rodent-dependent raptors have been almost exterminated (Kolman 2007). This very toxic thallium rodenticide has been banned in many countries but, unfortunately, is still used in some developing countries (Kazantzis 2000).

Apart from unintentional poisoning, the disturbing findings of persecution of raptors originating from Jordan and Israel primarily, have been detected at the Birding Center in Eilat, Israel. The brutal harm to these birds included feet bound together with wire or string, eyelids stitched open and wings partially clipped (Yosef 1996).

Indian Subcontinent

Prakash et al. (2003) were the first to alert the world to the most massive kill-off of birds of prey in history. Over 10 million birds are estimated to have died in just over a ten-year period.

In the 1990s on the Indian subcontinent, three species of vultures the white-rumped (*Gyps bengalensis*), Indian (*G. indicus*) and slender-billed (*G. tenuirostris*) vultures were brought to the brink of extinction. Ironically, a veterinary anti-inflammatory drug, diclofenac, turned out to be responsible (Oaks et al. 2004). This unintentional exposure resulted from the religious value placed on cattle there. Cattle are not slaughtered but mainly die naturally and are left for scavenging species, attracting the large numbers of vultures living there. Injured or ill cattle treated with diclofenac, die having diclofenac in all tissues. Exposed vultures die within 48 h with characteristic pathological lesions of visceral gout and massive renal tubular damage. Similar toxicity in other species including the African white-backed (*G. africanus*), Cape griffon (*G. coprotheres*), the Eurasian griffon (*G. fulvus*) and the Himalayan (*G. himalayensis*) vultures would occur from this and other anti-inflammatory drugs such as ketoprofen and aceclofenac (Swan et al. 2006; Naidoo et al. 2010; Das et al. 2011; Galligan et al. 2016).

Other contributing factors in this massive poisoning of vultures was the extremely low cost and therefore wide-scale use of diclofenac among veterinary clinics in Pakistan and India, exacerbated by the gregarious nature of vultures with a large number of birds feeding on each contaminated cattle carcass. Additionally, old-world vultures seem unusually sensitive to the drug compared with mammals and other birds (Rattner et al. 2008). Based on population modelling, contamination of as few as 1% of the total carcasses available for vultures could have caused this massive scale of toxicity (Green et al. 2007).

A 2013 survey of raptors in India reported that OC residues, mostly DDE, were below levels that cause deleterious effects (Dhananjayan and Muralidharan 2011).

Africa

As with the Middle East, despite extensive review of literature, there is a paucity of information from much of Africa except for the more developed countries such as South Africa and Kenya. A common problem with information about poisoning in raptors in Africa is that although the deaths of birds are reported, there are rarely supporting investigations into causes of the toxicities (Botha 2015; Virani et al. 2011; Thiollay 2007; Bridgeford 2001; Brown 1991a, b).

Birds of prey in Africa are being poisoned maliciously and accidentally. Intentional poisoning has been for harvesting birds for bush meat or traditional medicines (Roxburgh & McDougall 2012), and for their elimination linked to incorrect assumptions on their involvement in livestock killings. Vultures specifically are targeted by poachers to reduce the chance of their poaching activities being detected by game rangers (Groom et al. 2013; Ogada 2014). Vultures circle carcasses before landing for group feeding, so poachers bait the carcasses to kill as many vultures as possible to hide their nefarious activities. Vultures are also sometimes poisoned by some farmers who mistakenly believe that vultures prey upon their new born lambs or that vultures contaminate water holes by bathing in them and cause death in livestock.

In central Kenya, carbofuran poisoning has resulted in mass mortalities in vultures, in large part because this cheap cholinesterase-inhibiting pesticide is used so extensively that soil, water and plants are highly contaminated (Ogada and Keesing 2010). Likewise in South Africa, 49 vultures plus other raptors died following their accidental exposure to carbamates in baited carcasses meant to control jackal predation (Botha et al. 2015). Fenthion is another OP commonly used throughout Africa to control quelea (*Quelea quelea*), a seed-eating passerine responsible for major crop losses. Mcwilliam and Cheke (2004) provide an extensive summary of 30 reports of fenthion toxicity in raptors around Africa. Another OP, dichlorfos, used to control locusts in the Sahel region of Africa has induced mortalities in up to one in five exposed raptors (Keith and Bruggers 1998). Naidoo and Wolter (2016) found that the natural choline esterase activities in healthy Cape griffon vultures is substantially lower than other raptor species, resulting in them being unusually sensitive to these toxicants.

Strychnine, a poison which causes muscle rigidity and convulsions, with death generally due to respiratory failure, in Ethiopia is used for the illegal baiting of wild carnivores and stray dogs, with primary and secondary poisoning identified in Egyptian vultures (Arkumarev et al. 2014). The dog carcasses are illegally put into the city dumps, although official procedure requires them to be buried. Numerous reports of vulture and other raptor deaths from strychnine exposure are reported for South Africa (Allan 1989) and Namibia (Brown 1991a, b; Bridgeford 2001).

As with other areas in the worlds, in Africa most Pb toxicity occurs from exposure to lead bullets or fragments thereof. In Botswana, of almost 500 wild white-backed vultures blood sampled for Pb, 28% were exposed to Pb with a small number having toxic levels (Kenny et al. 2015). There were similar findings in South Africa in both Cape griffon and white backed vultures sampled around the country indicating widespread exposure to Pb (Naidoo et al. 2017). The physiological impact of chronic low Pb burdens in wild raptors is not known although studies show both hormetic (positive at low levels) and detrimental effects (Nain and Smits 2011; Naidoo et al. 2017).

While no formal reports of SGAR toxicity could be found, popular press blogs in Port Elizabeth, South Africa, implicated rodenticides in the deaths of booted eagles (*Aquila pennata*), wood owls (*Strix woodfordii*), rock kestrels (*Falco rupicolus*) and spotted eagle owls (*Bubo africanus*), while an online news report suggests that such SGARs were responsible for the death of barn owls in the KwaZulu-Natal province of South Africa, with the dead birds showing pathology consistent with anticoagulant toxicity. Zinc phosphide used for rodent control is also suspected in unintentional raptor deaths (Keith and Bruggers 1998).

Table 10.1 Blood lead concentrations in several vulture populations in Southern Africa

Species	Country	Status	*n*	Mean	SD	≥ 10 μg/dL
G. africanus	Botswana	Wild	477	10.6	11.36	147
G. africanus	South African	Wild	90	16.27	8.19	40.95
		Captive	6	19.67	7.94	5
G. coprotheres	South African	Wild	49	18.95	18.26	19
		Captive	35	11.71	4.97	18
G. africanus	Namibia	Wild	11	5.27	1.56	0
G. barbatus	South African	Wild	5	<3		0

Lead greater than 10 ug/dL exceed background environmental exposure; between 50 and 100 ug/dL indicates subclinical toxicity which typically results in reproductive abnormalities, whereas blood lead above 100 ug/dL causes clinical, usually lethal toxicity

On the small nation of Cape Verde off the Senegalese coast of Africa, compared with semi-scavenging and predatory raptors, scavengers are suffering more severe population declines and local extinctions (Hille and Collar 2011). Pesticides in common use there are OP and carbamate insecticides, DDT, strychnine and rodenticides – all of which are likely contributing to the precipitous decline in scavenger populations from secondary poisoning (Hille and Collar 2011). Scavengers would be most affected by the local practice of poisoning feral dogs with laced meat (Hazevoet et al. 1996), with the birds eating the bait or the poisoned carcasses.

Current levels of DDE in eggs or blood of raptors in Africa no longer appear to be associated with thinning of egg shells, egg death or embryonal death (Hollamby et al. 2004). It appears that pesticide poisoning is one of the main factors responsible for the major reduction of vulture populations in sub-Saharan Africa with additional problems from the ingestion of lead (Table 10.1).

Summary and Conclusions

Risk maps could provide a useful complement when classifying possible wildlife poisoning events in which toxicological analysis cannot provide conclusive results. However, it would be extremely difficult to create such maps because of the challenges inherent in working with wildlife that are found dead, such as unknown population sizes, inadequate remains for diagnosis, etc. (Guitart et al. 2010). Wildlife researchers are acutely aware of these problems. Published data confirm that most effects on wildlife are never observed, and much of observed mortality is not reported (Vyas 1999). Even delays in reporting or responses to reports of dead raptors, plus the existence of multiple stressors for any wildlife, distort any exposure–effect relationship and contribute further to uncertainty in determining the cause of death.

Although it would be appealing to be able to identify or link specific types of poisoning in raptors with particular countries or regions of the world, this would be fraught with risk, in large part because of the skewed information available to researchers.

Deliberate poisoning and persecution of raptors continues illegally around the world to eliminate perceived predators or for other 'cultural' reasons. Globally, insecticides, rodenticides, and Pb ammunition in the environment, secondarily, inflict heavy costs to birds of prey.

References

Allan DG (1989) Strychnine poison and the conservation of avian scavengers in the Karoo, South Africa. S Afr J Wildl Res 19(3):102–106

Allen GT, Veatch JK, Stroud RK, Vendel CG, Poppenga RH, Thompson L (1996) Winter poisoning of coyotes and raptors with furadan- laced carcass baits. J Wildl Dis 32:385–389

Arkumarev V, Dobrev V, Abebe YD, Popgeorgiev G, Nikolov SC (2014) Congregations of wintering Egyptian vultures *Neophron percnopterus* in afar, Ethiopia: present status and implications for conservation. Ostrich 85(2):139–145

Bahat O (2001) Conservation of threatened raptor populations in Israel. In Wings over Africa. In: Proceedings of the international seminar on bird migration: research, conservation, education and flight safety, pp 177–189

Bedrosian B, Craighead D, Crandall R (2012) Lead exposure in bald eagles from big game hunting: the continental implications and successful mitigation efforts. PLoS One 7(12):e51978

Berny P (2007) Pesticides and the intoxication of wild animals. J Vet Pharmacol Ther 30:93–100

Berny P, Vilagines L, Cugnasse JM, Mastain O, Chollet JY, Joncour G, Razin M (2015) VIGILANCE POISON: illegal poisoning and lead intoxication are the main factors affecting avian scavenger survival in the Pyrenees (France). Ecotoxicol Environ Saf 118:71–82

Blanco G, Frias O, Jimenez B, Gomez G (2003) Factors influencing variability and potential uptake routes of heavy metals in black kites exposed to emissions from a solid waste incinerator. Environ Toxicol Chem 22:2711–2718

Blanco G, Jimenez B, Frias O, Millan J, Davilla JA (2004) Contamination with nonessential metals from a solid-waste incinerator correlated with nutritional and immunological stress in prefledging black kites (*Milvus migrans*). Environ Res 94:94–101

Boreyko VE, Parnikoza IY (2012) Zinc phosphide – a threat to the steppe ecosystems of Ukraine. Steppe Bulletin 36: pp 25–28

Bortolotti GR, Smits JE, Bird DM (2003) Iris colour of American kestrels varies with age, sex and exposure to PCBs. Physiol Biochem Zool 76:99–104

Botha A, (2015) 65 endangered vultures killed in poisoning incident. *African Geographic*. available online on http://africageographic.com/blog/65-endangered-vultures-killed-in-poisoning-incident.

Botha CJ, Coetser H, Labuschagne L, Basson A (2015) Confirmed organophosphorus and carbamate pesticide poisonings in south African wildlife (2009–2014). J S Afr Vet Assoc 86(1):1329

Bridgeford P (2001) More vulture deaths in Namibia. Vultures News 44:22–26

Brown CJ (1991a) Declining martial *Polemaetus bellicosus* and tawny *Aquila rapax* eagle populations and causes of mortality on farmlands in Central Namibia. Biol Conserv 56(1):49–62

Brown CJ (1991b) An investigation into the decline of the bearded vulture *Gypaetus barbatus* in southern Africa. Biol Conserv 57(3):315–337

Christensen TK, Lassen P, Elmeros M (2012) High exposure rates of anticoagulant rodenticides in predatory bird species in intensively managed landscapes in Denmark. Arch Environ Contam Toxicol 63(3):437–444

Coeurdassier M, Riols R, Decors A, Mionnet A, David F, Quintaine T (2014) Unintentional wildlife poisoning and proposals for sustainable management of rodents. Conserv Biol 28(2):315–321

Das D, Cuthbert RJ, Jakati RD, Prakash V (2011) Diclofenac is toxic to the Himalayan vulture *Gyps himalayensis*. Bird Conservation International 21(1):72–75

Dell'Omo G, Constantine D, Wright J, Casagrande S, Shore R (2008) PCBs in the eggs of Eurasian kestrels indicate exposure to local pollution. AMBIO J Hum Environ 37(6):452–455

Dhananjayan V, Muralidharan S (2011) Distribution of persistent organochlorine chemical residues in blood plasma of three species of vultures from India. *Environ Monit Assess* 173(1):803–811

Eason CT, Spurr EB (1995) Review of the toxicity and impacts of brodifacoum on non-target wildlife in New Zealand. New Zeal J Zool 22(4):371–379

Eason CT, Murphy EC, Wright GR, Spurr EB (2002) Assessment of risks of brodifacoum to non-target birds and mammals in New Zealand. Ecotoxicology 11:35–48

Elliott JE, Langelier KM, Mineau P, Wilson LK (1996) Poisoning of bald eagles and red-tailed hawks by carbofuran and fensulfothion in the Fraser Delta of British Columbia, Canada. J Wildl Dis 32(3):486–491

Fildes K, Astheimer LB, Story P, Buttemer WA, Hooper MJ (2006) Cholinesterase response in native birds exposed to fenitrothion during locust control operations in eastern Australia. Environ Toxicol Chem 25:2964–2970

Fisher SA, Bortolotti GR, Fernie KJ, Smits JEG, Marchant TA, Drouillard KG, Bird DM (2001) Courtship behavior of captive Kestrels (*Falco sparverius*) exposed to polychlorinated biphenyls. Arch Environ Contam Toxicol 41:215–220

Fisher IJ, Pain DJ, Thomas VG (2006) A review of lead poisoning from ammunition sources in terrestrial birds. Biol Conserv 131(3):421–432

Fujita S, Chiba I, Ishizuka M, Hoshi H, Iwata H, Sakakibara A (2001) P450 in wild animals as a biomarker of environmental impact. Biomarkers 6(1):19–25

Galligan TH, Taggart MA, Cuthbert RJ, Svobodova D, Chipangura J, Alderson D (2016) Metabolism of aceclofenac in cattle to vulture-killing diclofenac. Conserv Biol 30(5):1122–1127

García-Fernández AJ, Calvo JF, Martínez-López E, María-Mojica P, Martínez JE (2008) Raptor ecotoxicology in Spain: a review on persistent environmental contaminants. AMBIO J Hum Environ 37(6):432–439

Gjershaug JO, Kålås JA, Nygård T, Herzke D, Folkestad AO (2008) Monitoring of raptors and their contamination levels in Norway. AMBIO J Hum Environ 37(6):420–424

Goldstein MI, Lacher TE, Zaccagnini ME, Parker ML, Hooper MJ (1999) Monitoring and assessment of Swainson's hawks in Argentina following restrictions on monocrotophos use, 1996–97. Ecotoxicology 8:215–224

Gombobaatar S, Sumiya D, Shagdarsuren O, Potapov E, Fox N (2004) Saker falcon (*Falco cherrug milvipes Jerdon*) mortality in Central Mongolia and population threats. Mong J Biol Sci 2(2):13–21

Green RE, Taggart MA, Senacha KR, Raghavan B, Pain DJ, Jhala Y, Cuthbert R (2007) Rate of decline of the oriental white-backed vulture population in India estimated from a survey of diclofenac residues in carcasses of ungulates. PLoS One 2(8):e686

Groom RJ, Gandiwa E, Gandiwa P, van der Westhuizen HJ (2013) A mass poisoning of white-backed and lappet-faced vultures in Gonarezhou National Park. NuSpace Institutional Repository 59(1):5–9

Guitart R, Sachana M, Caloni F, Croubels S, Vandenbroucke V, Berny P (2010) Animal poisoning in Europe. Part 3: wildlife. Vet J 183:260–265

Hazevoet CJ, Fischer S, Deloison G (1996) Ornithological news from the Cape Verde islands in 1995, including records of species new to the archipelago. Bull Zool Mus Univ Amsterdam 15:21–27

Helander B (1985) Reproduction of the white-tailed sea eagle in Sweden. Holarct Ecol 8:211–227

Helander B, Bignert A, Asplund L (2008) Using raptors as environmental sentinels: monitoring the White-Tailed Sea eagle *Haliaeetus albicilla* in Sweden. AMBIO J Hum Environ 37(6):425–431

Hernández M, Margalida A (2009) Assessing the risk of lead exposure for the conservation of the endangered Pyrenean bearded vulture (*Gypaetus barbatus*) population. Environ Res 109(7):837–842

Hille SM, Collar NJ (2011) Status assessment of raptors in Cape Verde confirms a major crisis for scavengers. Oryx 45(2):217–224

Hollamby S, Afema-Azikuru J, Sikarskie JG, Kaneene JB, Bowerman WW, Fitzgerald SD et al (2004) Mercury and persistent organic pollutant concentrations in African fish eagles, marabou storks, and Nile tilapia in Uganda. J Wildl Dis 40(3):501–514

http://www.save-vultures.org/save_latestnews.html#iran

Iwata H, Watanabe M, Kim EY, Gotoh R, Yasunaga G, Tanabe S, Masuda Y, Fujita S (2000) Contamination by chlorinated hydrocarbons and lead in Steller's sea eagle and White-Tailed Sea eagle from Hokkaido, Japan. In: First symposium on Stellar's and White-Tailed Sea eagles in East Asia. Wild Bird Society of Japan, Tokyo, pp 91–106

Karyakin IV (2010) Catastrophic consequences of deratization with use of bromadialone in Mongolia (2001–2003) for Russian bird populations. Pest Manag 1:20–26

Kazantzis G (2000) Thallium in the environment and health effects. Environ Geochem Health 22:275–280

Keith JO, Bruggers RL (1998) Review of hazards to raptors from pest control in Sahelian Africa. J Raptor Res 32(2):151–158

Kenntner N, Crettenand Y, Fünfstüc J, Janovsky M, Tataruch F (2007) Lead poisoning and heavy metal exposure of golden eagles (*Aquila chrysaetos*) from the European alps. J Ornithol 148(2):173–171

Kenny D, Reading R, Maude G, Hancock P, Garbett B (2015) Blood lead levels in White-Backed Vultures (*Gyps africanus*) from Botswana, Africa. *Vulture News 68*(25):25–31. p.e31

Kolman A (2007) Thallium sulfate, AcuteTox project, Available online: www.acutetox.eu/pdf_human_short/66-Thallium%20sulphate%20revised.pdf

Korovin VA (2014) Restoration of the little bustard population in the northern steppe trans-Urals. Biol Bull Russ Acad Sci 41:856

Krone O, Wille F, Kenntner N, Boertmann D, Tataruch F (2004) Mortality factors, environmental contaminants, and parasites of white-tailed sea eagles from Greenland. Avian Dis 48(2):417–424

Krone O, Stjernberg T, Kenntner N, Tataruch F, Koivusaari J, Nuuja I (2006) Mortality factors, helminth burden, and contaminant residues in white-tailed sea eagles (*Haliaeetus albicilla*) from Finland. AMBIO J Hum Environ 35(3):98–104

Kurosawa N (2000) Lead poisoning in Steller's sea eagles and white-tailed sea eagles. In: Ueta M, McGrady MJ (eds) First symposium on Steller's and white-tailed sea eagles in east Asia. Wild Bird Society of Japan, Tokyo, pp 107–109

Lambertucci SA, Donázar JA, Huertas AD, Jiménez B, Sáez M, Sanchez-Zapata JA, Hiraldo F (2011) Widening the problem of lead poisoning to a south-American top scavenger: lead concentrations in feathers of wild Andean condors. Biol Conserv 144(5):1464–1414

Langelier KM (1993) Barbiturate poisoning in twenty-nine bald eagles. *Raptor biomedicine*. University of Minnesota Press, Minneapolis, pp 231–232

Langford KH, Reid M, Thomas KV (2013) The occurrence of second generation anticoagulant rodenticides in non-target raptor species in Norway. Sci Total Environ 450:205–208

Ma M (2013) Government-sponsored falconry practices, rodenticides, and land development jeopardize golden eagles (*Aquila chrysaetos*) in Western China. J Raptor Res 47(1):76–79

MaMing R, Zhao XM, Xu GH, Caiwu J, Zhang T, Ding P, Xu F (2014) Raptor conservation and culture in the west of China. Ela J 3(1):23–29

Manning T, Ross GA, Symons R (2008) Environmental contaminants in white-bellied sea-eagles (*Haliaeetus Leucogaster*) found in Sydney, Australia. Australas J Ecotoxicol 14(1):21–30

Margalida A, Heredia R, Razin M, Hernández M (2008) Sources of variation in mortality of the bearded vulture *Gypaetus barbatus* in Europe. Bird Conserv Int 18(1):1–10

Martínez-Haro M, Mateo R, Guitart R, Soler-Rodríguez F, Pérez-López M, María-Mojica P, García-Fernández AJ (2008) Relationship of the toxicity of pesticide formulations and their commercial restrictions with the frequency of animal poisonings. Ecotoxicol Environ Saf 69:396–402

Martinez-Lopez E, Romero D, Maria-Mojica P, Martinez JE, Calvo JF, García-Fernández AJ (2009) Changes in blood pesticide levels in booted eagle (*Hieraaetus pennatus*) associated with agricultural land practices. Ecotoxicol Environ Saf 72(1):45–59

Mateo R, Cadenas R, Manez M, Guitart R (2001) Lead shot ingestion in two raptor species from Doñana, Spain. Ecotoxicol Environ Saf 48(1):6–10

Mateo-Tomás P, Olea PP, Sánchez-Barbudo IS, Mateo R (2012) Alleviating human–wildlife conflicts: identifying the causes and mapping the risk of illegal poisoning of wild fauna. J Appl Ecol 49:376–385

Mayr E (1982) The growth of biological thought: diversity, evolution and inheritance. The Belknap Press of Harvard University Press, Cambridge, MA, p 591

Mcwilliam AN, Cheke RA (2004) A review of the impacts of control operations against the red-billed quelea (*Quelea quelea*) on non-target organisms. Environ Conserv 31(02):130–137

Mendelssohn H, Paz U (1977) Mass mortality of birds of prey caused by azodrin, an organophosphorous pesticide. Biol Conserv 11:163–169

Mineau P, Fletcher MR, Glaser LC, Thomas NJ, Brassard C, Wilson LK et al (1999) Poisoning of raptors with organophosphorus and carbamate pesticides with emphasis on Canada, US and UK. J Raptor Res 33:1–37

Mooney N (1998) Status and conservation of raptors in Australia's tropics. J Raptor Res 32(1):64–73

Movalli P, Valvo ML, Pereira MG, Osborn D (2008) Organochlorine pesticides and polychlorinated biphenyl congeners in lanner *Falco biarmicus feldeggii* Schlegel chicks and lanner prey in Sicily, Italy. AMBIO: J Hum Environ 37(6):445–451

Muzinic J (2007) Poisoning of seventeen Eurasian griffons (*Gyps fulvus*) in Croatia. J Raptor Res 41(3):239–242

Naidoo V, Wolter K (2016) Serum and plasma cholinesterase activity in the cape griffon vulture (*Gyps coprotheres*). J Wildl Dis 52(2):369–372

Naidoo V, Wolter K, Cromarty D, Diekmann M, Duncan N, Meharg AA et al (2010) Toxicity of non-steroidal anti-inflammatory drugs to Gyps vultures: a new threat from ketoprofen. Biol Lett 6(3):339–341

Naidoo V, Wolter K, Botha CJ (2017) Lead ingestion as a potential contributing factor to the decline in vulture populations in southern Africa. Environ Res 152:150–156

Nain S, Smits JEG (2011) Subchronic lead exposure, immunotoxicity and disease resistance in Japanese quail. Ecotoxicol Environ Saf 74(4):787–792

Naldo JL, Samour JH (2004) Causes of morbidity and mortality in falcons in Saudi Arabia. J Avian Med Surg 18(4):229–241

Newton I, Wyllie I, Freestone P (1990) Rodenticides in British Barn owls (*Tyto alba*). Environ Pollut 68:101–117

Novotný L, Misík J, Honzlová A, Ondráèek P, Kuèa K, Vávra O et al (2011) Incidental poisoning of animals by carbamates in the Czech Republic. J Appl Biomed 9:157–161

Nygard T, Gjershaug JO (2001) The effects of low levels of pollutants on the reproduction of golden eagles in western Norway. Ecotoxicology 10:285–290

Oaks LJ, Gilbert M, Virani MZ, Watson RT, Meteyer CU et al (2004) Diclofenac residues as the cause of vulture population decline in Pakistan. Nature 427:630–633

Ogada DL (2014) The power of poison: pesticide poisoning of Africa's wildlife. Ann NY Acad Sci 1322(1):1–20

Ogada DL, Keesing F (2010) Decline of raptors over a three-year period in Laikipia, Central Kenya. J Raptor Res 44(2):129–135

Olsen J, Osgoode M (2006) Numbers of breeding little eagles in the Australian Capital Territory in 2006. Canberra Bird Notes 31(4):178–182

Olsen J, Debus SJS, Judge D (2013) Declining little eagles *Hieraaetus morphnoides* and increasing rabbit numbers near Canberra: is secondary poisoning by pindone the problem? Corella 37(2):33–35

Ostrowski S, Shobrak M (2001) Pesticide poisoning in free-ranging lappet-faced vulture (Torgos). Vet Rec 149:396–397
Prakash V, Pain DJ, Cunningham AA, Donald PF, Prakash N, Verma A et al (2003) Catastrophic collapse of Indian white-backed *Gyps bengalensis* and long-billed *Gyps indicus* vulture populations. Biol Conserv 109(3):381–390
Punch P (2001) A retrospective study of the success of medical and surgical treatment of wild Australian raptors. Aust Vet J 79:747–752
Rattner BA, Whitehead MA, Gasper G, Meteyer CU, Link WA, Taggart MA et al (2008) Apparent tolerance of Turkey vultures (*Cathartes aura*) to the non-steroidal anti-inflammatory drug diclofenac. Environ Toxicol Chem 27(11):2341–2345
Rodríguez B, Rodríguez A, Siverio F, Siverio M (2013) Causes of raptor admissions to a wildlife rehabilitation center in Tenerife (Canary Islands). J Raptor Res 44(1):30–39
Roxburgh L, McDougall R (2012) Vulture poisoning incidents and the status of vultures in Zambia and Malawi. Vulture News 62(1):33–39
Saggese MD, Quaglia AG, Lambertucci SA, Bo MS, Sarasola JH, Pereyra-Lobos RO, Maceda JJ (2009) Survey of lead toxicosis in free-ranging raptors from central Argentina. In: Watson RT, Fuller M, Pokras M, Hunt WG (eds) Ingestion of lead from spent ammunition: implications for wildlife and humans. The Peregrine Fund, Boise, pp 223–231
Saito K (2009) Lead poisoning of Steller's sea-eagle (*Haliaeetus pelagicus*) and white-tailed eagle (*Haliaeetus albicilla*) caused by the ingestion of lead bullets and slugs, in Hokkaido, Japan. In: Watson RT, Fuller M, Pokras M, Hunt WG (eds) Ingestion of lead from spent ammunition: implications for wildlife and humans. The Peregrine Fund, Boise
Sánchez-Barbudo IS, Camarero PR, Mateo R (2012) Primary and secondary poisoning by anticoagulant rodenticides of non-target animals in Spain. Sci Total Environ 420:280–288
Shilova SA (2015) Biocenotic consequences of small mammal population control in natural ecosystems. Russ J Ecol 46(2):175–180
Shlosberg A (2011) Monitoring wild birds for exposure to environmental toxicants. Israeli J Vet Med 66(2):48–50
Shlosberg A, Wu Q, Rumbeiha WK, Lehner A, Cuneah O, Hatzofe O et al (2012) Examination of Eurasian griffon vultures (*Gyps fulvus fulvus*) in Israel for exposure to environmental toxicants using dried blood spots. Arch Environ Contam Toxicol 62:502–511
Shobrak M (2003) Vultures in Saudi Arabia. Vulture News 48:20–23
Shore RF, Walker LA, Turk A, Wienburg CL, Wright J., Murk A, Wanless S (2006) Wildlife and pollution: 2003/04 annual report. JNCC report, no. 391. Joint Nature Conservation Committee, Peterborough, 91pp
Smits JE, Bortolotti GR (2001) Antibody mediated immunotoxicity in American kestrels (*Falco sparverius*) exposed to polychlorinated biphenyls. J Toxicol Environ Health A 62:101–110
Smits JEG, Fernie KJ, Bortolotti GR, Marchant TA (2002) Thyroid hormone suppression and cell mediated immunomodulation in American kestrels (*Falco sparverius*) exposed to PCBs. Arch Environ Contam Toxicol 43:338–344
Snoeijs T, Dauwe T, Pinxten R, Darras VM, Arckens L, Eens M (2005) The combined effect of lead exposure and high or low dietary calcium on health and immunocompetence in the zebra finch (*Taeniopygia guttata*). Environ Pollut 134:123–132
Swan GE, Cuthbert R, Quevedo M, Green RE, Pain DJ, Bartels P et al (2006) Toxicity of diclofenac to Gyps vultures. Biol Lett 2(2):279–282
Tavecchia G, Adrover J, Navarro AM, Pradel R (2012) Modelling mortality causes in longitudinal data in the presence of tag loss: application to raptor poisoning and electrocution. J Appl Ecol 49(1):297–305
Thiollay JM (2007) Raptor declines in West Africa: comparisons between protected, buffer and cultivated areas. Oryx 41(3):322–329
Thomas PJ, Mineau P, Shore RF, Champoux L, Martin PA, Wilson LK et al (2011) Second generation anticoagulant rodenticides in predatory birds: probabilistic characterisation of toxic liver concentrations and implications for predatory bird populations in Canada. Environ Int 37(5):914–920

Vandenbroucke V, Van Pelt H, De Backer P, Croubels S (2010) Animal poisonings in Belgium: a review of the past decade. Vlaams Diergen Tijds 79:259–268
Vergara P, Fargallo JA, Banda E, Parejo D, Lemus JA, García-Montijano M (2008) Low frequency of anti-acetylcholinesterase pesticide poisoning in lesser and Eurasian kestrels of Spanish grassland and farmland populations. Biol Conserv 141(2):499–505
Virani MZ, Kendall C, Njoroge P, Thomsett S (2011) Major declines in the abundance of vultures and other scavenging raptors in and around the Masai Mara ecosystem, Kenya. Biol Conserv 144(2):746–752
Vogler BR, Hoop R, Sinniger M, Albini S (2015) Intentional methomyl-poisoning of peregrine falcons (*Falco peregrinus*) in Switzerland. Eur J Wildl Res 61(4):611–615
Vyas N (1999) Factors influencing estimation of pesticide-related wildlife mortality. Toxicol Ind Health 15:187–192
Walker CH, Newton I (1999) Effects of cyclodiene insecticides on raptors in Britain – correction and updating of an earlier paper by Walker and Newton, Ecotoxicology 7, 185–189. Ecotoxicology 8:425–429
Walker LA, Shore RF, Turk A, Pereira MG, Best J (2008) The predatory bird monitoring scheme: identifying chemical risks to top predators in Britain. AMBIO J Hum Environ 37(6):466–471
Whitfield DP, David RA, McLeod DRA, Watson J, Fielding AH, Haworth PF (2003) The association of grouse moor in Scotland with the illegal use of poisons to control predators. Biol Conserv 114(2):157–163
Wildlife Disease Assoc (2016) www.wildlifedisease.org/wda/PUBLICATIONS/WDANewsletter
Wobeser G, Bollinger T, Leighton FA, Blakley B, Mineau P (2004) Secondary poisoning of eagles following intentional poisoning of coyotes with anticholinesterase pesticides in western Canada. J Wildl Dis 40:163–172
Xiaodi Y, Fox N (2003) China 2000. Falco: The newsletter of the Middle East Falcon Research Group 21: ISSN 1608–1544, pp 7–8
Xiaodi Y, Ziming W, Yanxia B (2001) The Saker Falcon in China is fighting for survival. Proceedings of the II international conference on the Saker Falcon and Houbara Bustard, Mongolia, 1–4 July 2000, pp 80–94
Xirouchakis S (2004) Causes of raptor mortality in Crete. Raptors Worldwide. World Working Group on Birds of Prey/MME. Budapest:849–860
Yom-Tov Y, Hatzofe O, Geffen E (2012) Israel's breeding avifauna: a century of dramatic change. Biol Conserv 147(1):13–21
Yosef R (1996) Raptor persecution in the Gulf of Elat region. Israeli J Zool 42(3):295–296
Zuberogoitia I, Martínez JA, Iraeta A, Azkona A, Zabala J, Jiménez B (2006) Short-term effects of the prestige oil spill on the peregrine falcon (*Falco peregrinus*). Mar Pollut Bull 52(10):1176–1181

Chapter 11
Lead Poisoning in Birds of Prey

Oliver Krone

A Very Brief History of Lead

Naturally lead (plumbum, Pb) is embedded in the earth's crust at a concentration of 0.016 g Pb/kg soil, making it a relatively rare metal. From there it is released into the environment by geochemical weathering, igneous processes and radioactive decay (Pattee and Pain 2003). Lead is probably the first metal used by mankind. It was known as opacifier and colourant for glazes and glasses since the fifth millennium B.C. But lead pigment has also been used in cosmetics as long ago as 4000 B.C. due to its softness and low melting point (327.5 °C); it is easily mined and moulded. Formed to coins and figures, lead played an important role in trading more than 4500 years ago in ancient Egypt. As a component of many metallic ores, lead was also considered as a by-product of mining precious metals such as silver. Cooking utensils have been made of lead, and lead piping was the mainstay of the water distribution system in the Roman Empire (Nriagu 1983). Since the Romans did not know sugar, they produced sapa, a syrup made of sweet fruits boiled in lead vessels. Sapa containing lead was used to sweeten drinks and meals. Lead poisoning from all these sources must have been a common disease in ancient Rome. Symptoms included colic, stillbirths, deformities and cases of brain damage. Although controversial (Scarborough 1984), high lead concentrations diagnosed in archaeological Roman bones arguably contributed to the fall of the Roman Empire (Gilfillan 1965). Described in antiquity, lead poisoning was no more mentioned in the literature until the Middle Ages, where it was then mentioned sporadically. Due to the increased use of lead in pottery, piping, shipbuilding, window making, arms industry, pigments and later book printing, lead poisoning reached epidemic dimensions during the period of industrialisation (Hernberg 2000). For millennia the main route of lead exposure was primarily via

O. Krone (✉)
Department of Wildlife Diseases, Leibniz Institute for Zoo and Wildlife Research, Berlin, Germany
e-mail: krone@izw-berlin.de

J. H. Sarasola et al. (eds.), *Birds of Prey*,
https://doi.org/10.1007/978-3-319-73745-4_11

occupation, but the introduction of leaded paint for residential use in the nineteenth century significantly increased lead accumulation in children (Bellinger 2004). Symptoms in children from lead paint recognised in Australia contributed largely to the understanding of childhood lead poisoning (Henretig 2006). European governments started to ban lead-based paints in the early 1900s, culminating in a ban by the League of Nations in 1922 (Gilbert and Weiss 2006).

At the beginning of the 1920s, lead was added to gasoline because of its potential to reduce engine knock and continued to be an additive to gasoline in most countries until the 1980s. Leaded gasoline was then phased out primarily because of its interference with catalytic converter operation, but increasing awareness of adverse health effects of lead may have supported the decision. This resulted in marked declines of lead levels worldwide since the 1980s. However, leaded gasoline remained available in selected markets until the early 1990s. Owing to its physical characteristics, lead-based alloys are widely used as solders in electric industry. In response to concerns for the environment and human health, the European Union has banned lead from electronic products in 2006 (Li et al. 2005). Lead from exhaust fumes still persists in soil and in dust in urban environments until today (Rossi 2008). Lead use continued in modern times with the production of lead acid batteries, solder, alloys, cable sheathing, pigments, rust inhibitors, ammunition, glazes and plastic stabilisers (WHO 1989). Other frequent and sporadic sources of lead accumulation and poisoning nowadays are still drinking water from lead pipes (Schock et al. 2008); lead paint, e.g. found in inexpensive children's toys, much of them imported from China (Aliyev et al. 2011); altered marihuana (Busse et al. 2008); and particles of lead-based hunting ammunition in game meat.

Based on the available knowledge on the toxicity of lead, the Joint FAO/WHO Expert Committee on Food Additives (JECFA)) proposed a provisional tolerable weekly intake (PTWI) of 25 μg of lead per kilogram of body weight for infants and children in 1986. This PTWI was reconfirmed by JECFA in 1993 and extended to all age groups. JECFA re-evaluated lead in 2010, finding that even exposure to low lead levels is associated with a wide range of pathological effects. JECFA therefore concluded that the PTWI could no longer be considered health protective, and it was withdrawn. JECFA reaffirmed this conclusion because of the neurodevelopmental effects of foetuses, infants and children which are most sensitive to lead (WHO 2011).

Sources of Lead in Birds of Prey

Oral ingestion of metallic lead particles is the most important route of lead intoxication in birds of prey. The source of lead is fragments from spent hunting ammunition, i.e. fragments of lead-based bullets or lead shot pellets ingested with food. Physical demonstration of shot or bullet fragments of metallic lead from post-mortem examinations directly indicates ingestion of lead. This has been proven further by comparing lead isotope signatures from tissue of raptors with isotope ratios from ammunition (Berny et al. 2015; Church et al. 2006; Fisher et al. 2006; Helander et al. 2009; Lambertucci et al. 2011; Madry et al. 2015; Mateo et al. 2001;

Pain et al. 2007; Scheuhammer and Templeton 1998) and feeding ecology (Nadjafzadeh et al. 2013). An increase in mortality or blood lead levels in birds of prey and other scavengers such as ravens correlating with hunting seasons has been demonstrated several times (Craighead and Bedrosian 2007; Kelly et al. 2014; Kramer and Redig 1997; Krone et al. 2009a; Cade 2007; Cruz-Martinez et al. 2012; Stauber et al. 2010; Strom et al. 2009).

Other sources of lead such as lead in paint are more relevant to captive birds of prey such as Egyptian vultures (*Neophron percnopterus*) and cinereous vultures (*Aegypius monachus*) poisoned with lead from paint dust and chips that were sanded off the steel aviary construction and contaminated the soil and subsequently the food put on the ground (Pikula et al. 2013).

Lead in Hunting Ammunition

When longbows and crossbows were replaced by early firearms in the sixteenth century, projectiles were already made of lead. With a specific gravity of 11.35 cm^3 and ready availability, lead had ideal ballistic characteristics to be used as bullets. While arquebuses and muskets were muzzle loaded, i.e. gunpowder and the projectile were inserted separately, later paper cartridges, also muzzle loaded, enabled an easier and faster reloading. The first fully metallic cartridge was patented by Houllier in France in 1846 (Kinard 2004). Later in the nineteenth century, function and design of the cartridges were improved, and a variety of different calibres developed. The typical rifle cartridge now consists of a metallic case containing a primer, gunpowder and the bullet. The bullet or projectile has a torpedo-like shape and is usually a composition of a lead core surrounded by a copper-alloy mantle with an open tip, so-called semi-jacketed bullet (Fig. 11.1). There are different bullet designs available on the market, but the semi-jacketed is in many countries the typical projectile used for hunting with rifles. The shooting distance is usually less than 100 metres but can reach up to 300 metres.

The wound ballistics describe the effect and behaviour of a projectile in the animal's body. Large game animals (dear, wild boar, wild sheep, ibex, chamois, etc.) are killed when the projectile hits vital organs. In traditional hunting, hunters target the heart of the animal. In the shot animal, the projectile penetrates the skin of the side of the animal, then the connective tissue and muscle and eventually the ribs, lung and heart and continues through the lung, muscle, ribs, connective tissue and skin again before leaving the body. On initial impact with the skin, the semi-jacketed bullet starts to mushroom; i.e. the torpedo-shaped projectile transforms into a mushroom-shaped projectile enlarging its diameter two times and enlarging the surface of the lead core. On its way through the tissues, this process continues, and little fragments of the lead core and sometimes also of the jacket are left behind, producing a cloud of lead particles around the wound channel (Fig. 11.2). Depending on the size, i.e. thickness of the animal, bones hit, calibre, shooting distance and angle, the bullet or its remaining parts (after fragmentation) can be encountered at some distance from the entry point and wound channel in the body.

Fig. 11.1 Hunting ammunition. (**a**) shotgun shell with pellets, (**b**) rifle cartridge with bullet

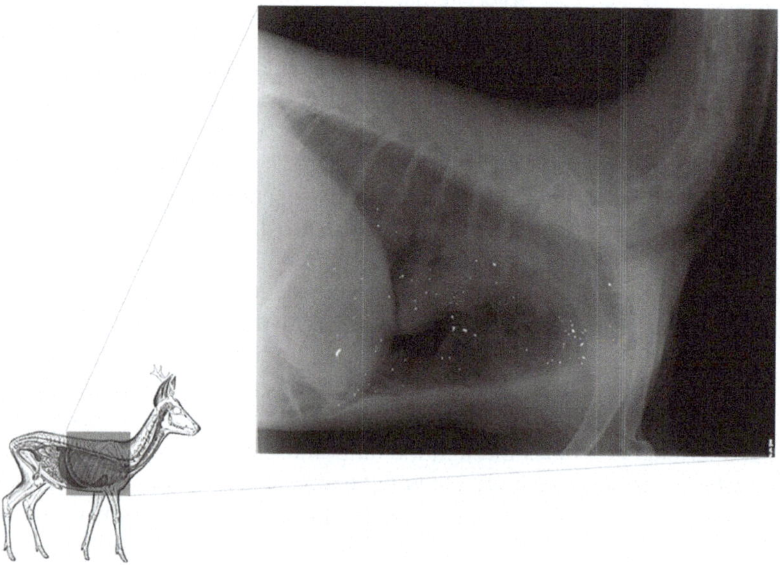

Fig. 11.2 Radiograph of roe deer (*Capreolus capreolus*) shot with a semi-jacketed lead core rifle bullet. The cloud of amorphous little lead fragments is located in the thorax of the deer

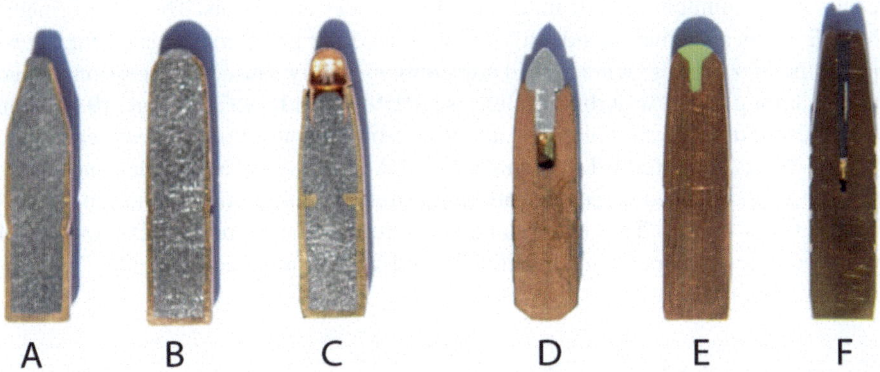

Fig. 11.3 Selection of three lead-based (**a**–**c**) and three lead-free rifle bullets (**d**–**f**) Material; shape and construction determine performance, movement and expanding behaviour in the target tissue. (**a**) Semi-jacketed lead-based bullet with thin mantle, (**b**) round nose lead-based semi-jacketed bullet, (**c**) semi-jacketed lead-based bullet with H-mantle and metallic cap covering the hollow tip, (**d**) lead-free fragmenting bullet with aluminium tip partly filling hollow point cavity, (**e**) lead-free (copper-based) deforming bullet with plastic tip filling hollow point cavity, (**f**) lead-free deforming bullet with iron pin

Originally produced for hunting at long distances, copper or brass bullets are faster and have a straighter trajectory than conventionally lead-based bullets. Copper or brass projectiles are solid and often contain a hollow tip (Fig. 11.3). This hole can be filled with a ballistic plastic or aluminium cap. Solids can be designed as pure deforming or fragmenting bullets. Deforming bullets usually do not fragment, whereas fragmenting bullets often have a defined number of predetermined breaking grooves. This construction allows the projectile to part into several, usually three to four nearly similar sized and shaped, fragments in addition to the remaining bottom part. Solids or lead-free (<1% lead contamination in the alloy) rifle bullets are gaining increasing acceptance among hunters who want to stop lead pollution of the environment, lead poisoning of birds of prey and scavengers and contamination of game meat.

In contrast to rifle ammunition, shotgun cartridges are larger in diameter, made of cardboard or plastic and contain several hundred small round pellets often between 2 and 4 mm in size in addition to a primer and gunpowder. This ammunition is used to shoot small game animals such as ducks, pigeons, pheasants, hares, rabbits, etc. The cluster of pellets increases in diameter the further it travels from the gun barrel, thus limiting the shooting/potential killing range to less than 50 metres. At larger distances or bad shot placements, small game animals may survive the shot but then carry lead pellets in their body. When shooting at birds flying in flocks, the shot may kill only the bird in the centre of the pattern, while surrounding birds get hit by fewer pellets. Traditionally lead-based shotgun ammunition is used for hunting waterbirds when flying. Birds shot with lead pellets can survive and carry lead pellets in their bodies and subsequently can become a source of lead poisoning to birds of prey and scavengers.

In addition to this animal welfare aspect and the contamination of prey for larger raptors, lead poisoning of waterbirds is a well-studied and understood problem. Intensive hunting of waterbirds with lead shot especially in shallow water bodies results in the unintentional death of mainly of ducks and swans. These birds ingest the pellets because they misidentify them as seeds or grit stones needed for digestion in the gizzard. The scale of lead poisoning mortality caused by ingestion of lead pellets among those waterbirds (Bellrose 1959; Friend 1987; Mudge 1983; Pain 1992) resulted in substantial policy attention from the outset of the treaty called the African-Eurasian Waterbird Agreement (AEWA). Its original action plan and subsequent resolution aimed to reduce lead-based shotgun ammunition for hunting waterfowl at water bodies. This is a great achievement to better protect 254 species of birds ecologically dependent on wetlands (http://www.unep-aewa.org).

Ingestion and Kinetics of Lead

Ingestion

Birds of prey ingest lead particles with their food when they hunt living prey previously shot or scavenge on carcasses or viscera left behind by hunters. Feeding experiments on white-tailed sea eagles (*Haliaeetus albicilla*) have shown that the likelihood of ingesting metallic particles is inversely correlated with the size of the particles (Nadjafzadeh et al. 2015). When lead particles (fragments of rifle bullets as well as lead shot pellets) are swallowed with food bites, they end up in the gizzard and stay there where they may be dissolved or a casting is produced of insoluble feathers, hairs, scales, claves, plant material, etc. in which the metallic particles might be embedded. Because of their weight and their size, especially small particles are trapped in the mucosal folds at the bottom of the gizzard. There they remain until fully dissolved or transported further through the pylorus into the intestine. In an experimental study on turkey vultures (*Cathartes aura*), Carpenter et al. (2003) had to redose the birds since some of them regurgitated or defecated lead shot pellet at a size of 4.5mm. The process of dissolution of lead particles in the bird's gizzard is not fully understood since the mechanisms are more complex than putting a piece of lead in a test tube with hydrochloric acid (HCl). In the bird's gizzard, a variety of organic processes enable the digestion of food (Denbow 2000) and are also responsible for the dissolution of lead particles.

Absorption

While lead salts such as lead acetate are rapidly dissolvable, metallic lead is less soluble in general. In the acid environment of a gizzard with pH values around 1.2–1.4, the gastric HCl produces a lead(II) chloride ($PbCl_2$) layer at the surface of every lead particle. Due to the motility of the gizzard and potential ingesta, this surface layer is repeatedly eroded and then built up again. This ongoing process makes lead(II) chloride permanent available. In warm conditions (body temperature) and in a concentrated hydrochloric acid, lead(II) chloride is highly soluble. In addition, lead binds to digestive enzymes at their sulphur sites because of the thiophilic character of lead. These altered enzymes may break apart and set lead sulphides free. Depending on the ingested food and the amount of swallowed saliva, other soluble lead salts are most certainly produced. There is no equilibrium between the solution and the metallic lead particles in the gizzard since the contents are transported further down the digestive system; therefore the dissolution process continues until the lead particles are fully dissolved. As food is digested, digestive liquid containing the finest particulates is allowed to pass into the intestine, where a different intestinal fluid contains bicarbonate ions and bile salts, and pancreatin is secreted (Martinez-Haro et al. 2009). In the intestine the same mechanisms and proteins responsible for the absorption of calcium enable the transportation of lead into the blood (Fullmer et al. 1985). Lead absorption occurs in the small intestine. Experimental studies on rats have shown that most lead is absorbed in the duodenum, followed by the jejunum and ileum. The presence of bile in the duodenum enhances the transport of lead across the intestinal mucosa epithelial cells (Conrad and Barton 1978). In general age, fasting status, nutrition (Thompson 2012), solubility, particle size, pH value (Ruby et al. 1992) and retention time (Stendell et al. 1979) strongly influence lead absorption from the gastrointestinal tract. The dietary intake of fat, calcium, iron, vitamin D and vitamin E also can influence absorption and susceptibility to lead intoxication (DeMichele 1984). In birds of prey, the low gastric pH values and the anatomy and physiology of the gizzard making it a one-way system for large indigestible items and thus the resulting long retention time of tiny metallic lead particles additionally support the absorption of lead.

Kinetics

Following absorption, a large proportion of lead is carried on erythrocyte membranes (60–90%, species dependent) with most of the remaining of lead being bound to protein or sulfhydryl compounds, with only a very small proportion found free in the serum (Thompson 2012). Elevation of blood lead concentration occurs within 24 h following ingestion of lead shot pellets (Hoffman et al. 1981). The half-life of lead in blood is several weeks to a few months in waterfowl (Pain 1996). The blood transports lead into the soft tissue and bones. Highest lead values can be found in the bone, followed by the kidney and liver with lower concentrations in the central nervous system and blood, and lowest in muscle tissue (Franson and Pain 2011). Depending on species and stage within the course of lead distribution in the organism, some have found higher kidney lead levels in birds of prey (Carpenter et al. 2003; Pattee et al. 2006), while others have reported higher liver values (Pattee et al. 1981; Kenntner et al. 2001, 2003; Kim et al. 1999; Krone et al. 2004, 2006; Saito et al. 2000). In soft tissues the half-life of lead is about 1-3 months in waterfowl (Pain 1996). Finally lead is stored in the bone matrix (Fig. 11.4) instead of calcium (Gangoso et al. 2009; Scheuhammer 1987). The bone marrow accumulates only small amounts of lead (Ethier et al. 2007). Higher lead concentrations found in bone versus liver tissue in the majority of birds (91%) have been interpreted as

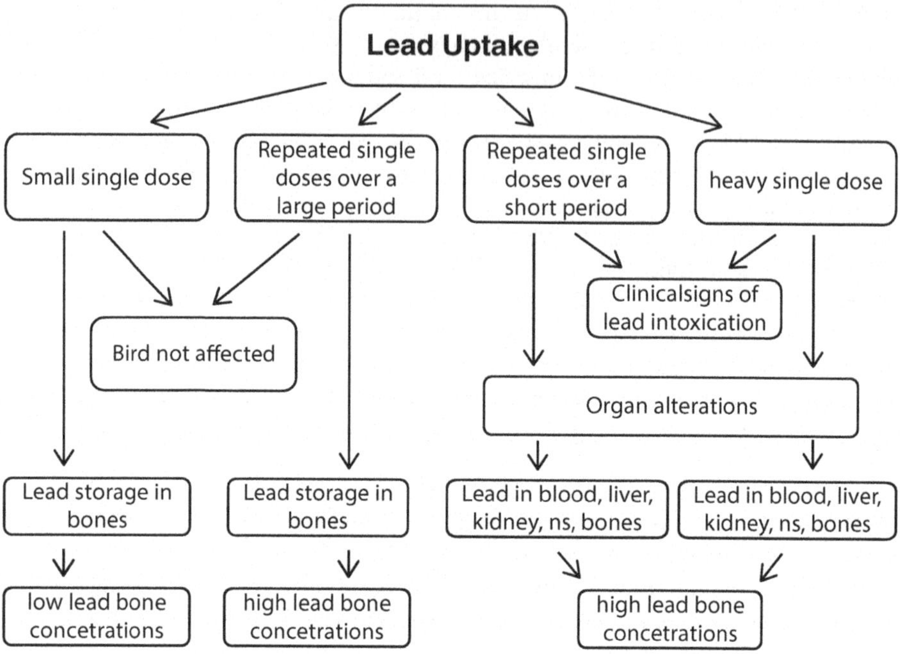

Fig. 11.4 Effects of amount and repeated uptake of lead on bones as final deposit (ns = neural system)

typical of environmental exposure to low lead levels over an extended period of time (Martin et al. 2008). Lead embedded in the bone matrix is relatively immobile but can be released in times of increased bone remodelling or calcium utilisation like egg production (Scheuhammer 1987; Pounds et al. 1991). Absorbed lead is excreted in urine and via the bile into faeces (Thompson 2012). Lead concentrations in bile are higher than in urine (Ishihara and Matsushiro 1986), but a high proportion of biliary lead is reabsorbed in the intestine (Ishihara et al. 1987). The higher calcium demand of young birds (i.e. nestlings) growing their body, especially bones requiring higher absorption rates, makes them more susceptible to lead accumulation. Hoffman et al. (1985) found more severe biochemical and haematological alterations in young American kestrels (*Falco sparverius*) than in adults. Females during egg laying also have a higher calcium demand resulting in an increased absorption of lead (Finley and Dieter 1978; Hutton and Goodman 1980). Species-specific differences in the accumulation and retention of lead can depend on different metabolic processes as suggested to explain results in experimentally dosed turkey vultures and red-tailed hawks (*Buteo jamaicensis*; Reiser and Temple 1981). Turkey vultures appear relatively tolerant to lead poisoning in comparison with other species dosed with lead shot (Pattee et al. 1981). Individual and species-related responses to lead intoxication also depend on genetic features. So far three polymorphic genes (ALAD, vitamin D receptor, haemochromatosis gene) have been recognised in potentially influencing the accumulation and toxicokinetics of lead in humans and animal models (Onalaja and Claudio 2000).

Molecular Interactions of Lead

The biomechanical mechanisms of lead toxicity are largely unknown, but at least some have been studied. At the molecular level, lead alters mainly calcium-mediated cellular processes. One known mechanism is the displacement of physiologically relevant metal ions (Ca^{2+} and Zn^{2+}) in proteins. Voltage-gated calcium channels are specifically targeted by lead (Atchison 2003). Pb^{2+} activates skeletal muscle troponin C, which regulates the contraction of striated muscles and calmodulin, which mediates smooth muscle contraction (Chao et al. 1990).

The calmodulin complex is also involved in memory, nerve growth and the immune response and mediates biological processes related to inflammation. Lead mimics calcium in the activation of protein kinase C at picomolar levels. Protein kinase C phosphorylates various cell membrane and transport proteins, regulating cellular growth and differentiation (Markovac and Goldstein 1988). Lead ions may alter normal synaptic activity by blocking presynaptic calcium channels resulting in a decrease in evoked release of neurotransmitters, but the alteration of the synaptic vesicle protein synaptotagmin I by lead is responsible for the increase in spontaneous release of neurotransmitters (Bouton et al. 2001). It is presumed that the ability of Pb^{2+} to bind opportunistically may extend to non-metalloproteins, thus increasing the number of potential target proteins whose function may be altered. Since lead replaces calcium, it interacts with proteins regulating cell division, apoptosis and intracellular signalling, as well as protein stability (Kirberger and Yang 2008).

Box (For Converting Values)

1 ppm = 1 μg/g or 1 mg/kg
1 ppm = 1 mg/l or 1 μg/ml or 100 μg/dl
1 ppm = 4,826 μmol/l (Pb molar mass = 207,2 g).
ww = wet weight
dw = dry weight

Conversion rates from ww to dw[1]:
Blood: 4.6x
Liver: 3.1x
Kidney: 4.3x
Bone: 1.2x

Pathology of Lead

Although lead occurs naturally in soil, human activities can result in the ubiquitous distribution of lead in the environment. Therefore, all animals can be exposed to certain levels of lead, but this background contamination does not result in clinically

[1] From Franson and Pain (2011)

relevant concentration. In smelting and mining areas, lead levels in animals can be elevated. Henny et al. (1991) found lead-induced biochemical changes in blood parameters but no effects on reproduction of ospreys (*Pandion haliaetus*) at a mining site in the USA.

Blood

Lead levels described as background values in blood are generally low, usually <20 μg/dl and frequently <10 μg/dl (Franson and Pain 2011). Lead exposure affects a number of enzymes resulting in impaired haem synthesis and an elevation or excretion of its metabolites. The inhibition of the delta-aminolevulinic acid dehydratase (ALAD)) is a reliable and sensitive indicator of lead exposure. ALAD strongly decreases after lead exposure and often remains depressed over an extended period in an otherwise apparently healthy bird (Franson et al. 1983). ALAD catalyses the second step in the porphyrin and haem biosynthesis. The inhibition of ALAD by lead results in anaemia primarily because it inhibits haem synthesis and shortens the lifespan of circulating red blood cells. In addition it stimulates the production of erythropoietin leading to inadequate maturation of erythrocytes from their progenitors.

Lead particles embedded in subcutaneous or muscle tissue also decrease levels of ALAD, indicating lead absorption into the blood (Repper et al. 1986). Erythrocyte protoporphyrin increases over time of lead exposure (Franson et al. 1986; Beyer et al. 1988; Rattner et al. 1989; Carpenter et al. 2003). Haematocrit often declines below 30–40% in experimentally dosed birds (Beyer et al. 1988; Carpenter et al. 2003; Hoffman et al. 1981, 1985). Blood lead concentrations greater than 1 ppm (wet weight) are indicative of lead toxicosis, and levels greater than 5 ppm (wet weight) are fatal (Franson 1996). Based on clinical experiences, Kramer and Redig (1997) classified eagles with blood lead concentrations between 0.2 and 0.6 ppm as subclinical lead exposure and birds with concentrations between 0.61 and 1.2 ppm as clinical (treatable) lead poisoned. Concentrations >1.2 ppm were invariably associated with mortality. Lead levels of more than 1.00 ppm in bald and golden eagles (*Haliaeetus leucocephalus* and *Aquila chrysaetos*) are considered as severely clinically poisoned, and recovery after treatment was rare (Stauber et al. 2010). Müller et al. (2007) treated white-tailed sea eagles with lead blood values between 0.4 and 1.5 ppm successfully and released them back to the wild.

Gross Pathology

Necropsy of lead-poisoned raptors often shows no significant or consistent gross lesions (Carpenter et al. 2003). Depending on the decomposition stage of the carcass, pathological alterations of organs are frequently found in white-tailed sea eagles and other birds of prey. These gross lesions consist of liver swelling, enlarged gall bladder with condensed bile fluids, enlarged bile ducts, reflux of bile into the

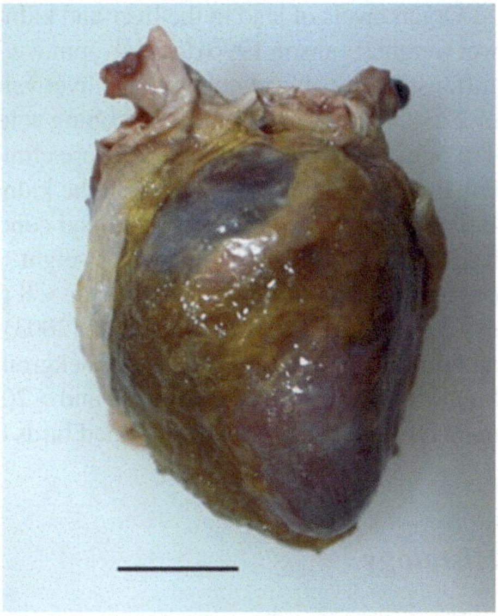

Fig. 11.5 Heart with removed pericard of a white-tailed sea eagle with acute lead intoxication showing fibrinous pericarditis

gizzard and fibrinous pericarditis. These lesions might not be considered pathognomonic for lead intoxication but should be recognised as indicative. Fibrinoid and myocardial necroses have also been detected infrequently in a variety of birds poisoned with lead including bald eagles, ducks, geese, swans, etc. (Beyer et al. 1988; Karstad 1971; Sileo et al. 1973; Pattee et al. 1981).

Acute lead intoxicated birds of prey are clinically affected and if not medically treated die from ingested lethal lead concentration. This may reflect exposure to a single lead dose, leading to death within a few days. Often pathological lesions are hardly recognisable with the bird in good alimentary condition. A fibrinous pericarditis is frequently found in acute lead-poisoned white-tailed sea eagles (Fig. 11.5).

Because of the recent uptake, radiography might reveal metallic particles in the digestive system. In a study on white-tailed sea eagles in Germany, 22.8% ($n = 390$) died due to lead intoxication from lead-based hunting ammunition. Fragments of semi-jacketed lead-based bullets were recovered from the gastrointestinal tract 5.6 times more often than lead shot (Krone et al. 2009a). As well as the potential for ingesting lethal levels of lead at one time point, repeated, episodic ingestion of sublethal levels of lead also results in clinically affected birds. In these chronically poisoned birds, lead triggers its adverse effects on the nervous, digestive, respiratory and haemopoetic system preventing the bird from hunting and finally dying from starvation. The highly emaciated birds frequently display the live and gall bladder lesions as mentioned above. The ingestion of low lead levels (a single dose or repeated uptake of small amounts) can be survived by birds without being severely affected as indicated by high lead levels in bones of those birds. Repeated uptake of small amounts as well as single ingestion of a large amount result can result in high lead bone levels. Therefore, a differentiation between chronic and acute lead exposure using bone concentrations alone without liver/kidney values is not possible. In fact lead concentrations in bones indicate long-term accumulation from both chronic and acute exposure.

Organ levels of lead in the liver and kidney thought to be lethal range from >5 ppm wet weight (Franson 1996) to >6–8 ppm ww (Friend and Franson 1999). Falconiformes suffer subclinical lead poisoning at liver values of 2 < 6 ppm ww and at kidney values of 2 < 4 ppm ww and clinical poisoning at lead levels of 6–10 ppm ww in the liver and 4–6 ppm ww in the kidney, and severe clinical poisoning occurs at levels of >10 ppm ww in the liver and > 6 ppm ww in the kidney (Franson and Pain 2011). Wayland and Bollinger (1999) considered liver lead concentrations greater than 30 ppm or kidney lead concentrations >20 ppm dry weight to be diagnostic of lead poisoning, while concentrations >6 ppm in the liver or > 8 ppm in the kidney indicated elevated lead exposure. Clark and Scheuhammer (2003) found values of <6 μg/g dw in the liver and/or kidney to be indicative for background lead levels, > 6 μg/g dw in the liver or kidney indicative for lead exposed and > 20 μg/g dw in the kidney or > 30 μg/g dw in the liver indicative for lead-poisoned birds of prey.

Histology

Bald eagles experimentally poisoned with lead showed renal tubular degeneration (nephrosis) and medial fibrinoid necrosis of small arteries in few cases but multifocal myocardial necrosis in all animals studied (Pattee et al. 1981). Liver lesions consisting of haemosiderin loading in hepatocytes and Kupffer's cells as well as intranuclear inclusion bodies in convoluted tubules in the kidneys seem to be characteristic for some bird species with lead intoxication (Kendall and Scanlon 1983; Pattee et al. 2006). In experimentally poisoned turkey vultures, haemosiderin was diagnosed in hepatocytes and Kupffer's cells. All treated vultures had some degree of acute renal tubule degeneration and some degree of status spongiosis in the white matter of the cerebrum, the cerebellum and the medulla oblongata (Carpenter et al. 2003). De Francisco et al. (2016) applied magnetic resonance imaging to identify areas in the brain affected by lead of live bald eagles. They found lesions in the forebrain, mid-brain, brainstem (pons and medulla) and cerebellum. These lesions corresponded to histological alterations such as acute focal to multifocal intraparenchymal haemorrhage, focal to multifocal pannecrosis of the neuropil (including glial and neuronal cells) and oedema without gitter cell infiltration (considered to be acute necrosis). Other lesions were characterised by vacuolar change and pallor of the neuropil with decrease or loss of differential staining in HE-stained sections. Mild intraparenchymal haemorrhage was present in one of these lesions. More severe lesions included necrosis and oedema as well as endothelial hypertrophy and infiltration with gitter cells (considered to be a subacute lesion), focal to multifocal pannecrosis of the neuropil with hypertrophy of capillary endothelial cells, infiltration by gitter cells and intraparenchymal haemorrhage with evidence of erythrophagocytosis (considered to be a chronic lesion). Pattee et al. (2006) also found perivascular and perineuronal oedema throughout the cerebrum. In the cerebellum they diagnosed spongiosis between the molecular and granular layer and pyknotic Purkinje cells in this area.

Clinical Symptoms of Lead Poisoning

Acutely lead-poisoned birds of prey are often found in good body condition with incoordination, blindness and other clinical signs indicative of CNS damage (Redig et al. 1980; Stauber et al. 2010) as well as profound, laboured and uncoordinated respirations accompanied by vocalisations (de Francisco et al. 2016). Weight loss is a typical response to chronically lead poisoning and has been reported in many avian species (Beyer et al. 1988; Janssen et al. 1986; Pattee et al. 1981). In general symptoms of birds with lead poisoning vary and can include behavioural changes, depression, lethargy, droopy wings, open-mouth breathing, anorexia, paralysis of the crop, oesophagus, proventriculus or gizzard, vomiting, diarrhoea, ataxia, paralysis of the legs or wings, amaurosis, convulsions, anaemia and emaciation (Lumeij 1985; Pattee et al. 2006). The green cloacal feathers are a result of intense green-coloured diarrhoea; in advanced phases there may be convulsions and even blindness (Labonde 1991). Lead-poisoned white-tailed sea eagles in their final stage often lay on the ground, or are unable to coordinate their body movements while standing, and show an open-mouth breathing indicating obvious problems supplying their body with sufficient oxygen and are blind. The birds are severely depressed and their reactions are depressed (Fig. 11.6). In chronic cases birds might fly to remote, less disturbed areas. This retreat allows them to concentrate their energy on curing a disease or overcome any kind of malady. In a case study, Krone et al. (2009c) reported about a white-tailed sea eagle with lead intoxication which flew to an island 11 days prior to death. The eagle significantly reduced its activity and movement on the island and finally died with lead values of 15.60 ppm (ww) in the

Fig. 11.6 Dying white-tailed sea eagle with lead intoxication showing typical symptoms of reduced flight distance, depression, droopy wings, weakness, uncoordinated movements, amygdaloid eyes, green-stained faeces and emaciation. Ten days before it was seen with much reduced flight distance but still able to fly. The adult, female eagle was equipped with a satellite transmitter to study habitat use and home range size 12 months before

liver and 7.352 ppm (ww) in the kidney. First changes in the rhythmic activity pattern were apparent 22 days before death indicating a 3-week-long case history.

Lead reduces the degree of bone mineralisation, with consequences to bone fragility (Gangoso et al. 2009). Sublethal chronic lead assimilation may favour secondary infections (Reiser and Temple 1981), higher mortality or reduced reproduction (Pain et al. 2009).

Bird of Prey Species Affected

Nearly all carnivorous raptor species, feeding on medium-sized to large prey, are susceptible to lead ingestion and subsequently to lead intoxication. These are raptors feeding on any game species or on animals considered as pests and therefore hunted, i.e. shot with lead-based ammunition. Hunted animals can contain fragments of hunting projectiles or lead shot pellets and might be killed or found dead by raptors.

At a population level, larger raptor species with naturally low mortality and reproduction are more affected than smaller species with higher reproduction rates. Among all birds of prey, obligate and facultative scavengers are more likely to ingest lead particles from spent ammunition than active hunters. Species with small global or local populations are more sensitive to additive mortality from lead intoxication such as the Spanish imperial eagle (*Aquila adalberti*; Pain et al. 2005; Mateo et al. 2001; Rodriguez-Ramos Fernandez et al. 2011), golden eagle (Bezzel and Fünfstück 1995; Jenni et al. 2015; Kenntner et al. 2007; Zechner et al. 2005), white-tailed sea eagle (Helander et al. 2009; Iwata et al. 2000; Kim et al. 1999; Kenntner et al. 2001, 2004; Krone et al. 2004, 2006, 2009b; Saito et al. 2000), Steller's sea eagle (*Haliaeetus pelagicus*; Iwata et al. 2000; Kim et al. 1999; Saito et al. 2000), red kites (*Milvus milvus*; Knott et al. 2009a), Egyptian vulture (*Neophron percnopterus*; Bounas et al. 2016), griffon vulture (*Gyps fulvus*; Garcia-Fernandez et al. 2005; Mateo et al. 1997), black vulture (Hernandez and Margalida 2008; Nam and Lee 2009), bearded vultures (*Gypaetus barbatus*; Berny et al. 2015), California condor (*Gymnogyps californianus*; Church et al. 2006; Finkelstein et al. 2010) and Andean condor (*Vultur gryphus*; Wiemeyer et al. 2017). Effort, costs, duration and susses of reintroduction programmes are adversely challenged by lead poisoning in California condor (Kelly et al. 2014), bearded vulture (Hernandez and Margalida 2009) and red kite (Pain et al. 2007). Meyer et al. (2016) found a reduced recovery rate of red kite when modelling the effects of mortality from lead poisoning. In his review, Mateo (2009) listed 14 birds of prey species found with lead poisoning in Europe. In addition to the above listed species from Europe, he also mentioned the honey buzzard (*Pernis apivorus*), marsh harrier (*Circus aeruginosus*), Eurasian sparrowhawk (*Accipiter nisus*), northern goshawk (*A. gentilis*), common buzzard (*Buteo buteo*) and the peregrine falcon (*Falco peregrinus*). Fisher et al. (2006) mentioned 59 terrestrial bird species including 12 birds of prey, 4 owl species and 3 new world vultures which have ingested lead or suffered lead poisoning. Raptors in captivity have been poisoned accidentally from lead in food carcasses, too. Pain et al. (2009) reported 33 raptor species (new and old world vultures, hawks, falcons, eagles and owls) exposed or poisoned by lead, of which five are globally threatened.

Solutions

To reduce the risk of lead exposure and subsequent intoxication, access to lead-contaminated food to raptors should be prevented. The two main sources of lead from hunting ammunition are lead shot pellets from shotguns and lead-based bullets from rifles.

Lead-Based Shotgun Pellets

Mass mortality of waterbirds from ingesting lead shot pellets resulted in significant work by intercontinental agreement (AEWA) obliging multiple countries to restrict the use of lead-based shotgun ammunition in wetlands to hunt waterfowl; however the problem for raptors is only partly addressed by this action. Compliance of hunters with national or regional regulations restricting the use of lead shot to hunt waterfowl in wetlands is variable and sometimes very poor. A game dealer survey conducted in England revealed poor compliance of hunters shooting waterfowl; 77% of 84 ducks have been illegally shot with lead (Cromie et al. 2015). However in other regions such as the Ebro Delta, Spain, noncompliance values declined from 27% in 2007–2008 to ≤2% in 2010–2011 following intensive education and enforcement (Mateo et al. 2014). For hunting other small game species such as hares, rabbits, grouse, partridge, pheasants, pigeons, martens, foxes, badgers, etc., lead shot ammunition is still allowed in most countries, except countries such as Denmark and the Netherlands (Kanstrup 2006). Shot small game animals, alive or dead, as mentioned previously, still constitute an important lead source for raptors. Only the total ban of lead-based shotgun ammunition and its substitution by lead-free shotgun pellets will impede further lead ingestion by raptors.

Lead-Based Rifle Bullets

To stop lead intoxication from lead-based rifle bullets, one possible solution is to bury gut piles, but this has been proven to be impractical on a large scale (Krone et al. 2009b). Moreover non-contaminated gut piles left behind by hunters constitute an important protein source for the scavenging raptor guild. As long as lead-based projectiles are used, these gut piles and unrecovered shot game animals constitute the most important source for lead intoxication, where large game hunting is practised. Therefore, only the use of lead-free projectiles for hunting has the potential to prevent lead fragments in prey and carrion that raptors are feeding on. Precision and performance of lead-free ammunition have been proven (Knott et al. 2009b; Trinogga et al. 2013), and wide product availability and prices comparable to lead-based ammunition have been demonstrated (Thomas 2013). Lead-free bullets, e.g. made of copper to hunt game animals, have been shown to be of no concern for human health pertaining to copper concentrations in meat portions (Irschik et al. 2013; Paulsen et al. 2015). Many key questions regarding the transition to use lead-free ammunition including communication of the lead poisoning problem and effectiveness of alternative lead-free ammunition, etc., have been answered by Thomas et al. (2015).

In general lead poisoning seems to be a simple problem for which a simple solution exists, but strong tradition and prejudice hamper the change to lead-free ammunition for hunting (Fisher et al. 2006). Often the discussion of the lead problem is perceived as an assault against hunting as a whole and not seen as a possibility to adjust hunting to the recent knowledge making it more sustainable. A combination of legislation paired with education and awareness raising has the greatest potential to overcome this problem. In addition distributing lead-free ammunition to hunters can be an effective measure to support the voluntary transition to non-toxic ammunition (Austin et al. 2012; Bedrosian et al. 2012). Despite the global scientific consensus on the health and environmental risks from lead exposure, major shooting, hunting and ammunition industry lobbyist organisations are opposing attempts to ban, or even restrict, the use of lead-based ammunition (Arnemo et al. 2016).

Perspective

It will take a considerable time before lead in ammunition is phased out and replaced by lead-free alternatives. Some progress has been made but mainly on a regional scale as in the USA, Japan and Germany. Even if some countries ban lead-based ammunition, others may not instal a legal basis in the short term. Raptors do not know political borders, that is why national bylaws are insufficient. Large-scale regulations, on continental level, will be more effective than regional ones. In 2014 parties to the Convention on Migratory Species adopted a resolution on poisoning with guidelines to phase out lead ammunition. For the European community, advisory bodies such as ECHA (European Chemicals Agency) and EFSA (European Food Safety Authority) are responsible for the protection of human health and the environment and need to take action. However under REACH responsibility is shifted from the authorities to industry and companies who have to demonstrate that a chemical can be handled safely without threatening human health or the environment. Many applications of lead have been restricted or regulated by the European Commission, for example, the use in electronics (European Union 2003) and in wheel balance weights (European Union 2013). Thus, it would be irrational allowing lead from other sources such as ammunition to contaminate human food, wildlife and the environment.

For environmental scientists, it will be important to continue monitoring the accumulation of lead as well as other pollutants in wildlife, especially raptors and the compliance of hunters not using lead-based ammunition in the future.

Acknowledgement I am grateful to I. Rottenberger from the German Institute for Experimentation and Testing of Hunting and Sporting Firearms (DEVA) for providing the cuts through the different bullet types and to R. Cromie from Wildfowl & Wetlands Trust (WWT), UK, for proofreading the manuscript.

References

Aliyev V, Bozalan M, Guvendik G, Soeylemezoglu T (2011) The potential health risk assessment of lead levels in children's toys. Toxicol Lett 205(Suppl 1):S255–S255

Arnemo JM, Andersen O, Stokke S, Thomas VG, Krone O, Pain DJ, Mateo R (2016) Health and environmental risks from lead-based ammunition: science versus socio-politics. EcoHealth 13(4):618–622

Atchison WD (2003) Effects of toxic environmental contaminants on voltage-gated calcium channel function: from past to present. J Bioenerg Biomembr 35:507–532

Austin W, Day K, Gatto A, Humphrey J, Parish C, Rodgers J, Sieg R, Smith B, Sullivan K, Young J (2012) A review of the third five years of the California condor reintroduction program in the Southwest (2007–2011). USFWS, Sacramento

Bedrosian B, Craighead D, Crandall R (2012) Lead exposure in bald eagles from big game hunting, the continental implications and successful mitigation efforts. PlosOne 7(12):e51978

Bellinger DC (2004) Lead. Pediatrics 113(4 Suppl):1016–1022

Bellrose FC (1959) Lead poisoning as a mortality factor in waterfowl populations. Ill Nat Hist Surv Bull 27:235–288

Berny P, Vilagines L, Cugnasse JM, Mastain O, Chollet JY, Joncour G, Razin M (2015) Vigilance poison: Illegale poisoning and lead intoxication are the main factors affecting avian scavenger survival in the Pyrenees (France). Ecotoxicol Environ Saf 118:71–82

Beyer WN, Spann JW, Sileo L, Franson JC (1988) Lead poisoning in six captive avian species. Arch Environ Contam Toxicol 17:121–130

Bezzel E, Fünfstück HJ (1995) Alpine Steinadler *Aquila chrysaetos* durch Bleivergiftungen gefährdet? J Ornithol 136:294–296

Bounas A, Ganoti M, Giannakaki E, Akrivos A, Vavylis D, Zorilla I, Saravia V (2016) First confirmed case of lead poisoning in the endangered Egyptian vulture (*Neophron percnopterus*) in the Balkans. Vulture News 70:22–29

Bouton CM, Frelin LP, Forde CE, Arnold Godwin H, Pevsner J (2001) J Neurochem 76:1724–1735

Busse F, Omidi L, Timper K, Leichtle A, Windgassen M, Kluge E, Stumvoll M (2008) Lead poisoning due to adulterated marijuana. N Engl J Med 358:1641–1642

Cade TJ (2007) Exposure of California condors to lead from spent ammunition. J Wildl Man 71(7):2125–2133

Carpenter JW, Pattee OH, Fritts SH, Rattner BA, Wiemeyer SN, Royle JA, Smith MA (2003) Experimental lead poisoning in turkey vultures (*Cathartes aura*). J Wildl Dis 39:96–104

Chao SH, Bu CH, Cheung WY (1990) Activation of troponin C by Cd^{2+} and Pb^{2+}. Arch Toxicol 64:490–496

Church ME, Gwiazda R, Risebrough RW, Sorenson K, Chamberlain CP, Farry S, Heinrich W, Rideout BA, Smith DR (2006) Ammunition is the principal source of lead accumulated by California condors re-introduced to the wild. Environ Sci Technol 40:6143–6150

Clark AJ, Scheuhammer AM (2003) Lead poisoning in upland-foraging birds of prey in Canada. Ecotoxicology 12:23–30

Conrad ME, Barton JC (1978) Factors effecting absorption and excretion of lead in the rat. Gastroenterology 74:731–740

Craighead D, Bedrosian B (2007) Blood lead levels of common ravens with access to big-game offal. J Wildl Man 72:240–245

Cromie R, Newth J, Reeves J, O'Brien M, Beckmann K, Brown M (2015) The sociological and political aspects of reducing lead poisoning from ammunition in the UK: why the transition to non-toxic ammunition is so difficult. In: Delahay RJ, Spray CJ (eds) Proceedings of the Oxford Lead Symposium. Lead ammunition: understanding and minimising the risk to human and environmental health. Edward Grey Institute, The University of Oxford, pp 104–124

Cruz-Martinez L, Redig PT, Deen J (2012) Lead from spent ammunition: a source of exposure and poisoning in bald eagles. Hum–Wildl Interact 6:94–104

De Francisco ON, Feeney D, Armién WA, Redig PT (2016) Correlation of brain magnetic resonance imaging of spontaneously lead poisoned bald eagles (*Haliaeetus leucocephalus*) with histological lesions: a pilot study. Res Vet Sci 105:236–242

DeMichele SJ (1984) Nutrition of lead. Comp Biochem Physiol 78:401–408
Denbow DM (2000) Gastrointestinal anatomy and physiology. In: Whittow GC (ed) Sturkie's avian physiology, 5th edn. Academic Press, San Diego, pp 299–325
Ethier ALM, Braune BM, Scheuhammer AM, Bond DE (2007) Comparison of lead residues among avian bones. Environ Pollut 145:915–919
European Union (2003) Directive 2002/96/EC of the European Parliament and of the council of 27 January 2003 on waste electrical and electronic equipment (WEEE). Off J Eur Union L37:24–39 http://eur-lex.europa.eu/legal-content/EN/TXT/HTML/?uri=CELEX:32002L0096&from=EN. Accessed 06 July 2017
European Union (2013) Commission directive 2013/28/EU of 17 may 2013 amending annex II to directive 2000/53/EC of the European Parliament and of the council on end-of-life vehicles. Off J Eur Union L135:14–18. http://eur-lex.europa.eu/legal-content/EN/TXT/HTML/?uri=CELEX:32013L0028&from=EN. Accessed 06 July 2017
Finkelstein ME, George D, Scherbinski S, Gwiazda R, Johnson M, Burnett J, Brandt J, Lawrey S, Pessier AP, Clark M, Wynne J, Grantham J, Smith DR (2010) Feather lead concentrations and 207Pb/206Pb ratios reveal lead exposure history of California condors (*Gymnogyps californianus*). Environ Sci Technol 44:2639–2647
Finley MT, Dieter MP (1978) Influence of laying on lead accumulation in bone of mallard ducks. J Toxicol Environ Health 4:123–129
Fisher IJ, Pain DJ, Thomas VG (2006) A review of lead poisoning from ammunition sources in terrestrial birds. Biol Conserv 131:421–432
Franson JC (1996) Interpretation of tissue lead residues in birds other than waterfowl. In: Beyer WN, Heinz GH, Redmon-Norwood AW (eds) Environmental contaminants in wildlife, interpreting tissue concentrations, SETAC Spec Publ Series. Lewis Publishers, CRC Press, Boca Raton, pp 265–279
Franson JC, Pain DJ (2011) Lead in birds. In: Beyer WN, Meador JP (eds) Environmental contaminants in biota: interpreting tissue concentrations, 2nd edn. CRC Press, Boca Raton, pp 563–593
Franson JC, Sileo L, Pattee OH, Moore JF (1983) Effects of chronic dietary lead in American kestrels (*Falco sparverius*). J Wildl Dis 19:110–113
Franson JC, Michael G, Perry MC, Moore JF (1986) Blood protoporphyrin for detecting lead exposure in canvasbacks. In: Feierabend JS, Russel AB (eds) Lead poisoning in wild waterfowl. A workshop. Proceedings of the symposium held 3–4 March 1984, Wichita, Kansas, Cooperative Lead Poisoning Control Information Program, National Wildlife Federation, Washington, pp 32–37
Friend M (1987) Lead poisoning. In: Friend M (ed) Field guide to wildlife diseases, vol 1. US Department of the Interior, Fish and Wildlife Service, Washington DC, pp 175–189
Friend M, Franson JC (1999) Field manual of wildlife diseases. General field procedures and diseases of birds. USGS, Biological Resource Division, Madison, p 426
Fullmer CS, Edelstein S, Wasserman RH (1985) Lead-binding properties of intestinal calcium-binding proteins. J Biol Chem 260:6816–6819
Gangoso L, Alvarez-Lloret P, Rodriguez-Navarro AAB, Mateo R, Hiraldo F, Donazar JA (2009) Long-term effects of lead poisoning on bone mineralization in vultures exposed to ammunition sources. Environ Pollut 157:569–574
Garcia-Fernandez AJ, Martinez-Lopez E, Romero D, Maria-Mojica P, Godino A, Jimenez P (2005) High levels of lead blood in griffon vultures (*Gyps fulvus*) from Cazorla Natural Park (Southern Spain). Environ Toxicol 20:459–463
Gilbert SG, Weiss B (2006) A rationale for lowering the blood lead action level from 10 to 2 mg/dL. Neurotoxicology 27:693–701
Gilfillan SC (1965) Lead poisoning and the fall of Rome. J Occup Med 7:53–60
Helander B, Axelsson J, Borg H, Holm K, Bignert A (2009) Ingestion of lead from ammunition and lead concentrations in white-tailed sea eagles (*Haliaeetus albicilla*) in Sweden. Sci Total Environ 407:5555–5563
Henny C, Blus LJ, Hoffman DJ, Grove RA, Hatfield JS (1991) Lead accumulation and osprey production near a mining site on the Coeur d'Alene river, Idaho. Arch Environ Contam Toxicol 21:415–424

Henretig FM (2006) Lead. In: Nelson LS, Lewin NA, Howland MA, Hoffmann RS, Goldfrank LR, Flomenbaum NE (eds) Goldfrank's toxicologic emergencies, 9th edn. McGraw-Hill Medical, New York, pp 1266–1283

Hernandez M, Margalida A (2008) Pesiticide abuse in Europe: effects on the cinereous vulture (*Aegypius monachus*) population in Spain. Ecotoxicology 17:264–272

Hernandez M, Margalida A (2009) Assessing the risk of lead exposure for the conservation of the endangered Pyrenean bearded vulture (*Gypaetus barbatus*) population. Environ Res 109:837–842

Hernberg S (2000) Lead poisoning in a historical perspective. Am J Ind Med 38(3):244–254

Hoffman DJ, Pattee OH, Wiemeyer SN, Mulhern B (1981) Effects of lead shot ingestion on δ-aminolevulinic acid dehydratase activity, hemaglobin concentration and serum chemistry in bald eagles. J Wildl Dis 17:423–431

Hoffman DJ, Franson JC, Pattee OH, Bunck CM, Murray HC (1985) Biochemical and haematological effects of lead ingestion in nestling American kestrels (*Falco sparverius*). Comp Biochem Physiol 80:431–439

Hutton M, Goodman GT (1980) Metal contamination of feral pigeons *Columba livia* from the London area: part 1—tissue accumulation of lead, cadmium and zinc. Environ Pollut A 22:207–217

Irschik I, Bauer F, Sager M, Paulsen P (2013) Copper residues in meat from wild artiodactyls hunted with two types of rifle bullets manufactured from copper. Eur J Wildl Res 59:129–136

Ishihara N, Matsushiro T (1986) Biliary and urinary excretion of metals in humans. Arch Environ Health 41:324–330

Ishihara N, Koizumi M, Yoshida A (1987) Metal concentrations in human pancreatic juice. Arch Environ Health 42:356–360

Iwata H, Watanabe M, Kim EY, Gotoh R, Yasunaga G, Tanabe S, Masuda Y, Fujita S (2000) Contamination by chlorinated hydrocarbons and lead in Stellar's sea eagle and white-tailed sea eagle from Hokkaido, Japan. In: Ueta M, McGrady MJ (eds) First symposium on Stellar's and White-Tailed Sea eagles in East Asia. Wild Bird Society of Japan, Tokyo, pp 91–106

Janssen DL, Oosterhuis JE, Allen JL, Anderson MP, Kelts DG, Wiemeyer SN (1986) Lead poisoning in free-ranging California condors. J Am Vet Med Assoc 189:1115–1117

Jenni L, Madry MM, Kraemer T, Kupper J, Naegeli H, Jenny H, Jenny D (2015) The frequency distribution of lead concentration in feathers, blood, bone, kidney and liver of golden eagles *Aquila chrysaetos*: insights into the modes of uptake. J Ornithol 156:1095–1103

Kanstrup N (2006) Non-toxic shot – Danish experience. In: Boere CG, Galbraith CA, Stroud DA (eds) Waterbirds around the world. The Stationery Office, Edinburgh, p 861

Karstad L (1971) Angiopathy and cardiopathy in wild waterfowl from ingestion of lead shot. Conn Med 35:355–360

Kelly TR, Grantham J, George D, Welch A, Brandt J, Burnet LJ, Sorenson KJ, Johnson M, Poppenga R, Moen D, Rasico J, Rivers JW, Battistone C, Johnson CK (2014) Spatiotemporal patterns and risk factors for lead exposure in endangered California condors during 15 years of reintroduction. Conserv Biol 28:1721–1730

Kendall RJ, Scanlon PF (1983) Histologic and ultrastructural lesions of mourning doves (*Zenaida macroura*) poisoned by lead shot. Poult Sci 62:952–956

Kenntner N, Tataruch F, Krone O (2001) Heavy metals in soft tissue of white-tailed eagles found dead or moribund in Germany and Austria from 1993 –2000. Environ Toxicol Chem 20:1831–1837

Kenntner N, Krone O, Altenkamp R, Tataruch F (2003) Environmental contaminants in liver and kidney of free-ranging northern goshawks in three regions of Germany. Arch Environ Contam Toxicol 45(1):128–135

Kenntner N, Oehme G, Heidecke D, Tataruch F (2004) Retrospektive Untersuchung zur Bleiintoxikation und Exposition mit potenziell toxischen Schwermetallen von Seeadlern *Haliaeetus albicilla* in Deutschland. Vogelwelt 125:63–75

Kenntner N, Cretten Y, Fünfstück HJ, Janovsky M, Tataruch F (2007) Lead poisoning and heavy metal exposure of golden eagles (*Aquila chrysaetos*) from the European alps. J Ornithol 148:173–177

Kim EY, Goto R, Iwata H, Masuda Y, Tanabe S, Fujita S (1999) Preliminary survey of lead poisoning of Steller's sea eagle (*Haliaeetus pelagicus*) and white-tailed sea eagle (*Haliaeetus albicilla*) in Hokkaido, Japan. Environ Toxicol Chem 18:448–451
Kinard J (2004) Pistols: an illustrated history of their impact. ABC-CLIO, USA
Kirberger M, Yang JJ (2008) Structural differences between Pb^{2+}- and Ca^{2+}-binding sites in proteins: implications with respect to toxicity. J Inorg Biochem 102:1901–1909
Knott J, Green R, Pain DJ (2009a) Lead poisoning in Red Kites *Milvus milvus*. In: David F (coord) Status, monitoring and trends, Statuts, suivis et tendances, Red Kite International symposium – October, 17th et 18th 2009 – Colloque international Milan royal, 17–18 October, 2009, Montbéliard, France, pp 106–108
Knott J, Gilbert J, Green RE, Hoccom DG (2009b) Comparison of the lethality of lead and copper bullets in deer control operations to reduce incidental lead poisoning: field trials in England and Scotland. Conserv Evid 6:71–78
Kramer JL, Redig PT (1997) Sixteen years of lead poisoning in eagles, 1980 – 1995: an epizootiologic view. J Raptor Res 32:327–332
Krone O, Wille F, Kenntner N, Boertmann D, Tataruch F (2004) Mortality factors, environmental contaminants, and parasites of White-Tailed Sea eagles from Greenland. Avian Dis 48:417–424
Krone O, Stjernberg T, Kenntner N, Tataruch F, Koivusaari J, Nuuja I (2006) Mortality, helminth burden and contaminant residues in white-tailed sea eagles from Finland. Ambio 35:98–104
Krone O, Kenntner N, Tataruch F (2009a) Gefährdungsursachen des Seeadlers (*Haliaeetus albicilla* L. 1758). Denisia 27: 139–146
Krone O, Kenntner N, Trinogga A, Nadjafzadeh N, Scholz F, Sulawa J, Totschek K, Schuck-Wersig P, Zieschank R (2009b) Lead poisoning in white-tailed sea eagles: causes and approaches to solutions in Germany. In: Watson RT, Fuller M, Pokras A, Hunt WG (eds) Ingestion of lead from spent ammunition: implications for wildlife and humans. The Peregrine Fund, Boise, pp 289–301
Krone O, Berger A, Schulte R (2009c) Recording movement and activity pattern of a White-Tailed Sea eagle (*Haliaeetus albicilla*) by a GPS datalogger. J Ornithol 150:273–280
Labonde J (1991) Avian toxicology. Vet Clinic North Am Small Anim Pract 21:1329–1342
Lambertucci SA, Donázar JA, Huertas AD, Jiménez B, Sáez M, Sanchez-Zapata JA, Hiraldo F (2011) Widening the problem of lead poisoning to a South-American to scavrenger: lead concentrations in feathers of wild Andean condors. Biol Conserv 144:1464–1471
Li Y, Moon K, Wong CP (2005) Electronics without lead. Science 308:1419–1420
Lumeij JT (1985) Clinicopathologic aspects of lead poisoning in birds: a review. Review papers. Vet Q 7:133–138
Madry MM, Kraemer T, Kupper J, Naegeli H, Jenny H, Jenni L, Jenny D (2015) Excessive lead burden among golden eagles in the Swiss alps. Environ Res Lett 10(3). https://doi.org/10.1371/journal.pone.0172232
Markovac J, Goldstein GW (1988) Picomolar concentrations of lead stimulate brain protein kinase C. Nature 334:71–73
Martin PA, Campbell D, Hughes K, McDaniel T (2008) Lead in the tissues of terrestrial raptors in southern Ontario, Canada, 1995–2001. Sci Total Environ 391:96–103
Martinez-Haro M, Taggart MA, Green AJ, Mateo R (2009) Avian digestive tract simulation to study the effect of grit geochemistry and food on Pb shot bioaccessibility. Environ Sci Technol 43:9480–9486
Mateo R (2009) Lead poisoning in wild birds in Europe and the regulations adopted by different countries. In: Watson RT, Fuller M, Pokras M, Hunt WG (eds) Ingestion of lead from spent ammunition: implications for wildlife and humans. The Peregrine Fund, Boise, pp 71–98
Mateo R, Molina R, Grifols J, Guitar R (1997) Lead poisoning in free ranging griffon vulture (*Gyps fulvus*). Vet Rec 140:47–48
Mateo R, Cadenas R, Manez M, Guitar R (2001) Lead shot ingestion in two raptor species from Donana, Spain. Ecotoxicol Environ Saf 48:6–10
Mateo R, Vallverdu-Coll N, Lopez-Antia A, Taggart MA, Martínez-Haro M, Guitart R, Ortiz-Santaliestra ME (2014) Reducing Pb poisoning in birds and Pb exposure in game meat consumers: the dual benefit of effective Pb shot regulation. Environ Int 63:163–168

Meyer CB, Meyer JS, Francisco AB, Holder J, Verdonck F (2016) Can ingestion of lead shot and poisons change population trends of three European birds: grey partridge, common buzzard, and red kite? PLoS One 11(1):e0147189. https://doi.org/10.1371/journal.pone.0147189 Accessed 06 July 2017

Mudge G (1983) The incidence and significance of ingested lead pellet poisoning in British waterfowl. Biol Conserv 27:333–372

Müller K, Altenkamp R, Brunnberg L (2007) Morbidity of free-ranging White-Tailed Sea eagles (*Haliaeetus albicilla*) in Germany. J Avian Med Surg 21:265–274

Nadjafzadeh M, Hofer H, Krone O (2013) The link between feeding ecology and lead poisoning in white-tailed eagles. J Wildl Manag 77:48–57

Nadjafzadeh M, Hofer H, Krone O (2015) Lead exposure and food processing in white-tailed eagles and other scavengers: an experimental approach to simulate lead uptake at shot mammalian carcasses. Eur J Wildl Res 61:763–774

Nam DH, Lee DP (2009) Abnormal lead exposure in globally threatened cinereous vultures (*Aegypius monachus*) wintering in South Korea. Ecotoxicology 18:225–229

Nriagu JO (1983) Lead & lead poisoning in antiquity. Wiley-Interscience Publication, New York

Onalaja AO, Claudio L (2000) Genetic susceptibility to lead poisoning. Environ Health Perspect 108:23–28

Pain DJ (1992) Lead poisoning in waterfowl: a review. In: Pain DJ (ed) Lead poisoning in waterfowl. Proceedings of the international waterfowl and wetlands research bureau workshop, Brussels, Belgium. I. WR.B. Special Publication 16. UK, p 7–13

Pain DJ (1996) Lead in waterfowl. In: Beyer WN, Heinz GH, Redmon-Norwood AW (eds) Environmental contaminants in wildlife – interpreting tissue concentrations. A Special Publication of SETAC, CRC Press, South Carolina, pp 251–264

Pain DJ, Meharg AA, Ferrer M, Taggart M, Penteriani V (2005) Lead concentrations in bone and feathers of the globally threatened Spanish imperial eagle. Biol Conserv 121:603–610

Pain DJ, Carter I, Sainsbury A, Shore R, Eden P, Taggart M, Konstantinos S, Walker L, Meharg A, Raab A (2007) Lead contamination and associated disease in captive and reintroduced red kites *Milvus milvus* in England. Sci Total Environ 376:116–127

Pain DJ, Fisher IJ, Thomas VG (2009) A global update of lead poisoning in terrestrial birds from ammunition sources. In: Watson RT, Fuller M, Pokras M, Hunt G (eds) Ingestion of lead from spent ammunition: implications for wildlife and humans. The Peregrine Fund, Boise, pp 99–118

Pattee OH, Pain DJ (2003) Lead in the environment. In: Hoffman DJ, Rattner BA, Burton GA Jr, Cairns J Jr (eds) Handbook of ecotoxicology, 2nd edn. Lewis, Boca Raton, pp 373–408

Pattee OH, Wiemeyer SN, Mulhern BM, Sileo L, Carpenter JW (1981) Experimental lead-shot poisoning in bald eagles. J Wildl Manag 45:806–810

Pattee OH, Carpenter JW, Fritts SH, Rattner BA, Wiemeyer SN, Royle JA, Smith MR (2006) Lead poisoning in captive Andean condors (*Vultur gryphus*). J Wildl Dis 42:772–779

Paulsen P, Sager M, Bauer F, Schumann-Irschik I (2015) Pilot study on metal contents in meat portions from wild game killed by "lead-free" rifle bullets. J Food Saf Food Qual 66:128–131

Pikula J, Hajkova P, Bandouchova H, Bednarova I, Adam V, Beklova M, Kral J, Ondracek K, Osickova J, Pohanka M, Sedlackova J, Skochova H, Sobotka J, Treml F, Kizek R (2013) Lead toxicosis of captive vultures: case description and responses to chelation therapy. BMC Vet Res 9:11

Pounds JG, Long GJ, Rosen JF (1991) Cellular and molecular toxicity of lead in bone. Environ Health Perspect 91:17–32

Rattner BA, Fleming WJ, Bunck CM (1989) Comparative toxicity of lead shot in black ducks (*Anas rubripes*) and mallards (Anas platyrhnchos). J Wildlife Dis 25:175–183

Redig PT, Stowe CM, Barnes DM, Arent TD (1980) Lead toxicosis in raptors. J Am Vet Med Assoc 177:941–943

Reiser MH, Temple SA (1981) Effects of chronic lead ingestion on birds of prey. In: Cooper JE, Greenwood AG (eds) Recent advances in the study of raptor diseases. Chiron Publications, Keighley, pp 21–25

Repper R, Drasch G, Grimm F, Kösters J (1986) Die Auswirkungen implantierter Schrotkugeln auf einige Parameter bei Tauben. Deutsche Veterinärmedizinische Gesellschaft e.V., Tagung der

Fachgruppe Geflügelkrankheiten in Verbindung mit dem Institut für Geflügelkrankheiten der Ludwig-Maximilians-Universität München, 5: 34–48

Rodriguez-Ramos Fernandez J, Höfle U, Mateo R, de Francisco ON, Abbott R, Acevedo P, Blanco JM (2011) Assessment of lead exposure in Spanish imperial eagle (*Aquila adalberti*) from spent ammunition in Central Spain. Ecotoxicology 20:670–681

Rossi E (2008) Low level environmental lead exposure – a continuing challenge. Clin Biochem Rev/Aust Assoc Clin Biochem 29(2):63–70

Ruby MV, Davis A, Kempton JH, Drexler JW, Bergstrom PD (1992) Lead bioavailability: dissolution kinetics under simulated gastric conditions. Environ Sci Technol 26:1242–1248

Saito K, Kurosawa N, Shimura R (2000) Lead poisoning in endangered sea-eagles in eastern Hokkaido through ingestion of shot sika deer. In: Lumeij JT, Remple DR, Redig PT, Lierz M, Cooper JE (eds) Raptor biomedicine III. Zoological Education Network, Florida, pp 163–166

Scarborough J (1984) The myth of lead poisoning among the romans: an essays review. J Hist Med Allied Sci 39(4):469–475

Scheuhammer AM (1987) The chronic toxicity of Aluminium, cadmium, mercury and lead in birds: a review. Environ Pollut 46:263–295

Scheuhammer AM, Templeton DM (1998) Use of stable isotope ratios to distinguish sources of lead exposure in wild birds. Ecotoxicology 7:37–42

Schock MR, Hyland RN, Welch MM (2008) Occurrence of contaminant accumulation in lead pipe scales from domestic drinking water distribution systems. Environ Sci Technol 42(12):4285–4291

Sileo L, Jones RN, Hatch RC (1973) The effect of ingested lead shot on the electrocardiogram of Canada geese. Avian Dis 17:308–313

Stauber E, Finch N, Talcott PA, Gay JM (2010) Lead poisoning of bald (*Haliaeetus leucocephalus*) and golden (*Aquila chrysaetos*) eagles in the US inland pacific northwest region – an 18-year retrospective study: 1981-2008. J Avian Med Surg 24:279–287

Stendell RC, Smith RI, Burnham KP Christensen RE (1979) Exposure of waterfowl to lead: a nationwide survey of residues in wing bones of seven species, 1972–1973. US Fish and Wildl Serv Spec Sci Rep Wildl 223:1–12

Strom SM, Langenberg JA, Businga NK, Batten JK (2009) Lead exposure in Wisconsin birds. In: Watson RT, Fuller M, Pokras M, Hunt WG (eds) Ingestion of spent lead ammunition: implications for wildlife and humans. The Peregrine Fund, Boise, Idaho, pp 194–201

Thomas VG (2013) Lead-free rifle ammunition: product availability, price, effectiveness, and role in global wildlife conservation. Ambio 42:737–745

Thomas VG, Kanstrup N, Gremse C (2015) Key questions and responses regarding the transition to use of lead-free ammunition. In: Delahay RJ, Spray CJ (eds) Proceedings of the Oxford lead symposium. Lead ammunition: understanding and minimising the risk to human and environmental health. Edward Grey Institute, The University of Oxford, Oxford, pp 125–134

Thompson LJ (2012) Chapter 31. Lead. In: Gupta RC (ed) Veterinary toxicology. Basic and clinical principles, 2nd edn. Elsevier, Amsterdam, pp 438–441

Trinogga A, Fritsch G, Hofer H, Krone O (2013) Are lead-free hunting rifle bullets as effective at killing wildlife as conventional lead bullets? A comparison based on wound size and morphology. Sci Total Environ 443:226–232

Wayland M, Bollinger T (1999) Lead exposure and poisoning in bald eagles and golden eagles in the Canadian prairie provinces. Environ Pollut 104:341–350

Wiemeyer GM, Perez MA, Bianchini LT, Sampietro L, Bravo GF, Jacome NL, Astore V, Lambertucci SA (2017) Repeated conservation threats across the Americas: high levels of blood and bone lead in the Andean condor widen the problem to a continental scale. Environ Pollut 220:672–679

World Health Organization (1989) Lead—environmental aspects. Environmental Health Criteria, No. 85. Geneva, Switzerland. http://www.inchem.org/documents/ehc/ehc/ehc85.htm. Accessed 06 July 2017

World Health Organization (2011) Lead in drinking-water. Background document for development of WHO Guidelines for Drinking-water Quality. Geneva, Switzerland. http://www.who.int/water_sanitation_health/dwq/chemicals/lead.pdf. Accessed 06 July 2017

Zechner L, Steineck T, Tataruch F (2005) Bleivergiftungen bei einem Steinadler (*Aquila chrysaetos*) in der Steiermark. Egretta 47: 157–158

Chapter 12
Raptor Electrocutions and Power Line Collisions

Duncan T. Eccleston and Richard E. Harness

Raptor Electrocutions

Introduction

Utility infrastructure is used by a wide range of bird species for roosting, foraging, hunting, nest building, and rearing young. In some cases, interactions benefit individuals and species by protecting them from ground predators, improving hunting success, or providing high value nesting and brood-rearing substrates (APLIC 2006; Ellis et al. 2009; Jenkins et al. 2013). Power poles also can cause avian mortality through electrocution (González et al. 2007). Electrocution occurs when a bird is killed by electric current passing through the body; electric shock injuries also may ultimately be fatal (Dwyer 2006; Kagan 2016). In a review of 110 studies, Lehman et al. (2007) showed that raptor electrocution is closely associated with 1,000–60,000 V (1–60 kV) electric distribution electric systems and occurs on nearly every continent.

The US power grid was initially developed in the 1880s and subsequently expanded into rural areas, where it was frequently encountered by raptor species. Eagle electrocution was first documented in 1922 (Hallinan 1922 in Hunting 2002), and hawk electrocution first recorded in 1933 (Manville 2005). Although avian electrocution on power lines has been documented in a wide variety of taxa (Hunting 2002), most North American studies have shown that raptor species comprise a high proportion of electrocuted birds, especially in rural areas (Harness and Wilson 2001; Lehman et al. 2010). Raptors are especially susceptible to electrocution because of their large size and their predilection for using power poles for perching and nesting (Hunting 2002). In the Western United States, raptors comprise 96% of electrocutions recorded by utilities (Harness and Wilson 2001). In other parts of the world,

D. T. Eccleston (✉) · R. E. Harness
EDM International Inc, Fort Collins, CO, USA
e-mail: deccleston@edmlink.com

J. H. Sarasola et al. (eds.), *Birds of Prey*,
https://doi.org/10.1007/978-3-319-73745-4_12

electrocution also is an important component of anthropogenic mortality for raptors, even where raptors comprise a smaller proportion of documented electrocutions (Gerdzhikov and Demerdzhiev 2009).

Scope of the Problem

Electrocution has been studied extensively in the United States (49 studies), Spain (16 studies), and South Africa (11 studies) and also has been documented in Canada, Mexico, the United Kingdom, Italy, France, Germany, Norway, Hungary, Slovakia, Bulgaria, Russia, Kazakhstan, Australia, Israel, and Sudan (Lehman et al. 2007). Since 2007, other English-language sources have documented raptor electrocution in China (Dixon et al. 2013), Mongolia (Harness et al. 2008), India (Harness et al. 2013), Morocco (Amezian et al. 2015), New Zealand (Fox and Wynne 2010), the Dominican Republic (Spencer 2016), and Argentina (Ibarra and De Lucca 2015; Galmes et al. (in press)).

By its nature, raptor electrocution is difficult to study; Lehman et al. (2007) provides a good introduction to advantages and drawbacks of various methodologies and examples of studies that have employed them. Differences in methodology and reporting conventions make it difficult to directly compare electrocution rates between studies. For systems with no protective raptor electrocution retrofitting or minimal retrofitting, electrocution rates reported worldwide have ranged from minimum estimates of 0.03–0.04 deaths/pole/year (Kemper et al. 2013; Ledger 1984 as calculated by Lehman et al. 2010) to maximum estimates of 0.32–0.35 deaths/pole/year (Cartron et al. 2005 and Ferrer et al. 1991, as calculated by Lehman et al. 2010). Studies with the highest rates generally were associated with grounded steel and concrete poles and crossarms, which are more hazardous to raptors than ungrounded wood poles and crossarms (Lehman et al. 2010; Harness 2000a, b; Janss and Ferrer 1999b).

Systems with grounded poles and grounded metal crossarms (Fig. 12.1) are of particular interest because they have been implicated in population and subpopulation declines among multiple species of special conservation concern. For example, electrocution has placed subpopulations of the Bonelli's eagle (*Aquila fasciata*) at risk of extinction (Hernández-Matías et al. 2015), and electrocution was a critical driver of population declines in the Spanish Imperial Eagle *(Aquila adalberti)* (López-López et al. 2011). Angelov et al. (2013) and Donázar et al. (2002) attributed population declines of the endangered (Bird Life International 2017) Egyptian vulture (*Neophron percnopterus*) in east Africa and the Canary Islands, respectively, to electrocution. Harness et al. (2008) and Dixon et al. (2013) expressed concern that high rates of electrocution observed in Mongolia could impact populations of the endangered (BirdLife International 2017) saker falcon (*Falco cherrug*, in this book Chap. 17).

Avian electrocution rates are probably high in developing countries. Developing countries may lack basic legal protections for birds and other wildlife (Purev-Ochir,

Fig. 12.1 Turkey vulture electrocution on a grounded pole with a grounded metal crossarm; note ineffective wooden elevated perch

pers. comm. 2016), and where legal protections exist, awareness (Boshoff et al. 2011), reporting mechanisms, and enforcement (Juvvadi, pers. comm. 2016) may be emergent or nonexistent. Additionally, many of the electrical systems in developing countries are constructed of grounded poles and grounded metal crossarms (e.g., Dixon et al. 2013; Angelov et al. 2013; Goroshko 2011). A bird perched on a grounded metal crossarm need only touch one wire to be injured or killed.

In the developing countries of Sudan and Morocco, Angelov et al. (2013) and Amezian et al. (2015), respectively, documented electrocutions of large wintering raptors occurring at such high rates that they were responsible for declines in breeding populations elsewhere. These studies suggest that in those countries, resident populations of other raptor species also may have been strongly impacted by electrocution but without documentation. Given the lack of avian electrocution documentation from many countries with large raptor populations, it is likely that specific areas of the world where raptor electrocution is problematic have not yet been identified. Most of the electrical grid expansion in the coming decades will occur in developing areas. If these electrical systems are not built to minimize avian electrocutions, worldwide rates of raptor electrocution could rise substantially.

Hazardous Poles

Overhead distribution power lines are defined as lines with less than 60,000 volts or 60 kV (APLIC 2006). Lines at 60 kV or higher are termed transmission lines and are implicated with fewer electrocutions because they require greater separation between wires and between wires and grounded surfaces. Distribution power poles support one, two, or three differently energized conductor wires, termed phases.

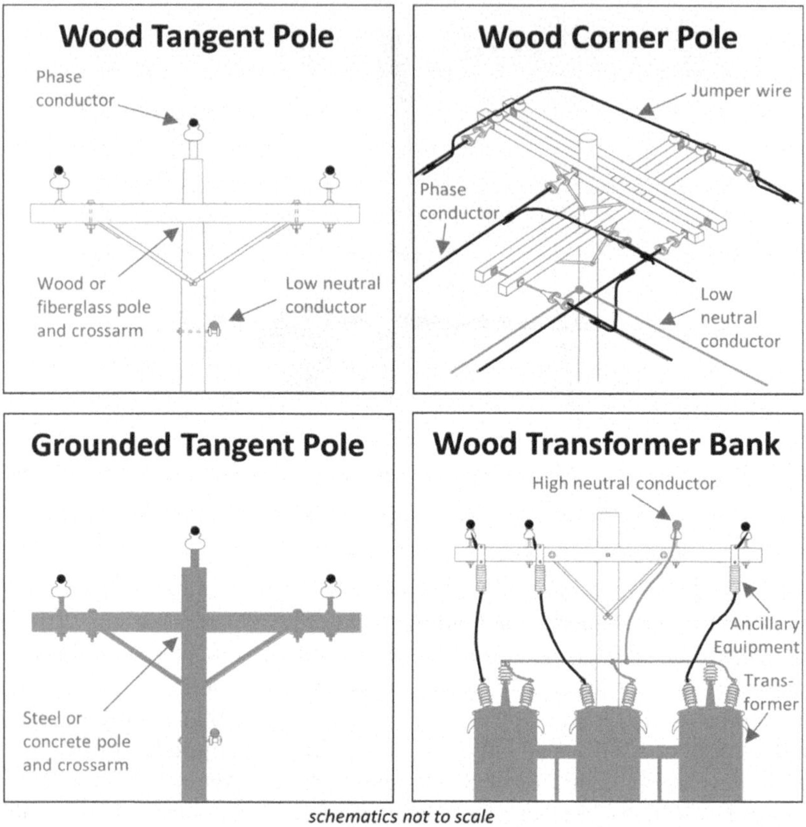

Fig. 12.2 Four common distribution designs showing energized (black), grounded (gray), and non-grounded and non-energized or non-conductive areas in white

Distribution power lines also may have a grounded neutral wire to provide a return path for electricity. Electrocution or shock injury occurs when a bird simultaneously contacts two differently energized phases or one energized phase and one grounded area (APLIC 2006). Figure 12.2 illustrates four common distribution designs and introduces key vocabulary; from top left, clearances generally decrease in a clockwise direction.

Avian electrocution risk is distributed unevenly among distribution poles (Tintó et al. 2010). Electrocution risk has been associated with both pole design elements and habitat elements. Some pole configurations pose a greater risk to raptors; Harness and Wilson (2001) found that three-phase transformer banks in rural habitats were responsible for 53% of detected raptor electrocutions in the Western United States but comprised less than 3% of all poles (Fig. 12.3). Certain "lethal poles" can individually kill large numbers of birds: on a single site visit, Ibarra and De Lucca (2015) located 18 electrocuted black-chested buzzard-eagles (*Geranoaetus melanoleucus*) beneath one distribution pole near a garbage dump in Argentina;

Fig. 12.3 Swainson's hawk at risk on an unprotected three-phase transformer bank

Demerdzhiev et al. (2009) found 9 suspected electrocutions beneath a single pole within an Audubon Important Bird Area in southern Bulgaria; and Harness et al. (2008) documented a single pole responsible for at least 34 mortalities in Mongolia.

A conceptual model illustrates avian electrocution risk as the product of pole hazard and avian exposure (Fig. 12.4, Dwyer et al. 2016a), with each factor being independent of the other. Pole hazard is an index of the relative danger associated with any given pole configuration, which is determined by the number, spacing, and arrangement of conductors, neutral wires, equipment and jumper wires, and the type and extent of grounding (e.g., Tintó et al. 2010; Kemper et al. 2013; Dwyer et al. 2013). As the number and proximity of potentially lethal contacts increase, the pole becomes intrinsically more hazardous. Avian exposure describes the level of pole utilization by species susceptible to electrocution. As avian exposure increases, the number of opportunities for a bird to concurrently touch lethal contacts also rises. Because long-term avian exposure cannot be ascertained during a limited number of field visits, it is usually assessed using proxy variables such as surrounding habitat, proximity to nests or bird use areas, pole vantage, prey base, whitewash, and prior mortalities (e.g., Ferrer et al. 1991; Benson 1981; Guil et al. 2011) that suggest or demonstrate pole utilization. Electrocution risk is greatest for a hazardous pole with a high level of avian exposure.

In developing a statistical model of electrocution risk, Dwyer et al. (2013) compared 213 known avian electrocution poles in Southern California with a comparable number of randomly selected poles based on 14 attributes describing the pole and setting. A statistically rigorous process was used to develop over one thousand candidate models and select the best one. The final model used four variables to predict electrocution risk. The three pole hazard variables included the number of

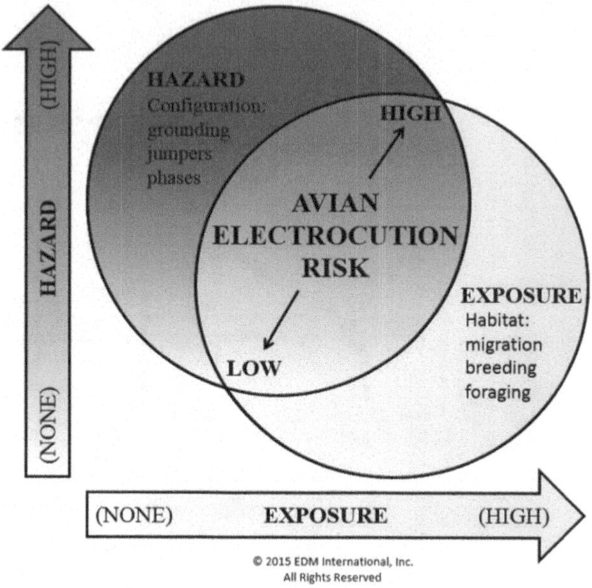

Fig. 12.4 Conceptual model of avian electrocution risk. (After Dwyer et al. 2016a)

energized primary phase conductors, the number of energized jumper wires, and the presence of a high grounded contact; the fourth variable was favorable habitat specific to the species studied. The authors believed number of jumpers was a meaningful predictor of electrocution because it served as an index of general pole complexity, effectively integrating the hazards posed by the jumper wires themselves with hazards posed by associated pole equipment.

Pole Configurations and Materials

Specific designs associated with raptor electrocution vary by study area, but common themes have emerged. The most determinative design factor appears to be crossarm grounding. Birds perched on a grounded metal crossarm may be electrocuted or shocked if they touch only one energized phase wire (Fig. 12.5), in comparison to the two contacts to precipitate electrocution on an ungrounded wood or fiberglass crossarm (Janss and Ferrer 1999b). Not only does a grounded metal crossarm substantially increase the likelihood that a large perching raptor will be electrocuted, it also places small raptors like the European kestrel (*Falco tinnunculus*), which are not large enough to touch two conductors simultaneously, at risk of phase to ground electrocution (Bayle 1999; Janss 2000). In Siberia, Goroshko (2011) discovered an average of 0.14 carcasses per pole at 491 concrete poles with grounded metal crossarms, but no carcasses beneath 260 wood poles with ungrounded crossarms; a similar result was observed in Mongolia by Harness et al. (2008).

Studies of systems with grounded poles and metal crossarms have frequently attributed variability in hazard to the insulator type. For example, Demerdzhiev

Fig. 12.5 Small raptors such as American kestrels can be at risk on poles with metal-grounded crossarms

(2014) recorded a per pole carcass discovery rate of 0.03–0.04 for suspended insulator configurations, 0.07–0.08 for configurations with vertical post insulator mounted on metal brackets, and 0.19 for terminal configurations with bare jumper wires routed through areas where they were likely to be contacted by perching birds. This relative ranking of pole hazard reflects the ease with which a perching bird would be expected to make contact with an energized wire. Suspended conductors are less hazardous because energized phase wires are mostly or completely isolated from a bird perched on a grounded metal crossarm (Fig. 12.6). In contrast, conductors mounted above the crossarm on vertical post or pin insulators are frequently encountered by heads, bodies, and wings when small to large raptors land, perch, or depart from the crossarm. Similarly, Guzmán and Castaño (1998) found that on power lines comprised primarily of poles with suspended insulators, 97% of the mortalities were discovered beneath the <20% of poles having vertical post insulators or terminal configurations with bare jumpers above the crossarm. Similarly, terminal poles with jumpers routed below the crossarm typically are associated with fewer electrocutions (Harness et al. 2008; Dixon et al. 2013).

In Western India, Harness et al. (2013) found that whether the pin insulator was short (~14 cm) or tall (~46 cm) (Harness and Juvvadi 2013) was a statistically significant predictor of whether a pole would be associated with an avian mortality. This minor detail underscores the lethal nature of the grounded poles: so many mortalities occurred when the pin was short, that increasing vertical separation between the grounded pole and the energized conductor significantly reduced avian electrocution. The longer pin insulator precluded smaller species from making a phase to ground contact. The presence of a transformer also was a significant predictor of electrocution (Harness and Juvvadi 2013). This result was unique because

Fig. 12.6 Eagle perched on a crossarm fitted with suspended insulators

studies of systems with grounded crossarms generally do not mention the role of pole-mounted equipment.

For systems using non-grounded wood or fiberglass crossarms, the presence of pole-mounted equipment exerts a strong influence over the pole hazard. In Alberta, Canada, Kemper et al. (2013) found that three-phase transformer banks were associated with a significantly higher electrocution rate than any other pole type, echoing the 2001 findings of Harness and Wilson in a study of the Western United States and Canada. Pole-mounted equipment introduces additional energized and grounded contacts, frequently in close proximity to one another. Frequently, equipment is connected to additional protective equipment such as fuses and lightning surge arresters with bare jumper wires. Thus, relative to a comparable pole without equipment, such configurations have many more potentially lethal contacts. The close proximity of these jumpers also allows smaller birds to become electrocuted.

Vulnerability

Hunting (2002) compiled a list of 64 raptor species known to have been electrocuted, worldwide. Size and behavior are the primary factors that determine raptor species' susceptibility to electrocution. Janss (2000) used size, weight, and indices of wing morphology reflective of behavior to develop a discriminant model successfully predicting species susceptible to electrocution (Fig. 12.7). The minimum size for electrocution to occur depends on the separation provided by the distribution system. An individual is potentially at risk of electrocution if the vertical separation is smaller than its head-to-foot length (above a grounded perch) or tail-to-foot length (below the perch). An individual also is at risk of electrocution if the

Fig. 12.7 Golden eagle electrocution

horizontal separation is less than the distance between the outermost skin components with wings spread (APLIC 2006).

In the United States, more golden eagle (*Aquila chrysaetos*) electrocutions are reported than any other species (Lehman et al. 2007); not only is the species among the largest on the continent, it also inhabits areas where commanding perches are at a premium and has a predilection for still hunting (Hunting 2002). Harness and Wilson (2001) found that in the Western United States, the golden eagle is the most reported electrocuted eagle species, the red-tailed hawk (*Buteo jamaicensis*) is the most electrocuted hawk, and the great horned owl (*Bubo virginianus*) is the most electrocuted owl. Species that avoid perching on power poles have very little exposure to electrocution. Accipiter species are rarely electrocuted because, within their forested habitat, natural perches are both preferred and widely available (Benson 1980; O'Neil 1988; Switzer 1977).

Contributing Factors

Researchers have identified numerous factors contributing to variability in electrocution rates. Factors identified in the literature as contributing to electrocution risk include sex, age, seasonal activity, precipitation, prevailing winds, perch availability, prey base, social activities, nesting, and excretion. Each is discussed briefly below.

Sex Sexual dimorphism is common in raptors, with females generally larger than males (Newton 2010), which may increase females' risk of electrocution (Fox and

Wynn 2010). Ferrer and Hiraldo (1992) found that 78% of electrocuted Spanish Imperial Eagles were female, an imbalance that exacerbated the population level effects of electrocution and hampered the species' recovery. While a sex-related size difference was one factor explaining higher electrocution rates for female Harris' hawks (*Parabuteo unicinctus*), behavior differences meant that females also were exposed to power poles more frequently than males (Dwyer 2009).

Age Disproportionately high electrocution rates in juvenile and subadult raptors have been widely observed in the United States, especially in eagles. For example, a review of utility raptor electrocution data found that 94% of electrocuted eagles were juveniles or subadults, though pole utilization by non-adult age classes was only 62% (EPRI 1982). Similar results were documented by Boeker and Nickerson (1975) and Nelson and Nelson (1977). Nelson and Nelson (1976) used video analysis to determine that immature birds had not yet developed the skills used by adult birds to maintain controlled flight as they approached power poles. Elevated rates of electrocution in non-adult age classes also have been observed in other species and regions (González et al. 2007; Ferrer et al. 1991; Ferrer and Hiraldo 1992).

Seasonal Activity Seasonal increases in electrocutions have been attributed to fledging, when many inexperienced birds first encounter come utility poles (Harness and Wilson 2001). Lasch et al. (2010) attributed a late summer increase in electrocution rates in Kazakhstan to fledging. Demerdzhiev (2014) also found raptor electrocution rates to be elevated during the fledging season, as well as during migration, perhaps due to temporarily increased populations and exposure. Barbazyuk et al. (2010) observed increased eagle electrocutions during migration in southwestern Russia. Other work suggests that in some areas, electrocution rates are highest in the winter months due to increased humidity and precipitation (Nelson and Nelson 1977), although an alternative explanation is that reduced winter scavenging allows carcasses to persist for longer and be recovered at a higher rate (Harness and Wilson 2001; Dixon et al. 2013). While seasonal effects may be seen locally, the timing and magnitude vary with site-specific factors.

Precipitation Precipitation can have a strong influence on electrocution (Benson 1981). Contact between feathers and conductors may be ten times more likely to result in an electrocution if the feathers are wet than if they are dry (Nelson 1979, 1980). In some climates, seasonal variation in precipitation may drive variability in electrocution rates (Nelson and Nelson 1977; Gonzáles et al. 2007). Even single-phase wood poles considered avian safe under dry conditions may cause raptor electrocutions (Eccleston, unpublished data) if precipitation causes the generally non-conductive wood to become temporarily grounded (Hamerstrom et al. 1974 in Hunting 2002).

Wind High winds, especially when gusty, can complicate power pole landings and departures by blowing raptors off course and into hazardous areas. These hazards are exacerbated when the wind direction is parallel to the crossarm. Benson (1981) found approximately twice as many carcasses at poles with crossarms parallel or diagonal to the prevailing wind, a result predicted by Nelson and Nelson (1976).

Perch Availability In deserts (Donázar et al. 2002), grasslands (Harness et al. 2008), or other habitats with few available natural perches (Lehman et al. 2010), raptors are more likely to perch on power poles. Guil et al. (2015) found that regional tree cover was inversely proportional to electrocution rates in Spain. Phipps et al. (2013) showed that movements of Cape vultures (*Gyps coprotheres*), which prefer to perch, roost, and forage in the vicinity of power poles, are highly influenced by the location and extent of the power grid; the study documented the species' range expanding to reflect the growth of the transmission network into areas of Southern Africa previously lacking suitable perches.

Prey Base Because certain raptors use power poles to facilitate still hunting, specific poles offering a superior vantage point are utilized at a higher rate than nearby poles (Hunting 2002; Tintó et al. 2010); high surrounding slopes (Guil et al. 2011) are an indicator of pole vantage. Guil et al. (2011) showed that prey abundance was a significant predictor of electrocution and hypothesized that local raptor concentrations increased because of high prey availability. A pole in an area with excellent food availability for scavengers (Ibarra and DeLucca 2015) or an advantage in hunting (Olendorf 1972) may function as an ecological "trap" if the pole itself is hazardous (Guil et al. 2015).

Social Activities Communal roosting, a behavior characteristic of many vulture species (Ledger and Hobbs 1999) and certain other raptor species (Dwyer and Mannan 2007), can increase both the likelihood of an electrocution and the number of birds that are killed in a single incident (Leger and Hobbs 1999; Janss and Ferrer 1999b). Recommended avian clearances are based on the dimensions of a single bird (APLIC 2006); when multiple birds are tightly packed into a perch, electrocutions may occur even on poles providing "avian-friendly" separation (Dwyer and Mannan 2007). Electrocutions may be more prevalent during the breeding season, as courtship and territorial defense result in aggressive behaviors that increase raptors' exposure to electrocution through distraction and carelessness and coupled birds that together are larger than either individual (Dehmerdzhiev 2014; Dickerman 2003).

Nesting Power pole nesting has been documented in a large number of species, particularly raptors. Although power pole nesting may be beneficial to raptors in many cases (Jenkins et al. 2013), it is not beneficial under all circumstances (Hunting 2002). One potential consequence is the potential for wet nesting material to cause electrocution and/or fire (Ledger and Hobbs 1999). When carrying sticks and prey items to the nest site, a raptor's agility is reduced and effective size is increased, which may result in elevated exposure to electrocution (Benson 1980). Nesting platforms in safer locations can reduce the likelihood of raptor electrocution (Vanderburgh 1993). Nesting in the vicinity of power lines also may increase adults'

(Dwyer 2009) and fledglings' (Dwyer and Mannan 2007) exposure to hazardous poles, compounding the likelihood of electrocution; Guzmán and Castaño (1998) found an inverse correlation between the likelihood of a pole being associated with a raptor electrocution and the distance to a nest.

Excretion A streamer is a jet of semiliquid bird excrement that can exceed 60 inches in length (Burnham 1995; van Rooyen et al. 2002; Zhou et al. 2009). An outage can occur when a streamer simultaneously contacts differently energized components or if the streamer fills the air gap between them and initiates arcing (van Rooyen et al. 2003). Though rare, electrocution may occur if a streamer is launched from a grounded perch (Harness and Kratz 2007; Demerdzhiev 2014).

Prevention and Mitigation

Separation, insulation, and redirection are the strategies typically employed to prevent electrocution, in descending order of preference (Fig. 12.8, Dwyer et al. 2016b):

- Separation reduces electrocution hazard by providing avian-friendly clearances between potential contacts. Separation is preferred because it is a permanent, design-based strategy that does not require ongoing maintenance. However, separation is practical primarily for new construction and rebuilds and cannot be used to address hazards posed by equipment, especially components having both energized and grounded areas, like transformers (Harness 2004, 2007; APLIC 2006).
- Insulation uses dielectric materials to strategically cover energized or grounded contacts in order to provide avian-friendly clearances (EPRI 2013). Wildlife covers provide a level of insulation protection for birds and other wildlife but do not make energized equipment safe for human contact (APLIC 2006). Although many products initially will function well, some are more durable than others, which may degrade rapidly in response to environmental stresses. Before selecting wildlife protective products to install on overhead electrical distribution systems, it is useful to reference IEEE standard 1656™, "Guide for Testing the Electrical, Mechanical, and Durability Performance of Wildlife Protective Devices on Overhead Power Distribution Systems Rated up to 38 kV." This standard provides test recommendations for animal electrocution mitigation products (IEEE 2010).
- Redirection refers to the use of perch discouragers and alternative perches to shift birds toward safer locations. This strategy was recommended (APLIC 1996) in North America until recently (APLIC 2006). Today, redirection is the least preferred avian protection approach and is particularly difficult when used on grounded metal crossarms (Janss and Ferrer 1999a, b). Studies have shown some existing discouragers cannot reliably exclude a determined raptor (Prather and

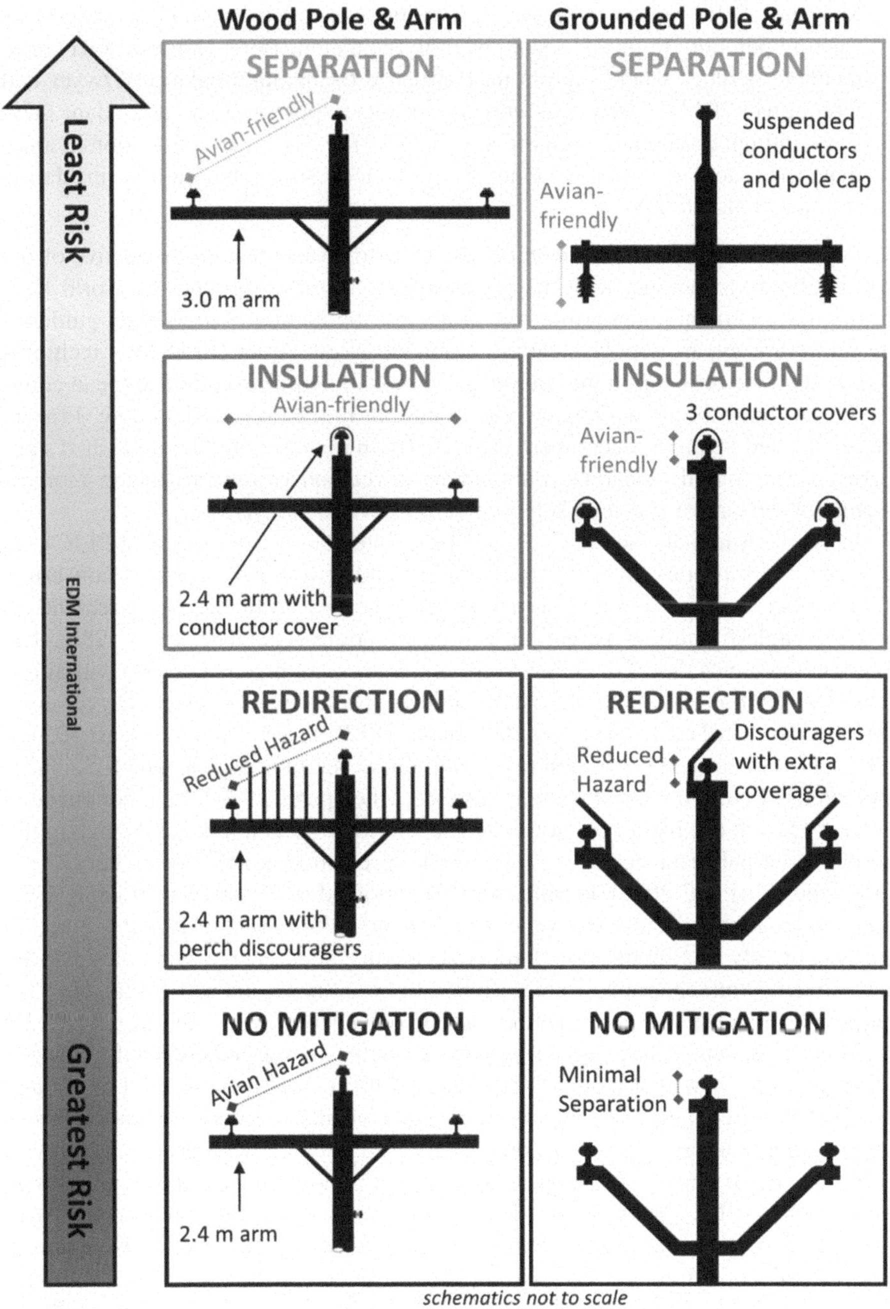

Fig. 12.8 Pole hazard is greatest for poles with no mitigation and least for poles with avian-friendly clearances achieved through insulation or separation; left, wood pole with an ungrounded wood crossarm; right, concrete pole with a grounded metal crossarm. (After Dwyer et al. 2016b)

Messmer 2010) and may, in some cases, increase exposure to electrocution by shifting perching to unsafe locations (Guyonne et al. 1999, Harness 2001) or to unstable perches where raptors may struggle to maintain balance (Dwyer and Doloughan 2014). Perch discouragers are most appropriate as redundant measures in high hazard areas (Kemper et al. 2013) or as primary protection in areas that, for practical reasons, cannot be protected using separation or insulation (Dwyer et al. 2016b).

Separation and/or insulation should be used to ensure spacing protective of the species likely to interact with the pole. Although utilities around the world have begun to implement avian protection strategies, the literature offers little guidance on minimum avian-friendly clearances outside North America. In fact, technical standards for electrocution mitigation are "not yet mature" even in European countries that have been worldwide electrocution mitigation leaders like France, Germany, and Spain (Schuerenberg et al. 2010). In regions without established standards, avian-friendly clearances should be developed to reflect spacing requirements for the largest species likely to perch (Dwyer et al. 2015).

In North America, the Avian Power Line Interaction Committee (APLIC), an association of utilities, industry trade groups, and government agencies concerned about avian electrocution and collision, used the best available measurements to develop eagle-friendly construction recommendations (Olendorff et al. 1981). The approach assumes that if the largest bird in the area can perch on a pole with minimal electrocution risk, then the risk to smaller species is even lower (Dwyer et al. 2016b). For distribution lines in eagle habitat, APLIC recommended at least 102 cm (40 inches) of vertical separation to accommodate perching eagles, which can measure up to 71 cm (28 inches) from head to toe, when perched. APLIC recommended at least 152 cm (60 inches) of horizontal separation to accommodate the maximum carpal-to-carpal distance for a golden eagle, presumed at the time to be 137 cm (54 inches) (APLIC 2006). Utilities were encouraged to adjust avian-friendly standards in locations where larger (e.g., California condor, *Gymnogyps californianus*) species, or only smaller ones, are likely to come in contact with power poles. APLIC clearance recommendations were designed to be intuitive and self-explanatory for engineers, linemen, and others without specialized experience with birds.

Many researchers have urged system operators to use avian-friendly designs when building new distribution lines, especially in rural areas (e.g., Ledger and Hobbs 1999; Lehman 2001). For systems that typically use wood or other types of ungrounded poles and crossarms, only minor design modifications are necessary to achieve avian-friendly separation, such as longer crossarms mounted slightly lower on the pole (APLIC 2006; Benson 1981). For systems depending on grounded steel or concrete poles and crossarms, more fundamental design changes may be required. For example, operators in Germany have begun to build distribution lines with suspension insulators (Schürenberg et al. 2010) instead of arm-mounted pin and post insulators because the suspended insulators effectively isolate the conductors from perching birds.

Existing power poles posing an elevated electrocution risk to birds can be modified to reduce risk. The effectiveness of retrofitting has been tested only in a modest number of locations (Lehman et al. 2007; Hunting 2002; Tintó et al. 2010), though some studies have reported high rates of success. After distribution lines were modified to reduce electrocutions using insulation and updated insulators, both Tintó et al. (2010) and Dehmerzhiev (2014) found zero carcasses beneath hundreds of retrofitted poles, where dozens of electrocuted carcasses previously had been discovered. Benson (1981) found that four line segments rebuilt with longer crossarms to meet APLIC spacing recommendations did not cause any mortalities. Matsyna et al. (2010) found only 1 carcass beneath more than 400 retrofitted poles. Dwyer (2004) documented a 74% reduction in electrocutions after transformer poles near active Harris' hawk nests were retrofitted. Retrofitting the most hazardous poles in Doñana National Park and Andalucía reduced Spanish Imperial Eagle electrocutions by 97% and 67%, respectively, and has been credited with the species' recovery in the area (López-López et al. 2011).

However, retrofitting has not been a panacea. Gonzalez et al. (2007) noted that within a decade of installation, heat shrink tape had substantially deteriorated (Fig. 12.9), resulting in eight electrocutions. As a result, Guil et al. (2011) urged operators to focus on durable structural retrofits like eliminating overarm insulators and extending dead-end insulators, rather than adding insulation that would require regular monitoring and eventual replacement. Lehman et al. (2010) found that retrofitting 7,424 high-risk poles, or 13% of all poles in the service territory, reduced the overall electrocution rate by 47% but did not eliminate the issue, perhaps because of a heavy reliance on redirection (Harness, pers. comm. 2017). A number of studies have suggested that retrofitting may have unintended consequences. For example,

Fig. 12.9 Example of insulating tape coming loose from a retrofitted structure in the Mongolian steppe

though perch discouragers may reduce electrocutions (Garrett 1993), they also may encourage birds to perch in close proximity to conductors, thereby increasing electrocution risk (Harness et al. 2008). Bayle (1999) reported multiple examples of failed retrofitting products and strategies and urged extensive testing, preferably with trained or captive birds, prior to widespread implementation.

It is possible to design and/or modify lines to significantly reduce electrocution rates. There are numerous published guidelines on how to remedy these issues and scores of commercially available mitigation products. EPRI (2013) includes a comprehensive list of manufacturers and an extensive list of avian electrocution mitigation products. When installing such products, results of testing performed by manufacturers should be reviewed and compared (EPRI 2014). Implementation testing should be carried by each operator to ensure specific products are compatible with local equipment and work practices (EPRI 2016). Finally, products need to be installed correctly. A raptor electrocution on a retrofitted pole represents a missed conservation opportunity and a waste of resources. Dwyer et al. (2016b) reviewed 52 retrofitted poles associated with 56 electrocutions in the Western United States and determined that retrofitting failures can be minimized through careful device selection, proper installation, and training.

Raptor Collisions

In North America, bird collisions with wires were first documented in 1876, when Coues (1876) noted numerous bird collisions with rural telegraph wire as he rode a horse between Denver, Colorado, and Cheyenne, Wyoming. Today, collisions with power lines are a widely documented cause of avian mortality (APLIC 2012), and the worldwide scope of the problem is largely unknown. Rioux et al. (2013) estimated that collisions with transmission lines across Canada range from 2.5 million to 25.6 million birds per year, incorporating detection, scavenging, and crippling biases. Hunting (2002) estimated annual bird collisions with transmission lines in the United States (USA) are from tens of thousands to 174 million birds. Using a sensitivity analysis, Loss et al. (2014) estimated that between 8 and 57 million birds are killed annually by transmission power line collision in the United States. These projected mortality figures could be underestimates because they do not include collision mortality due to distribution lines, which are more abundant across the landscape. The potential collision impacts to birds of prey from renewable energy transmission linkages are discussed by Dwyer et al. (in this book Chap. 13).

Bird of prey collisions have been documented globally (Harness et al. 2003), but in general raptors are good fliers with excellent eyesight and are not typically prone to power line collision (Fig. 12.10) (APLIC 2012). Exceptions are raptors having high wing loading (APLIC 2012), such as California condors, which have been susceptible to power line collisions in pre- and post-reintroduction efforts (Snyder and Snyder 2000; Meretsky et al. 2001; Snyder 2007). However even raptors without high wing loading can collide with wires under certain circumstances, such as

Fig. 12.10 Raptors are good fliers with excellent eyesight and are not typically susceptible to power line collision

when visibility is restricted or high winds make flight more difficult. Olendorff and Lehman (1986) documented 88 raptor wire collisions including species from the orders Falconiformes, Accipitriformes, and Strigiformes. The most commonly reported species were peregrine falcons (*Falco peregrinus*; $n = 24$), bald eagles (*Haliaeetus leucocephalus*; $n = 15$), golden eagles ($n = 9$), osprey (*Pandion haliaetus*; $n = 7$), and red-tailed hawks ($n = 7$). Falcons and eagles were the groups most likely to collide with distribution power lines. The authors speculated that when birds of prey engage in courtship flights, defend a territory, or are preoccupied pursuing prey, they are at elevated collision risk.

Habitat may also play an important role in raptor power line collision risk. NorthWestern Energy reported 21 golden eagle carcasses mid-span under distribution lines located near Roundup, Montana. In one event cluster, three golden eagles collided with two adjacent spans near an active golden eagle nest and less than a mile from a prairie dog town (Harness et al. 2003). The authors suggested these spans bisected a regular hunting/travel corridor. Mañosa and Real (2001) similarly reported transmission lines near Bonelli's eagle nesting territories in Spain created elevated power line collision risk. At the Aberdeen Proving Ground on the upper Chesapeake Bay, Maryland, Mojica et al. (2009) documented 21 bald eagle carcasses under distribution lines, each showing blunt force trauma characteristic of collisions. They also reported that landscape factors such as proximity to shoreline had a significant impact on the location of bald eagle collisions with power lines. Additional studies have shown certain African vulture (Boshoff et al. 2011) and eagle species (Rollan et al. 2010) to be more vulnerable to power line collision in foraging habitats.

Fig. 12.11 Bald eagle carcass under a distribution power line after a fly-in

Raptor collisions with power lines often result in neck, head, and beak trauma and broken and/or damaged wings (Olendorff and Lehman 1986). Raptors may also collide with the wires with such force that one conductor is pushed toward another, resulting in both collision and electric shock trauma (Harness et al. 2003). Such an event is termed a "fly-in" (Fig. 12.11). Even where APLIC-recommended wire spacing is present, large birds such as eagles can occasionally fly into wires with enough force to cause a fly-in (Harness et al. 2003).

Proposed solutions to avian collisions include modifying habitats surrounding lines, burying lines, removing overhead shield wires, and marking wires to increase their visibility to birds. Modifying surrounding habitats can reduce an area's attractiveness to birds but is largely impractical because electric utilities rarely control land use outside the right of way. Burying lines is impractical on a large scale because it significantly increases construction costs (Johnson 2004; Hall 2012). Eliminating overhead shield wires from transmission lines is also not practical because doing so leaves conductors unprotected from lightning, negatively impacting reliability (APLIC 2012). Thus, the primary strategy for reducing avian-power line collision on existing lines is using marking products to make wires more visible to birds (Beaulaurier 1981).

There are few studies specific to raptor-power line collisions and mitigation; wire marking is a tool that typically reduces, but does not eliminate, collision risk in other avian species. Wire-marking devices increase the line profile, particularly for smaller-diameter wires. In studies of waterfowl and cranes, wire marking reduced collisions by up to 61% (Morkill 1990; Morkill and Anderson 1991; Brown and Drewien 1995). The most robust wire-marking studies have normalized the number of collision fatalities by the number of bird crossings. These studies report that wire marking reduces mortality by 55% to 94%, with a mean reduction of 78% (Barrientos et al. 2011).

Raptor collision with power lines can be a key regulatory concern where new lines bisect known raptor migration corridors. Chelan County Public Utility District (Chelan PUD) used radar and human observers to monitor raptor migration near the top of Burch Mountain in central Washington during fall 2005. Surveys were used to assess collision potential for raptor fall migrants crossing a proposed 230 kV transmission line (Pope et al. 2006). The 230 kV line was constructed in 2006 as an H-Frame with one optical ground wire (OGW) and one overhead static wire (OHS). At spans with high passage rates of migrating raptors, the OGW and OHS were marked with Bird Flight Diverters with 30-foot staggered spacing. In 2007, Chelan PUD repeated the raptor migration study with the transmission line in place. In 753 observer hours over 45 days, 1,598 raptors were observed with no collisions (Pope, pers. comm. 2014).

In a pre- and post-construction study of a major raptor migration corridor in Pennsylvania, spans of a 500 kV transmission line were fitted with Swan Flight Diverters and instrumented with Bird Strike Indicators (EDM International, Inc., Fort Collins, CO) to remotely monitor collisions. The line segment crossing the ridgeline also was monitored with visual observers during the day. Post-construction, observers recorded 4,482 safe raptor crossings and no collisions with the new line; raptors responded to the new line by flying higher than before construction (Luzenski et al. 2016). However, wire marking does not always reduce the risk of raptor collision. A wire marker used at the Aberdeen Proving Ground in Maryland did not adequately mitigate bald eagle collisions at one location, and the distribution power line was subsequently undergrounded (Mojica et al. 2009; Watts et al. 2015).

Two classes of marking devices exist. "Passive" marking devices do not have moving parts (Figs. 12.12 and 12.13). These products include the Bird Flight Diverter (BFD), Swan Flight Diverter (SFD), Spiral Vibration Damper (SVD), FireFly High Wind diverter, Bird Flight Diverter, and Avian Flight Diverter.

"Active" marking devices have moving parts that spin or flutter in the wind. These products include the FireFly, BirdMark, and CROCFAST Bird Diverters which also have glow in the dark and reflective stickers attached (Fig. 12.14). The Overhead Warning Light (OWL) diverter is a nocturnal wire marker with solar-powered LEDs and moving parts to increase nocturnal line visibility (Fig. 12.15).

Although numerous wire-marking studies have been conducted, few studies have compared efficacy among marking devices. Differences in study sites, species, and methodologies preclude comparing device effectiveness across multiple studies (Barrientos et al. 2011). Sporer et al. (2013) monitored the BirdMark, SFD, and FireFly HW and concluded that collectively these devices reduced collisions; however, small sample sizes precluded direct comparisons between the three devices. Also these studies were conducted primarily on non-raptor species.

When marking lines, staggering markers on adjacent wires maximizes visibility and minimizes costs. For example, to give the appearance of 4.6-meter (15-foot) spacing on a three-phase distribution line, diverters on each wire should be spaced 13.7 meters (45 feet) apart, with the first diverter 4.6, 9.1, or 13.7 meters (15, 30, or

Fig. 12.12 Passive coil-type wire markers: Bird Flight Diverter (left), Swan Flight Diverter (right)

Fig. 12.13 Passive wire markers: FireFly High Wind (left), Bird Flight Diverter (middle), Avian Flight Diverter (right)

45 feet) from the pole (x.9). For transmission lines with two overhead static wires, the diverters on each wire would be placed 9.1 meters (30 feet) apart and staggered (Fig. 12.16).

Bird species should be considered when selecting a product. Unfortunately, many collisions occur during low light, and some marking devices have been shown to be ineffective at night (Murphy et al. 2016). Several glowing marker products

Fig. 12.14 Active wire markers: FireFly (left), BirdMark (middle left), CROCFAST Static (middle right) and CROCFAST Dynamic Diverter (right)

Fig. 12.15 Illuminated wire markers: Overhead Warning Light (left) and Raptor Clamp LED Diverter (right), night images shown below

have been introduced to the market to address nocturnal collision. However, little is currently known about the properties of diurnal and nocturnal devices and how they relate to avian vision. Using UV light has been proposed as a mitigation tool to assist birds in recognizing and avoiding objects at night. Researchers are beginning to test the use of ultraviolet (UV) color and light to modify bird behavior and reduce collisions with anthropogenic objects. While it is unclear whether raptors can see in the UV spectrum, several studies indicate UV color on objects can provide contrast to assist birds in recognizing objects before collision (Blackwell et al. 2012; Fernández-Juricic et al. 2011). UV markings on windows can reduce collisions only if the reflecting background has high UV contrast to the window markings (Håstad and Ödeen 2014).

Line markers can be installed on energized distribution wires, grounded distribution neutral wires, and grounded transmission overhead shield wires. From an engineering perspective, wire marking is not an option for all power lines. Devices that

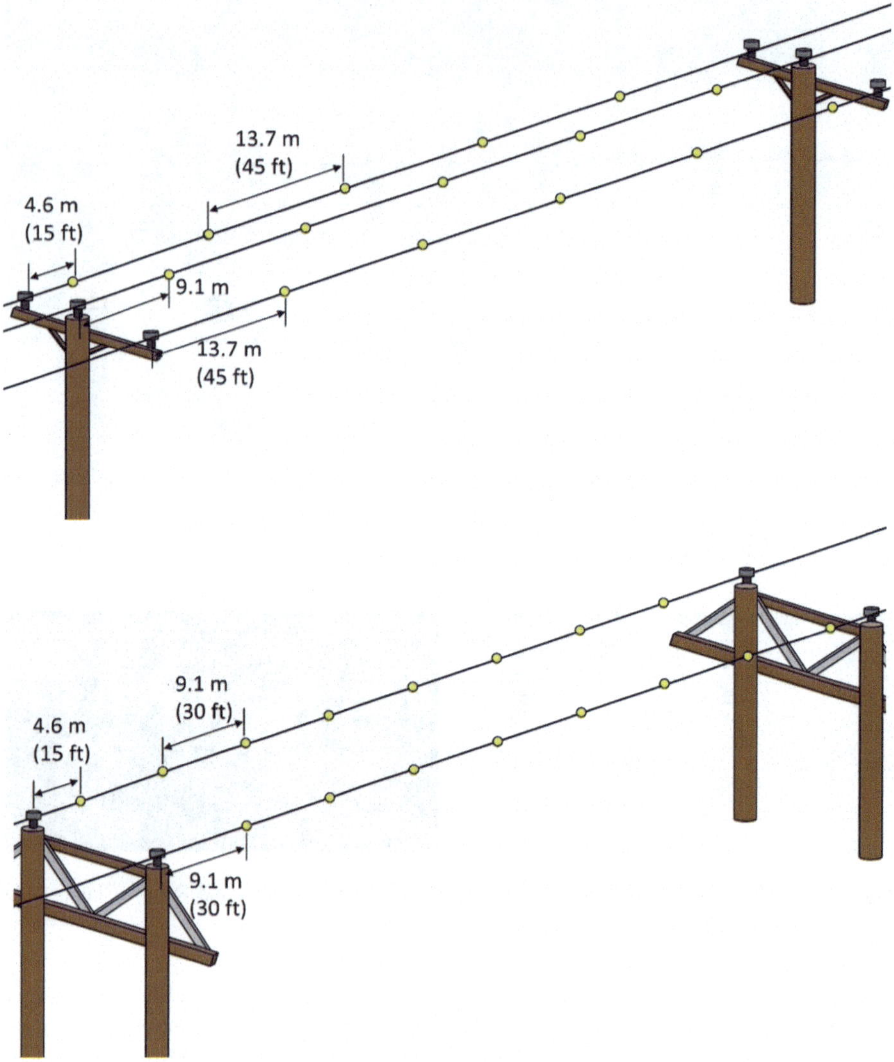

Fig. 12.16 A) Marker placement for three wires in two planes. B) Marker placement for two overhead static wires in one plane (conductors omitted for clarity)

enlarge the wire may cause increased wind or ice loading, increased line tension and stress loads, and an elevated risk of wire breaks and power outages. Devices may also abrade and damage the conductors. Line markers typically are not installed on energized transmission wires greater than 115 kV because of the resulting corona discharge (Hurst 2004), which produces electromagnetic interference and audible noise.

When selecting devices, utility engineers should consider long-term effectiveness, durability, and potential effects to line operation. If optical ground wires are to be marked, the utility should contact the vendor to ensure marking the wires will

Fig. 12.17 Installing Swan Flight Diverters with a helicopter

not void any warrantees. Prior to installing diverters, engineers should perform a structural analysis on the line where marking is proposed. If the static wire or poles cannot safely support the additional load plus appropriate safety factors for all design conditions, then alternate solutions must be evaluated. Alternate solutions may include reducing the number of diverters or delaying installation until that particular segment of line is rebuilt (for routine replacement or capacity upgrade). Worker safety, line access, and installation methods also must be considered when selecting devices and the installation approach (Fig. 12.17). It is also possible to install certain wire markers using drones (Lobermeier et al. 2015). As collision mitigation continues to advance, utility personnel should periodically review new technologies and deployment strategies to ensure the most effective approach is implemented whenever a collision risk is addressed.

Conclusions

As a significant conservation challenge, raptor mortality on power lines is an issue that deserves global attention. Doing so will require the cooperative efforts of utilities, governments, and biologists. Only utilities can carry out avian-friendly retrofits and line marking and begin building power lines that are protective of large birds. Electrocutions associated with outages directly cost utilities in lost revenue, equipment damage or loss, and outage response labor (Reynolds et al. 2014). Additional costs are incurred when an avian electrocution starts a wildfire (Harness and Wilson 2001; Lehman and Barrett 2002; Tintó et al. 2010). A number of utilities have made a business case for addressing avian electrocutions in order to improve system reliability. However, if less than 10% of avian electrocutions trigger an outage (Dwyer 2004; Kemper et al. 2013), the business case may be restricted to the most vulnerable utility assets, such as substations (Heck and Sutherland 2013), which each serves a large customer base.

Government is the only entity that can set minimum conservation standards for industry, and standards are meaningless without enforcement (Suazo 1998). Since the federal prosecution of the Moon Lake Electric Association (Melcher and Suazo 1999) in 1999 for violations of the Migratory Bird Treaty Act and the Bald and Golden Eagle Protection Act, electrocution (APLIC 2006) and collision (Mojica et al. 2009) mitigation activities by utilities have increased substantially in the United States. In Russia, the ecological damage associated with electrocutions is calculated as a function of the number and species of mortalities (Barbuzyuk et al. 2010; Goroshko 2011) and may be levied as a fine. Avian protection standards for power lines in Europe, Germany, Spain, and the Catalonia region of Spain have led to significant progress in avian mitigation (Prinsen et al. 2011; Scheurenberg et al. 2010), including undergrounding all distribution lines in the Netherlands (Bayle 1999). The possibility of fines resulting from avian electrocutions also has led to large ongoing investments in retrofitting projects in Hungary (Antal 2010 in Prinsen 2011). In addition to its regulatory role, government can, and frequently does, fund research that informs conservation priorities and strategies (e.g., Dwyer et al. 2013; Lehman 2001).

Regardless of the legal ramifications and financial penalties assessed, utilities always will face financial pressures to accomplish more with less; funds for avian protection will never be unlimited. Biologists can help utilities target available budgets to ensure maximum conservation impact. Recent studies have resulted in validated models for predicting poles with the greatest likelihood of causing electrocution (Tintó et al. 2010; Dwyer et al. 2013; Harness et al. 2013). In the aggregate, these models demonstrate the predictive power of both habitat and pole elements, but to some unknown extent, each model reflects the vagaries of the ecology and electrical system used to develop it. Models identifying movement corridors of raptor species can be overlaid with line and pole data to further inform risk of collision and electrocution in high-use areas (Watts et al. 2015; Mañosa and Real 2001). Similar efforts representing a broader range of habitats and distribution systems will improve conservation outcomes by improving the applicability of model results and ensuring the most high-risk poles are addressed first (Tintó et al. 2010). To date, statistical models predicting collision risk have not been developed; such models would help appropriately focus collision mitigation. Additional work with species susceptible to electrocution and collision and further development and testing of avian protection products also will improve conservation outcomes.

Leaders in Spain have demonstrated that utilities, government, and biologists can make a strikingly positive impact on raptor conservation (López-López et al. 2011) by working together. The Spanish Imperial Eagle remains a unique success story in which the risks posed by power lines were effectively managed to save a sensitive species. To protect raptors worldwide in coming decades, all interested parties must work together to create similar successes across landscapes on an ever-expanding power grid.

Acknowledgments We'd like to thank EDM International, Inc. for supporting the effort to develop this section. Special thanks to reviewers and co-workers Libby Mojica, Melissa Landon, and Dr. James Dwyer. Bryon Robb helped develop figure 2 and figure 8. We also wish to thank the

Electric Power Research Institute, California Energy Commission, National Rural Electric Association, US Fish and Wildlife Service, and the Avian Power Line Interaction Committee for sponsoring ongoing research that contributed greatly to the development of this chapter.

References

Amezian M, Irizi A, Errati A, Loran H, El Khamlichi R, Morandini V, González DG, Garrido JR (2015) Spanish imperial eagles and other eagles found electrocuted in Morocco and proposition of correction measures. Available via figshare. https://doi.org/10.6084/m9.figshare.1613292. Accessed 12 Dec 2016

Angelov I, Hashim I, Oppel S (2013) Persistent electrocution mortality of Egyptian vultures Neophron percnopterus over 28 years in East Africa. Bird Conserv Int 23:1–6. https://doi.org/10.1017/S0959270912000123

Antal M (2010) Policy measures to address bird interactions with power lines – a comparative case study of four countries. Ostrich 81:217–223

Avian Power Line Interaction Committee (APLIC) (1994) Mitigating bird collisions with power lines: the state of the art in 1994. Edison Electric Institute, Washington, DC

Avian Power Line Interaction Committee (APLIC) (1996) Suggested practices for raptor protection on power lines: the state of the art in 1996. Edison Electric Institute/Raptor Research Foundation, Washington, DC

Avian Power Line Interaction Committee (APLIC) (2006) Suggested practices for avian protection on power lines: the state of the art in 2006. Edison Electric Institute, APLIC, and the California Energy Commission, Washington, DC/Sacramento

Avian Power Line Interaction Committee (APLIC) (2012) Reducing avian collisions with power lines: the state of the art in 2012. Edison Electric Institute and APLIC, Washington, DC

Avian Power Line Interaction Committee and U.S (2005) Fish and Wildlife Service (APLIC and USFWS) Avian protection plan (APP) guidelines

Barbazyuk EV, Bakka SV, Barashkova AN, Semenov AR, Smelyanskiy IE (2010) The outcomes of preliminary monitoring for death of raptors and other bird species through electrocution in the eastern Orenburg District, Russia. Raptors Conserv 20:40–47

Barrientos R, Alonso JC, Ponce C, Palacín C (2011) Meta-analysis of the effectiveness of marked wire in reducing avian collisions with power lines. Conservation. Biol 25(5):893–903

Bayle P (1999) Preventing birds of prey problems at transmission lines in western Europe. J Raptor Res 33:43–48

Beaulaurier DL (1981) Mitigation of bird collisions with transmission lines. Bonneville Power Administration, Portland

Benson PC (1980) Large raptor electrocution and power pole utilization: a study in six western states. Raptor Res 14:125–126

Benson, PC (1981) Large raptor electrocution and power line utilization: a study in six western states. PhD dissertation, Brigham Young University

BirdLife International (2017) International union for conservation of nature and natural resources red list for birds. http://www.birdlife.org. Accessed 18 Jan 2017

Blackwell BF, DeVault TL, Seamans TW, Lima SL, Baumhardt P, Fernández-Juricic E (2012) Exploiting avian vision with aircraft lighting to reduce bird strikes. J Appl Ecol 49(4):758–766

Boeker EL, Nickerson PR (1975) Raptor electrocutions. Wildl Soc Bull 3(2):79–81

Boshoff AF, Minne JC, Tambling CJ, Michael MD (2011) The impact of power line-related mortality on the cape vulture (Gyps coprotheres) in a part of its range, with an emphasis on electrocution. Bird Conserv Int 21:311–327

Brown WM, Drewien RC (1995) Evaluation of two power line markers to reduce crane and waterfowl mortality. Wildl Soc Bull 23:217–227

Burnham JT (1995) Bird streamer flashovers on FPL transmission lines. IEEE Trans Power Delivery 10:970–977

Cartron J-LE, Harness RE, Rogers RC, Manzano-Fischer P (2005) Impact of concrete power poles on raptors and ravens in northwestern Chihuahua, Mexico. In: Cartron J-L, Ceballos G, Felger RS (eds) Biodiversity, ecosystems, and conservation in northern Mexico. Oxford University Press, New York, pp 357–369
Coues E (1876) The destruction of birds by telegraph wires. Am Nat 10:734–736
Demerdzhiev DA (2014) Factors influencing bird mortality caused by power lines within special protected areas and undertaken conservation efforts. Acta Zool Bulg 66(2):411–423
Demerdzhiev DA, Stoychev SA, Petrov TH, Angelov ID, Nedyalkov NP (2009) Impact of power lines on bird mortality in southern Bulgaria. Acta Zool Bulg 61(2):175–183
Dickerman RW (2003) Talon-locking in the red-tailed hawk. J Raptor Res 37:176
Dixon A, Maming R, Gungaa A, Purev-Ochir G, Batbayar N (2013) The problem of raptor electrocution in Asia: case studies from Mongolia and China. Bird Conserv Int 23:520–529. https://doi.org/10.1017/S0959270913000300
Donázar JA, Palacios CJ, Gangoso L, Ceballos O, González MJ, Hiraldo F (2002) Conservation status and limiting factors in the endangered population of Egyptian vulture (Neophron percnopterus) in the Canary Islands. Biol Conserv 107:89–97
Dwyer JF (2004) Investigating and mitigating raptor electrocution in an urban environment. MS Thesis, University of Arizona, Tucson
Dwyer JF (2006) Electric shock injuries in a Harris's hawk population. J Raptor Res 40(3):193–199
Dwyer JF (2009) Raptor electrocution: a case study on ecological traps, sinks, and additive mortality. J Nat Resour Life Sci Educ 38:93–98. https://doi.org/10.4195/jnrlse.2009.0002u
Dwyer JF, Doloughan K (2014) Testing systems of avian perch deterrents on electric power distribution poles. Hum-Wildl Interact 7:39–54
Dwyer JF, Mannan RW (2007) Preventing raptor electrocutions in an urban environment. J Raptor Res 41:59–267
Dwyer JF, Harness RE, Donohue K (2013) Predictive model of avian electrocution risk on overhead power lines. Conserv Biol 28(1):159–168. https://doi.org/10.1111/cobi.12145
Dwyer JF, Kratz GE, Harness RE, Little SS (2015) Critical dimensions of raptors on electric utility poles. J Raptor Res 49(2):210–216
Dwyer JF, Harness RE, Gerber BD, Landon MA, Petersen P, Austin DD, Woodbridge B, Williams GE, Eccleston D (2016a) Power pole density informs spatial prioritization for mitigating avian electrocution. J Wildl Manag 80:634–642. https://doi.org/10.1002/jwmg.1048
Dwyer JF et al (2016b) Avian electrocutions on incorrectly retrofitted power poles. Final report submitted to the USFWS, Western Golden Eagle Conservation Team (in press)
Electric Power Research Institute (EPRI) (1982) Prevention of golden eagle electrocution. EPRI EA-2680. Palo Alto
Electric Power Research Institute (EPRI) (2013) Effectiveness and durability of animal-caused outage prevention products. EPRI 3002001185, Palo Alto
Electric Power Research Institute (EPRI) (2014) Animal bushing cover testing: current practices and opportunities to improve comparability and transparency. EPRI 3002003574, Palo Alto
Electric Power Research Institute (EPRI) (2016) Conductor cover selection: an animal outage device selection process to maximize installation ease, system compatibility, and overall performance. EPRI 3002006333, Palo Alto
Ellis DH, Craig T, Craig E, Postupalsky S, Larue CT, Nelson RW, Henny CJ, Watson J, Millsap BA, Dawson JW, Cole KL, Martin EM, Margalida A, Kung P (2009) Unusual raptor nests around the world. J Raptor Res 43:175–198
Fernández-Juricic E, Gaffney J, Blackwell BF, Baumhardt P (2011) Bird strikes and aircraft fuselage color: a correlational study. Hum-Wildl Interact 5(2):224–234
Ferrer M, Hiraldo F (1992) Man-induced sex-biased mortality in Spanish imperial eagles. Biol Conserv 60:57–60
Ferrer M, De La Riva M, Castroviejo J (1991) Electrocution of raptors on power lines in southwestern Spain. J Field Ornithol 62:181–190
Fox N, Wynne C (2010) The impact of electrocution on the New Zealand falcon (Falco novaeseelandiae). Notornis 57:71–74

Galmes MA, Sarasola JH, Grande JM, Vargas FH (in press) Electrocution risk for the endangered crowned solitary eagle and other birds in semiarid landscapes of central Argentina. Bird Conserv International. https://doi.org/10.1017/S0959270917000272
Garrett MG (1993) PacifiCorp's program for managing birds on power lines- a case study. In: Huckabee JW (ed) Proceedings of the avian interactions with utility structures international workshop, Miami, September 1992, p 18–1–18-5
Gerdzhikov GP, Demerdzhiev DA (2009) Data on bird mortality in "Sakar" IBA (BG021), caused by hazardous power lines. Ecol Balk 1:67–77
González LM, Margalida A, Mañosa S, Sánchez R, Oria J, Molina JI, Caldera J, Aranda A, Prada L (2007) Causes and spatio-temporal variations of non-natural mortality in the vulnerable Spanish imperial eagle Aquila adalberti during a recovery period. Oryx 41(4):495–502. https://doi.org/10.1017/S0030605307414119
Goroshko OA (2011) Bird electrocution in the Daurian steppe (south-eastern trans-Baikal region), Russia. Raptors Conserv 21:84–99
Guil F, Fernández-Olalla M, Moreno-Opo R, Mosqueda I, Gómez ME, Aranda A, Arrendondo Á, Guzmán J, Oria J, González LM, Margalida A (2011) Minimising mortality in endangered raptors due to power lines: the importance of spatial aggregation to optimize the application of mitigation measures. PLoS One 6(11):e28212
Guil F, Colomer MA, Moreno-Opo R, Margalida A (2015) Space-time trends in Spanish bird electrocution rates from alternative information sources. Glob Ecol Conserv 3:379–388
Guyonne FE, Ferrer J, Ferrer M (1999) Mitigation of raptor electrocution on steel power poles. Wildl Soc Bull 27(2):263–273
Guzmán J, Castaño JP (1998) Electrocucion de rapaces en lineas electricas de distribucion en Sierra Morena Oriental y Campo de Montiel. Ardeola 45(2):161–169
Hall KL (2012) Out of sight, out of mind: an updated study on the undergrounding of overhead power lines. Edison Electric Institute, Washington, DC
Hallinan T (1922) Bird interference on high tension electric transmission lines. Auk 39:573
Hamerstrom FN Jr, Harrell BE, Olendorff RR (eds) (1974) Management of raptors – proceedings of the conference on raptor conservation techniques, Fort Collins, March 1973. Part 4. Raptor Res Foundation, Inc, Raptor Res Rpt no 2
Harness RE (2000a) Steel distribution poles – environmental implications. In: Chancellor RD, Meyburg B-U (eds) Raptors at risk. Hancock House, Blaine, pp 751–755
Harness RE (2000b) Raptor electrocutions and distribution pole types. North American Wood Pole Council Technical Bulletin-Raptor-10-00
Harness RE (2001) Effectively retrofitting power lines to reduce raptor mortality. In: The leading edge: meeting environmental challenges for siting and reliability, 44th annual IEEE Rural Electric Power Conference, Louisville, 7–9 May 2000
Harness RE (2004) Bald Eagle Haliaeetus leucocephalus electrocutions in Alaska and Florida – a comparison of retrofitting measures. In: Chancellor RD, Meyburg B-U (eds) Raptors worldwide. Proceedings of the 6th world conference on birds of prey and owls, Budapest, 2003
Harness RE (2007) Mitigation. In: Bird DM, Bildstein KL (eds) Raptor research and management techniques. Hancock House, Surrey/British Columbia, pp 365–382
Harness RE, Juvvadi PR (2013) Bird electrocutions in Western Rajasthan, India. 2013 Proceedings of the IEE Rural Electric Power Conference, Stone Mountain, Georgia, April–May 2013, p A4–1 to A4–5
Harness RE, Kratz G (2007) Bird streamers as the probable cause of a bald eagle electrocution. Colo Birds 41(2):86–90
Harness RE, Wilson KR (2001) Utility structures associated with raptor electrocutions in rural areas. Wildl Soc Bull 29:612–623
Harness RE, Milodragovich S, Schomburg J (2003) Raptors and power line collisions. Colo Birds 37(3):118–122
Harness RE, Gombobaatar S, Yosef R (2008) Mongolian distribution power lines and raptor electrocutions. Papers presented at the 52nd annual conference, Charleston, 27–30 April 2008
Harness RE, Juvvadi PR, Dwyer JF (2013) Avian electrocutions in western Rajasthan, India. J Raptor Res 47:352–364

Håstad O, Ödeen A (2014) A vision physiological estimation of ultraviolet window marking visibility to birds. PeerJ. https://doi.org/10.7717/peerj.621
Heck N, Sutherland T (2013) Use of custom-fitted cover-ups to reduce power outages and mitigate avian electrocutions. T&D Magazine Nov-Dec p 21–25
Hernández-Matías A, Real J, Parés F, Pradel R (2015) Electrocution threatens the viability of populations of the endangered Bonelli's eagle (Aquila fasciata) in southern Europe. Biol Conserv 191:110–116
Hunting K (2002) A roadmap for PIER research on avian collisions with power lines in California. Technical report P500–02-071F. California Energy Commission, Sacramento
Hurst N (2004) Corona testing devices used to mitigate bird collisions. EDM International, California Energy Commission, PIER Energy-Related Environmental Research. #500 04 086F
Ibarra J, De Lucca E (2015) Águilas Moras (Geranoaetus melanoleucus), víctimas de electrocución en luján de cuyo, Mendoza, Argentina. Nótulas Faunísticas 176 Segunda Serie 176:1–7
Institute of Electrical and Electronics Engineers (IEEE) (2010) Guide for testing the electrical, mechanical, and durability performance of wildlife protective devices on overhead power distribution systems rated up to 38kV. New York
Janss GFE (2000) Avian mortality from power lines: a morphologic approach of a species-specific mortality. Biol Conserv 95(3):353–359
Janss GFE, Ferrer M (1999a) Avian electrocution on power poles: European experiences. In: Ferrer M, Janss GFE (eds) Birds and power lines: collision, electrocution, and breeding. Quercus Editores, Madrid, pp 145–164
Janss GFE, Ferrer M (1999b) Mitigation of raptor electrocution on steel power poles. Wildl Soc Bull 27:263–273
Jenkins AR, de Goede KH, Sebele L, Diamond M (2013) Brokering a settlement between eagles and industry: sustainable management of large raptors nesting on power infrastructure. Bird Conserv Int 23:232–246. https://doi.org/10.1017/S0959270913000208
Johnson B (2004) Out of sight, out of mind? A study on the costs and benefits of undergrounding overhead power lines. Edison Electric Institute, Washington, DC
Kagan RA (2016) Electrocution of raptors on power lines: a review of necropsy methods and findings. Vet Pathol 53(5):1030–1036. https://doi.org/10.1177/0300985816646431
Kemper CM, Court GS, Black JA (2013) Estimating raptor electrocution mortality on distribution power lines in Alberta, Canada. J Wildl Manag 77:1342–1352
Lasch U, Zerbe S, Lenk M (2010) Electrocution of raptors at power lines in Central Kazakhstan. Waldökologie, Landschaftsforschung und Naturschutz 9:95–100
Ledger JA (1984) Engineering solutions to the problem of vulture electrocutions on electricity towers. Certif Eng 57:92–95
Ledger JA, Hobbs JCA (1999) Raptor use and abuse of powerlines in southern Africa. J Raptor Res 33(1):49–52
Lehman RN (2001) Raptor electrocution on power lines: current issues and outlook. Wildl Soc Bull 29:804–813
Lehman RN, Barrett JS (2002) Raptor electrocutions and associated fire hazards in the Snake River Birds of Prey National Conservation Area. Technical bulletin No. 02–7, Idaho Bureau of Land Management, Boise
Lehman RN, Kennedy PL, Savidge JA (2007) The state of the art in raptor electrocution research: a global review. Biol Conserv 136:159–178
Lehman RN, Kennedy PL, Savidge JA, Harness RE (2010) Raptor electrocution rates for a utility in the intermountain western United States. J Wildl Manag 74(3):459–470. https://doi.org/10.2193/2008-332
Lobermeier S, Moldenhauer M, Peter CM, Slominski L, Tedesco RA, Meer MV, Dwyer JF, Harness RE, Stewart AH (2015) Mitigating avian collision with power lines: a proof of concept for installation of line markers via unmanned aerial vehicle. J Unmanned Veh Syst 3(4):252–258
López-López P, Ferrer M, Madero A, Casado E, McGrady M (2011) Solving man-induced large-scale conservation problems: the Spanish imperial eagle and power lines. PLoS One. https://doi.org/10.1371/0017196

Loss SR, Will T, Marra PP (2014) Refining estimates of bird collision and electrocution mortality at power lines in the United States. PLoS One. https://doi.org/10.1371/0101565
Luzenski J, Rocca CE, Harness RE, Cummings JL, Austin DD, Landon MA, Dwyer JF (2016) Collision avoidance by migrating raptors encountering a new transmission power line. Condor 118(2):402–410
Mañosa S, Real J (2001) Potential negative effects of collisions with transmission lines on a Bonelli's eagle population. J Raptor Res 35(3):247–252
Manville AM II (2005) Birds strikes and electrocutions at power lines, communications towers, and wind turbines: state of the art and state of the science—next steps toward mitigation. U.S. Forest Service General Technical Report PSW-GTR-191, Washington, DC
Matsyna AI, Matsyna EL, Matsyna AA, Grishutkin GF, Spiridonov SN (2010) Estimating the efficiency of bird protection activities on power lines 6-10 kV in the national park of Smolny, Russia. Raptors Conserv 20:35–39
Melcher C, Suazo LR (1999) Raptor electrocutions: the unnecessary losses continue. J of the Colo Field Ornithol 33:221–224
Meretsky VJ, Snyder NFR, Beissinger SR, Clendenen DA, Wiley JW (2001) Demography of the California condor: implications for reestablishment. Conserv Biol 14:957–967
Mojica EK, Watts BD, Paul JT, Voss ST, Pottie J (2009) Factors contributing to bald eagle electrocutions and line collisions on Aberdeen proving ground, Maryland. J Raptor Res 43:57–61
Morkill AE (1990) Effectiveness of markers in reducing Sandhill Crane collisions with power lines. MS Thesis, University of Wyoming, Laramie
Morkill AE, Anderson SH (1991) Effectiveness of marking powerlines to reduce Sandhill crane collisions. Wildl Soc Bull 19:442–449
Murphy RK, Mojica EK, Dwyer JF, McPherron MM, Wright GD, Harness RE, Pandey AK, Serbousek KL (2016) Crippling and nocturnal biases in as study of Sandhill crane (Grus canadensis) collisions with a transmission line. Waterbirds 39:312–317
Nelson MW (1979) Power lines progress report on eagle protection research. Idaho Power Company, Boise
Nelson MW (1980) Update on eagle protection practices. Idaho Power Company, Boise
Nelson MW, Nelson P (1976) Power lines and birds of prey. Idaho Wildl Rev 28:3–7
Nelson MW, Nelson P (1977) Power lines and birds of prey. In: Chandler RD (ed) Proceedings of world conference on birds of prey, International Council for Bird Preservation, Vienna, 1975
Newton I (2010) Population ecology of raptors. Poyser, London
O'Neil T (1988) An analysis of bird electrocutions in Montana. J Raptor Res 22:27–28
Olendorff RR (1972) Eagles, sheep and power lines. Colo Outdoors 21(1):3–11
Olendorff RR, Lehman RN (1986) Raptor collisions with utility lines: an analysis using subjective field observations. Final report. Pacific Gas and Electric Co., Research and Development, San Ramon
Olendorff RR, Miller AD, Lehman RN (1981) Suggested practices for raptor protection on power lines: the state of the art in 1981. Raptor Research Report No. 4. Raptor Research Foundation, University of Minnesota, St. Paul
Phipps WL, Wolter K, Michael MD, MacTavish LM, Yarnell RW, Farwig N (2013) Do Power Lines and Protected Areas Present a Catch-22 Situation for Cape Vultures (Gyps coprotheres)?. PLoS ONE 8 (10):e76794
Platt C (2005) Patterns of raptor electrocution mortality on distribution power lines in southeast Alberta. MS Thesis, University of Alberta
Pope VR, Fielder PC, Cordell KA, Harness RE, Hamer TE (2006) Pre-construction evaluation of collision potential for fall migrating raptors with a transmission line across Burch Mountain, Chelan County, Washington. Public Utility District No. 1 of Chelan County, EDM International, and Hamer Environmental, Wenatchee, Fort Collins, Mount Vernon
Prather PR, Messmer TA (2010) Raptor and corvid response to power distribution line perch deterrents in Utah. J Wildl Manag 74(4):796–800. https://doi.org/10.2193/2009-204
Prinsen H, Boere G, Píres N, Smallie J (2011) Review of the conflict between migratory birds and electricity power grids in the African-Eurasian region. Tenth meeting of the conference of the parties Bergen, 20–25 November 2011

Reynolds B, Dupre R, Eccleston D, "To protect and deter," T&D World Magazine. [Online]. Sept 2014, pp. 54-58. Available: http://tdworld.com/overhead-transmission/protect-and-deter.
Rioux S, Savard J-PL, Gerick AA (2013) Avian mortalities due to transmission line collisions: a review of current estimates and field methods with an emphasis on applications to the Canadian electric network. Avian Conserv Ecol 8(2):7
Rollan A, Real J, Bosch R, Tintó A, Hernandez-Matias A (2010) Modeling the risk of collision with power lines in Bonelli's eagle Hieraaetus fasciatus and its conservation implications. Bird Conserv Int 20:279–294
Schürenberg B, Schneider R, Jerrentrup H (2010) Implementation of recommendation no. 110/2004 on minimising adverse effects of above- ground electricity transmission facilities (power lines) on birds. Convention on the conservation of European wildlife and natural habitats. Strasbourg
Snyder NFR (2007) Limiting factors for wild California condors. In: Mee A, Hall LS (eds) California condors in the 21st century. Nuttall Ornithological Club and the American Ornithologists' Union, Washington, DC, pp 9–33
Snyder N, Snyder H (2000) The California condor, a saga of natural history and conservation. Academic Press, London
Spencer R (2016) Ridgway's Hawks in the Dominican Republic. Presentation at the international conference on overhead lines, Fort Collins, 4–7 April 2016
Sporer MK, Dwyer JF, Gerber BD, Harness RE, Pandey AK (2013) Marking power lines to reduce avian collisions near the Audubon National Wildlife Refuge, North Dakota. Wildl Soc Bull 37:796–804
Suazo LR (1998) Powerlines and raptors: using regulatory influence to prevent electrocutions. Paper presented at fifth world conference on birds of prey and owls. Midrand, South Africa
Switzer F (1977) Saskatchewan Power's experience. Blue Jay 35:259–260
Tintó A, Real J, Mañosa S (2010) Predicting and correcting electrocution of birds in Mediterranean areas. J Wildl Manag 74:1852–1862
Van Rooyen C, Vosloo H, Harness RE (2002) Eliminating bird streamers as a cause of faulting on transmission lines. Proceedings of Rural Electric Power Conference, Colorado Springs
Van Rooyen C, Vosloo H, Harness RE (2003) Watch the birdie! IEEE Ind Appl Mag 9:55–60
Vanderburgh DC (1993) Manitoba hydro accommodates osprey. Blue Jay 51:173–177
Watts BD, Mojica EK, Paxton BJ (2015) Using Brownian bridges to assess potential interactions between bald eagles and electrical hazards within the upper Chesapeake Bay. J Wildl Manag 79:435–445
Zhou Y, Hu H, Xu X, Cui J, Yuan Y, Li Z, Chen Y, Yu D, Nie Q, Liang X, Guan Z (2009) High-risk region of bird streamer flashover in high-voltage transmission lines. J Electrost 67:311–315

Chapter 13
Impact of Renewable Energy Sources on Birds of Prey

James F. Dwyer, Melissa A. Landon, and Elizabeth K. Mojica

Introduction

Renewable energy, defined as energy generated from natural processes that are replenished over time (Johnson and Stephens 2011), is increasingly important in global energy portfolios. This chapter begins by reviewing reasons for shifting from fossil fuels to renewable energy, including reasons which have nothing to do with environmental concerns but are nevertheless driving advances in the renewable sector. The chapter then focuses on birds of prey, describing actual and potential direct and indirect mortality, habitat loss, avoidance, and displacement resulting from the development and operation of renewable energy facilities. The chapter considers renewable energy facilities themselves, including wind, biofuel, solar, hydro, geothermal, and oceanic energy sources. Transmission connections linking renewable facilities to the existing electric transmission grid are considered, as are potential offsite impacts where the materials used to construct renewable infrastructure are mined and manufactured. The chapter closes with a discussion of mitigation strategies designed to reduce or compensate for negative impacts for birds of prey and a discussion of potential benefits of renewable energy facilities for birds of prey. The latter are important to understand when evaluating the overall balance of costs and benefits of renewable energies on birds of prey.

Knowledge of the connections between global conflicts and international dependencies on fossil fuels is important in understanding how macroeconomic forces independent of environmental concerns drive the advancement of renewable energy technologies. Because "green" initiatives may not in fact be grounded in environmental concerns, but be grounded instead in economics and national interests, potential negative environmental impacts of renewables and their high initial investment costs may carry little weight in the overall discussion, a paradox not readily apparent without consideration of the context of global competition over traditional energy reserves.

J. F. Dwyer (✉) · M. A. Landon · E. K. Mojica
EDM International Inc., Fort Collins, CO, USA
e-mail: jdwyer@edmlink.com

J. H. Sarasola et al. (eds.), *Birds of Prey*,
https://doi.org/10.1007/978-3-319-73745-4_13

Fig. 13.1 (**a**) A pump designed to extract liquid and gas fossil fuels from terrestrial deposits; note great horned owl (*Bubo virginianus*) nest and whitewash. (**b**) Collection facility for traditional liquid and gas fossil fuels from terrestrial deposits. (**c**) Transport (left) and collection (right) of traditional fossil fuels, (**d**) Traditional coal-burning electricity generation station

Fossil fuels have been the primary energy source for developing and developed nations since the Industrial Revolution of the early 1800s when coal began to be used to power steam-driven machines and energy-intensive metallurgic and chemical processes. Emissions from these machines and processes were recognized almost immediately as harmful, triggering early environmental responses to protect urban air and water. From the late 1800s through the early twenty-first century, fossil fuels remained the primary solution to global energy needs as petroleum and natural gas products made the storage and use of chemical energy more efficient and economical (Fig. 13.1).

The resulting dependence of national and international economies on fossil fuels has created two fundamental problems. The first is a globally ubiquitous reliance on fossil fuels often derived from outside national boundaries. This reliance can place less developed nations with large reserves at the center of conflicts for control of those reserves and can place more developed nations without large reserves at the mercy of nations with reserves. Shifting energy sources from fossil fuels to renewables offers nations the ability to achieve energy independence.

The second fundamental problem created by the global reliance on fossil fuels is the impact of combustion products on the global climate. Greenhouse gases released during combustion of fossil fuels are contributing to global climate changes. Shifting energy sources from fossil fuels to renewables offers nations the ability to achieve energy independence and offers potential environmental benefits. These benefits are not without their own potential costs however, and it is those potential costs, as exerted on birds of prey populations, that are discussed here.

Effects at Renewable Facilities

Potential effects to birds of prey at renewable facilities include direct mortality and indirect effects resulting from habitat loss, avoidance, and displacement. Direct mortality is defined as death occurring as an immediate consequence of an interaction between a bird of prey and a component of renewable infrastructure. For example, a golden eagle (*Aquila chrysaetos*) killed when struck by a rotating wind turbine blade or killed when colliding with the suspended high-voltage wires of a transmission power line connecting a renewable facility to the electric grid. Habitat loss is defined as occurring when the landscape occupied by birds of prey is converted to non-habitat, for example, the displacement of prey species resulting from conversion of hunting habitat to a mirror field for a solar plant or the removal of a nest tree when creating an agricultural monoculture for biofuel production. Avoidance and displacement are similar processes occurring at different scales. Both occur when habitat persists, but is no longer used. Avoidance is defined as a shift in use of specific portions of a renewable facility, not the entire site (Band et al. 2007). Displacement occurs when an entire site is abandoned (Band et al. 2007).

These effects rarely occur in isolation but are instead likely additive, co-occurring with one another and with other anthropogenic and natural agents of mortality. Additive effects can be problematic, even at low rates, because most birds of prey are k-selected species with relatively little annual reproduction and breeding often delayed during multiple years of maturation. Population persistence for many bird of prey species requires individual breeding adults to produce young over an entire lifetime. Mortality of breeding adults can have substantial effects on the population (Bellebaum et al. 2013). For example, at some sites, griffon vultures (*Gyps fulvus*) and red kites (*Milvus milvus*) cannot maintain stable local populations with additive mortality from wind farms (Carrete et al. 2009; Bellebaum et al. 2013).

Wind Resource Areas

Direct effects of wind energy facilities (Fig. 13.2) on birds of prey involve mortality occurring when rotating turbine blades strike birds in flight. Impacts are largely species-specific. Directly affected species are characterized by low-altitude flight when gliding on local winds and on thermal and orographic lifts (Katzner et al. 2012; de Lucas et al. 2008). Because wind turbines are designed and specifically placed to harvest the kinetic energy in some of these same winds, low-altitude flight behaviors largely dictate risk by placing birds of prey and rotating turbine blades

Fig. 13.2 (**a**) A wind resource area in desert habitat; note substation under construction in the background will provide a connection from the wind resource facility to the existing transmission power line network. (**b**) A wind resource area above agricultural fields, potentially facilitating both wind energy and biofuel production. (**c**) Close view of a solar field illustrating the bare and leveled earth (non-habitat) typical of such facilities. (**d**) Wide view of a solar field, illustrating fencing and bare earth designed to limit attractiveness as habitat and illustrating associated distribution and transmission lines

together in the same airspace. Hunting in these airspaces has been hypothesized to hinder the ability of a bird of prey to recognize turbines as a flight hazard (Orloff and Flannery 1992; Smallwood et al. 2009), so species habituated to hunting within wind resource areas can be at higher risk of collision. Collision risk can also increase along flight corridors where large numbers of migrating birds of prey funnel along narrow ridges and coastlines supporting wind energy facilities (Barrios and Rodriguez 2004; Katzner et al. 2012; de Lucas et al. 2012) or where communal roosts occur near wind resource areas (Carrete et al. 2012). Intraspecific and interspecific interactions during flight also increase risk for collision because birds of prey can be distracted and less likely to recognize flight hazards (Dahl et al. 2013; Smallwood et al. 2009).

Though at least 34 bird of prey species have been documented in collisions with wind turbines, population-level impacts from direct effects are unknown for most species (Beston et al. 2016); only griffon vultures (Carrete et al. 2009), red kites (Bellebaum et al. 2013) and golden eagles (USFWS 2013) are currently known to be at risk of population-level effects from these collisions.

Species-specific behaviors also drive indirect effects of wind resource areas. Species avoiding or displaced by wind resource areas tend not to be affected by direct mortality but may abandon breeding territories (Dahl et al. 2013), shift local space use (Walker et al. 2005), or decrease in local abundance (Garvin et al. 2011; de Lucas et al. 2004). Some species show avoidance behaviors for individual turbine structures by adjusting flight paths to fly between or around turbines (Cabrera-Cruz and Villegas-Patraca 2016; Hull and Muir 2013; de Lucas et al. 2004) or adjust altitude to fly over turbines in their path (Johnston et al. 2014; de Lucas et al. 2004). There is limited evidence of net population loss in birds of prey from avoidance or displacement attributable to wind resource areas, but effects could be important for threatened species when considered with direct effects (Martínez et al. 2010).

Biofuels

Biofuels primarily describe energy resources developed from agriculture and most often describe production by industrial farms focused on extracting the greatest possible crop yields per acre. Yields are maximized by eliminating as many non-producing inclusions as possible and by promoting maximum growth through regular inputs of synthetic chemicals. Eliminating inclusions requires conversion of potential nest groves and bird of prey hunting habitat to cropland. Chemical inputs regularly consist of fertilizers to maximize crop yields, and pesticides, rodenticides, and herbicides, to protect monoculture crops from competing organisms in the environment. Collectively, these processes contribute to agricultural intensification which has been at least partly responsible for declines in farmland bird populations (Campbell et al. 1997; Uden et al. 2015).

Meeting increasing demand for ethanol requires increasing cropland in production, and consequently, the development footprint of biofuels is expected to be one of the fastest growing of all renewable energy sources in the next two decades (Johnson and Stephens 2011). Impacts of biofuel energy production on birds of prey occur primarily due to indirect effects triggered by the loss of breeding and foraging habitats when stands of trees used for nesting and open spaces used for hunting are converted to biofuel monocultures. Indirect effects include habitat loss, decreases in prey abundance, and potential biochemical effects from exposure to toxic chemicals. Direct effects are generally limited to rare occurrences of nestling mortality when nest trees are removed during breeding seasons, though exposure to bioaccumulating chemicals may also have effects that have not yet been identified.

Solar Facilities

Solar energy facilities also have the potential to impact birds of prey. Direct effects most often include electrocution on collection power lines, collisions with mirrors, and thermal trauma in solar flux fields (Kagan et al. 2014; McCrary et al. 1986). Electrocution can occur when a bird of prey simultaneously contacts two differently energized conductors or an energized conductor and a path to ground (APLIC 2006, in this book Chap. 12). Collisions occur when birds apparently mistake reflections of the sky in mirrors as the sky itself and attempt to fly through a mirror, perhaps in pursuit of prey.

Solar flux fields are the areas of concentrated light surrounding the collection tower(s) at thermal solar plants. Mirrors are used at these facilities to concentrate solar energy on a single area where water within a container is heated to produce steam which powers a generator. The air around the collection tower can reach 500–800 °C (McCrary et al. 1986; Diehl et al. 2016). Damage to feathers occurs at 160 °C (Wendelin et al. 2016), so flight through a solar flux field can result in burns to feathers and tissues, causing immediate mortality or limiting or eliminating the ability to fly, depending on individual exposure. Unlike other renewable energy technologies like wind turbines, which are relatively benign when not operational, solar flux fields can be dangerous to birds even when solar flux fields are not focused on collection towers (Wendelin et al. 2016). This can occur because mirrors in standby positions often focus solar energy just above collection towers. Heat in these standby positions can be intense enough to harm birds.

Morbidity and mortality of birds of prey in solar flux fields appear relatively rare, but when cases do occur, taxonomic patterns are emerging. Specifically, falcon (Falconiformes) species may be more susceptible, apparently because falcons are attracted to hunt aerial prey concentrated near collection towers (WEST 2016). Alternatively, in both active and standby positions, warm air rising above collection towers may attract buteos and vultures seeking thermal air currents to power flight, and these birds may inadvertently enter solar flux zones regardless of the presence or absence of potential prey.

Indirect effects of solar energy facilities include habitat loss, displacement, and avoidance (Hernandez et al. 2014). Unlike wind energy facilities where some of these effects might be temporary, with birds returning after construction, solar facilities eliminate habitat from within the facility, creating a flat bare earth-scape unattractive for hunting or nesting by birds of prey. Habitat loss at solar energy facilities is generally greater per megawatt generated than at wind facilities because wind resource areas retain most of the habitat below turbines, whereas solar facilities cover much of the facility in mirror arrays. Birds of prey and other wildlife species also may avoid habitats in and around solar facilities as a result of increased human activity and habitat alteration (DeVault et al. 2014).

Other Renewable Facilities

Other renewable energy sources include geothermal, hydroelectric, and oceanic. There are no substantial direct mortality effects to birds of prey documented for these energy sources. Geothermal power stations use heat energy from within the earth's crust to generate electrical energy. Facility footprints are similar to those of liquid and gas fossil fuel extraction facilities, with impacts to birds of prey limited to indirect effects resulting from disturbance during construction and operation. Roads to extraction wells increase habitat fragmentation (Jones and Pejchar 2013), impacting edge-sensitive species. Geothermal emissions often contain vaporized toxins which, while less than coal burning plants, release toxins into the air including hydrogen sulfide, carbon dioxide, ammonia, methane, and boron, mercury, and other heavy metals (Kagel et al. 2007), so indirect effects could also include reactions to toxic emissions.

Hydroelectric and oceanic renewable energy facilities use the energy of flowing rivers or tides to turn turbines and generate electricity. Hypothetically, aquatic hunters like osprey (*Pandion haliaetus*) could become entrapped in the machinery of hydroelectric or oceanic renewable energy infrastructure, but neither of these potential agents of mortality has yet been documented. This indicates that even if mortality occurs, levels are sufficiently low to preclude population impacts. Indirect effects likely do occur, though are not necessarily negative. Construction of reservoirs to store water for a hydroelectric dam floods and destroys bottomland habitats used as nest sites by some bird of prey species, but this habitat loss may be offset by creation of new reservoirs with far more shoreline hunting and nesting habitat than existed previously.

Effects of Transmission Linkages

Renewable facilities are connected to the existing electric system through construction of new transmission lines (Fig. 13.3), termed connections, interconnections, links, or linkages (hereafter interconnections). These interconnections have the potential to create avian collision and habitat fragmentation concerns well away from, but directly attributable to, renewable energy facilities. Post-construction environmental impacts of renewable energy infrastructure are generally considered only within the footprint of renewable energy facilities, but may not include the associated interconnections even though transmission lines are associated with avian collision mortalities (Bevanger 1998; Loss et al. 2014; Rogers et al. 2014). Because renewable interconnections have not yet been thoroughly studied with respect to potential impact to birds of prey, this section summarizes knowledge of potential impacts of transmission lines in general.

Direct effects of power lines on birds occur through mortality caused by electrocution and collision (Bevanger 1998; Loss et al. 2014). Electrocution is limited mostly to distribution lines (<69 kV) where clearances are minimal and birds can simultaneously contact multiple energized components or energized and grounded components (APLIC 2006, in this book Chap. 12). Transmission clearances designed to prevent electrical energy from arcing across conductors generally include separations greater than birds can bridge with extended wings, though there are exceptions on certain configurations used for lower transmission voltages (69–138 kV). Because electrocution is generally of little concern at the transmission voltages used in renewable energy interconnections, and because detailed

Fig. 13.3 Transmission line issues: (**a**) Transmission line bisecting a water source used by birds as a movement corridor. (**b**) Numerous transmission lines within a transmission corridor. (**c**) Overhead shield wires are less visible than conductors. (**d**) Transmission line partially obscured by fog

discussion of avian electrocution is available elsewhere in this book (in this book Chap. 12), this chapter does not address avian electrocutions.

Avian collision mortality is an ongoing global concern (Sporer et al. 2013; Rioux et al. 2013; Loss et al. 2014), though most research on the topic is not bird-of-prey-specific. Collisions involving transmission lines occur when a flying bird hits suspended wires, most often at night. Transmission lines are typically constructed with relatively thin overhead shield wires at the top and thicker energized conductors below. Birds appear to adjust flight altitudes upward to avoid large-diameter energized wires and then collide with smaller, less visible overhead shield wires (Murphy et al. 2016; Ventana Wildlife Society 2009; Martin and Shaw 2010). Transmission lines do not pose consistent risk. Rather, collision risk varies as a function of avian species and populations in the area of a given line, the surrounding habitat, and the line design (Bevanger and Brøseth 2004; Mojica et al. 2009; Rollan et al. 2010). Among birds, factors affecting collision risk include size, maneuverability, and flocking behavior (Jenkins et al. 2011; APLIC 2012). Transmission lines bisecting daily movement corridors, such as those located between roosting and foraging sites, also have been most associated with avian collisions (Bevanger and Brøseth 2004; APLIC 2012), with risk exacerbated during low-light, fog, and other inclement weather conditions (APLIC 2012; Hüppop and Hilgerloh 2012).

Birds of prey are at relatively low risk for power line collisions in general (SAIC 2000; Rioux et al. 2013), though large raptors with high wing loading and poor in-flight maneuverability like bustard species and condor species are collision prone.

In part, collision risk is low for birds of prey because they tend to fly diurnally during good weather (Ligouri 2005) and appear to detect and avoid transmission lines (Pope et al. 2006; Luzenski et al. 2016). Though risk for birds of prey is low compared to some other avian groups, collisions involving birds of prey do occur (Olendorff and Lehman 1986; Rollan et al. 2010, in this book Chap. 12). For example, California condors (*Gymnogyps californianus*) have collided with power lines (Snyder 2007), the Ventana Wildlife Society (2009) documented collisions by a northern harrier (*Circus cyaneus*) and a white-tailed kite (*Elanus leucurus*), and Mojica et al. (2009) documented multiple carcasses of bird of prey species (bald eagle (*Haliaeetus leucocephalus*), osprey, and owls) under distribution lines. Studies have shown certain African birds of prey are vulnerable to colliding with lines in foraging habitats (Boshoff et al. 2011; Rollan et al. 2010). Peregrine falcons can be at risk because they attain high speeds when pursuing prey near the ground (Olendorff and Lehman 1986). Mañosa and Real (2001) documented both collisions of breeding Bonelli's eagle (*Hieraaetus fasciatus*) and high turnover rates of pairs nesting within 1 km of power lines in Catalonia, Spain. González et al. (2007) documented infrequent collision as a cause of mortality in a study examining 267 records of nonnatural mortality of the Spanish imperial eagle (*Aquila adalberti*).

Indirect effects of transmission lines on birds of prey are not well studied but are likely low following initial disturbance and acclimation during and following construction given the fact that many birds of prey readily nest on or near transmission lines. Transmission lines can create corridors for human incursion into otherwise natural landscapes because maintenance access roads and rights-of-way may be used for recreational activities (hiking, running, mountain biking, cross-country skiing, all-terrain vehicles, etc.). Some bird of prey species respond negatively to recreational human traffic (Steidl and Anthony 1996), but no firm connection has yet been established to confirm widespread impacts with respect to power lines.

Power lines generate strong electromagnetic fields, UV discharges, and acoustic signatures which can affect animal health and behavior (Phernie et al. 2000; Tyler et al. 2014). Recent research suggests that avoidance by reindeer (*Rangifer tarandus*) may be linked to their ability to detect ultraviolet light emitted by transmission lines (Tyler et al. 2014). At least some birds also see in the ultraviolet spectrum (Lind et al. 2014), but the potential implications of this for indirect effects have not been investigated in birds of prey (in this book Chap. 12).

Offsite Effects

Offsite effects are indirect by definition. The natural resources used in constructing renewable infrastructure are typically harvested from areas well beyond the boundaries of renewable project sites. This has the potential to shift some of the environmental costs of renewable energy away from project sites where resources are used, to mine and factory sites where resources are extracted and processed. Consequently, offsite mining should be considered when developing a comprehensive understanding of potential impacts of renewable energy sources on birds of prey.

Effects of mines on birds of prey are site-specific and species-specific. For example, peregrine falcons and gyrfalcons (*Falco rusticolus*) breeding near two diamond mines in Northwest Territories, Canada, showed no difference in nest occupancy or breeding success as a function of distance from mine footprints, despite those footprints expanding during the study (Coulton et al. 2013). In contrast, prairie falcons (*Falco mexicanus*) in New Mexico appeared to avoid an entire mountain range where mining and blasting for various minerals was common but did nest in two adjacent ranges with similar habitats but less mining activity (Bednarz 1984). Mild responses to the vibration and noise associated with mining may derive from the occurrence of such natural events as thunder and landslides (Holthuijzen et al. 1990), with which birds of prey are presumably familiar both individually and over evolutionary time. Across studies, with few exceptions, evidence of disturbance by mining activity seems isolated and in some cases can be offset by relocating birds of prey nests prior to the advance of mine operations (McKee 2007). However, at least some mine sites likely included nesting territories prior to initiation of mining activities. In these cases, productivity from directly affected territories likely was reduced at least while affected individuals sought alternate nest sites. Even these impacts may be minimized, however, with measures specifically designed to support birds of prey populations, for example, through installation during reclamation of permanent structures designed to serve as nest substrates (Harshbarger 1997) and through the use of unreclaimed anthropogenic cliffs used for nesting (Moore et al. 1997). Mines also are associated with environmental pollution. Mining and smelting can lead to increased levels of lead in ospreys and American kestrels (*Falco sparverius*) nesting downstream (Henny et al. 1991, 1994) and in Eurasian eagle owls (*Bubo bubo*; Espin et al. 2014), though to our knowledge, definitive links to survival or productivity specifically related to mine sites have not been established. Though reductions in nesting attempts or productivity appear minimal overall, spills, pollution, and sedimentation from mine sites may have effects that are difficult to link conclusively to evidence of impacts specifically affecting birds of prey.

Though mining does have deleterious ecological consequences, and some examples involving birds of prey can be identified, overall it appears that offsite indirect impacts are either small or difficult to quantify and isolate (Anderson et al. 2008). Regardless of potential effects associated with renewable infrastructure, mined materials would also be necessary for fossil fuel extraction, which renewable energy facilities are designed to replace. That being so, it appears that indirect effects of extractive industries on birds of prey are minimal and offset by equivalent needs across energy sources.

Mitigation

Renewable energy facilities have the potential to bring together ecologically novel combinations of juxtaposed land covers like water bodies in deserts, prominent features like tall perches where none existed naturally, potential risks to wildlife like electrocution and mirror collisions, and potentially, unique combinations of species drawn to these features from their respective native habitats. Consequently, the removal and addition of biotic and abiotic materials at renewable energy facilities may require novel mitigation strategies applied to microclimates and biological communities which may not occur naturally. The rotor-swept zones of wind resource areas and the heated-air zones of solar tower collection areas have no natural analogues and thus no evolutionary context preparing wildlife for the risks encountered in these areas.

It should be incumbent on those creating these new landscapes, to also provide new and effective mitigation. With regard to mitigation of bird of prey mortalities at wind resource areas, innovative techniques are being developed to compensate for mortality at the renewable sites by mitigating the electrocution of birds of prey elsewhere (Fig. 13.4), creating a net benefit overall (USFWS 2013).

Wind energy facilities can also adjust turbine operations to prevent collisions by curtailing operations when birds of prey are flying within the wind resource area, and by increasing minimum operational wind speeds to wind speeds above those within which birds of prey generally choose to fly (USFWS 2013). At solar facilities with collection towers, successful mitigation involves spreading the aim points of mirrors apart to reduce the peak flux value to <4 kW/m^2 when the facility is in standby mode and not actively producing power (Multiagency Avian-Solar Collaborative Working Group 2016). For both wind resource areas and solar facilities, direct and indirect effects may be minimized by siting facilities away from

Fig. 13.4 Retrofitted power poles: (**a**) Insulation on center wire. (**b**) Insulation on connecting wires and on switches. (**c**) Insulation on connecting wires and on energized components of equipment. (**d**) Installation of insulation on equipment. (See in this book Chap. 12 for additional technical details on electrocution of birds of prey)

concentrated populations of birds of prey at migration, foraging, or roosting sites. Collisions involving birds of prey and transmission interconnections can be mitigated by marking transmission lines to increase their prominence to approaching birds of prey so lines can be avoided (in this book Chap. 12).

Unlike compensation programs for wind and solar energy, which are still in their infancy, compensation programs for biofuel monocultures are well established within a general framework of minimizing agricultural impacts to natural systems to the extent practical. Mitigation for biofuel monocultures may be achieved through existing mitigation programs, such as the US Department of Agriculture's Conservation Reserve Program which enables farmers to remove environmentally sensitive land from agricultural production in exchange for an annual payment. These types of programs tend to be successful if three obstacles can be overcome. First, because participation is voluntary, individual decisions may be influenced by the value of the payment compared to the value of potential crop yields. This mitigation strategy may lose effectiveness if demands for biofuels, and other crops competing in the market place for the same land, result in crop profits per acre that are greater than payments (Johnson and Stephens 2011). Second, compensation may undermine an individual's sense of responsibility for the land (Ramsdell et al. 2016), potentially resulting in a reduced sense of stewardship over the long term and enabling landowners to justify conversion of natural habitats if compensation programs terminate. Third, compensation programs may not be practical in developing countries lacking the necessary financial or political resources. Despite the potential obstacles involved in compensation-based mitigation programs, these solutions are nevertheless the best currently available, at least in areas like the USA where most arable farmland is privately owned and decisions affecting land use are primarily market driven. Though not necessarily focused on bird of prey concerns, these approaches often result in habitat patches that can contain hunting habitat or potential nest sites, creating focal locations which allow bird of prey populations to persist within areas dominated by agriculture.

Siting new facilities in previously disturbed habitat like nonproductive agricultural fields also can reduce impacts to birds from loss of breeding and foraging habitat (Pearce et al. 2016). Birds of prey can be intentionally displaced from solar projects when nesting sites are destroyed during construction. Burrowing owls (*Athene cunicularia*) have been successfully translocated to new breeding sites away from solar facilities (Multiagency Avian-Solar Collaborative Working Group 2016).

Benefits to Birds of Prey

Birds of prey also can benefit from renewable energy facilities and transmission linkages, primarily through provision of new nesting opportunities (Fig. 13.5) since birds of prey routinely nest on transmission structures. For example, bald eagles and osprey regularly nest on utility structures (Buehler 2000; Poole et al. 2002).

Fig. 13.5 (**a**) A golden eagle (*Aquila chrysaetos*) departing a transmission tower, potentially benefitting through hunting opportunities and, simultaneously, potentially at risk of collision with transmission wires. (**b**) A golden eagle roosting atop a transmission pole. (**c**) A golden eagle nest on a transmission tower. (**d**) An osprey (*Pandion haliaetus*) nest on a transmission H-frame structure

Other species nesting on utility structures include ferruginous hawks (*Buteo regalis*; Gilmer and Wiehe 1977), hobbies (*Falco subbuteo*; Puzović 2008), common kestrels (*Falco tinnunculus*; Krueger 1998), greater kestrels (*Falco rupicoloidesa*; Ledger and Hobbs 1999), martial eagles (*Polemaetus bellicosus*; Jenkins et al. 2013), prairie falcons (Roppe et al. 1989), lanner Falcons (*Falco biarmicus*; Ledger and Hobbs 1999), upland buzzards (*Buteo hemilasius*; Ellis et al. 2009), Swainson's hawks (*Buteo swainsoni*; James 1992), tawny eagles (*Aquila rapax*; Jenkins et al. 2013), black eagles (*Aquila verreauxii*; Jenkins et al. 2013), African hawk eagles (*Hieraaetus fasciatus*; Ledger and Hobbs 1999), and white-backed vultures (*Gyps africanus*, Ledger and Hobbs 1999). Though none of these were on renewable interconnections, the consistency between transmission structures in general and transmission structures supporting renewable interconnections specifically indicates that nesting is likely. Nesting habitat can also be created from mines providing new nest substrates for cliff-nesting birds of prey like peregrine falcons (Moore et al. 1997). Habitat conversion for dams and agriculture can also increase food availability for birds of prey because dams and reservoirs create aquatic habitat and provide abundant year-round food resources for birds of prey including water snakes (Tingay et al. 2010), waterbirds (Mukherjee and Wilske 2006; Mwaura et al. 2002), and stunned or dead fish flowing through dam spillways or turbines (Sánchez-Zapata et al. 2016).

Integrated vegetation management techniques employed in rights-of-way management for renewable energy interconnections can also play an important role in maintaining and improving habitat for wildlife (Ball 2012; Rogers 2016). These activities could create hunting habitat for birds of prey or be used as migration corridors (Denoncour and Olson 1982).

Other indirect benefits may also be important. The fundamental motivators of shifting global economies from fossil fuels to renewable energies are national energy independence and reduction of greenhouse gas emissions. Energy independence is perhaps irrelevant to birds of prey, but reduction of greenhouse gas emissions and global climate change do have substantial potential benefits for birds of prey. Global climate change is associated with increased frequency and intensity of weather events. Late spring and high-intensity weather events can directly impact the productivity and survival of birds of prey. For example, breeding success is negatively correlated with precipitation during nesting in peregrine falcons (Anctil et al. 2014; Burke et al. 2015). Survival of peregrines migrating south from the Artic is negatively correlated with climatic events suggesting the species is vulnerable to weather events along the migration route (Franke et al. 2011). Reduced impacts of climate change in general will likely reduce weather-related impacts on nesting birds of prey.

Conclusions

Ultimately, the large, widely dispersed territories of most birds of prey minimize the population impacts of either direct or indirect effects at most renewable energy facilities, transmission interconnections, or mines. This is because even if a specific territory is affected by a renewable energy facility, through habitat loss, for example, the effect is unlikely to have a population-level effect. There are exceptions however. For example, collisions involving migrating or wintering birds of prey with wind turbines can result in impacts dispersed throughout breeding ranges, and large-scale biofuel monocultures can result in elimination of habitat patches far larger than a single territory. These two areas of renewable energy advancement in particular warrant ongoing consideration, mitigation, and monitoring as renewable energy facilities expand into the habitats of birds of prey.

Acknowledgments We thank R.E. Harness for insights which guided this writing and EDM International, Inc. (Fort Collins, CO) for providing funding and logistical support.

References

Anctil A, Franke A, Bêty J (2014) Heavy rainfall increases nestling mortality of an arctic top predator: experimental evidence and long-term trend in peregrine falcons. Oecologia 174:1033–1043

Anderson DW, Suchanek TH, Eagles-Smith CA, Cahill TM Jr (2008) Mercury residues and productivity in osprey and grebes from a mine-dominated ecosystem. Ecol Appl 18(8 Supplement):A227–A238

Avian Power Line Interaction Committee (APLIC) (2006) Suggested practices for avian protection on power lines: the state of the art in 2006. Edison Electric Institute, APLIC/The California Energy Commission, Washington, DC/Sacramento, CA

Avian Power Line Interaction Committee (APLIC) (2012) Reducing avian collisions with power lines: the state of the art in 2012. Edison Electric Institute/APLIC, Washington, DC

Ball S (2012) Capitalizing on conservation: the ecological benefits of transmission line rights-of-way. In: Evans JM, Goodrich-Mahoney JW, Mutrie D, Reinemann J (eds) Proceedings of the 9th international symposium on environmental concerns in right-of-way management, Champaign

Band W, Madders M, Whitfield DP (2007) Developing field and analytical methods to assess avian collision risk at wind farms. In: de Lucas M, Janss GFE, Ferrer M (eds) Birds and wind farms: risk assessment and mitigation. Quercus, Madrid, pp 259–275

Barrios L, Rodriguez A (2004) Behavioural and environmental correlates of soaring bird mortality at on-shore wind turbines. J Appl Ecol 41:2–81

Bednarz JC (1984) The effect of mining and blasting on breeding prairie falcon (*Falco mexicanus*) occupancy in the Caballo Mountains, New Mexico. Raptor Res 8:16–19

Bellebaum J, Korner-Nievergelt F, Dürr T, Mammen U (2013) Wind turbine fatalities approach a level of concern in a raptor population. J Nat Conserv 21:394–400

Beston JA, Diffendorfer JE, Loss SR, Johnson DH (2016) Prioritizing avian species for their risk of population-level consequences from wind energy development. PLoS One. https://doi.org/10.1371/0150813

Bevanger K (1998) Biological and conservation aspects of bird mortality caused by electricity power lines: a review. Biol Conserv 86:67–76

Bevanger K, Brøseth H (2004) Impact of power lines on bird mortality in a subalpine area. Biodivers Conserv 27:67–77

Boshoff AF, Minnie JC, Tambling CJ, Michael MD (2011) The impact of power line-related mortality on the cape vulture (*Gyps coprotheres*) in a part of its range, with an emphasis on electrocution. Bird Conserv Int 21:311–327

Buehler DA (2000) Bald Eagle (*Haliaeetus leucocephalus*). In: Rodewald PG (ed) The birds of North America. Cornell Lab of Ornithology, Ithaca; Retrieved from the birds of North America: https://birdsna.org/Species-Account/bna/species/baleag. Accessed 16 Jan 2017

Burke BJ, Clarke D, Fitzpatrick A, Carnus T, McMahon BJ (2015) Population status and factors affecting the productivity of peregrine falcon *falco peregrinus* in county Wicklow, Ireland, 2008–2012. Biol Environ Proc R Ir Acad 115B:115–124

Cabrera-Cruz SA, Villegas-Patraca R (2016) Response of migrating raptors to an increasing number of wind farms. J Appl Ecol 53:1667–1675

Campbell LH, Avery MI, Donald PF, Evans AD, Green RE, Wilson JD (1997) A review of the indirect effects of pesticides on birds. Joint nature conservation committee report 227. Peterborough, UK

Carrete M, Sánchez-Zapata JA, Benítez JR, Lobón M, Donázar JA (2009) Large scale risk-assessment of wind-farms on population viability of a globally endangered long-lived raptor. Biol Conserv 142:2954–2961

Carrete M, Sánchez-Zapata JA, Benítez JR, Lobón M, Montoya F, Donázar JA (2012) Mortality at wind-farms is positively related to large-scale distribution and aggregation in griffon vultures. Biol Conserv 145:102–108

Coulton DW, Virgl JA, English C (2013) Falcon nest occupancy and hatch success near two diamond mines in the Southern Arctic, Northwest Territories. Avian Conserv Ecol 8:14

Dahl EL, May R, Hoel PL, Bevanger K, Pedersen HC, Røskaft E, Stokke BG (2013) White-tailed Eagles (*Haliaeetus albicilla*) at the Smøla wind-power plant, Central Norway, lack behavioral flight responses to wind turbines. Wildl Soc Bull 37:66–74

Denoncour JE, Olson DP (1982) Raptor utilization of power line rights-of-way in New Hampshire. In: Crabtree AF (ed) Proceedings of the 3rd international symposium on environmental concerns in rights-of-way management, San Diego, 1982

DeVault TL, Seamans TW, Schmidt JA, Belant JL, Blackwell BF (2014) Bird use of solar photovoltaic installations at US airports: implications for aviation safety. Landsc Urban Plan 122:122–128

Diehl RH, Valdez EW, Preston TM, Wellik MJ, Cryan PM (2016) Evaluating the effectiveness of wildlife detection and observation technologies at a solar power tower facility. PLoS One. https://doi.org/10.1371/0158115

Ellis DH, Craig T, Craig E, Postupalsky S, LaRue CT, Nelson RW, Henny CJ, Watson J, Millsap BA, Dawson JW, Cole KL, Martin EM, Margalida A, Kung P (2009) Unusual raptor nests around the world. J Raptor Res 43:175–198

Espin S, Martinez-Lopez E, Leon-Ortega M, Martinez JE, Garcia-Fernandez AJ (2014) Oxidative stress biomarkers in Eurasian eagle owls (*Bubo bubo*) in three different scenarios of heavy metal exposure. Environ Res 131:134–144

Franke A, Therrien J-F, Descamps S, Bêty J (2011) Climatic conditions during outward migration affect apparent survival of an arctic top predator, the peregrine falcon *Falco peregrinus*. J Avian Biol 42:544–551

Garvin JC, Jennelle CS, Drake D, Grodsky SM (2011) Response of raptors to a windfarm. J Appl Ecol 48:199–209

Gilmer DS, Wiehe JM (1977) Nesting by ferruginous hawks and other raptors on high voltage powerline towers. Prairie Nat 9:1–10

González LM, Margalida A, Mañosa S, Sánchez R, Oria J, Molina JI, Caldera J, Aranda A, Prada L (2007) Causes and spatio-temporal variations of non-natural mortality in the vulnerable Spanish imperial eagle (*Aquila adalberti*) during a recovery period. Oryx 41:495–502

Harshbarger RM (1997) Reclamation planning for sensitive species in southwest Wyoming. In: 14th annual meeting of the American society of mining & reclamation, Austin, 10–15 May 1997

Henny CJ, Blus LJ, Hoffman DJ, Grove RA, Hatfield JS (1991) Lead accumulation and osprey production near a mining site on the Coeur d'Alene River, Idaho. Arch Environ Contam Toxicol 21:415–424

Henny CJ, Blus LJ, Hoffman DJ, Grove RA (1994) Lead in hawks, falcons and owls downstream from a mining site on the Coeur-d'Alene River, Idaho. Environ Monit Assess 29:267–288

Hernandez RR, Easter SB, Murphy-Mariscal ML, Merstre FT, Tavassoli M, Allen EB, Barrows CW, Belnap J, Ochoa-Hueso R, Ravi S, Allen MF (2014) Environmental impacts of utility-scale solar energy. Renew Sustain Energy Rev 29:766–779

Holthuijzen AMA, Eastland WG, Ansell AR, Kochert MN, Williams RD, Young LS (1990) Effects of blasting on behavior and productivity of nesting prairie falcons. Wildl Soc Bull 18:270–281

Hull CL, Muir SC (2013) Behavior and turbine avoidance rates of eagles at two wind farms in Tasmania, Australia. Wildl Soc Bull 37:49–58

Hüppop O, Hilgerloh G (2012) Flight call rates of migrating thrushes: effects of wind conditions, humidity and time of day at an illuminated offshore platform. J Avian Biol 43:85–90

James PC (1992) Urban-nesting of Swainson's hawks in Saskatchewan. Condor 94:773–774

Jenkins AR, Shaw JM, Smallie JJ, Gibbons B, Visagie R, Ryan PG (2011) Estimating the impacts of power line collisions on Ludwig's Bustards *Neotis ludwigii*. Bird Conserv Int 21:303–310

Jenkins AR, de Goede KO, Sebele L, Diamond M (2013) Brokering a settlement between eagles and industry: sustainable management of large raptors nesting on power infrastructure. Bird Conserv Int 23:232–246

Johnson GD, Stephens SE (2011) Wind power and biofuels: a green dilemma for wildlife conservation. In: Nagle DE (ed) Energy development and wildlife conservation in western North America. Island Press, Washington, DC

Johnston NN, Bradley JE, Otter KA (2014) Increased flight altitudes among migrating golden eagles suggest turbine avoidance at a Rocky Mountain wind installation. PLoS One. https://doi.org/10.1371/0093030

Jones NF, Pejchar L (2013) Comparing the ecological impacts of wind and oil & gas development: a landscape scale assessment. PLoS One. https://doi.org/10.1371/0081391

Kagan RA, Viner TC, Trail PW, Espinoza EO (2014) Avian mortality at solar energy facilities in southern California: a preliminary analysis. http://alternativeenergy.procon.org/sourcefiles/avian-mortality-solar-energy-ivanpah-apr-2014.pdf. Accessed 10 Dec 2016
Kagel A, Bates D, Gawell K (2007) A guide to geothermal energy and the environment. Geothermal Energy Association. http://geoenergy.org/pdf/reports/AGuidetoGeothermalEnergyandtheEnvironment10.6.10.pdf. Accessed 10 Dec 2016
Katzner T, Brandes D, Miller T, Lanzone M, Maisonneuve C, Tremblay JA, Mulvihill R, Merovich GT Jr (2012) Topography drives migratory flight altitude of golden eagles: implications for on-shore wind energy development. J Appl Ecol 49:1178–1186
Krueger TE Jr (1998) The use of electrical transmission pylons as nesting sites by the Kestrel *Falco tinnunculus* in north-east Italy. In: Meyburg BU, Chancellor RD, Ferrero JJ (eds) Holarctic birds of prey. The World Working Group on Birds of Prey and Owls, Berlin, pp 141–148
Ledger JA, Hobbs JCA (1999) Raptor use and abuse of powerlines in Southern Africa. J Raptor Res 33:49–52
Ligouri J (2005) Hawks from every angle: how to identify raptors in flight. Princeton University Press, Princeton
Lind O, Mitkus M, Olsson P, Kelber A (2014) Ultraviolet vision in birds: the importance of transparent eye media. Proc R Soc Lond B Biol Sci. https://doi.org/10.1098/rspb.2013.2209
Loss SR, Will T, Marra PP (2014) Refining estimates of bird collision and electrocution mortality at power lines in the United States. PLoS One. https://doi.org/10.1371/0101565
de Lucas M, Janss GFE, Ferrer M (2004) The effects of a wind farm on birds in a migration point: the Strait of Gibraltar. Biodivers Conserv 13:395–407
de Lucas M, Janss GFE, Whitfield DP, Ferrer M (2008) Collision fatality of raptors in wind farms does not depend on raptor abundance. J Appl Ecol 45:1695–1703
de Lucas M, Ferrer M, Bechard MJ, Muñoz AR (2012) Griffon vulture mortality at wind farms in southern Spain: distribution of fatalities and active mitigation measures. Biol Conserv 147:184–189
Luzenski J, Rocca CE, Harness RE, Cummings JL, Austin DD, Landon MA, Dwyer JF (2016) Collision avoidance by migrating raptors encountering a new electric power transmission line. Condor 118:402–410
Mañosa S, Real J (2001) Potential negative effects of collisions with transmission lines on a Bonelli's eagle population. J Raptor Res 35:247–252
Martin GR, Shaw JM (2010) Bird collisions with power lines: failing to see the way ahead? Biol Conserv 143:2695–2702
Martínez JE, Calvo JF, Martínez JA, Zuberogoitia I, Cerezo E, Manrique J, Gómez GJ, Nevado JC, Sánchez M, Sánchez R, Bayo J, Pallarés A, González C, Gómez JM, Pérez P, Motos J (2010) Potential impact of wind farms on territories of large eagles in southeastern Spain. Biodivers Conserv 19:3757–3767
McCrary MD, McKernan RL, Schreiber RW, Wagner WD, Sciarrotta TC (1986) Avian mortality at a solar energy power plant. J Field Ornithol 57:135–141
McKee G (2007) Wildlife mitigation techniques at surface coal mines in northeast Wyoming. In: Barnhisel RI (ed) National meeting of the American society of mining and reclamation, Gillette, 2007
Mojica EK, Watts BD, Paul JT, Voss ST, Pottie J (2009) Factors contributing to bald eagle electrocutions and line collisions on Aberdeen Proving Ground, Maryland. J Raptor Res 43:57–61
Moore NP, Kelly PF, Lang FA, Lynch JM, Langton SD (1997) The peregrine *Falco peregrinus* in quarries: current status and factors influencing occupancy in the Republic of Ireland. Bird Study 44:176–181
Mukherjee A, Wilske B (2006) Importance of wetlands for conservation of bird life in the dry lands of western India. In: Boere GC, Galbraith CA, Stroud DA (eds) Waterbirds around the world. The Stationary Office, Edinburgh, pp 303–304

Multiagency Avian-Solar Collaborative Working Group (2016) Avian-solar science coordination plan. http://blmsolar.anl.gov/program/avian-solar/docs/Final_Avian-Solar_Science_Coordination_Plan.pdf. Accessed 10 Dec 2016

Murphy RK, Mojica EK, Dwyer JF, McPherron MM, Wright GD, Harness RE, Pandey AK, Serbousek KL (2016) Crippling and nocturnal biases in as study of Sandhill Crane (*Grus canadensis*) collisions with a transmission line. Waterbirds 39:312–317

Mwaura F, Mavuti KM, Wamicha WN (2002) Biodiversity characteristics of small high-altitude tropical man-made reservoirs in the Eastern Rift Valley, Kenya. Lakes Reserv Res Manag 7:1–12

Olendorff RR, Lehman RN (1986) Raptor collisions with utility lines: an analysis using subjective field observations. http://ulpeis.anl.gov/documents/dpeis/references/pdfs/Olendorff_and_Lehman_1986.pdf. Accessed 16 January 2017

Orloff S, Flannery A (1992) Wind turbine effects on avian activity, habitat use, and mortality in Altamont Pass and Solano County wind resource areas, 1989–1991. http://www.energy.ca.gov/windguidelines/documents/2006–12-06_1992_FINAL_REPORT_1989–1991.PDF. Accessed 23 Dec 2016

Pearce D, Strittholt J, Watt T, Elkind EN (2016) A path forward: identifying least-conflict solar PV development in California's San Joaquin Valley. https://www.law.berkeley.edu/wp-content/uploads/2016/05/A-PATH-FORWARD-May-2016.pdf. Accessed 16 Jan 2017

Phernie K, Bird DM, Dawson RD, Lague PC (2000) Effects of electromagnetic field on the reproductive success of American Kestrel. Physiol Biochem Zool 73:60–65

Poole AF, Bierregaard RO, Martell MS (2002) Osprey (*Pandion haliaetus*). In: Rodewald PG (ed) The birds of North America. Cornell Lab of Ornithology, Ithaca; Retrieved from the birds of North America: https://birdsna.org/Species-Account/bna/species/osprey. Accessed 16 Jan 2017

Pope VR, Fielder PC, Cordell KA, Harness RE, Hamer TE (2006) Pre-construction evaluation of collision potential for fall migrating raptors with a transmission line across Burch Mountain, Chelan County, Washington. Public Utility District No. 1 of Chelan County, EDM International, and Hamer Environmental, Wenatchee WA, Fort Collins CO, Mount Vernon WA

Puzović S (2008) Nest occupation and prey grabbing by saker falcon (*Falco cherrug*) on power lines in the province of Vojvodina (Serbia). Arch Biol Sci 60:271–277

Ramsdell CP, Sorice MG, Dwyer AM (2016) Using financial incentives to motivate conservation of an at-risk species on private lands. Environ Conserv 43:34–44

Rioux S, Savard J-PL, Gerick AA (2013) Avian mortalities due to transmission line collisions: a review of current estimates and field methods with an emphasis on applications to the Canadian electric network. Avian Conserv Ecol. https://doi.org/10.5751/ACE-00614-080207

Rogers T (2016) Impacts of vegetation management practices on animal, plant, and pollinator habitats. In: Doucet J (ed) Proceedings of the 11th international symposium on environmental concerns in right-of-way management, Forest Lake, MN, 2016

Rogers AM, Gibson MR, Pockette T, Alexander JL, Dwyer JF (2014) Scavenging of migratory bird carcasses in the Sonoran Desert. Southwest Nat 59:542–547

Rollan A, Real J, Bosch R, Tintó A, Hernández-Matías A (2010) Modelling the risk of collision with power lines in Bonelli's eagle *Hieraaetus fasciatus* and its conservation implications. Bird Conserv Int. https://doi.org/10.1017/S0959270910000250

Roppe JA, Siegel SM, Wilder SE (1989) Prairie falcon nesting on transmission towers. Condor 91:711–712

Sánchez-Zapata JA, Clavero M, Carrete M, DeVault TL, Hermoso V, Losada MA, Polo MJ, Sánchez-Navarro S, Pérez-García JM, Botella F, Ibáñez C, Donázar JA (2016) Effects of renewable energy production and infrastructure on wildlife. In: Mateo R, Arroyo B, Garcia JT (eds) Current trends in wildlife research. Springer International Publishing, Cham, pp 97–123

Science Applications International Corporation (SAIC) (2000) Avian collision at transmission lines associated with the Hells Canyon complex. Technical report appendix E.3.2–20. https://www.idahopower.com/pdfs/Relicensing/hellscanyon/hellspdfs/techappendices/Wildlife/e32_20.pdf. Accessed 16 Jan 2017

Smallwood KS, Rugge LM, Morrison ML (2009) Influence of behavior on bird mortality in wind energy developments. J Wildl Manag 73:1082–1098
Snyder NFR (2007) Limiting factors for wild California condors. In: Mee A, Hall LS (eds) California condors in the 21st century. The Nuttall Ornithological Club/The American Ornithologists' Union, Cambridge, MA/Washington, DC, pp 9–33
Sporer MK, Dwyer JF, Gerber BD, Harness RE, Pandey AK (2013) Marking power lines to reduce avian collisions near the Audubon National Wildlife Refuge, North Dakota. Wildl Soc Bull 37:796–804
Steidl RJ, Anthony RG (1996) Responses of bald eagles to human activity during the summer in interior Alaska. Ecol Appl 6:482–491
Tingay RE, Nicoll MAC, Whitfield DP, Visal S, McLeod DRA (2010) Nesting ecology of the grey-headed fish-eagle at Prek Toal, Tonle Sap Lake, Cambodia. J Raptor Res 44:165–174
Tyler N, Stokkan K-A, Hogg C, Nellemann C, Vistnes A-I, Jeffery G (2014) Ultraviolet vision and avoidance of power lines in birds and mammals. Conserv Biol 28:630–631
Uden DR, Allen CR, Mitchell RB, McCoy TD, Guan Q (2015) Predicted avian responses to bioenergy development scenarios in an intensive agricultural landscape. GCB Bioenergy 7:717–726
United States Fish and Wildlife Service (USFWS) (2013) Eagle conservation plan guidance: module 1 – land-based wind energy, version 2. https://www.fws.gov/migratorybirds/pdf/management/eagleconservationplanguidance.pdf Accessed 16 Jan 2017
Ventana Wildlife Society (2009) Evaluating diverter effectiveness in reducing avian collisions with distribution lines at San Luis National Wildlife Refuge Complex, Merced County, California. http://www.ventanaws.org/pdf/condor_reports/CEC-500-2009-078.pdf. Accessed 16 Jan 2017
Walker D, McGrady M, McCluskie A, Madders M, McLeod DRA (2005) Resident golden eagle ranging behaviour before and after construction of a windfarm in Argyll. Scott Birds 25:24–40
Wendelin T, Ho CK, Sims C (2016) Development of tools, training, and outreach to address solar glare and flux-related avian impact. http://blmsolar.anl.gov/program/avian-solar/docs/Avian-Solar_CWG_May_2016_Workshop_Slides.pdf. Accessed 15 Dec 2016
Western EcoSystems Technology, Inc. (WEST) (2016) Ivanpah solar electric generating system avian and bat monitoring plan, 2014–2015 annual report and two year comparison. http://docketpublic.energy.ca.gov/PublicDocuments/07-AFC-05C/TN212042_20160630T145041_ISEGS_Avian_and_Bat_Monitoring_Plan_20142015.pdf. Accessed 15 Dec 2016

Part III
Raptor Conservation

Chapter 14
Use of Drones for Research and Conservation of Birds of Prey

David Canal and Juan José Negro

Introduction

In the last two decades, unmanned aircraft systems (UASs) have experienced an exponential development. Originally conceived for military use, technological advances and a dramatic reduction of prices are leading to widespread use of UASs in environmental disciplines including remote sensing, ecology, wildlife management or environmental monitoring (Chabot and Bird 2015; Linchant et al. 2015; Christie et al. 2016).

UASs have been used in wildlife management and conservation, among others, to monitor terrestrial (e.g. chimpanzees (van Andel et al. 2015), elephants and rhinoceroses (Vermeulen et al. 2013), ungulates (Barasona et al. 2014), black bears (Ditmer et al. 2015)) and marine mammals (e.g. dugongs (Hodgson et al. 2013), whales (Koski et al. 2015)) and birds (e.g. raptors (Canal et al. 2016; Junda et al. 2016), bird colonies (Sardá-Palomera et al. 2012; Rümmler et al. 2015; Sardà-Palomera et al. 2017)), as an antipoaching tool (Mulero-Pázmány et al. 2014b) and for tracking tagged wildlife (Cliff et al. 2015; Dos Santos et al. 2015). In the field of agriculture and the environment, the development of UAS has been so revolutionary that different reviews are already available on particular applications (e.g. see Zhang and Kovacs 2012; Colomina and Molina 2014; Shahbazi et al. 2014 and references therein).

The advantages of UASs compared to conventional aviation go beyond security by removing occupants from aircrafts – accidents on plane and helicopters are a leading cause of work-related mortality among field biologists in North America (Sasse 2003). First, UASs can fly preprogrammed GPS tracks to collect repeatable

D. Canal (✉) · J. J. Negro
Department of Evolutionary Ecology, Doñana Biological Station (EBD-CSIC), Seville, Spain
e-mail: davidcanal@ebd.csic.es

J. H. Sarasola et al. (eds.), *Birds of Prey*,
https://doi.org/10.1007/978-3-319-73745-4_14

aerial data at very large temporal and spatial resolution. They can fly over areas of difficult access and/or cover vast areas, becoming a cost-effective solution. Second, the ample range of available models, varying widely in size (from insect-sized to full-sized aircraft) and designs (review in Vergouw et al. 2016), makes UASs highly versatile. As an example of their diversity, Van Blyenburgh (2011) catalogued near 1500 different UAS models a few years ago. This figure is much higher nowadays.

UASs may be classified according to multiples characteristics (e.g. size, autonomy, payload; Vergouw et al. 2016), but probably the most simple and interesting classification for wildlife research is to distinguish between fixed and rotary wing devices (although hybrid systems are steadily appearing in the market). Fixed-wing UASs have a higher payload capacity and autonomy, which allow them to fly long distances, often beyond visual line of sight (BVLOS). The applications requesting BVLOS are subjected to legal considerations in a majority of countries, and this has to be taken seriously when operating the UAS. Flights are usually performed at a higher speed and altitude than their rotary UAS counterparts. A landing strip may also be required for certain models. Rotary-wing UASs are typically smaller, with lower autonomy (typically, lower than 30–40 min) and payload capacity than fixed wings. The great advantages of rotary UASs are their vertical take-off and landing coupled with the high manoeuvrability and capacity to fly at very low speed and hover in the air. As a third important characteristic, UASs – both fixed and rotary wings – can be equipped (taking into account payload limitations) with a broad range of sensors (one or several simultaneously), including, among others, video and still photo cameras (e.g. at the visible wavelength and infrared), microphones, biological and chemicals sensors or meteorological sensors (reviewed in Vergouw et al. 2016), making UASs particularly useful in collecting a wide range of research data. Finally, multiple UASs may be flown concurrently, communicating among them and accomplishing tasks collaboratively, thus expanding their possibilities and applications. As a curiosity, in 2015, within the project Drone 100 (https://www.aec.at/futurelab/en/project/drone-100), a 100-UAS squadron equipped with LED lights illuminated the night sky in Tornesch (Germany) while flying in accord with a full orchestra playing Beethoven's Fifth Symphony, which meant a Guinness World Record.

Raptors are generally hard to study. They often nest in remote areas or in rather inaccessible places such as treetops, rocky outcrops, coastal cliffs or power pylons. Their population densities are low, and excepting a few colonial species, territorial couples are sparsely distributed even in favourable habitats, where their home ranges often cover many square kilometres. To monitor movements and productivity of raptors, biologists have resorted in the last years to high-tech devices, including VHS radio tracking, GPS monitoring and photo or video recordings, and, more recently, to the use of UAS (mostly to check nest contents on hard-to-climb locations). We will review current use of UAS in raptor research and monitoring (Table 14.1), we will foresee future potential applications and, last, we will comment on key issues and limitations that, aside from the technological questions outlined above – autonomy, payload capacity and sensor attached – must be considered for the successful use of UASs in raptor studies.

Table 14.1 List of works using UAS in raptor research

Species	Common name	Aim of the study	Reference
Falco naumanni	Lesser kestrel	Habitat selection	Rodríguez et al. (2012)
Haliaeetus pelagicus	Steller's sea eagle	Nest surveying	Potapov et al. (2013)
Pandion haliaetus	Osprey	Nest surveying	Junda et al. (2015)
Haliaeetus leucocephalus	Bald eagles		
Buteo regalis	Ferruginous hawks		
Buteo jamaicensis	Red-tailed hawks		
Pandion haliaetus	Osprey	Parental nest defence to a novel aerial intruder	Junda et al. (2016)
Haliaeetus leucocephalus	Bald eagles		
Buteo regalis	Ferruginous hawks		
Buteo jamaicensis	Red-tailed hawks		
Milvus migrans	Black kite	Test if nest decoration acts as a signal of individual quality	Canal et al. (2016)

UAS for Wildlife Population Monitoring and Breeding Success

A straightforward application of UASs in wildlife monitoring is the use of multirotor UASs to check the content of open nests (nesting status, clutch size, brood size and nestling development) when it cannot be made appropriately with optical equipment and the nest is located in places of difficult access, such as cliffs (golden eagles and different vulture species, for instance) or the tree canopy (see specific examples below and Fig. 14.1). In such situations, the use of UAS is a relative inexpensive, faster and safer alternative to either climbing or the use of piloted aircrafts (Potapov et al. 2013; Weissensteiner et al. 2015; Junda et al. 2015). This approach has been successfully employed to survey nests of Steller's sea eagle (*Haliaeetus pelagicus*; Potapov et al. 2013), osprey (*Pandion haliaetus*), bald eagles (*Haliaeetus leucocephalus*), ferruginous hawks (*Buteo regalis*) and red-tailed hawks (*Buteo jamaicensis*; Junda et al. 2015) as well as of non-raptorial avian species such as white storks (*Ciconia ciconia*) (Mulero-Pázmány et al. 2014a; Weissensteiner et al. 2015). Further, through personal communications, we are aware this is a routinely approach used by several research groups to survey hard-to-reach nests; thus additional examples concerning other species will accumulate soon in the literature.

Monitoring and surveillance of wildlife populations is possibly the most popular application of UASs in ecology (reviewed in Linchant et al. 2015). UASs offer a great opportunity to substantially improve the survey process by being a cost-effective method, with lower disturbance and higher accuracy than traditional approaches, particularly in remote or difficult-to-access areas. In birds, a majority of examples to date have focused on surveying breeding colonies with UASs, either using fix wing or multirotor systems equipped with cameras operating in the visible wavelength (Sardá-Palomera et al. 2012; Grenzdörffer 2013; Goebel et al. 2015; Sardà-Palomera et al. 2017). The UASs are flown over the target colonies to photograph them, and birds and/or their nests are counted directly on the photographs or using

Fig. 14.1 Nest surveying using the traditional climbing approach (left) and an unmanned aerial vehicle equipped with a camera (right). The traditional method involves at least one but usually two experienced persons (the belayer and the climbing person) and specific equipment. This approach is time consuming, which implies more disturbance for the studied species, is risky for the climber, and not all the nests can be reached. By contrast, the unmanned aerial vehicle can be operated after a short training, it does not require long preparation previously the flight, and only a small patch without high vegetation is needed for take-off and landing

object recognition software. If the colonies are large, the post-processing of the images may involve the construction of photomosaics. When flying over colonies, it is essential to do it at a sufficient altitude in order not to disturb the birds. Sea gulls and herons, typically, stand flying UAS at an altitude of 100 metres above ground level. Using the right photo equipment, nests and their contents may still be visible and discernible at such flight altitude.

Although a majority of raptor species are territorial, UASs could be used to survey semicolonial raptors that nest on structures able to be detected from the air such as ospreys (*Pandion haliaetus*; Poole 1989). Further, for raptors nesting on artificial linear structures, UASs may be used both to monitor the status of the nests and to record the type of pylon according to structural design, or even electrocution risk (Mulero-Pázmány et al. 2014a). Alternatively, UASs might be utilized to monitor raptors breeding at low densities in remote areas when nests are detectable from the air. An example may be the endangered crowned solitary eagle (*Buteogallus coronatus*) in South America, a large snake-eating eagle which nests on treetops at a very low density (Ferguson-Lees and Christie 2001; Fandiño and Pautasso 2014). Irrespective of the case study – i.e. numerous individuals in a small area or a vast area inhabited by few individuals – the use of UASs for wildlife surveying typically implies the collection of a large number of images, and this number increases as the overlap of the pictures increases. Consequently, the manual examination of all the images may be a time-consuming and exhaustive task. Although several authors have developed different methods (e.g. Chabot and Bird 2012; Grenzdörffer 2013; Christiansen et al. 2014; Linchant et al. 2015 and references herein), the automatic detection and count of individuals and/or nests may be challenging, particularly in sandy areas or when species/nests do not contrast well with the environment. Researchers should also pay attention to local legal restrictions of BVLOS operations.

As an alternative to still cameras, UASs can be equipped with thermal cameras to detect animals at night. Flights may be also performed mounting compact video cameras, when a high resolution is not necessary or when ambient noise and/or behaviour wants to be recorded. The use of UASs equipped with thermal cameras is becoming very popular due to its perceived potential in conservation as an anti-poaching system. The detection of wildlife concealed within agricultural fields is also a promising UAS application since increased working widths and speeds of harvesters kill or injure a large number of animals breeding in farmlands. This is important for animal welfare, but also because carcases in decomposition may cause health hazards and great economic losses. Israel (2012) pioneered the use of a multirotor equipped with a thermal camera to detect roe deer (*Capreolus capreolus*) fawns concealed in meadows. Fawns were frequent victims of harvesting machines because, as a defensive strategy, they remain hidden and still during the harvest. Regarding potential applications in raptor research, one may envision the utility of UASs equipped with thermal cameras to locate nests of ground-dwelling raptors of the harrier group (*Circus* spp.) in extensive agricultural fields. For such aim, an UAS multirotor should be the preferred platform in most of the situations, although the final election of the UAS model will depend on the surface to be surveyed and the resolution required for target detection. Beyond legal restrictions concerning nocturnal flights, the major drawback of this approach is related to the analysis of the data. Manual processing of a large number of images to detect nests is a time-consuming task, whereas the implementation of an automatic recognition system is a challenge (Chabot and Bird 2012; Grenzdörffer 2013; Christiansen et al. 2014; Linchant et al. 2015) as, for example, inanimate objects that are heated by the sun may be confused with concealed animals (Christiansen et al. 2014). In this respect, flights performed right before sunset would contribute to increase detection efficiency.

UAS for Wildlife Tracking and Location

The deployment of electronic devices for wildlife is widely extended and has multiple applications; it allows to record the movements of the marked individuals, serves to locate their feeding and breeding areas or permits registering multiple activity and physiological parameters. VHF radio tagging and GPS logging devices are nowadays two of the most common wildlife marking systems, but tracking and (re)capturing animals maybe challenging and are both time-consuming and expensive (Cliff et al. 2015; Dos Santos et al. 2015).

When radio tracking animals, the individual is equipped with a small transmitter that emits a signal in the very high frequency (VHF) band (a unique frequency for each marked animal) and the researcher uses a directional antenna – the VHF receiver – to locate the animal. By orientating the antenna in multiple directions, the researcher monitors the signal and, based on the intensity of the received signals, locates the direction of the tagged animal. This operation is performed on three different and sufficiently spaced locations, and the position of the VHF tag is obtained by triangulation. The tracking process tends to be time consuming because, for example, when tracking large predators with large home ranges or in the phase of juvenile dispersal, the points needed to triangulate the signal may be separated by dozens of kilometres. Further, fading effects, resulting from obstacles (e.g. terrain or vegetation), hamper the tracking process and translate into a greater investment of resources (Körner et al. 2010; Cliff et al. 2015). Employing aircrafts to track (air to ground) wildlife maybe a solution to reduce fading effects, but it typically implies significant logistic costs and added risks to the surveyors. Regarding logging devices, they often have a reduced wireless transmission capability (the smaller the device, the lower the transmission). Thus, to retrieve the information stored in these devices, one needs to be near the tagged animals, which first implies to know their location. Although other devices such as satellite or GSM transmitters allow the wireless transmission of the data, these devices are very expensive relative to VHF or data-logger devices.

Under this scenario, the use of UASs to automatically track and locate animals is a promising solution by having a great potential to reduce cost and efforts in this research field. Broadly, the UAS-based tracking system would involve one or various UASs, equipped with a VHF receiver. Flying over/near the area containing the tagged animals, the UAS would be able to receive the VHF signals, process them for triangulation and geolocation and, ideally, reach the position of the animals to photograph/film them or retrieve the data from the attached loggings. However, although exciting, this approach is still in the initial stages. Initial work has focused on developing the technique's conceptual and mathematical backbone (Soriano et al. 2005; Posch and Sukkarieh 2009; Körner et al. 2010; Cliff et al. 2015), but the development of an efficient UAS-based tracking system is a major technological challenge because, among others issues, the incidence of noise caused by the electronic systems of the UASs hampers the ideal reception of the signal, leading to uncertainty in the location of the tag. The performed research promises good results of tag location using complex methods of detection and/or large receivers and antennas (Soriano et al. 2005; Posch and Sukkarieh 2009; Körner et al. 2010; Cliff et al. 2015). For example, Cliff et al. (2015) have recently demonstrated it is possible to autonomously locate radio-tagged small birds by a multirotor UAS, but their refinement to be employed in small UASs and large distances is still a challenge. Aside technical shortcomings, the regulation of the airspace may still be the major obstacle for the employment of tracking systems based on UASs, as this approach will frequently imply BVLOS operations, which, as we have stated above, are severely restricted in most countries.

UAS for Habitat Sensing

UASs constitute an invaluable tool to retrieve habitat and environmental data when high temporal or spatial resolution is required (Chabot and Bird 2015). As some illustrative examples, Chabot et al. (2014) showed that UAS-based images may reveal fine-scale details of wetland habitat, otherwise non-detected with conventional methods. This was shown to be important for the conservation of the threatened least bittern (*Ixobrychus exilis*). Breckenridge et al. (2011) and Beck et al. (2014) utilized UASs to acquire high-resolution images and characterize sagebrush habitats associated with the endangered greater sage grouse (*Centrocercus urophasianus*), whereas Barasona et al. (2014) determined the landscape predictors of ungulate abundance and the associated risk of tuberculosis infection with the assistance of UAS imagery. The type of environmental information extracted from these approaches would be interesting to apply in raptor research, although, to our knowledge, this has not been attempted yet.

Recent works in spatial ecology goes a step further by using UASs to complement other techniques. In this regard, Mulero-Pázmány et al. (2015) operated a fixed-wing UAS as a complement of GPS-GSM collars to predict cattle (*Bos taurus*) distribution, whereas Rodríguez et al. (2012) combined the use of GPS data loggers with a fixed-wing UAS to study habitat selection in a colonial small falcon, the lesser kestrel (*Falco naumanni*; Fig. 14.2). After downloading the information from tagged kestrels, the researchers programmed the UAS to overflow and take photographs of the route done by the kestrels shortly after their flight. This allowed the researchers to study the availability of different habitat types along the bird flightpath at near-real time in a highly dynamic landscape. As these authors suggested, this approach may be particularly interesting to study habitat selection and/or foraging behaviour in small animals where other portable but heavier devices are not feasible. In the future, when technological advances (and airspace restrictions) allow it, one may envision a scenario where UASs followed GPS-bearing animals at a safe distance to record, among other things, environmental parameters such as temperature, air pressure as well as wind direction and speed.

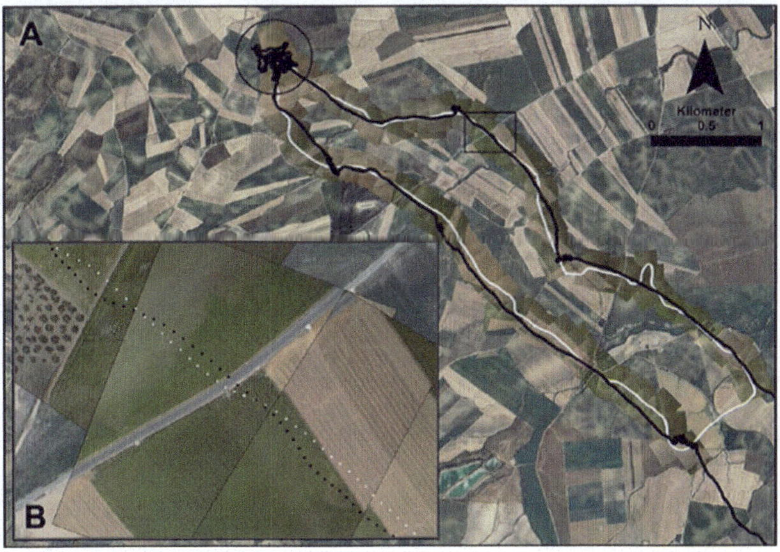

Fig. 14.2 (**a**) Foraging flight of a tagged lesser kestrel. Images were obtained by an unmanned aerial vehicle programmed to overflow (white track) and photograph the route done by the kestrel (black track) shortly after their flight. (**b**) Enlarged image of the lesser kestrel hunting area. (Figure was taken from Rodríguez et al. (2012))

UAS for Investigations on Behavioural Ecology

Beyond the most widespread UAS applications found in the literature and that we have reported above, UASs may be used to study behavioural traits or social interactions in raptors. Although still scant, some recent examples are worth mentioning. Junda et al. (2016) investigated parental nest defence response to a UAS, acting as a novel object, in four raptors species (osprey, bald eagle, ferruginous hawk and red-tailed hawk) at different stages of the breeding cycle (egg incubation and nestling stage). They documented the aggressive behaviour of raptors (flight behaviour and vocalizations) while a UAS, equipped with a GoPro, was progressively approaching the nests. In another study (Canal et al. 2016) on black kite (*Milvus migrans*) nest ornamentation, we utilized a hexacopter to take pictures of the nest platforms, with and without an experimentally placed decoration, from different altitudes and distances simulating the perspective of a flying and approaching, prospecting black kite intruder (Fig. 14.3). The images were later shown to human volunteers in detection trials to determine whether detection rates varied in relation to the nest decoration and distance. Our study complemented previous works in the species and confirmed that nest decoration is a signal to enhance nest visibility that acts as a reliable threat informing to trespassing conspecifics on the individual quality of the signaller. Conspecifics might thus assess nest decoration and minimize aggressions from the territory owners, which would contribute to the evolutionary maintenance of this communication system.

Fig. 14.3 Hexacopter taken photographs around a black kite nests with and without an experimentally placed decoration. Images were taken at different altitudes and distances simulating the perspective of a flying and approaching intruder to test whether detection rates varied according to nest decoration and distance. (Figure taken from Canal et al. (2016))

UASs have been also utilized in songbird surveys (Wilson et al. 2017), although the application of this approach in raptor research seems to be limited to highly vocal species such as the kestrels or the chimango caracara (*Phalcoboenus chimango*). In particular, Wilson et al. (2017) operated a quadcopter with a suspended lightweight recorder and tested the feasibility of the system to estimate songbird abundance in relation to traditional ground-based surveys. Detection rates using the UAS were generally comparable to standard methods. However, the system underestimated species with a low-frequency song, and species richness was generally lower in relation to standard methods, possibly due to the noise produced by the own UAS. Thus, although promising, this approach still requires technical developments to be reliably employed in the field. Finally, we envision that UAS may be operated to study predator-prey interactions, social dynamics (e.g. monitoring birds of different species during migration; Chabot and Bird 2012) or registering behavioural actions, such as direct attacks, escape reactions or foraging bouts of predators on their prey.

More Foreseen Applications of UAS

In this section, we outline applications of UASs that, although not strictly related to scientific research, are indirectly related to the ecological role of raptors. For instance, UASs have been operated to deter nuisance wildlife to avoid bird collisions with aircrafts in airports or to disperse birds causing damage to crops. Although regular commercial UASs were used in those tasks initially, nowadays, and aiming to increase the deterring efficacy, it is possible to use UAS models resembling birds of prey. Further, to increase the efficacy, UASs may be equipped with additional scare devices and, among others, they may emit predator calls, loud noises or the sounds of gun shots. However, despite these innovations, the long-term efficacy of this system is possibly limited because the target wildlife likely will get habituated to the unnatural, repetitive movements (and/or sounds) of currents UASs. Thus, future computational development should focus on improving UAS movements, including relative random movements, to resemble the real flight pattern and behaviour of actual raptors. It remains to be seen whether this approach is cost-effective, since a pilot is currently needed to manage the UAS. Finally, major limitations are due to airspace regulations (BVLOS operations) and ethical concerns over the effects of this unspecific approach on target and not-target species.

In the realm of recreational use, UAS are being utilized in falconry as a complement to exercise and train birds of prey. In this technique, on the one hand, a multirotor UAS carrying bait (which in turn is attached to a parachute) is placed at the desired altitude, whereas, on the other hand, a training raptor is released and attracted to the bait. The animal is in fact trained to approach the UAS. Once the raptor is near the UAS, the bait is detached by remote control from the UAS and captured by the raptor. In successive flights, the UAS can be hovered progressively at higher altitudes and distances, becoming a useful complement to train the birds of prey for future hunting situations. A refinement of the approach is the employment of UASs resembling other birds of prey or just as surrogates of actual prey. UAS shaped and painted as bustards and houbaras are used by Middle East falconers due to the scarcity of actual prey (e.g. see http://www.rofalconry.com/page.php?id=4).

Considerations on the Limits of UAS Utilization

In this section, we outline several issues and potential limits that should be considered before working with UASs.

Training, Field Area and Weather Conditions

Although small UASs have become progressively more user-friendly, working with professional models (e.g. DJI-S1000 octocopter or fixed-wing UAS) requires a relatively long learning curve and periodic training, including a minimum knowledge of the mechanisms and electronic devices to solve potential setbacks. This is particularly important when missions are going to be carried out in remote areas where the opportunities for external technical assistance are non-existent. Dealing with drones often means investing a significant fraction of the time fixing a range of technical problems. Thus, for example, it is not infrequent that the vegetation or stones cause small damages during the landing of fixed-wing UAS in the field. Further, it is also essential to count with a power generator to charge batteries in the field or carry multiple batteries to avoid unsurmountable setbacks.

Weather conditions are another important factor determining the success of the planned missions. Wind speed is the most obvious weather-related factor affecting UAS performance. Different models support different maximum wind speeds, so the researcher must check the typical wind conditions in the research area before acquiring and/or operating a UAS. Wind type is also of great concern, i.e. if it is constant or gusty, since the latter may destabilize the UAS and provoke a crash. The pilot should also pay attention to uneven terrain – which may cause gusty winds. Plus very low-flight altitudes leave little margin for corrections and the hazards increase. When operating in extreme cold conditions as, for example, in high altitudes or in polar regions, the researcher should take into account that the cold may reduce drastically the performance of the batteries and also affect other electronic devices of the UAS.

Finally, one may consider whether the use of UAS is really advantageous in relation to alternative methods. For example, if the objective is to detect marsh harrier nests in relatively small areas, it may be faster to observe the behaviour of adults and then perform traditional eye searches in the areas supposedly containing the nests than flying a UAS and process manually all the acquired images.

Ethical Considerations

Birds of prey are all sensitive species. Some of them are endangered with extinction and others have vulnerable populations. Therefore, maximum care must be exerted to avoid disturbance or to cause damage to individuals or their nests. Mulero-Pázmány et al. (2017) have recently listed a number of guidelines to be followed when dealing with protected wildlife or in surveys conducted in nature reserves. To summarize, these authors propose to use UAS in good working order, with full batteries at the start of any mission, and in optimal weather conditions to minimize crashes or incidents that may delay operations (see above). The pilot should be licensed and of proven ability. Avoid training and practice involving wild raptors. These guidelines incorporate two specific pieces of advice directly addressed to raptor research and monitoring that we reproduce here:

1. If the UAS flight purpose is nest inspection, approach the nest when adults are away and at times when unprotected eggs/chicks are safe. If an adult approaches in aggressive attitude towards the UAS or the pilot/observers, cancel the operation and return the UAS to base as soon as possible to pack and leave the area.
2. If the flights are performed around raptor territories that previously have shown aggressive behaviour, fly at daytimes when the temperature is lower, such as early morning or late evening, because diurnal raptors are less prone to fly in such conditions. If, however, incubating or brooding adults leave the nest, consider cancelling the mission if small nestlings are suspected or actually observed at the nest.

References

Barasona JA, Mulero-Pázmány M, Acevedo P, Negro JJ, Torres MJ, Gortázar C, Vicente J (2014) Unmanned aircraft systems for studying spatial abundance of ungulates: relevance to spatial epidemiology. PLoS One 9:e115608. Available from http://dx.plos.org/10.1371/journal.pone.0115608

Beck JL, Terrance Booth D, Kennedy CL (2014) Assessing greater sage-grouse breeding habitat with aerial and ground imagery. Rangel Ecol Manag 67:328–332. Available from http://www.bioone.org/doi/abs/10.2111/REM-D-12-00141.1

Breckenridge RP, Dakins M, Bunting S, Harbour JL, White S (2011) Comparison of unmanned aerial vehicle platforms for assessing vegetation cover in sagebrush steppe ecosystems. Rangel Ecol Manag 64:521–532. Available from http://www.bioone.org/doi/abs/10.2111/REM-D-10-00030.1. Accessed 8 Sept 2014

Canal D, Mulero-Pázmány M, Negro JJ, Sergio F (2016) Decoration increases the conspicuousness of raptor nests. PLoS One 11:e0157440. Available from http://dx.plos.org/10.1371/journal.pone.0157440

Chabot D, Bird DM (2012) Evaluation of an off-the-shelf unmanned aircraft system for surveying flocks of geese. Waterbirds 35:170–174. Available from http://www.bioone.org/doi/abs/10.1675/063.035.0119

Chabot D, Bird DM (2015) Wildlife research and management methods in the 21st century: where do unmanned aircraft fit in? In: Proceedings of Robotics Science and Systems XI 3:137–155. Available from http://www.nrcresearchpress.com/doi/full/10.1139/juvs-2015-0021

Chabot D, Carignan V, Bird DM (2014) Measuring habitat quality for least bitterns in a created wetland with use of a small unmanned aircraft. Wetlands 34:527–533
Christiansen P, rild SKA, yholm JRN, Karstoft H (2014) Automated detection and recognition of wildlife using thermal cameras. Sensors (Basel, Switzerland) 14:13778–13793
Christie KS, Gilbert SL, Brown CL, Hatfield M, Hanson L (2016) Unmanned aircraft systems in wildlife research: current and future applications of a transformative technology. Front Ecol Environ 14:241–251
Cliff OM, Fitch R, Sukkarieh S, Saunders DL, Heinsohn R (2015) Online localization of radio-tagged wildlife with an autonomous aerial robot system. Robotics: Science and Systems (RSS) XI: 42
Colomina I, Molina P (2014) Unmanned aerial systems for photogrammetry and remote sensing: a review. ISPRS J Photogramm Remote Sens 92:79–97. International Society for Photogrammetry and Remote Sensing, Inc. (ISPRS). Available from https://doi.org/10.1016/j.isprsjprs.2014.02.013
Ditmer MA, Vincent JB, Werden LK, Tanner JC, Laske TG, Iaizzo PA, Garshelis DL, Fieberg JR (2015) Bears show a physiological but limited behavioral response to unmanned aerial vehicles. Curr Biol 25:2278–2283. Elsevier Ltd. Available from http://linkinghub.elsevier.com/retrieve/pii/S0960982215008271
Dos Santos GAM et al (2015) Small unmanned aerial vehicle system for wildlife radio collar tracking. In: Proceedings – 11th IEEE international conference on Mobile Ad Hoc and Sensor Systems, MASS 2014:761–766
Fandiño B, Pautasso A (2014) Distribution, natural history and conservation of Harpyhaliaetus coronatus (Birds: accipitridae) in central-East Argentina. Nat Neotrop 44:41–42
Ferguson-Lees J, Christie DA (2001) Raptors of the world. Helm Identification Guides, London
Goebel ME, Perryman WL, Hinke JT, Krause DJ, Hann NA, Gardner S, LeRoi DJ (2015) A small unmanned aerial system for estimating abundance and size of Antarctic predators. Polar Biol 38:619–630
Grenzdörffer GJ (2013) UAS-based automatic bird count of a common gull colony. Int Arch Photogramm Remote Sens XL:4–6
Hodgson A, Kelly N, Peel D (2013) Unmanned aerial vehicles (UAVs) for surveying marine fauna: a dugong case study. PLoS One 8:1–15
Israel M (2012) A Uav-based roe deer fawn detection system. ISPRS – International Archives of the Photogrammetry, Remote Sens Spatial Inf Sci XXXVIII-1:51–55. Available fromhttp://www.int-arch-photogramm-remote-sens-spatial-inf-sci.net/XXXVIII-1-C22/51/2011/
Junda J, Greene E, Bird DM (2015) Proper flight technique for using a small rotary-winged drone aircraft to safely, quickly, and accurately survey raptor nests. J Unmanned Veh Syst 3:222–236. Available from http://www.nrcresearchpress.com/doi/10.1139/juvs-2015-0003
Junda JH, Greene E, Zazelenchuk D, Bird DM (2016) Nest defense behaviour of four raptor species (osprey, bald eagle, ferruginous hawk, and red-tailed hawk) to a novel aerial intruder – a small rotary-winged drone. J Unmanned Veh Syst 4:217–227. Available from http://www.nrcresearchpress.com/doi/10.1139/juvs-2016-0004
Körner F, Speck R, Göktoğan AH, Sukkarieh S (2010) Autonomous airborne wildlife tracking using radio signal strength. In: IEEE/RSJ 2010 international conference on Intelligent Robots and Systems, IROS 2010 – Conference Proceedings, pp 107–112
Koski W, Gamage G, Davis A, Mathews T, LeBlanc B, Ferguson S (2015) Evaluation of UAS for photographic re-identification of bowhead whales, Balaena mysticetus. J Unmanned Veh Syst 3:22–29
Linchant J, Lisein J, Semeki J, Lejeune P, Vermeulen C (2015) Are unmanned aircraft systems (UASs) the future of wildlife monitoring? A review of accomplishments and challenges. Mammal Rev 45:239–252
Mulero-Pázmány M, Negro JJ, Ferrer M (2014a) A low cost way for assessing bird risk hazards in power lines: fixed-wing small unmanned aircraft systems. J Unmanned Veh Syst 2:5–15

Mulero-Pázmány M, Stolper R, van Essen LD, Negro JJ, Sassen T (2014b) Remotely piloted aircraft systems as a rhinoceros anti-poaching tool in Africa. PLoS One 9:e83873. Available from http://www.pubmedcentral.nih.gov/articlerender.fcgi?artid=3885534&tool=pmcentrez&rendertype=abstract. Accessed 26 May 2014

Mulero-Pázmány M, Barasona JÁ, Acevedo P, Vicente J, Negro JJ (2015) Unmanned aircraft systems complement biologging in spatial ecology studies. Ecol Evol 5:4808–4818

Mulero-Pázmány M, Jenni-Eiermann S, Strebel N, Sattler T, Negro JJ et al (2017) Unmanned aircraft systems as a new source of disturbance for wildlife: a systematic review. Plos One 12(6):e0178448. https://doi.org/10.1371/journal.pone.0178448

Poole A (1989) Ospreys: a natural and unnatural history. Cambridge University Press, Cambridge

Posch A, Sukkarieh S (2009) UAV based search for a radio tagged animal using particle filters. In: Australasian Conference on Robotics and Automation ACRA. Available from http://www.araa.asn.au/acra/acra2009/papers/pap136s1.pdf

Potapov ER, Utekhina IG, McGrady MJ, Rimlinger D (2013) Usage of UAV for surveying Steller's sea eagle nests. Raptors Conserv 27:253–260

Rodríguez A, Negro JJ, Mulero M, Rodríguez C, Hernández-Pliego J, Bustamante J (2012) The eye in the sky: combined use of unmanned aerial systems and GPS data loggers for ecological research and conservation of small birds. PLoS One 7:e50336. Available from http://dx.plos.org/10.1371/journal.pone.0050336

Rümmler M-C, Mustafa O, Maercker J, Peter H-U, Esefeld J (2015) Measuring the influence of unmanned aerial vehicles on Adélie penguins. Polar Biol:1–6. Available from http://link.springer.com/10.1007/s00300-015-1838-1

Sardá-Palomera F, Bota G, Viñolo C, Pallarés O, Sazatornil V, Brotons L, Gomáriz S, Sardà F (2012) Fine-scale bird monitoring from light unmanned aircraft systems. Ibis 154:177–183. Available from http://doi.wiley.com/10.1111/j.1474-919X.2011.01177.x

Sardà-Palomera F, Bota G, Padilla N, Brotons L, Sardà F (2017) Unmanned aircraft systems to unravel spatial and temporal factors affecting dynamics of colony formation and nesting success in birds. J Avian Biol 48:1273–1280

Sasse DB (2003) Job-related mortality of wildlife workers in the United States, 1937-2000. Wildl Soc Bull 31:1015–1020

Shahbazi M, Théau J, Ménard P (2014) Recent applications of unmanned aerial imagery in natural resource management. GIScience Remote Sens 51:339–365

Soriano P, Caballero F, Ollero A (2005) RF-based particle filter localization for wildlife tracking by using an UAV. In: 4th international symposium on Robotics

van Andel AC, Wich SA, Boesch C, Koh LP, Robbins MM, Kelly J, Kuehl HS (2015) Locating chimpanzee nests and identifying fruiting trees with an unmanned aerial vehicle. Am J Primatol 77:1122–1134

van Blyenburgh P (2011) UAS: the global perspective 2011/2012. Blyenburgh & Co., Paris

Vergouw B, Nagel H, Bondt G, Custers B (2016) Drone technology: types, payloads, applications, frequency spectrum issues and future developments. In: Custers B (ed) The future of drone use. T.M.C. Asser Press, pp 21–45. Available from http://link.springer.com/10.1007/978-94-6265-132-6

Vermeulen C, Lejeune P, Lisein J, Sawadogo P, Bouché P (2013) Unmanned aerial survey of elephants. PLoS One 8. https://doi.org/10.1371/journal.pone.0054700

Weissensteiner MH, Poelstra JW, Wolf JBW (2015) Low-budget ready-to-fly unmanned aerial vehicles: an effective tool for evaluating the nesting status of canopy-breeding bird species. J Avian Biol 46:425–430. Available from http://doi.wiley.com/10.1111/jav.00619

Wilson AM, Barr J, Zagorski M (2017) The feasibility of counting songbirds using unmanned aerial vehicles. Auk 134:350–362. Available from http://www.bioone.org/doi/10.1642/AUK-16-216.1

Zhang C, Kovacs JM (2012) The application of small unmanned aerial systems for precision agriculture: a review. Precis Agric 13:693–712

Chapter 15
Conservation Genetics in Raptors

Begoña Martínez-Cruz and María Méndez Camarena

Introduction

Biodiversity is being depleted worldwide at unprecedented rates, due to direct or indirect human actions. These biodiversity losses affect all three fundamental and interrelated levels of biodiversity: ecosystems, species and genetic levels. Conservation genetics emerged in the 1970s as a discipline committed to preserve *genetic diversity* and minimize the risk of extinction of threatened populations by using genetic tools.

During these last decades, researchers in conservation of endangered species, including raptors, have made a strong effort in developing molecular markers that could be interrogated in their species of concern to understand the ecological and evolutionary processes going on, to describe the genetic patterns present and to unambiguously answer questions that could not be solved by other means. As technology improved and decreased costs, a succession of molecular markers was developed and applied, and among those, mitochondrial DNA fragments and microsatellites have been the markers of choice in the last two decades. The mitochondrial DNA narrates the maternal side of the story, while microsatellites are normally localized within the nuclear genome and tell the story of the whole population. In the last years, the decrease in prices is opening the door of genomics to the conservation of biodiversity. The use of all these molecular tools has led to important contributions in conservation at the individual, population and species level of raptors that we review in this chapter, with a special focus on endangered populations. It does not pretend to be an exhaustive review of all the work done in

B. Martínez-Cruz (✉)
Department of Integrative Ecology, Doñana Biological Station (EBD-CSIC), Seville, Spain
e-mail: begona.martinez.cruz@ebd.csic.es

M. Méndez Camarena
Department of Conservation Biology, Doñana Biological Station (EBD-CSIC), Seville, Spain

J. H. Sarasola et al. (eds.), *Birds of Prey*,
https://doi.org/10.1007/978-3-319-73745-4_15

conservation genetics of raptors, as this will largely exceed the number of pages dedicated. Otherwise, it is intended to give a general overview on how genetic tools may assist in the conservation of populations and species.

Molecular Markers and DNA

Molecular Markers in Conservation

Molecular markers reflect heritable differences in homologous DNA sequences among individuals due to nucleotide changes, rearrangements, base pair insertions or deletions, variation in the number of tandem repeats, etc. They are ubiquitous, stably inherited, show two or more *alleles* per marker and are detectable in all tissues at all ages. They differ in their mode of inheritance, either *dominant* or *codominant*. They are amplifiable by different techniques and are useful to answer different questions. Table 15.1 synthesizes the most common molecular markers used in conservation, with their advantages and inconveniences. Among them, microsatellites and the mitochondrial DNA control region fragment have been the most widely used markers in the last two decades.

Microsatellites

Microsatellites are short tandem repeats (STR) of one to six nucleotide motives in the DNA sequence. They are abundant and widespread in the genome of organisms, generally in non-coding regions. They are codominant and have a high *mutation* rate. The variations in the number of repetitions of the motives determine the different sizes of the fragments or alleles (*polymorphisms*). These polymorphisms can be easily amplified by *PCR* using specific fluorescently labelled *primers* designed to match at the flanking regions, which should be highly conserved to avoid amplification failures (*null alleles*). The alleles can be differentiated by high-resolution *electrophoresis*, normally in an automatic sequencer. The combination of different alleles at different loci found in an individual conforms an exclusive mark that can be used (1) at the individual level to identify individuals, migrants, hybrids, conduct paternity tests, etc.; (2) at the population level because the frequency of each allele is characteristic of a population and these frequencies can be used to define the assignment of individuals to those populations, detecting population structure, migration rates, *inbreeding*, etc.; and (3) at the species level because the type and frequency of each allele can also be used to assign individuals to species or detect cryptic species.

Since 2000 a total of 466 microsatellite markers have been developed and published for 29 raptor species worldwide belonging to 21 genus (Table 15.2), 308 of these microsatellites isolated in diurnal raptors and 158 in nocturnal raptors. Most of the studies have performed cross-amplifications of the microsatellites developed

Table 15.1 Molecular markers used in conservation

Marker	Description	Advantages	Inconvenients
Allozymes	The differences in amino acids change the charge in a protein, so different variants of the protein can be distinguished by gel electrophoresis	They are codominant. It is a simple, low-cost technique. It was the first marker highly used in natural populations, especially in plants	This technique masks information: silent substitutions in the nucleotides do not translate into a different protein, this variation remaining hidden
RFLPs (restriction fragment length polymorphism) (DNA fingerprinting)	Some restriction enzymes are able to recognize a determined series of four to six nucleotides in the DNA molecule, a restriction site, and cut the DNA at this level. The different lengths of resulting fragments can be separated by gel electrophoreses and posterior molecular hybridization to DNA sequences fluorescently or radioactively labelled. If a mutation or an indel occurs at one site, the enzyme will not cut there	Highly polymorphic markers, results in a pattern specific of individuals, a DNA *fingerprint*. These characteristics made them the markers of choice in forensic studies	The technique requires a high amount of not degraded DNA, and the process is long and complex; consequently, they were replaced by PCR-based approaches
RAPD (random amplified polymorphic DNA)	Is a PCR based on a random amplification of the genome using primers of around ten bases length. The primers will randomly match different complementary sequences of the genome and, therefore, generate different amplification products. Variantions in the genomes of the individuals determine whether there is an amplification product, by matching the primer with the sequence, or not. Fragments can be separated by electrophoresis. The resulting data are scored as presence-absence of polymorphism	The advantage is that no previous information is needed on the DNA sequence and that the number of loci amplified can be increased just adding primer pairs	They are dominant markers and the repeatability is not high
AFLPs (amplified fragment length polymorphism)	The DNA is cut with restriction enzymes followed by ligation to adaptors of a known sequence. A subset of these fragments is selected for amplification. Fragments can be separated by electrophoresis. The resulting data are scored as presence-absence of polymorphism	The repeatability is higher than for RAPDs	They are dominant markers

(continued)

Table 15.1 (continued)

Marker	Description	Advantages	Inconvenients
Microsatellites or STRs (short tandem repeats)	Microsatellites are repetitions in tandem of a small motif of one to six nucleotides length. They are abundant and widespread in the genome, mainly in non-coding regions. Using primers designed to match the flanking regions of the repeat motif, you can amplify the fragment containing the microsatellite. The difference in the number of repetitions of the motif constitutes the different alleles and therefore the polymorphism. They can be separated by electrophoresis, and the length of the alleles can be detected by automatic sequencers	Microsatellites are very polymorphic codominant markers. They are very short, between 100 and 300 bp, which makes them suitable to work with degraded DNA. They have been the markers of choice for the study of wild populations in the last two decades	Previous knowledge of the sequence is needed in order to design the primers
Amplification of single nucleotide polymorphisms (SNPs)	These are based on the polymorphism on one nucleotide in a sequence. Once this polymorphism has been detected by sequencing fragments of DNA, different kinds of techniques can be used to amplify this single base, either one by one (e.g. using TaqMan probes) or in battery (SNP chip)	These are codominant markers	The costs of elaboration of SNP chips are very high, out of reach for the majority of non-model species studies
Sequence fragments (pre-genomics era; Sanger)	The data is not a marker but the reading of the sequence of nucleotides in a DNA fragment. Fragments are PCR amplified and read a posteriori using an automatic sequencer	This is the most informative kind of data, given that all the information in the sequence of the DNA fragment is collected	The classical (Sanger) sequencing is expensive, limiting the amount of samples that can be processed
Sequencing (genomics era)	The data is the reading of the sequence of nucleotides, either from the whole genome or from extensive regions of DNA	The standardization of whole genomes or extensive regions of DNA amplification has been made possible, thanks to the development of the next-generation sequencing (NGS) technology. Although still expensive, prices are dropping and its use is now getting extended	Studies on population/ conservation genetics of non-model species, based on whole genomes, are still scarce

Table 15.2 Microsatellite markers developed in raptor species

Genus	Species	No. of loci[a]	N alleles	He[b]	References	Cross-species amplification[c]	Year published
Accipiter	*A. gentilis*	10	3–7	0,49–0,88	Topinka and May (2004)	*A. striatus*, *A. cooperii*, *Haliaeetus leucocephalus*, *B. jamaicensis*, *B. swainsoni*, *F. sparverius*	2004
	A. gentilis	13	3–7	0,49–0,88	Haughey et al. (2016)	*A. cooperii*, *A. striatus*	2016
Aquila	*A. adalberti*	18	2–7	0,29–0,77	Martínez-Cruz et al. (2002)	*A. heliaca*, *A. nipalensis*, *A. chrysaetos*, *Hieraaetus pennatus*, *Haliaeetus vociferoides*	2002
	A. heliaca	7	2–5	0,24–0,68	Busch et al. (2005)	*A. nipalensis*, *Haliaeetus albicilla*	2005
	A. audax	20	2–14	0,02–0,90	Austin et al. (2014)		2014
Athene	*A. cunicularia*	7	3–19	0,07–0,71	Korfanta et al. (2002)	None	2002
	A. cunicularia	11	5–25	0,52–0,94	Macías-Duarte et al. (2010)	None	2009
	A. cunicularia	18	2–11	0,04–0,87	Faircloth et al. (2010)	None	2010
Bubo	*B. bubo*	7	3–12	0,21–0,86	Isaksson and Tegelstrom (2002)	None	2002
	B. bubo	8	2–9	0,34–0,85	Kleven et al. (2013)	None	2013
	B. bubo	10	5–13	0,73–0,84	León-Ortega and González-Wangüemert (2015)	None	2015
Buteo	*B. buteo*	11	2–17	0,11–0,93	Johnson et al. (2005)	*A. chrysaetos* (two markers)	2005
	B. swainsoni	25	6–49	0,44–0,96	Hull et al. (2007)	*B. lineatus*, *B. jamaicensis*, *B. albicaudatus*, *B. platypterus*, *B. regalis*, *B. lagopus*, *B. galapagoensis*	2007

(continued)

Table 15.2 (continued)

Genus	Species	No. of loci[a]	N alleles	He[b]	References	Cross-species amplification[c]	Year published
Circus	*C.* pigargus	16	4–12	0,48–0,86	Janowski et al. (2014)	None	2014
Falco	*F. rusticolus*	8	2–7	0,11–0,73	Nesje and Roed (2000)	*F. peregrinus*, *F. columbarius*, *F. subbuteo*, *F. tinnunculus*, *A. chrysaetos*, *Pandion haliaetus*	2000
	F. peregrinus	12	2–11	0,12–0,89	Nesje et al. (2000)	*F. rusticolus*, *F. columbarius*, *F. subbuteo*, *F. tinnunculus*, *A. chrysaetos*, *Pandion haliaetus*	2000
	F. naumanni	2	23–172	0,77–0,99	Ortego et al. (2007)	None	2007
	F. naumanni	14	3–20	0,29–0,92	Padilla et al. (2009)	*F.* tinnunculus	2009
Gymnogyps	*G. californianus*	17	2–5	0,16–0,67	Romanov et al. (2009)	None	2009
Gypaetus	*G. barbatus*	14	2–13	0,07–0,90	Gautschi et al. (2000)	*Neophron percnopterus*, *Aegypius monachus*, *Gyps fulvus*	2000
Gyps	*G. fulvus*	5	2–9	0,39–0,87	Mira et al. (2002)	*Hieraaetus fasciatus*	2002
Haliaeetus	*H. albicilla*	14	2–8	0,03–0,81	Hailer et al. (2005)	None	2005
	H. vociferoides	8	2–4	0,05–0,66	Tingay et al. (2007)	*Haliaeetus vocifer*, *Haliaeetus albicilla*	2007
	H. leucocephalus	3	2–3	0,22–0,51	Tingay et al. (2007)	*Haliaeetus vocifer*, *Haliaeetus albicilla*	2007
Harpia	*H. harpyja*	10	2–6	0,05–0,76	Coser et al. (2014)	*Morphnus guianensis*	2014
Hieraaetus	H. fasciatus	15	2–8	0,10–0,86	Mira et al. (2005)	None	2005
Neophron	*N. percnopterus*	18	2–11	0,03–0,89	Agudo et al. (2008)	*N. percnopterus percnopterus*, *N. percnopterus majorensis*	2008

(continued)

Table 15.2 (continued)

Genus	Species	No. of loci[a]	N alleles	He[b]	References	Cross-species amplification[c]	Year published
Ninox	*N. strenua*	14	2–8	0,21–0,82	Hogan et al. (2007)	None	2007
Otus	*Otus elegans*	12	5–25	0,65–0,90	Hsu et al. (2003)	None	2003
	Otus elegans	6	5–8	0,62–0,76	Hsu et al. (2006a, b)	*Ninox scutulata, Otus lettia, O. spilocephalus, O. sunia*	2006
Pandion	*P. haliaetus*	13	3–10	0,36–0,88	Dawson et al. (2015)	None	2015
Spilornis	*S. cheela*	24	2–8	0,11–0,84	Hsu et al. (2013)	None	2013
Spizaetus	*S. nipalensis*	11	2–6	0,27–0,80	Hirai and Yamazaki (2010)	None	2010
Strix	*S. occidentalis*	7	6–16	0,70–0,84	Thode et al. (2002)	None	2002
	S. nebulosa	37	2–7	0,13–0,86	Hull et al. (2008a, b)	*S. occidentalis, S. varia*	2008
Tyto	*T. alba*	21	2–31	*0,16–0,95*	Burri et al. (2008)	None	2008

[a]Only polymorphic markers are reported
[b]He reported is obtained in the individuals genotyped in the marker released paper
[c]We report the cross species amplification performed in the marker released paper

in closely related species. Due to *ascertainment bias*, we expect lower levels of diversity when using microsatellites in a different species from the one they were design for. Despite this lower diversity, they have proven to be variable enough to successfully enable population and conservation genetic studies for many raptor species and not only the focal one. Most of the studies have been focused on threatened and/or small populations, most of them living in the Palaearctic or in the Nearctic (Table 15.2).

Mitochondrial DNA

Mitochondrial DNA (mtDNA hereafter) has been profusely used to study the genetic diversity of species (in this book Chap. 1). mtDNA is a maternally inherited genome enclosed in the mitochondria and does not recombine, which means that the entire molecule works as a single locus. mtDNA genomes are abundant: while there is only a nuclear genome, there are hundreds of mitochondrial genomes in each cell. This implies that when DNA is extracted from a sample, there are much more mtDNA genomes than nuclear DNA molecules. Consequently, it allows the use of minute amounts of DNA to obtain good mtDNA sequencing results, making the mitochondria a very interesting marker when dealing with non-invasive or historical/ancient samples. The existence of well-conserved genes in the mitochondria across species allowed the design of universal primers (Kocher et al. 1989) that amplify the entire control region in most vertebrates. Otherwise, the gene order of the mtDNA in birds is different (Desjardins and Morais 1990), a characteristic that has been confirmed in raptors (i.e. (Roques et al. 2004). The majority of conservation genetic studies have sequenced the mitochondrial control region or D-loop, as it has been shown to be the most variable region of the mtDNA. The mtDNA has proven to be a useful marker in the description of population structure, the detection of hybridization and *introgression*, species delimitation, etc.

The Transition to Conservation Genomics

Conservation genetic studies using "classical" markers have two main limitations. First, most of the conservation genetics studies have made use of neutral markers (as microsatellites) under the assumption that they represented also the genetic diversity at non-neutral regions. Second, only a small number of markers or small sequences are used, representing a minute fraction of the genome. This situation is changing as we have entered the era of genomes. Next-generation sequencing (NGS) technologies developed at the dawn of the twenty-first century have marked "a now and then" in genetic studies. NGS allows the rapid collection of enormous amounts of data, including neutral and non-neutral loci, either across long, unlinked portions of the genome or whole genomes. This new technology enables the detection and characterization of functional genes directly involved in the adaptation to new conditions. Elevated economical costs made that this technology was reserved to human and model species studies, but at present prices have dropped and are still dropping, and the use of NGS in conservation genetics studies of non-model species is already a reality, including raptors (Martínez-Cruz et al. in preparation).

At present the *reference genomes* of several raptors have been assembled de novo (Table 15.3). A reference genome is necessary as mould against which the *reads* from *re-sequenced* individuals would be aligned. Building a genome de novo is the most expensive and arduous step to perform any population genomics study. Fortunately, although with limitations, the reference genome from a closely related species can be used to answer important questions relevant in conservation. One of the most common approaches is to use the information in the genome to design a battery of SNPs (single nucleotide polymorphism) (see Table 15.1). Those are single nucleotide changes in the DNA sequence and are abundant along the genome. The combined use of hundreds to hundreds of thousands SNPs is a powerful tool to answer a high number of questions in population and conservation genetics.

Table 15.3 Complete genomes of raptors published at present

Species	Genome size (Gb)	Depth	No. of scaffolds	Total length of scaffolds (Gb)	N50 contig (kb)	N50 scaffold (kb)	No. of protein coding genes	Mean length genes (kb)	Refs.	DOI
Bald eagle	1,4	88X	346419	1,26	10	670	16526	19	Gilbert et al. (2014a)	https://doi.org/10.5524/101040
Barn owl	1,6	27X		1,14	13	51	13613	13,8	Gilbert et al. (2014b)	https://doi.org/10.5524/101039
Peregrine falcon	1,2	107X	7021		28,6	3890	16263		Bruford et al. (2014)	https://doi.org/10.5524/101006
Turkey vulture		25X		1,17	12	35	13534	10,8	Gilbert et al. (2014c)	https://doi.org/10.5524/101021
White-tailed eagle		26X		1,14	20	56	13831	14,2	Gilbert et al. (2014d)	https://doi.org/10.5524/101027
Golden eagle	1,28	39X	42,926			1747	16,571	25	Doyle et al. (2014)	https://doi.org/10.1371/journal.pone.0095599

Where to Obtain the DNA for Studies on Raptors

Fresh Samples

DNA can be extracted from almost all kinds of tissues. In birds, in general, blood is the preferred source to obtain DNA from living specimens. Thanks to the medium to big body size of raptors species, blood can be easily extracted in the field from the wing vein, the jugular or the leg. It can be stored at room temperature in ethanol or in a standard buffer for transportation until DNA is extracted in the laboratory. The erythrocytes are the most abundant cell type in the blood. In birds, contrary to mammals, the erythrocytes maintain the nucleus in the maturity; thus a small amount of blood yields good amounts of DNA. Plucked feathers contain also enough blood and tissue to obtain moderate to high DNA quantity. Either bleeding or feather plucking are invasive collection methods, as they require capturing the bird plus extracting the blood with a syringe or tear tissue. Other preferred source of DNA is muscle that in the case of raptors is obtained only when the individual is dead (in other phyla, like mammals, muscle samples are sometimes obtained by biopsy from living individuals, or in reptiles a small piece of the tail or a finger is often sampled without killing the animal). Otherwise given that conservation genetics deals with threatened populations, there is an obvious interest in applying alternative, non-invasive methods for obtaining DNA from biological remnants that individuals leave behind. Stuffed individuals from museums or private collections represent an invaluable source of information as well (Casas-Marce et al. 2012; Shaffer et al. 1998; Tingley and Beissinger 2009).

DNA from Non-invasive Sampling

Conservation genetics studies have a special concern about using non-invasive sampling methods in order to avoid any perturbation to the individuals in endangered or elusive species. In the case of birds, most researchers are putting their effort in non-invasively collected shed feathers ((Gausterer et al. 2013; Horvath et al. 2005; Segelbacher 2002) and more), although other sources are also useful (e.g. eggshells (Chemnick et al. 2000)). Shed feathers can be easily collected on the ground, on perching sites, nests, etc., without interfering with the individual. At this step, contamination should be avoided by using plastic gloves at collection. Feathers can be stored separately in plastic or paper bags until their arrival to the laboratory. The simplicity of the procedure allows the sampling of many individuals with a small effort and makes easy to incorporate it to the routine of bird monitoring professionals in nature. Of course, a difficult access to sampling areas can make this simple procedure not so simple and even impossible for certain species.

Typically DNA is extracted from the tip and/or the clot (Horvath et al. 2005), discarding the rest of the feather. The main problem derived from these non-invasively collected samples is the small amount of DNA present in them. Additionally, DNA is degraded and divided in chunks due to natural processes of degradation. Low amounts of degraded DNA may result in technical problems in the PCR amplification step, which may fail. As already mentioned, higher amounts of mtDNA in the samples make this genome easier to amplify than the nuclear. Additionally, amplifications with nuclear codominant markers as microsatellites or molecular sex determination may result in genotyping errors due to *allelic dropout*, and special measures must be taken to avoid erroneous results (Taberlet et al. 1996; Taberlet et al. 1999). Basically, those measures imply that the PCR amplifications are repeated a number of times until the probability of a genotyping error is minimized. This protocol of repetitions raises the price and time invested by several times, limiting the number of samples that can be processed. And still it is possible that some samples could not be genotyped.

Besides this, there is the danger of potential contamination, mainly linked to low amounts of DNA. To avoid it, in addition to the explained precautions taken in the field, these samples must be processed in a dedicated clean room, and the researcher must wear adequate clothing and gloves and use always positive and negative controls.

Despite all these technical issues, moulted, shed feathers are being used successfully in the monitoring of raptor populations (Rudnick et al. 2005), sex determination (Rudnick et al. 2005; Vili et al. 2009) and hybridization studies (Vali et al. 2010), among others.

DNA from Museum Individuals

Conservation biologists have usually assumed that low levels of genetic diversity can result either from recent population declines or from recent interruption in gene flow. These studies commonly analyse contemporary genetic patterns to make inferences about the conservation status of declining populations assuming that past genetic diversity was always higher as compared to current diversity. Many times levels were compared to those in a closely related, non-endangered species or populations. This assumption has important implications for understanding the effects of genetic erosion on populations and species viability and the conservation measures needed.

Historical samples and the advances on molecular techniques make possible a direct assessment of changes over time and provide a window into evolution (Holmes et al. 2016) and answers on how populations respond to environmental changes. Additionally, in some cases museum collections may be the simplest way of accessing to the information, if sampling in the wild at present becomes hampered by too expensive expeditions, the political situation of a country or even bureaucracy. In the extreme case of extinct species or populations, museum collections may represent the only source for obtaining any genetic information.

The extension of stuffed animal collections in some of the biggest museums in the world represents an enormous potential for the study of multiple species. Raptors are not the exception and researchers are making use of this potential. Diverse studies with historical specimens of raptors have extracted DNA from the skin, feathers or footpad (D'Elia et al. 2016; Godoy et al. 2004; Martinez-Cruz et al. 2007, Méndez et al. under review). The technical caveats are similar to those encountered with non-invasive sampling. The amount of DNA from stuffed individuals is scarce, and DNA quality is low due to the natural postmortem process. Therefore, the applicability has been limited to certain genetic markers. Nevertheless, the recent advances in new-generation sequencing technologies are allowing the sequencing of complete **genomes** from museum specimens although still at high costs not affordable for most budgets. However the rapid and pronounced decrease in prices that these technologies have experienced since the first human genome was sequenced suggests that they will soon become accessible for more modest projects. As for the case of non-invasive material, special care must be taken when handling these samples from stuffed individuals in order to avoid contamination. This implies working in exclusively dedicated laboratories using strict protocols of handling, contamination control and authentication of the data.

In spite of all these technical limitations, studies conducted on museum specimens have provided very useful insights to improve management and conservation plans on raptor populations. In this sense, researchers have been able to evaluate the genetic impact of population reduction (D'Elia et al. 2016; Martinez-Cruz et al. 2007, Méndez et al. under review), disentangle the evolutionary history of species (Martinez-Cruz et al. 2007; Nebel et al. 2015), assess on reintroductions (D'Elia et al. 2016; Godoy et al. 2004), detect hybridization (Nittinger et al. 2007) and make clear recommendations for population reinforcement in declining areas by detecting candidate source populations (Méndez et al. under review).

Applications in Raptor's Conservation

Gender and Individual Identification

Although most raptor species show reversed size dimorphism, determining the sex by morphometric traits is many times difficult, as many species are apparently *monomorphic*. This difficulty is absolute when we are dealing with remains such as feathers. Molecular techniques overcome these limitations with a simple PCR *assay*, based on the fact that raptor females are *heterogametic* (ZW), while males are not (ZZ).

Gender determination is necessary to understand multiple aspects of species biology and to take management actions. Molecular sexing helped to point at the scarcity of males as the main factor slowing the recovery of the Norfolk Island Boobook owl (Double and Olsen 1997). In captive breeding an early gender determination of chicks allows the adequate management of the captive population, while unambiguously assigning the sex avoids pairing of two birds of the same sex. Those reasons are behind the systematic molecular sexing in the condor recovery programme (Chemnick et al. 2000).

Molecular determination of gender in raptors as vultures, eagles and others has been used increasingly over the past years (Garcia et al. 2009; Chang et al. 2008; Ellegren 1996; Fridolfsson and Ellegren 1999; Ghorpade et al. 2012; Griffiths et al. 1996; Griffiths et al. 1998; Ito et al. 2003; Katzner et al. 2014; Rudnick et al. 2005; Sacchi et al. 2004). It is reliable, easy, quick and a relatively inexpensive method and can be performed even in situ in the field, with the help of a voltage inverter to adapt a thermo-block to a car lighter (Centeno-Cuadros et al. 2017).

In the wild, the application of genetic tools to the unambiguous identification of individuals from remains allows the investigation of important aspects in the demography, ecology and evolution of species, as well as the elaboration of population censuses, the follow-up of individuals in space, the monitoring of populations and also the estimation of population sizes by using a strategy analogous to capture-recapture methods among others. For example, Bulut et al. (2016) used non-invasive sampling to evaluate the space use of a white-tailed eagle population over 3 years, revealing it was closer to a semicolonial behaviour than to the classical territorial and colonial model expected. The analyses of shed feathers from consecutive years showed surprisingly high rates of loss and turnover in adult breeding populations of eastern imperial eagles in Kazakhstan and Central Europe (Rudnick et al. 2005; Vili et al. 2013). The demography and ecology of nonbreeding eastern imperial eagle population in Kazakhstan were unveiled using shed feathers (Rudnick et al. 2008). By sampling in roosting sites, they found that eastern imperial eagles shared roost with other raptors and that those roosts are big and found that reserve in Kazakhstan may serve as a critical refugee for pre-adults and itinerant floaters (Rudnick et al. 2008). Shed feathers from adults, in addition to blood samples from chicks in the nest, were also employed to study site fidelity and dispersal in gyrfalcon populations in Alaska (Booms et al. 2011). This study revealed high territorial but low nest-site fidelity (22%) and a mean annual turnover rate of 20%.

Parentage and Kinship Information in Conservation Genetics

Information on parentage is essential to study the impact of inbreeding, to verify pedigrees used in genetic management of threatened species and to determine the effective size of populations. Frequently, parentage cannot be determined from direct behavioural observations in species where females copulate with many males during their fertile periods, or in secretive species. Molecular markers can assist effectively in this task. If mother, offspring and putative fathers are genotyped for enough number of loci, positive paternity assignments can be made with high probabilities (Fleischer 1996). Otherwise, knowledge of mating systems can be used to assess the conservation risks in management strategies.

Molecular methods have evidenced that birds are rarely monogamous (Griffith et al. 2002). However, in raptors extra-pair paternity (EPP) levels differ among species. Relatively low levels of EPP were found in northern goshawks (Gavin et al. 1998), lesser kestrel (Alcaide et al. 2005), griffon vulture (Le Gouar et al. 2011) or

Swainson's hawks in California (Briggs and Collopy 2012). It was absent or nearly absent in black vulture (Decker et al. 1993), in the flammulated owls (Arsenault et al. 2002), in the eastern imperial eagle (*Aquila heliaca*) (Rudnick et al. 2005), in the Lanus scops owl (Hsu et al. 2006a), in the tawny owl (Saladin et al. 2007), among female boreal owls (Koopman et al. 2007), in the powerful owl (Hogan and Cooke 2010) and in the Ridgway's hawk (Woolaver et al. 2013). Determination of paternity also gives insight in how new ecological conditions affect the reproductive behaviour in raptors. For instance, colonization of urban areas seems not to be affecting mating fidelity in burrowing owls (Rodriguez-Martinez et al. 2014) and peregrine falcons (*Falco peregrinus*) (Caballero et al. 2016), but high levels of EPP found in a Cooper's hawk population and the mating behaviour in spotted towhees were suggested to respond to urban, human-altered environments (Rosenfield et al. 2015; Smith et al. 2016).

Due to the negative consequences for the viability and the fitness of individuals, inbreeding avoidance is key in the conservation of wildlife. It requires a good knowledge of *kin relationships* among individuals in a population, in order to apply adequate management policies both in captivity and in the wild. The most common management strategy to maintain the highest amount of genetic diversity in captive breeding prioritizes the reproduction between individuals with the lowest kinship. Kinship between pairs of individuals is calculated based on the pedigrees, with the assumption that the founders are not related. Researchers and managers use the subsequent pedigree to assess the genetic status of populations and select the individuals to be reintroduced or translocated.

A correct use of the information gathered from molecular pedigrees in the bearded vulture (*Gypaetus barbatus*) captive breeding programme resulted in a smaller kinship coefficient among the captive pairs compared to random expectations, suggesting that inbreeding had been successfully avoided in the captive breeding population (Loercher et al. 2013). In the case of the California condor (*Gymnogyps californianus*), the reconstruction of pedigrees in the captive population allowed the researchers to minimize the incidence of the lethal chondrodystrophy allele present in the population (Ralls and Ballou 2004; Ralls et al. 2000).

Identifying Populations at Risk of Genetic Erosion

When a population becomes small due to a demographic decline and/or fragmentation, the level of inbreeding increases, and *genetic drift*, which is a stochastic process, becomes the predominant evolutionary force shaping the genetic pool of the population, at the expenses of natural selection. Inbreeding and genetic drift lead to loss of genetic variation, the raw material for evolution, thus reducing the population's ability to cope with changing environmental conditions. Genetic drift may cause the loss of long-term acquired genetic variants responsible of local adaptations. At the individual level, the genetic consequences of diversity loss are the increase in inbreeding that occurs even if mating is random, because the mean kinship among individuals rises and the probability for close relatives to cross is high. As individuals become less heterozygous, there is an accumulation of deleterious alleles in homozygosity that might affect negatively the individual fitness (*inbreeding depression*). For all these reasons, the estimation levels of genetic diversity in the populations, the evaluation of risks derived from its decline and the proposal of management actions in order to increase it are of critical importance in conservation.

On the other hand, the fragmentation of populations leads to the accumulation of differentiations among the resulting subpopulations due to the differential action of drift in each fragment and the decrease of *gene flow* among them. This process results in a heterogeneous distribution of the genetic diversity, giving rise to a genetic structure. The degree of connectivity among subpopulations depends on the migration rate, which can buffer the detrimental effects of genetic diversity loss and accumulation of inbreeding in each fragment. Molecular markers can be used to describe the genetic structure of populations and use this to estimate the levels of gene flow or the time since isolation.

As said before, the more reliable way to evaluate how drift has affected a population is contrasting the genetic diversity at present with this in the historical population before the decline. This approach is not always possible due to the difficulty of accessing the appropriate samples, which come generally from museums. In spite of this difficulty, there are nice examples of this approach in the scientific literature. In an extreme case of population endangerment, the Mauritius kestrel (*Falco punctatus*) recovered from a single breeding pair without the addition of new genetic variation, after years of extremely low population size (Groombridge et al. 2000). As expected, the authors found the genetic variation extant in the population predating the decline to be very high with respect to the variation at present and comparable to that in sister nonthreatened continental species. Compared to the historical population, the Mauritius kestrel had lost more than 50% of its genetic diversity. Another interesting example is the highly endangered Spanish imperial eagle (*Aquila adalberti*), which suffered a strong population decline and fragmentation during the twentieth century that brought the species to the brink of extinction. Mitochondrial but not nuclear genetic diversity was lower with respect to the historical population, and also to its sister species, the eastern imperial eagle (*A. heliaca*), due to shifts in allelic frequencies (Martinez-Cruz et al. 2004, 2007). Otherwise, the comparison with the historical population showed that the current pattern of genetic structure, with the population of Doñana National Park genetically differentiated from the rest, is the direct result of the decline and fragmentation in the last century (Martinez-Cruz et al. 2007). At present there is an ongoing reintroduction programme that tries to establish a stepping stone population to reconnect the Doñana population with the rest, as it was in historical times. In the same way, habitat fragmentation and deterioration is one of the factors affecting the decline of the harpy eagle (*Harpia harpyja*) in Central and South America (Vargas et al. 2006). Banhos et al. (2016) studied historical and current samples to investigate how this dynamic had affected the genetic diversity of populations of this near-threatened species in Brazilian tropical forests. Their results showed that the genetic diversity of the harpy eagle had decreased with intense deforestation, in contrast with a previous study showing high levels of mitochondrial genetic diversity at a bigger geographical scale (Lerner et al. 2009). Otherwise the authors did not suggest any genetic action as they stated that conservation measures leading to the preservation of the habitat would help to preserve the genetic diversity.

Evaluating whether drift had shaped not only the neutral but also the adaptive variation is important to understand the real impact of losing genetic variation in small populations (Hedrick and Kalinowski 2000). Functional loci such as genes of the *major histocompatibility complex (MHC)* have become widely used models for the investigation of *adaptive variation* and natural selection because they constitute an essential component of the immune system of individuals. They have a crucial role in fighting off pathogens, which is directly related to individual fitness and survival. Agudo et al. (2011) found evidences of a major role of genetic drift in shaping the spectrum and frequencies of MHC alleles in isolated populations of the Egyptian vulture (*Neophron percnopterus*). However, they also found that the negative effects of drift might be counteracted by an evolutionary mechanism of coevolution of gene duplicates. In any case, those populations showed a positive correlation between functional MHC genetic diversity and breeding success, suggesting a negative effect of inbreeding (Agudo et al. 2012). The study of the genetic diversity in the MHC of the endangered Blakiston's fish owl (*Bubo blakistoni*) population showed low genetic diversity levels and high levels of population differentiation and declining levels of MHC diversity with time in several populations. This pattern was the result of the past decline and fragmentation of the populations (Kohyama et al. 2015).

Estimate the Size of the Effective Population

Low levels of diversity imply low *effective population size* (N_e). In small populations the N_e is a critical piece of information for its correct management as the rate of genetic drift is proportional to this parameter and not to the census size (N_c). Minimising drift and the consequent losses of genetic diversity requires maintaining a large N_e. Genetic tools allow the estimation of N_e, and results show that it is usually low, with a mean N_e/N_c ratio of 0.32 in birds (Frankham 1995). Additionally genetic methods allow the inference of historical patterns of population expansion or decline, even if the historical population is not sampled. This historical perspective helps to understand the strength of drift suffered by the population, which can be useful for conservation efforts even if the inference goes back to evolutionary times. The demographic history of the formerly critically endangered Seychelles kestrel (*Falco araea*) was investigated using historical and current samples (Groombridge et al. 2009). Their results showed that the species, now common, had gone through a severe genetic bottleneck by the middle of the twentieth century, declining from a $N_e = 378$ individuals to as low as $N_e = 8$ at present. Although now the species counts hundreds individuals, the N_e estimation reflects the period during which the population was at very low sizes. These results revealed that the Seychelles kestrel has suffered a decline nearly as severe as its relative the Mauritius kestrel (Groombridge et al. 2009). Similarly, with a census size around 200 birds, N_e of for the Canary Islands Egyptian vulture population has been estimated in approximately 38 individuals (Agudo et al. 2010). Researchers showed that the species had suffered a strong decline around 2500 years ago, coincident with the arrival of humans and cattle to the island that would have made possible the establishment of a group of Egyptian vulture in the island. The historical N_e estimate of more than 21,000 individuals would reflect that of the ancestral population before the split (Agudo et al. 2010). The most recent census of the highly endangered Madagascar fish eagle (*Haliaeetus vociferoides*) showed a population of 222 adults, and the N_e estimate was 24 individuals, and in comparison with other *Haliaeetus* species, levels of diversity were extremely low (Johnson et al. 2009). The authors found that the population showed signals of a strong decline that occurred hundreds to thousands years ago, suggesting that the low levels of diversity were the result of an ancient bottleneck and that the population have had low genetic diversity since then. As a consequence, they put the conservation priority in the maintenance of habitat requirements and the reduction of anthropic pressures more than in the genetic issues. A higher rate N_e/N_c was found in the British Golden eagle (*Aquila chrysaetos*), with a N_e estimated in 437 individuals in a population of 422 breeding pairs (Bourke et al. 2010). As in the case of the Egyptian vulture, the split from an ancestral population and the colonization of an island by a group (a founder effect) 22 ky ago was the interpretation given by the authors for the strong population decline. As in the case of the Madagascar fish eagle, the fact that the population could have been viable at low levels of diversity for such a long period of time made the researchers suggest limited need to take actions to increase the levels of genetic diversity. The critically endangered hawk (*Buteo ridgwayi*), endemic to the island of Hispaniola in the Caribbean, presents relatively high level of heterozygosity, inbreeding and relatedness within the population, indicative of a bottleneck (Hull et al. 2008a; Woolaver et al. 2013). Together with habitat destruction and anthropic pressures, these genetic factors represented a further threat to the species. They suggested translocations and the initiation of a captive population with the most divergent individuals as founders, as a safeguard of the remaining genetic diversity. On the contrary, the history of the white-tailed eagle (*Haliaeetus albicilla*) shows a different pattern. In spite of the dramatic declines experienced during the last century, the populations have retained significant levels of genetic diversity after recovery (Hailer et al. 2006; Ponnikas et al. 2013). Interestingly, long generation time and a high philopatry could have acted as a buffer against loss of genetic diversity, leading to a shorter effective time of the experienced bottleneck.

Relating Genetic Erosion to Human Pressures: Landscape Genetics

Landscape genetics is an approach for understanding how landscape characteristics determines gene flow and local adaptation (Manel and Holderegger 2013). This understanding is crucial, not only for improving ecological knowledge but also for managing properly the genetic diversity of threatened and endangered populations. This new approaches analyse spatial genetic data without requiring discrete populations in advance. Moreover, the analysis involves the detection of genetic discontinuities and the correlation of these with landscape features.

Several examples of the applicability of landscape genetics have been done in birds (e.g. Mendez et al. 2014); unfortunately there are less in raptors, probably due to their high mobility. Alcaide et al. (2009a, 2009b) integrated capture-recapture and molecular data to understand the consequences of habitat fragmentation on population connectivity in lesser kestrels (*Falco naumanni*) in Western Europe. Habitat fragmentation has increased genetic differentiation between lesser kestrel populations. Their genetic data challenge traditional assumptions associating philopatry with the emergence of fine-scale genetic structuring. Even in high mobile species, restricted dispersal can generate weak but significant genetic differentiation, raising concerns about genetic divergence in small and isolated demes. Di Maggio et al. (2015) analysed the role of human impact in promoting or disrupting the genetic structure in lesser kestrels in two subpopulations in Sicily. Colony features as size and human disturbance negatively affected the relatedness among chicks form different nest. The core of the agricultural plain, with the higher level of human disturbance, maintained colonies with the less related individuals. In contrast, the marginal colonies, less disturbed, maintained more-related individuals. Given the high philopatry of this species, the results are consistent with the disruption of colony fidelity related to the an intensification of agricultural practices.

Hybridization

Habitat changes due to climate change and other anthropological pressures have resulted in distribution changes leading to contact and interbreeding of many species or populations formerly separated. Hybridization and posterior *introgression* may result in all individuals being hybrids to different degrees, which would imply a de facto extinction of the indigenous groups, and so are of major concern in the conservation of species. Hybrids could be morphologically very similar to one or the other species, making it difficult to study the processes going on and to take decisions on the management strategy (i.e. the extirpation of non-pure individuals from the wild). Genetic markers have shown to be the most accurate tools to investigate the boundaries between different taxonomic entities (including the degree of hybridization or introgression between groups).

Northern spotted owls hybridize with expanding populations of barred owls, producing viable offspring although with potentially reduced fitness (Haig et al. 2004). Several molecular markers were developed specifically to detect hybrids and to estimate the extension of the ongoing hybridization process (Funk et al. 2007; Haig et al. 2004) that can be used in the protection strategy of the endangered spotted owl (Haig et al. 2016). Vali et al. (2010) applied molecular markers to study the hybridization and introgression between the vulnerable greater spotted eagle and the common lesser spotted eagle sympatric populations in Eastern Europe. They showed that hybridization was occurring in all studied populations and revealed that it was sex biased (favouring crosses between lesser spotted females and greater spotted males) and introgression had a preferential sense, from greater to lesser spotted eagles. Researchers could verify that in some cases hybridization preceded the disappearance of greater spotted eagle populations, showing the potential danger of this process for the conservation of the vulnerable greater spotted eagle.

Units for Conservation

The delineation of conservation units is crucial for the correct management of populations. The scientific literature harbours different definitions of what should be considered a conservation unit, but the most commonly used and accepted are evolutionary significant units (ESUs) and management units (MUs). ESUs are defined by significant genetic differentiation in the nuclear genome and reciprocal monophyly for the mtDNA (Moritz 1999). Ideally, the definition of an ESU must take into account the historical situation (Crandall et al. 2000). MUs are a less stringent definition, being demographically independent populations evidenced in divergence of allele frequencies (Moritz 1999). Defining ESUs and MUs is of paramount importance, as different units within populations may be adapted to different conditions, and managing them as a single unit could result in losses of diversity important for present and future adaptive evolutionary potential. Conversely, a separate management of populations that compound a single unit would lead, for instance, to unnecessary costs. In an attempt to verify the classification of the Osprey (*Pandion haliaetus*) into subspecies, Monti et al. (2015) identified four different ESUs worldwide that did not overlap exactly with the four previously recognized subspecies. Concerned about the abundant reintroduction programmes of this species (up to 25 in Europe and the United States in the last decades as indicated by the authors), they recommended a separated management of the four units, avoiding translocation/reintroduction of individuals from one unit to the other. The divergent origin of bearded vulture (*Gypaetus barbatus*) captive population, a mix of individuals from Asia and the Pyrenees, made (Gautschi 2001) warn about a potential outbreeding depression in case of crossbreeding of the two populations. Although phylogenetic analyses revealed the existence of two differentiated mitochondrial lineages, one in the eastern and the other in the western populations, evidences of old secondary contact and ecological exchangeability suggested the management of the species as a single unit (Godoy et al. 2004). Indeed, at present management plans of the bearded vulture alpine population include creating a corridor to connect the Alpine and the Pyrenean population (Hegglin et al. 2010). Another big raptor, the Black vulture, shows also two differentiated genetic lineages in Europe, one in the Iberian Peninsula and a second in the Balkans (Poulakakis et al. 2008). Lack of data made that recent and historical ecological exchangeability could not be ruled out, and thus, as a preventive measure, the separate management in two management units was advised.

Genetic Monitoring in Conservations Projects

Many populations have recovered from severe declines, thanks to the conservation measures taken, or from reintroductions in places where they were extirpated. The genetic follow up of those populations is however scarce, although it may be necessary to help making important management decisions and ensure the long-term viability of the population. The primary goal of any recovery effort is to produce a long-term self-sustaining population, and these genetic measures provide a metric by which we can monitor our progress and assist taking management decisions.

For instance, a genetic survey of the reintroduced Alpine bearded vulture (*Gypaetus barbatus*) population revealed that the genomes of founders were not equally represented in the wild and suggested that genetic diversity was low (Loercher et al. 2013). The authors recommended additional releases from the captive population in order to increase those parameters. Ewing et al. (2008) conducted a study on the genetic health of a reintroduced population of Mauritius kestrel (*Falco punctatus*). They verified high rate of both genetic diversity losses and increase in inbreeding that could affect the viability of the population and thus claimed for conservation strategies leading to reduce the impact of *genetic erosion.* A genetic monitoring on recovered peregrine falcon populations (*Falco peregrinus*) in North America revealed the success of the recovery programmes, showing a stable or growing population with no indication of recent decline in genetic diversity, and highlighted the importance of conducting temporal sampling of the populations to do a follow-up of the recovery (Johnson et al. 2010). Le Gouar et al. (2008) analysed genetic diversity and structure from native and reintroduced Griffon vulture (*Gyps fulvus*) populations successfully restored in Southern Europe. They found that genetic diversity was similar in all populations, but they detected signs of recent isolation in one native population. The reintroduced population showed a high immigration rate that limited the losses of genetic diversity. Johnson et al. (2008) analysed populations of oriental white-backed vultures (*Gyps bengalensis*) in south Asia. These species have declined over 95% since the mid-1990s due to feeding on livestock carcasses treated with diclofenac, an anti-inflammatory fatal to Gyps vultures. They documented temporal genetic diversity in the remaining population, during 6 years in the last colony of Pakistan, and concluded that a much larger captive population size is required than currently maintained to prevent further loss of genetic diversity. They suggested that before this species is extinct in the wild, it is crucial that additional individuals are included in the captive population.

Conservation Genetics in Action: The California Condor

The story of the California condor, *Gymnogyps californianus*, is one of success in conservation: the dramatic decline and posterior recovery of an almost extinct species, thanks to strong commitment and drastic conservation actions. The California condor is one of the most threatened birds of prey in the world. The species is classified as critically endangered by the UICN red list of endangered species (Birdlife International 2016). The California condor is the only remnant of its genus, within the Cathartidae family. With a wingspan up to 3 m and a weight up to 12 kg, condors are the largest North American land birds. They are long-living birds, reaching up to 60 years. Condors are scavengers that eat a high amount of carrion, preferably from large mammals.

Adams and collaborators gave a good overview to the history and the genetics of the species (Adams and Villablanca (2007) and references therein). The California condor declined rapidly throughout its historic range during the nineteenth and twentieth centuries due to habitat fragmentation, direct persecution, specimen collection and lead poisoning. They inhabited most of the US territory before the arrival of Europeans, but by the middle of the twentieth century, its population was confined to Southern California. Conservation measures started in 1930 mainly focused to preserving the condor habitat, but the species continued its decline until only 22 condors remained in the world in 1982. A programme of captive breeding started in 1982, using eggs and chicks removed from the wild. However, after a severe crash in the already decimated remnant population, scientists and managers took the drastic decision of capturing all free individuals and integrated them into the captive breeding programme in a desperate attempt of saving the species from extinction. The trapping of the last individual in 1987, rendering the species extinct in the wild, was not exempt of controversy. Fourteen individuals founded the captive population, and the first reintroduction took place in 1992: eight individuals reared in captivity were released free in Southern California. The efforts have shown to be really successful, and in 2016 the population of California condors counted more than 250 individuals in the wild in California, Arizona, Utah at the United States and Baja California, in Mexico, and more than 150 more are in captivity ((Walters et al. 2010), California Condor Recovery Program https://www.fws.gov/cno/es/calcondor/Condor.cfm).

The genetic management of the California condor has an important role in this history and constitutes a good example for other threatened species of raptors, and vertebrates in general. Together with demographic, genetic information is taken into consideration into the final decisions about the number of individuals and which ones are released into the wild (Ralls and Ballou 2004; Ralls et al. 2000). In relation to genetic questions, two mandatory questions were addressed since the beginning of the programme: to ensure a correct sex determination of the individuals and to know as accurately as possible the genealogical relations among individuals.

Sex determination based on molecular techniques was of paramount importance for the managing of the individuals in captivity in a monomorphic species like the condor. The reason is very simple: in monomorphic species, with conspicuous sexual organs, it would be easy to put in the cage two individuals of the same sex to breeding. The time loss would be terrible and additionally frustrating. The first cell-based molecular method used to determine the sex of individuals was a long-lasting, laborious cytogenetic method, and only a limited number of samples could be processed simultaneously (Chemnick et al. 2000). Molecular sexing was a significant improvement for gender determination, demanding a much smaller sample amount, quicker and that could be applied to non-invasive samples as eggshells (Chemnick et al. 2000). Molecular sex determination became systematically used in the captive programme.

A second subject of concern in the management of a captive population implies the knowledge of the pedigree, the genealogical relationship between the individuals. It is of high importance to assess the crosses among breeders, in order to retain the maximum amount of genetic diversity present in the captive population, to minimize inbreeding while avoiding outbreeding, in the measure of the possibilities. The first step was determining the relationships of the individuals in captivity, in order to avoid inbreeding whenever possible to prevent any potential loss of fitness. Using multilocus DNA fingerprinting to study the nuclear genome, the fourteen individuals retained to start the captive breeding programme were found to belong to three distinct clans, where individuals in a clan were more related to each other than to individuals of the other clans (Geyer et al. 1993). The levels of remnant genetic diversity were further investigated using allozymes and mtDNA D-loop sequences. As expected in a population that suffered a strong bottleneck, the genetic diversity turned out to be very low. In a study performed with allozymes in 1988, 24 out of 31 enzyme loci investigated were monomorphic (Corbin and Nice 1998). In the same line, only three mitochondrial control region (CR) haplotypes were present in the founders, which were uniquely defined by four polymorphic sites (D'Elia et al. 2016). Moreover, and as expected in such a small population, the frequencies of these three haplotypes shifted during the 25 years since the total population was kept in captivity in 1987, translating into a loss of haplotype diversity (Adams and Villablanca 2007). Otherwise, Ralls and Ballou (2004) had estimated that the founders contained >90% of the wild base population heterozygosity. They also found that, thanks to a correct management based on mean kinship among the individuals, almost 99% of it had been retained in the current population, with alleles from the founders well represented in the current captive breeding pools and in the wild.

The question still remained whether the California condor low genetic diversity was the result of the heavy population crash during the twentieth century or, on the contrary, it was characteristic of the species. Recently, a work suggested that low levels were the result of the population bottleneck: the study of historical museum specimens has shown that the species lost >80% of the mitochondrial haplotypes in the last two centuries (D'Elia et al. 2016). Additionally, this study did not found any geographical distribution of diversity, pointing to high levels of gene flow within a historical *panmictic* population across the entire range, and the authors concluded with the appropriateness of reintroducing the species in the totality of the historical range.

Up to this point, we have only made reference to neutral diversity. But for a species of conservation concern, it is very important to know whether the circumstances that brought it to endangerment have affected functional diversity. Unfortunately for the condor, the founders carried an autosomal recessive mutation (Ralls et al. 2000) that if expressed in homozygosity resulted in the development of the chondrodystrophy, a lethal form of dwarfism. The frequency of that allele in the population was estimated to be around 9%, which represents a very high frequency for a lethal allele, and was the result of the founder effect that occurred in the population. In a situation like this, there are two possible options to eliminate or minimize the possibility of the manifestation of chondrodystrophy: either eliminating all the possible carriers from the breeding programme or avoiding crosses between potential carriers. The first option would completely eradicate the malformation (assuming that all the potential carriers were properly identified), but this strategy implied reducing the number of breeders to the half. Starting a population from 14 founders was already a situation at the limit (Witzenberger and Hochkirch 2011), so the possibility of eliminating breeders was not a good option. By implementing the second strategy, although there was still the risk of appearance of the lethal phenotype, the probability of occurrence was minimized, while the genetic diversity remnant in the genomes of potential carriers was also preserved. This was the strategy chosen and the results obtained during all these years indicate that it was a correct decision. This example illustrates how the maintenance of the genetic diversity in small populations might be a trade-off that should be carefully evaluated and executed.

At present, genomic tools are being incorporated in the management of the California condor population. The complete genomes from 36 individuals have been sequenced, and about 4 millions of SNPs have been identified (Ryder and collaborators, abstract for the Plant and Animal Genome conference XXIV, January 9–13 2016, San Diego, California https://pag.confex.com/pag/xxiv/webprogram/Paper19697.html). The selection of the breeding pairs is made in base of their kinship relationships based on the complete nuclear genomes of all the founders and the pedigree information on their descendants (D'Elia et al. 2016). Researchers are in active search of an easy-to-handle set of markers that identify carriers of the lethal chondrodystrophy to incorporate this information in the population management strategy.

Glossary

Adaptive variation Changes in the genomes that improve the fitness of individuals to its environments. It is due to natural selection.

Allele Each of the alternative forms of a gene, locus or nucleotidic sequence

Allelic dropout Failure of a microsatellite allele to amplify in some of the PCRs reactions. It is a common error when dealing with low quality and quantity of DNA.

Ascertainment bias Systematic bias in the measure of a frequency associated to how the data has been collected. For instance, ascertainment biases occur when a marker is amplified in a species other than the one it was developed from.

Assay An investigative procedure used to assess or measure the presence, amount or functionality of an allele.

Codominant In codominant inheritance, the two alleles in a locus are expressed and make a different variant of the protein.

Electrophoresis A separating method for proteins or DNA fragments in a gel based on their net charge, size or shape.

Effective population size The size of an ideal population that experiences genetic drift at the rate of the population in question.

Gene A part of the genome that can be transcribed to yield a protein.

Genetic diversity Genetic variation in a population or species. It is a mean of the heterozygous individuals for a locus in a population.

Genetic drift Evolutionary stochastic process through which allelic frequencies vary from one generation to the next due to random sampling. In small populations genetic drift reduces genetic diversity by eliminating low-frequency alleles and driving others to fixation, ignoring selection.

Gene flow Transfer of genetic variation from one population to another thanks to migrants.

Genetic erosion Loss of genetic diversity and increase in inbreeding, especially in small populations.

Genome Collective term to refer to all the genetic information in an organism. In eukaryotes we consider nuclear, mitochondrial and plastidic as differentiated genomes.

Heterogametic sex Sex in which the two sexual chromosomes are different.

Inbreeding Increase in homozygosis due to non-random mating of closely related individuals. Small populations where all individuals are related due to the small size and drift can also have high levels of inbreeding in spite of a random-mating system.

Inbreeding depression Reduction of fitness (reproduction or survival) due to the increase of homozygosity caused by inbreeding.

Introgression Incorporation of genes from one population or species to another through hybridization with fertile offspring, which in turn hybridize with parental populations or species.

Kin relationship/kinship The probability that one allele from one individual is identical by descent to one allele of another individual.

Major Histocompatibility Complex (MHC) A large family of loci involved in the immune system and the fight to diseases.

Molecular marker A molecular marker is any trait that can be used as a marker of genetic variation. This variation can exist within or among individuals. A marker is codominant when the expression of heterozygote phenotypes differs from either homozygote genotype. For instance, microsatellites are codominant because they can distinguish a heterozygote (two bands on a gel, two alleles detected by the sequencer, etc.) from a homozygote (single band on a gel).

Monomorphic A locus in which only one allele is present, in contrast to polymorphic (see below).

Mutation A change in the DNA sequence.

Null allele Allele that cannot be detected by PCR due to a mutation in the flanking region that prevents the matching of the oligonucleotide and consequently its amplification.

PCR (polymerase chain reaction) Consist in making replicate copies (amplify) specific fragments of the DNA sequences.

Panmictic population A population with random mating.

Primer A small sequence of nucleotides (18–24 in the case of microsatellite markers), forming a single-stranded chain to which additional nucleotides can be added by a polymerase enzyme. The primer anneals to the DNA of the organism of interest and the enzyme starts copying the template.

Polymorphism The existence of two or more variants or alleles at a given DNA locus. To be considered a polymorphism, an allele must occur at a minimum frequency of 1%.

Read Ultimately is the sequence of a section of a unique fragment of DNA produced with next-generation sequencing techniques. NGS produces thousands of millions of reads that are mapped to a reference genome to reconstruct the genome or fragment of the genome of the individual that has been re-sequenced.

Reference genome Is a database of the sequence of nucleic acids as a representative of the genome of a particular species. It can be assembled from a single or several individuals.

Re-sequencing It the amplification of the genome of an individual to determine its genomic variations in relation to the reference genome. The sequence generated is aligned to that reference and mined for SNPs, CNVs, genomic rearrangements, etc.

References

Adams MS, Villablanca FX (2007) Consequences of a genetic bottleneck in California condors: a mitochondrial DNA perspective. In: Mee A, Hall LS (eds) California condors in the 21st century. Nuttall Ornithological Club and American Ornithologist's Union, Cambridge, pp 35–55

Agudo R, Roques S, Galarza JA, Rico C, Hiraldo F, Donazar JA (2008) Isolation and characterization of 18 microsatellite loci in the Egyptian vulture (Neophron percnopterus). Conserv Genet 9(5):1345–1348. https://doi.org/10.1007/s10592-007-9486-9

Agudo R, Rico C, Vila C, Hiraldo F, Antonio Donazar J (2010) The role of humans in the diversification of a threatened island raptor. BMC Evol Biol 10. https://doi.org/10.1186/1471-2148-10-384

Agudo R, Alcaide M, Rico C, Lemus JA, Blanco G, Hiraldo F, Donazar JA (2011) Major histocompatibility complex variation in insular populations of the Egyptian vulture: inferences about the roles of genetic drift and selection. Mol Ecol 20(11):2329–2340. https://doi.org/10.1111/j.1365-294X.2011.05107.x

Agudo R, Carrete M, Alcaide M, Rico C, Hiraldo F, Donazar JA (2012) Genetic diversity at neutral and adaptive loci determines individual fitness in a long-lived territorial bird. Pro R Soc B-Biol Sci 279(1741):3241–3249. https://doi.org/10.1098/rspb.2011.2606

Alcaide M, Negro JJ, Serrano D, Tella JL, Rodriguez C (2005) Extra-pair paternity in the Lesser Kestrel Falco naumanni: a re-evaluation using microsatellite markers. Ibis 147(3):608–611. https://doi.org/10.1111/j.1474-919x.2005.00429.x

Alcaide M, Serrano D, Negro JJ, Tella JL, Laaksonen T, Mueller C, Gal A, Korpimaki E (2009a) Population fragmentation leads to isolation by distance but not genetic impoverishment in the philopatric Lesser Kestrel: a comparison with the widespread and sympatric Eurasian Kestrel. Heredity 102(2):190–198. https://doi.org/10.1038/hdy.2008.107

Alcaide M, Serrano D, Tella JL, Negro JJ (2009b) Strong philopatry derived from capture-recapture records does not lead to fine-scale genetic differentiation in lesser kestrels. J Anim Ecol 78(2):468–475. https://doi.org/10.1111/j.1365-2656.2008.01493.x

Arsenault DP, Stacey PB, Hoelzer GA (2002) No extra-pair fertilization in Flammulated Owls despite aggregated nesting. Condor 104(1):197–202. https://doi.org/10.1650/0010-5422(2002)104[0197:nepfif]2.0.co;2

Austin JJ, Olivier L, Nankervis D, Brown WE, Gardner MG, Burridge CP (2014) Twenty microsatellite loci for population and conservation genetic studies of the wedge-tailed eagle (Aquila audax). Aust J Zool 62(3):235–237. https://doi.org/10.1071/zo14030

Banhos A, Hrbek T, Sanaiotti TM, Farias IP (2016) Reduction of genetic diversity of the Harpy Eagle in Brazilian tropical forests. PLoS One 11(2). https://doi.org/10.1371/journal.pone.0148902

BirdLife International (2016) Gymnogyps californianus. The IUCN red list of threatened species 2016: e.T22697636A93626406. https://doi.org/10.2305/IUCN.UK.2016-3.RLTS.T22697636A93626406.en. Downloaded on 03 Aug 2017

Booms TL, Talbot SL, Sage GK, McCaffery BJ, McCracken KG, Schempf PF (2011) Nest-site fidelity and dispersal of gyrfalcons estimated by noninvasive genetic sampling. Condor 113(4):768–778. https://doi.org/10.1525/cond.2011.100178

Bourke BP, Frantz AC, Lavers CP, Davison A, Dawson DA, Burke TA (2010) Genetic signatures of population change in the British golden eagle (Aquila chrysaetos). Conserv Genet 11(5):1837–1846. https://doi.org/10.1007/s10592-010-0076-x

Briggs CW, Collopy MW (2012) Extra-pair paternity in Swainson's Hawks. J Field Ornithol 83(1):41–46. https://doi.org/10.1111/j.1557-9263.2011.00354.x

Bruford MW, Cao C, Chen Y, Chen Y, Dixon A, Fox NC, Gao S, He J, Hou H, Hu L, Li G, Liao S, Liu Y, Luo Q, Muller MG, Ni P, Pan S, Wang J, Wang J, Wang Z, Xia J, Xu P, Yin Y, Yue Z, Zhan X (2014) Genomic data from the peregrine falcon (Falco peregrinus). GigaScience Database. http://dx.doi.org/10.5524/101006.

Bulut Z, Bragin EA, DeWoody JA, Braham MA, Katzner TE, Doyle JM (2016) Use of noninvasive genetics to assess nest and space use by white-tailed eagles. J Raptor Res 50(4):351–362

Burri R, Antoniazza S, Siverio F, Klein A, Roulin A, Fumagalli L (2008) Isolation and characterization of 21 microsatellite markers in the barn owl (Tyto alba). Mol Ecol Resour 8(5):977–979. https://doi.org/10.1111/j.1755-0998.2008.02121.x

Busch JD, Katzner TE, Bragin E, Keim P (2005) Tetranucleotide microsatellites for aquila and haliaeetus eagles. Mol Ecol Notes 5(1):39–41. https://doi.org/10.1111/j.1471-8286.2004.00823.x

Caballero IC, Bates JM, Hennen M, Ashley MV (2016) Sex in the city: breeding behavior of urban Peregrine Falcons in the Midwestern US. PLoS One 11(7). https://doi.org/10.1371/journal.pone.0159054

Casas-Marce M, Revilla E, Fernandes M, Rodriguez A, Delibes M, Godoy JA (2012) The value of hidden scientific resources: preserved animal specimens from private collections and small museums. Bioscience 62(12):1077–1082. https://doi.org/10.1525/bio.2012.62.12.9

Centeno-Cuadros A, Abbasi I, Nathan R (2017) Sex determination in the wild: a field application of loop-mediated isothermal amplification successfully determines sex across three raptor species. Mol Ecol Resour 17(2):153–160. https://doi.org/10.1111/1755-0998.12540

Chang H-W, Cheng C-A, Gu D-L, Chang C-C, Su S-H, Wen C-H, Chou Y-C, Chou T-C, Yao C-T, Tsai C-L, Cheng C-C (2008) High-throughput avian molecular sexing by SYBR green-based real-time PCR combined with melting curve analysis. BMC Biotechnol 8. https://doi.org/10.1186/1472-6750-8-12

Chemnick LG, Kumamoto AT, Ryder OA (2000) Genetic analyses in support of conservation efforts for the California condor Gymnogyps californianus. Int Zoo Yearb 37:330–339. https://doi.org/10.1111/j.1748-1090.2000.tb00738.x

Corbin K, Nice CC (1998) Genetic variation of California condors. In: Abstracts of symposia and papers to be presented at the 56th annual spring meeting of the Academy, Macalester College, St. Paul, Minnesota, 29–30 April, 1988. Journal of Minnesota Academy of Science 53:27

Coser R, Martinez JG, Nunes M, de Assuncao EN, Astolfi-Filho S, Banhos A, Sanaiotti TM, Hrbek T, Farias IP (2014) Development of microsatellite markers for the near threatened eagles Harpia harpyja and Morphnus guianensis using next-generation sequencing. Conserv Genet Resour 6(4):999–1001. https://doi.org/10.1007/s12686-014-0268-7

Crandall KA, Bininda-Emonds ORP, Mace GM, Wayne RK (2000) Considering evolutionary processes in conservation biology. Trends Ecol Evol 15(7):290–295. https://doi.org/10.1016/s0169-5347(00)01876-0

Dawson DA, Kleven O, dos Remedios N, Horsburgh GJ, Kroglund RT, Santos T, Hewitt CRA (2015) A multiplex microsatellite set for non-invasive genotyping and sexing of the osprey (Pandion haliaetus). Conserv Genet Resour 7(4):887–894. https://doi.org/10.1007/s12686-015-0497-4

Decker MD, Parker PG, Minchella DJ, Rabenold KN (1993) Monogamy in black vultures – genetic evidence from DNA fingerprinting. Behav Ecol 4(1):29–35. https://doi.org/10.1093/beheco/4.1.29

D'Elia J, Haig SM, Mullins TD, Miller MP (2016) Ancient DNA reveals substantial genetic diversity in the California condor (Gymnogyps californianus) prior to a population bottleneck. Condor 118:703–714

Desjardins P, Morais R (1990) Sequence and gene organization of the chicken mitochondrial genome – a novel gene order in higher vertebrates. J Mol Biol 212(4):599–634. https://doi.org/10.1016/0022-2836(90)90225-b

Di Maggio R, Mengoni C, Mucci N, Campobello D, Randi E, Sarà M (2015) Do not disturb the family: roles of colony size and human disturbance in the genetic structure of lesser kestrel. J Zool 295(2):115

Double M, Olsen P (1997) Simplified polymerase chain reaction (PCR)-based sexing assists conservation of an endangered owl, the Norfolk Island boobook Ninox novaeseelandiae undulata. Bird Conserv Int 7(3):283–286

Doyle JM, Katzner TE, Bloom PH, Ji Y, Wijayawardena BK, DeWoody JA (2014) The Genome Sequence of a Widespread Apex Predator, the Golden Eagle (Aquila chrysaetos). PLoS ONE 9(4): e95599. https://doi.org/10.1371/journal.pone.0095599

Ellegren H (1996) First gene on the avian W chromosome (CHD) provides a tag for universal sexing of non-ratite birds. Proc Royal Soc B-Biol Sci 263(1377):1635–1641. https://doi.org/10.1098/rspb.1996.0239

Ewing SR, Nager RG, Nicoll MAC, Aumjaud A, Jones CG, Keller LF (2008) Inbreeding and loss of genetic variation in a reintroduced population of Mauritius kestrel. Conserv Biol 22(2):395–404. https://doi.org/10.1111/j.1523-1739.2008.00884.x

Faircloth BC, Title A, Tan K, Welty J, Belthoff JR, Gowaty PA (2010) Eighteen microsatellite loci developed from western burrowing owls (Athene cunicularia hypugaea). Conserv Genet Resour 2(1):167–171. https://doi.org/10.1007/s12686-010-9214-5

Fleischer RC (1996) Application of molecular methods to the assessment of genetic mating systems in vertebrates. In: Ferraris JD, Palumbi SR (eds) Molecular zoology: advances, strategies, and protocols. Wiley-Liss, New York

Frankham R (1995) Effective population size / adult population size ratios in wildlife – a review. Genet Res 66(2):95–107

Fridolfsson AK, Ellegren H (1999) A simple and universal method for molecular sexing of non-ratite birds. J Avian Biol 30(1):116–121. https://doi.org/10.2307/3677252

Funk WC, Mullins TD, Forsman ED, Haig SM (2007) Microsatellite loci for distinguishing spotted owls (Strix occidentalis), barred owls (Strix varia), and their hybrids. Mol Ecol Notes 7(2):284–286. https://doi.org/10.1111/j.1471-8286.2006.01581.x

Garcia CB, Insausti JA, Gil JA, de Frutos A, Alcantara M, Gonzalez J, Cortes MR, Bonafonte JI, Arruga MV (2009) Comparison of different procedures of DNA analysis for sex identification in the endangered bearded vulture (Gypaetus barbatus). Eur J Wildl Res 55(3):309–312. https://doi.org/10.1007/s10344-008-0239-y

Gausterer C, Stein C, Pichler C, Probst R (2013) Molecular identification of traces from the White-tailed Sea Eagle. Forensic Sci Med Pathol 9(2):231–237. https://doi.org/10.1007/s12024-012-9370-x

Gautschi B (2001) Conservation genetics of the bearded vulture (Gypaetus barbatus). University of Zurich, Zurich

Gautschi B, Tenzer I, Muller JP, Schmid B (2000) Isolation and characterization of microsatellite loci in the bearded vulture (Gypaetus barbatus) and cross-amplification in three Old World vulture species. Mol Ecol 9(12):2193–2195. https://doi.org/10.1046/j.1365-294X.2000.105321.x

Gavin TA, Reynolds RT, Joy SM, Leslie D, May B (1998) Genetic evidence for low frequency of extra-pair fertilizations in northern Goshawks. Condor 100(3):556–560. https://doi.org/10.2307/1369724

Geyer CJ, Ryder OA, Chemnick LG, Thompson EA (1993) Analysis of relatedness in the California condors, from DNA fingerprints. Mol Biol Evol 10(3):571–589

Ghorpade PB, Gupta PK, Prakash V, Cuthbert RJ, Kulkarni M, Prakash N, Das A, Sharma AK, Saini M (2012) Molecular sexing of threatened Gyps vultures: an important strategy for conservation breeding and ecological studies. Spring 1. https://doi.org/10.1186/2193-1801-1-62

Gilbert MP, Howard JT, Jarvis ED, The Avian Genome Consortium, Warren W, Wilson RK, Zhang G (2014a) Genomic data of the Bald eagle (Haliaeetus leucocephalus). GigaScience Database. https://doi.org/10.5524/101040

Gilbert MP, Jarvis ED, Li B, Li C, The Avian Genome Consortium, Wang J, Zhang G (2014b) Genomic data of the Barn owl (Tyto alba). GigaScience Database. https://doi.org/10.5524/101039

Gilbert MP, Howard JT, Jarvis ED, Li B, Li C, The Avian Genome Consortium, Wang J, Zhang G (2014c) Genomic data of the Turkey Vulture (Cathartes aura). GigaScience Database. https://doi.org/10.5524/101021

Gilbert MP, Jarvis ED, Li B, Li C, The Avian Genome Consortium, Wang J, Zhang G (2014d) Genomic data of the White tailed eagle (Haliaeetus albicilla). GigaScience Database. https://doi.org/10.5524/101027

Godoy JA, Negro JJ, Hiraldo F, Donazar JA (2004) Phylogeography, genetic structure and diversity in the endangered bearded vulture (Gypaetus barbatus, L.) as revealed by mitochondrial DNA. Mol Ecol 13(2):371–390. https://doi.org/10.1046/j.1365-294X.2003.02075.x

Griffith SC, Owens IPF, Thuman KA (2002) Extra pair paternity in birds: a review of interspecific variation and adaptive function. Mol Ecol 11(11):2195–2212. https://doi.org/10.1046/j.1365-294X.2002.01613.x

Griffiths R, Daan S, Dijkstra C (1996) Sex identification in birds using two CHD genes. Proc Royal Soc B-Biol Sci 263(1374):1251–1256. https://doi.org/10.1098/rspb.1996.0184

Griffiths R, Double MC, Orr K, Dawson RJG (1998) A DNA test to sex most birds. Mol Ecol 7(8):1071–1075. https://doi.org/10.1046/j.1365-294x.1998.00389.x

Groombridge JJ, Jones CG, Bruford MW, Nichols RA (2000) Conservation biology – 'Ghost' alleles of the Mauritius kestrel. Nature 403(6770):616–616. https://doi.org/10.1038/35001148

Groombridge JJ, Dawson DA, Burke T, Prys-Jones R, Brooke MdL, Shah N (2009) Evaluating the demographic history of the Seychelles kestrel (Falco araea): genetic evidence for recovery from a population bottleneck following minimal conservation management. Biol Conserv 142(10):2250–2257. https://doi.org/10.1016/j.biocon.2009.04.026

Haig SM, Mullins TD, Forsman ED, Trail PW, Wennerberg L (2004) Genetic identification of spotted owls, barred owls, and their hybrids: legal implications of hybrid identity. Conserv Biol 18(5):1347–1357. https://doi.org/10.1111/j.1523-1739.2004.00206.x

Haig SM, Miller MP, Bellinger R, Draheim HM, Mercer DM, Mullins TD (2016) The conservation genetics juggling act: integrating genetics and ecology, science and policy. Evol Appl 9(1):181–195. https://doi.org/10.1111/eva.12337

Hailer F, Gautschi B, Helander B (2005) Development and multiplex PCR amplification of novel microsatellite markers in the White-tailed Sea Eagle, Haliaeetus albicilla (Aves : Falconiformes, Accipitridae). Mol Ecol Notes 5(4):938–940. https://doi.org/10.1111/j.1471-8286.2005.01122.x

Hailer F, Helander B, Folkestad AO, Ganusevich SA, Garstad S, Hauff P, Koren C, Nygard T, Volke V, Vila C, Ellegren H (2006) Bottlenecked but long-lived high genetic diversity retained in white-tailed eagles upon recovery from population decline. Biol Lett 2(2):316–319. https://doi.org/10.1098/rsbl.2006.0453

Haughey CL, Sage GK, DeGange GR, Sonsthagen SA, Talbot SL (2016) Development of novel microsatellite markers for the Northern Goshawk (Accipiter gentilis) and their utility in cross-species amplification. Avian Biol Res 9(3):195–199. https://doi.org/10.3184/175815516x14667737479433

Hedrick PW, Kalinowski ST (2000) Inbreeding depression in conservation biology. Annu Rev Ecol Syst 31:139–162. https://doi.org/10.1146/annurev.ecolsys.31.1.139

Hegglin D, Frey H, Terrasse M (2010) La strategia della Fondazione per la Conservazione degli Avvoltoi nel progetto di reintroduzione del gipeto sulle Alpi. Infogipeto 27:2

Hirai M, Yamazaki T (2010) Isolation and characterization of eleven microsatellite loci in an endangered species, Mountain Hawk-Eagle (Spizaetus nipalensis). Conserv Genet Resour 2:113–115. https://doi.org/10.1007/s12686-010-9226-1

Hogan FE, Cooke R (2010) Insights into the breeding behaviour and dispersal of the Powerful Owl (Ninox strenua) through the collection of shed feathers. Emu 110(2):178–184. https://doi.org/10.1071/mu09116

Hogan F, Burridge C, Cooke R, Norman J (2007) Isolation and characterization of microsatellite loci to DNA fingerprint the Powerful Owl (Ninox strenua). Mol Ecol Notes 7(6):1305–1307. https://doi.org/10.1111/j.1471-8286.2007.01864.x

Holmes MW, Hammond TT, Wogan GOU, Walsh RE, Labarbera K, Wommack EA, Martins FM, Crawford JC, Mack KL, Bloch LM, Nachman MW (2016) Natural history collections as windows on evolutionary processes. Mol Ecol 25(4):864–881. https://doi.org/10.1111/mec.13529

Horvath MB, Martinez-Cruz B, Negro JJ, Kalmar L, Godoy JA (2005) An overlooked DNA source for non-invasive genetic analysis in birds. J Avian Biol 36(1):84–88. https://doi.org/10.1111/j.0908-8857.2005.03370.x

Hsu YC, Severinghaus LL, Lin YS, Li SH (2003) Isolation and characterization of microsatellite DNA markers from the Lanyu scops owl (Otus elegans botelensis). Mol Ecol Notes 3(4):595–597. https://doi.org/10.1046/j.1471-8286.2003.00523.x

Hsu YC, Li SH, Lin YS, Philippart MT, Severinghaus LL (2006a) High frequency of extra-pair copulation with low level of extra-pair fertilization in the Lanyu scops owl Otus elegans botelensis. J Avian Biol 37(1):36–40. https://doi.org/10.1111/j.2006.0908-8857.03687.x

Hsu YC, Li SH, Lin YS, Severinghaus LL (2006b) Microsatellite loci from Lanyu scops owl (Otus elegans botelensis) and their cross-species application in four species of strigidae. Conserv Genet 7(1):161–165. https://doi.org/10.1007/s10592-005-5477-x

Hsu H-H, Ding S-T, Chang Y-Y, Chao M-C, Tsao H-S, Chan F-T, Hsu C-C, Yuan H-W, Wang P-H (2013) Development of 24 new microsatellite markers in the Crested Serpent Eagle (Spilornis cheela hoya). Conserv Genet Resour 5(2):417–420. https://doi.org/10.1007/s12686-012-9817-0

Hull JM, Tufts D, Topinka JR, May B, Ernest HB (2007) Development of 19 microsatellite loci for Swainson's hawks (Buteo swainsoni) and other buteos. Mol Ecol Notes 7(2):346–349. https://doi.org/10.1111/j.1471-8286.2006.01604.x

Hull JM, Anderson R, Bradbury M, Estep JA, Ernest HB (2008a) Population structure and genetic diversity in Swainson's Hawks (Buteo swainsoni): implications for conservation. Conserv Genet 9(2):305–316. https://doi.org/10.1007/s10592-007-9342-y

Hull JM, Keane JJ, Tell LA, Ernest HB (2008b) Development of 37 microsatellite loci for the great gray owl (Strix nebulosa) and other Strix spp. owls. Conserv Genet 9(5):1357–1361. https://doi.org/10.1007/s10592-007-9489-6

Isaksson M, Tegelstrom H (2002) Characterization of polymorphic microsatellite markers in a captive population of the eagle owl (Bubo bubo) used for supportive breeding. Mol Ecol Notes 2(2):91–93. https://doi.org/10.1046/j.1471-8286.2002.00156.x

Ito H, Sudo-Yamaji A, Abe M, Murase T, Tsubota T (2003) Sex identification by alternative polymerase chain reaction methods in Falconiformes. Zool Sci 20(3):339–344. https://doi.org/10.2108/zsj.20.339

Janowski (2014) Development of new microsatellite (STR) markers for Montagu's harrier (Circus pygargus) via 454 shot-gun pyrosequencing. Open Ornithol J 7:11–18

Johnson PCD, Fowlie MK, Amos W (2005) Isolation of microsatellite loci from the common buzzard, Buteo buteo (Aves : Accipitridae). Mol Ecol Notes 5(2):208–211. https://doi.org/10.1111/j.1471-8286.2005.00878.x

Johnson JA, Gilbert M, Virani MZ, Asim M, Mindell DP (2008) Temporal genetic analysis of the critically endangered oriental white-backed vulture in Pakistan. Biol Conserv 141(9):2403–2409. https://doi.org/10.1016/j.biocon.2008.07.001

Johnson JA, Tingay RE, Culver M, Hailer F, Clarke ML, Mindell DP (2009) Long-term survival despite low genetic diversity in the critically endangered Madagascar fish-eagle. Mol Ecol 18(1):54–63. https://doi.org/10.1111/j.1365-294X.2008.04012.x

Johnson JA, Talbot SL, Sage GK, Burnham KK, Brown JW, Maechtle TL, Seegar WS, Yates MA, Anderson B, Mindell DP (2010) The use of genetics for the management of a recovering population: temporal assessment of migratory peregrine falcons in North America. PLoS One 5(11). https://doi.org/10.1371/journal.pone.0014042

Katzner TE, Jackson DS, Ivy J, Bragin EA, DeWoody A (2014) Variation in offspring sex ratio of a long-lived sexually dimorphic raptor, the Eastern Imperial Eagle Aquila heliaca. Ibis 156(2):395–403. https://doi.org/10.1111/ibi.12139

Kleven O, Dawson DA, Gjershaug JO, Horsburgh GJ, Jacobsen K-O, Wabakken P (2013) Isolation, characterization and predicted genome locations of Eurasian eagle-owl (Bubo bubo) microsatellite loci. Conserv Genet Resour 5(3):723–727. https://doi.org/10.1007/s12686-013-9891-y

Kocher TD, Thomas WK, Meyer A, Edwards SV, Paabo S, Villablanca FX, Wilson AC (1989) Dynamics of mitochondrial DNA evolution in animals- amplification and sequencing with conserved primers. Proc Natl Acad Sci U S A 86(16):6196–6200. https://doi.org/10.1073/pnas.86.16.6196

Kohyama TI, Omote K, Nishida C, Takenaka T, Saito K, Fujimoto S, Masuda R (2015) Spatial and temporal variation at major histocompatibility complex class IIB genes in the endangered Blakiston's fish owl. Zoological Lett 1:13–13

Koopman ME, McDonald DB, Hayward GD (2007) Microsatellite analysis reveals genetic monogamy among female boreal owls. J Raptor Res 41(4):314–318. https://doi.org/10.3356/0892-1016(2007)41[314:margma]2.0.co;2

Korfanta NM, Schable NA, Glenn TC (2002) Isolation and characterization of microsatellite DNA primers in burrowing owl (Athene cunicularia). Mol Ecol Notes 2(4):584–585. https://doi.org/10.1046/j.1471-8286.2002.00326.x

Le Gouar P, Rigal F, Boisselier-Dubayle MC, Sarrazin F, Arthur C, Choisy JP, Hatzofe O, Henriquet S, Lecuyer P, Tessier C, Susic G, Samadi S (2008) Genetic variation in a network of natural and reintroduced populations of Griffon vulture (Gyps fulvus) in Europe. Conserv Genet 9(2):349–359. https://doi.org/10.1007/s10592-007-9347-6

Le Gouar P, Sulawa J, Henriquet S, Tessier C, Sarrazin F (2011) Low evidence for extra-pair fertilizations in two reintroduced populations of Griffon vulture (Gyps fulvus). J Ornithol 152(2):359–364. https://doi.org/10.1007/s10336-010-0593-x

León-Ortega M, González-Wangüemert M (2015) Characterization of 10 new tetranucleotide microsatellite markers for the European eagle owl, Bubo bubo: useful tools for conservation strategies. Biochem Syst Ecol 63:109–112. https://doi.org/10.1016/j.bse.2015.10.003

Lerner HRL, Johnson JA, Lindsay AR, Kiff LF, Mindell DP (2009) It's not too late for the harpy eagle (Harpia harpyja): high levels of genetic diversity and differentiation can fuel conservation programs. PLoS One 4(10). https://doi.org/10.1371/journal.pone.0007336

Loercher F, Keller L, Hegglin D (2013) Low genetic diversity of the reintroduced bearded vulture (Gypaetus barbatus) population in the Alps calls for further releases. Paper presented at the 5th symposium for research in protected areas, 10–12 June 2013

Macias-Duarte A, Conway CJ, Munguia-Vega A, Culver M (2010) Novel microsatellite loci for the burrowing owl Athene cunicularia. Conserv Genet Resour 2(1):67–69. https://doi.org/10.1007/s12686-009-9150-4

Manel S, Holderegger R (2013) Ten years of landscape genetics. Trends Ecol Evol 28(10):614–621. https://doi.org/10.1016/j.tree.2013.05.012

Martínez-Cruz B, David VA, Godoy JA, Negro JJ, O'Brien SJ, Johnson WE (2002) Eighteen polymorphic microsatellite markers for the highly endangered Spanish imperial eagle (Aquila adalberti) and related species. Mol Ecol Notes 2(3):323–326. https://doi.org/10.1046/j.1471-8278.2002.00231.x

Martinez-Cruz B, Godoy JA, Negro JJ (2004) Population genetics after fragmentation: the case of the endangered Spanish imperial eagle (Aquila adalberti). Mol Ecol 13(8):2243–2255. https://doi.org/10.1111/j.1365-294X.2004.02220.x

Martinez-Cruz B, Godoy JA, Negro JJ (2007) Population fragmentation leads to spatial and temporal genetic structure in the endangered Spanish imperial eagle. Mol Ecol 16(3):477–486. https://doi.org/10.1111/j.1365-294X.2007.03147.x

Mendez M, Voegeli M, Tella JL, Godoy JA (2014) Joint effects of population size and isolation on genetic erosion in fragmented populations: finding fragmentation thresholds for management. Evol Appl 7(4):506–518. https://doi.org/10.1111/eva.12154

Mira S, Billot C, Guillemaud T, Palma L, Cancela ML (2002) Isolation and characterization of polymorphic microsatellite markers in Eurasian vulture Gyps fulvus. Mol Ecol Notes 2(4):557–558. https://doi.org/10.1046/j.1471-8278.2002.00314.x

Mira S, Wolff K, Cancela ML (2005) Isolation and characterization of microsatellite markers in Bonelli's eagle (Hieraaetus fasciatus). Mol Ecol Notes 5(3):493–495. https://doi.org/10.1111/j.1471-8286.2005.00967.x

Monti F, Duriez O, Arnal V, Dominici J-M, Sforzi A, Fusani L, Gremillet D, Montgelard C (2015) Being cosmopolitan: evolutionary history and phylogeography of a specialized raptor, the Osprey Pandion haliaetus. BMC Evol Biol 15. https://doi.org/10.1186/s12862-015-0535-6

Moritz C (1999) Conservation units and translocations: strategies for conserving evolutionary processes. Hereditas 130(3):217–228. https://doi.org/10.1111/j.1601-5223.1999.00217.x

Nebel C, Gamauf A, Haring E, Segelbacher G, Villers A, Zachos FE (2015) Mitochondrial DNA analysis reveals Holarctic homogeneity and a distinct Mediterranean lineage in the Golden eagle (Aquila chrysaetos). Biol J Linn Soc 116(2):328–340. https://doi.org/10.1111/bij.12583

Nesje M, Roed KH (2000) Microsatellite DNA markers from the gyrfalcon (Falco rusticolus) and their use in other raptor species. Mol Ecol 9(9):1438–1440. https://doi.org/10.1046/j.1365-294x.2000.00999-4.x

Nesje M, Roed KH, Lifjeld JT, Lindberg P, Steen OF (2000) Genetic relationships in the peregrine falcon (Falco peregrinus) analysed by microsatellite DNA markers. Mol Ecol 9(1):53–60. https://doi.org/10.1046/j.1365-294x.2000.00834.x

Nittinger F, Gamauf A, Pinsker W, Wink M, Haring E (2007) Phylogeography and population structure of the saker falcon (Falco cherrug) and the influence of hybridization: mitochondrial and microsatellite data. Mol Ecol 16(7):1497–1517. https://doi.org/10.1111/j.1365-294X.2007.03245.x

Ortego J, Gonzalez EG, Sanchez-Barbudo I, Aparicio JM, Cordero PJ (2007) Novel highly polymorphic loci and cross-amplified microsatellites for the lesser kestrel Falco naumanni. Ardeola 54(1):101–108

Padilla JA, Parejo JC, Salazar J, Martinez-Trancon M, Rabasco A, Sansinforiano E, Quesada A (2009) Isolation and characterization of polymorphic microsatellite markers in lesser kestrel (Falco naumanni) and cross-amplification in common kestrel (Falco tinnunculus). Conserv Genet 10(5):1357–1360. https://doi.org/10.1007/s10592-008-9711-1

Ponnikas S, Kvist L, Ollila T, Stjernberg T, Orell M (2013) Genetic structure of an endangered raptor at individual and population levels. Conserv Genet 14(6):1135–1147. https://doi.org/10.1007/s10592-013-0501-z

Poulakakis N, Antoniou A, Mantziou G, Parmakelis A, Skartsi T, Vasilakis D, Elorriaga J, De La Puente J, Gavashelishvili A, Ghasabyan M, Katzner T, McGrady M, Batbayar N, Fuller M, Natsagdorj T (2008) Population structure, diversity, and phylogeography in the near-threatened Eurasian black vultures Aegypius monachus (Falconiformes; Accipitridae) in Europe: insights from microsatellite and mitochondrial DNA variation. Biol J Linn Soc 95(4):859–872. https://doi.org/10.1111/j.1095-8312.2008.01099.x

Ralls K, Ballou JD (2004) Genetic status and management of California condors. Condor 106(2):215–228. https://doi.org/10.1650/7348

Ralls K, Ballou JD, Rideout BA, Frankham R (2000) Genetic management of chondrodystrophy in California condors. Anim Conserv 3:145–153. https://doi.org/10.1111/j.1469-1795.2000.tb00239.x

Rodriguez-Martinez S, Carrete M, Roques S, Rebolo-Ifran N, Tella JL (2014) High urban breeding densities do not disrupt genetic monogamy in a bird species. PLoS One 9(3). https://doi.org/10.1371/journal.pone.0091314

Romanov MN, Tuttle EM, Houck ML, Modi WS, Chemnick LG, Korody ML, Mork EMS, Otten CA, Renner T, Jones KC, Dandekar S, Papp JC, Da Y, Green ED, Magrini V, Hickenbotham MT, Glasscock J, McGrath S, Mardis ER, Ryder OA, Progra NCS (2009) The value of avian genomics to the conservation of wildlife. BMC Genomics 10. https://doi.org/10.1186/1471-2164-10-s2-s10

Roques S, Godoy JA, Negro JJ, Hiraldo F (2004) Organization and variation of the mitochondrial control region in two vulture species, Gypaetus barbatus and Neophron percnopterus. J Hered 95(4):332–337. https://doi.org/10.1093/jhered/esh047

Rosenfield RN, Sonsthagen SA, Stout WE, Talbot SL (2015) High frequency of extra-pair paternity in an urban population of Cooper's Hawks. J Field Ornithol 86(2):144–152. https://doi.org/10.1111/jofo.12097

Rudnick JA, Katzner TE, Bragin EA, Rhodes OE, Dewoody JA (2005) Using naturally shed feathers for individual identification, genetic parentage analyses, and population monitoring in an endangered Eastern imperial eagle (Aquila heliaca) population from Kazakhstan. Mol Ecol 14(10):2959–2967. https://doi.org/10.1111/j.1365-294X.2005.02641.x

Rudnick JA, Katzner TE, Bragin EA, DeWoody JA (2008) A non-invasive genetic evaluation of population size, natal philopatry, and roosting behavior of non-breeding eastern imperial eagles (Aquila heliaca) in central Asia. Conserv Genet 9(3):667–676. https://doi.org/10.1007/s10592-007-9397-9

Sacchi P, Soglia D, Maione S, Meneguz G, Campora M, Rasero R (2004) A non-invasive test for sex identification in Short-toed Eagle (Circaetus gallicus). Mol Cell Probes 18(3):193–196. https://doi.org/10.1016/j.mcp.2004.01.002

Saladin V, Ritschard M, Roulin A, Bize P, Richner H (2007) Analysis of genetic parentage in the tawny owl (Strix aluco) reveals extra-pair paternity is low. J Ornithol 148(1):113–116. https://doi.org/10.1007/s10336-006-0109-x

Segelbacher G (2002) Noninvasive genetic analysis in birds: testing reliability of feather samples. Mol Ecol Notes 2(3):367–369. https://doi.org/10.1046/j.1471-8286.2002.00180.x

Shaffer HB, Fisher RN, Davidson C (1998) The role of natural history collections in documenting species declines. Trends Ecol Evol 13(1):27–30. https://doi.org/10.1016/s0169-5347(97)01177-4

Smith SB, McKay JE, Murphy MT, Duffield DA (2016) Spatial patterns of extra-pair paternity for spotted towhees Pipilo maculatus in urban parks. J Avian Biol 47(6):815–823. https://doi.org/10.1111/jav.00931

Taberlet P, Griffin S, Goossens B, Questiau S, Manceau V, Escaravage N, Waits LP, Bouvet J (1996) Reliable genotyping of samples with very low DNA quantities using PCR. Nucleic Acids Res 24(16):3189–3194. https://doi.org/10.1093/nar/24.16.3189

Taberlet P, Waits LP, Luikart G (1999) Noninvasive genetic sampling: look before you leap. Trends Ecol Evol 14(8):323–327

Thode AB, Maltbie M, Hansen LA, Green LD, Longmire JL (2002) Microsatellite markers for the Mexican spotted owl (Strix occidentalis lucida). Mol Ecol Notes 2(4):446–448. https://doi.org/10.1046/j.1471-8286.2002.00267.x

Tingay RE, Dawson DA, Pandhal J, Clarke ML, David VA, Hailer F, Culver M (2007) Isolation of 22 new Haliaeetus microsatellite loci and their characterization in the critically endangered Madagascar fish-eagle (Haliaeetus vociferoides) and three other Haliaeetus eagle species. Mol Ecol Notes 7(4):711–715. https://doi.org/10.1111/j.1471-8286.2007.01690.x

Tingley MW, Beissinger SR (2009) Detecting range shifts from historical species occurrences: new perspectives on old data. Trends Ecol Evol 24(11):625–633. https://doi.org/10.1016/j.tree.2009.05.009

Topinka JR, May B (2004) Development of polymorphic microsatellite loci in the northern goshawk (Accipiter gentilis) and cross-amplification in other raptor species. Conserv Genet 5(6):861–864. https://doi.org/10.1007/s10592-004-1973-7

Vali U, Dombrovski V, Treinys R, Bergmanis U, Daroczi SJ, Dravecky M, Ivanovski V, Lontkowski J, Maciorowski G, Meyburg BU, Mizera T, Zeitz R, Ellegren H (2010) Widespread hybridization between the Greater Spotted Eagle Aquila clanga and the Lesser Spotted Eagle Aquila pomarina (Aves: Accipitriformes) in Europe. Biol J Linn Soc 100(3):725–736

Vargas JD, Whitacre D, Mosquera R, Albuquerque J, Piana R, Thiollay JM, Marquez C, Sanchez JE, Lezama-Lopez M, Midence S, Matola S, Aguilar S, Rettig N, Sanaiotti T (2006) Status and current distribution of the Harpy Eagle (Harpia harpyja) in Central and South America. Ornitol Neotrop 17(1):39–55

Vili N, Horvath Marton B, Kovacs S, Jozef C, Hornung E, Kalmar L (2009) Alternative sampling methods in avian genetic studies: sexing, microsatellites based individual identification and mtDNA analyses of eastern imperial eagles (Aquila heliaca). Magyar Allatorvosok Lapja 131(7):426–435

Vili N, Szabo K, Kovacs S, Kabai P, Kalmar L, Horvath M (2013) High turnover rate revealed by non-invasive genetic analyses in an expanding eastern imperial eagle population. Acta Zool Acad Sci Hung 59(3):279–295

Walters JR, Derrickson SR, Fry DM, Haig SM, Marzluff JM, Wunderle JM Jr (2010) Status of the California condor (Gymnogyps californianus) and efforts to achieve its recovery. Auk 127(4):969–1001. https://doi.org/10.1525/auk.2010.127.4.969

Witzenberger KA, Hochkirch A (2011) Ex situ conservation genetics: a review of molecular studies on the genetic consequences of captive breeding programmes for endangered animal species. Biodivers Conserv 20(9):1843–1861. https://doi.org/10.1007/s10531-011-0074-4

Woolaver LG, Nichols RK, Morton ES, Stutchbury BJM (2013) Population genetics and relatedness in a critically endangered island raptor, Ridgway's Hawk Buteo ridgwayi. Conserv Genet 14(3):559–571. https://doi.org/10.1007/s10592-013-0444-4

Chapter 16
Conservation Status of Neotropical Raptors

José Hernán Sarasola, Juan Manuel Grande, and Marc Joseph Bechard

Introduction

Humans exert the greatest threats for animal conservation. Effects of habitat loss and perturbations as a result of human activities have become global and affect all corners of the earth through intensification of agriculture, development of infrastructures, spread of invasive species, pollution, and overexploitation of resources. Birds of prey have declined throughout the world, and some species have become severely threatened as a result of all these activities. Diurnal birds of prey are the predominant apex avian predators for most natural ecosystems. Besides the role they play as top predators of food webs in most habitats, raptors are among the most susceptible species to the negative effects of habitat transformations and human perturbations in part due to their large habitat requirements, low population densities, and turnover rates (Donázar et al. 2016). All these features make birds of prey especially susceptible to the variety of human pressures on their populations and habitats.

The Neotropical realm extends from tropical Mexico to Central and South America including the Caribbean archipelagoes and adjacent islands (Udvardy 1975) covering 18 million km^2 which corresponds to 12% of the world's land surface. This region includes some of the most surprising and renowned landscapes on the planet that have inspired the work of biologist and naturalist for centuries. From the Amazonian rainforests to Patagonian steppes, the diversity of habitats on

J. H. Sarasola (✉) · J. M. Grande
Centro para el Estudio y Conservación de las Aves Rapaces en Argentina (CECARA), Universidad Nacional de La Pampa (UNLPam) & Consejo Nacional de Investigaciones Científicas y Técnicas de Argentina (CONICET), Santa Rosa, La Pampa, Argentina
e-mail: sarasola@exactas.unlpam.edu.ar

M. J. Bechard
Raptor Research Center, Department of Biological Sciences, Boise State University, Boise, ID, USA

J. H. Sarasola et al. (eds.), *Birds of Prey*,
https://doi.org/10.1007/978-3-319-73745-4_16

the Neotropical region is also in conjunction with a great biodiversity of insects, vertebrates, and plants. The Neotropics also host one of the greatest raptor communities holding almost a third of the currently recognized species in this group (del Hoyo et al. 1994; Fergusson-Lees and Christie 2001), including in this list some of the most singular and flagship raptor representatives, such as the Andean condor (*Vultur gryphus*) and the harpy eagle (*Harpia harpyja*). In addition, four out of five orders of diurnal birds of prey are represented in the Neotropics with the exception of the monospecific and African exclusive Sagittariidae. Several of the raptor species in this region are also listed as globally threatened by the International Union for Conservation of Nature (IUCN). Conversion and destruction of natural habitats, active persecution, poisoning, and other human-related causes are among the threats that these species face in a region where human population growth rate is among the most accelerated of the world.

Important efforts have been made in the past aiming to assess the conservation status of Neotropical raptors at different spatial scales, ranging from regions and countries (Wiley 1986; Bierregaard 1998; Bildstein et al. 1998) to more local biomes or habitats (Alvarez-López and Kattan 1995; Thiollay 2009; Trejo and Ojeda 2015). Bierregaard (1995) made the first comprehensive review and assessment on the conservation status and future research needs for birds of prey in Central and South America. Because it was based on a continental rather than biogeographic scale, that overview did not include some Neotropical raptor species from the Caribbean region, which are mostly island endemic species and are among the most threatened in this realm. The spatial extent of the assessment also did not include areas of Mexico that do not belong to this realm, but the species in this region are very similar to those found in most of the Neotropics. Nevertheless, this assessment gave the most accurate picture of the conservation status of raptors in the Neotropical region.

Here we examine and review information on the diversity and conservation status of raptors in the Neotropics highlighting the most important causes threatening species in this region and gaps and future research needs with the aim of preserving their populations.

Data Source and Analysis

We reviewed information on the conservation status of birds of prey in the Neotropics using species as the lower taxonomic level and hence as the basic conservation unit following IUCN (Birdlife International 2017, but see in this book Chaps. 1 and 15 for futher considerations about units of conservation). Subspecies or races are sometimes considered as the primary level of biodiversity. However, the spatial extent of our assessment and variety of species involved precluded an analysis below the species level. However, it is important to remark that the number of subspecies is very high in some Neotropical raptors. That is the case, for example, of the turkey vulture (*Cathartes aura*), with more than a dozen recognized subspecies that vary in important ecological traits such as migratory status and behavior (del

Hoyo et al. 1994). For some groups of species, and particularly for tropical forest-dwelling hawks in the Accipitridae, there are also an important number of unresolved or controversial species classifications that need consideration. This is the case of sharp-shinned hawk (*Accipiter striatus*) for which several former races are now considered as distinct subspecies for some authors and all of them facing severe threats, particularly on insular systems (Wiley 1986). Due to the lack of consensus on the systematic and taxonomic status of each species, subspecies, or races, and also because we required a unique and authorized source with regard to their conservation status, we followed the species lists and conservation criteria for each of them determined by Birdlife International (2017). We adopted these criteria favoring a single species although there are several potential and yet unresolved distinct taxonomic units. Any uncertainty on the proper classification involved only globally threatened taxa.

We also considered IUCN as the authority for the conservation categorization for each species in this assessment (Birdlife International 2017). This source also defines criteria for other aspects of the biology of species, such as migratory and endemism status, which is based on periodic reviews of the available information conducted by experts on the species or topics. Thus, and according with this criteria, we considered as migrant those raptor species performing regular movements irrespective of their extent (i.e., local, regional, or continental) or the fraction of the global population involved (i.e., whole or partially migrant). Thus we did not distinguish between short- or long-distance migrant species. For country endemic species, we considered those whose breeding range is restricted to a single country but not to a single or particular habitat.

The assessment of the conservation status of Neotropical raptors was based on the most up-to-date information on threats, population status, and trends of species at global levels (Birdlife International 2017). Some species, however, may differ on their categorization and conservation status when examined at smaller spatial scales (e.g., country or region level). Such species categorizations, however, were sometimes defined based on small local or marginal populations usually located at the borders of their distribution range where they face particular threats. For the same reasons explained below, such categorizations where not considered in this analysis. For each raptor species categorized as having global conservation concern, either globally threatened (i.e., critically endangered, endangered, vulnerable) or near threatened, we examined species-habitat associations and main threats. For species-habitat association, we considered six main habitat types: forests, savannas, shrublands, grasslands, deserts, and wetlands. This information was gathered for each species according to the occurrence of these habitats among each species' range. Thus, each species was either assigned to a single habitat, when it was the single and dominant habitat across the entire species' range, or assigned to as many habitats up to a maximum of six according to the habitats included on the species' distribution range. After gathering this information, a habitat score was constructed by summing all the habitat types for each species and considered as a proxy to characterize each species as a habitat-generalist or habitat-specialist.

Also following Birdlife International (2017), we examined the most important threats identified for each species. For this, we searched for the main threats defined by IUCN for species of conservation concern in the Neotropical region.

Species Richness and Conservation Status

A total of 97 raptor species are found in the Neotropical region belonging to three out of the four orders of diurnal raptors: Accipitriformes, Falconiformes, and Cathartiformes. The most important representation in species richness belongs to the Accipitriformes with 66 species from 2 families: the highly diverse Accipitridae and the monospecific Pandionidae. The second more represented order among Neotropical raptors is Falconiformes with 25 species followed by Cathartiformes with 6 species.

Approximately 25% of the Neotropical diurnal raptor species are considered to be of conservation concern and categorized either as critically endangered (3 species), endangered (4 species), vulnerable (5 species), or near threatened (12 species) by the IUCN. This last category includes species that are on the brink of being in the globally threatened category. Thus, the number of species globally threatened includes a total of 12 raptors (Table 16.1) belonging mostly to the Accipitridae (97%, 11 species). None of the New World vulture species in the Cathartiformes are included among the globally threatened species in this region, and only the Andean condor (*Vultur gryphus*) is considered as near threatened and hence as species of conservation concern.

Populations of roughly 80% of threatened or near-threatened raptor species in the Neotropics are experiencing declining trends, and only 16% of these species are considered as having stable populations (Table 16.1). Furthermore, all except two of the threatened species (83%), the Galapagos hawk (*Buteo galapagoensis*) and the rufous-tailed hawk (*Buteo ventralis*), show decreasing population trends. Only the grey-bellied hawk (*Accipiter poliogaster*) in this group seems to be experiencing a population recovery and an increasing population trend.

Overall, the conservation status of Neotropical raptors during the period 1988 to 2016 worsened when we considered the variation in time in the categorization for those species currently considered of conservation concern (Fig. 16.1). In 1988 only three categories were considered when classifying the conservation status of the species, with the current categorization scheme starting in 1994. For 1988, 21 out of 24 species currently considered of conservation concern were listed. For the period 1994–2000, not a single raptor species was categorized at the highest conservation risk. As soon as the categorization was updated in 2000, the white-collared kite (*Leptodon forbesi*) and the Ridgway's hawk (*Buteo ridgwayi*) were listed for the first time as critically endangered species. The rearrangement of the classification system also caused some species to be momentarily downgraded to a lower risk category and a return to their initial categorization status. All of the species considered initially to be of least concern were progressively updated to riskier categories.

Table 16.1 Globally threatened (*CR* critically endangered, *EN* endangered, *VU* vulnerable) and near-threatened (NT) raptor species in the Neotropical region following Birdlife International (2017)

Common name	Scientific name	Status	Region	Population size[a]	Population trend	Area[b]	Country endemic[c]	Landmass type
White-collared kite	*Leptodon forbesi*	CR	SA	50–249	Decreasing	15,000	Yes	Continent
Cuban kite	*Chondrohierax wilsonii*	CR	CRB	50–249	Decreasing	4,100	Yes	Island
Ridgway's hawk	*Buteo ridgwayi*	CR	CRB	240–260	Decreasing	570	Yes	Island
Black-and-chestnut eagle	*Spizaetus isidori*	EN	SA	250–999	Decreasing	4,830,000	No	Continent
Crowned solitary eagle	*Buteogallus coronatus*	EN	SA	250–999	Decreasing	6,590,000	No	Continent
Grey-backed hawk	*Pseudastur occidentalis*	EN	SA	250–999	Decreasing	64,400	No	Continent
Gundlach's hawk	*Accipiter gundlachi*	EN	CRB	700–900	Decreasing	102,000	Yes	Island
Plumbeous forest falcon	*Micrastur plumbeus*	VU	SA	6000–15,000	Decreasing	115,000	No	Continent
Plumbeous hawk	*Cryptoleucopteryx plumbea*	VU	SA-CA	10,000–19,000	Decreasing	656,000	No	Continent
White-necked hawk	*Buteogallus lacernulatus*	VU	SA	2500–9999	Decreasing	1,040,000	Yes	Continent
Galapagos hawk	*Buteo galapagoensis*	VU	SA	270–330	Stable	29,600	Yes	Island
Rufous-tailed hawk	*Buteo ventralis*	VU	SA	250–999	Stable	1,070,000	No	Continent
Striated caracara	*Phalcoboenus australis*	NT	SA	1000–2499	Stable	154,000	No	Island
Orange-breasted falcon	*Falco deiroleucus*	NT	SA-CA	20,000–49,999	Decreasing	16,300,000	No	Continent
Crested eagle	*Morphnus guianensis*	NT	SA-CA	670–6700	Decreasing	15,600,000	No	Continent
Harpy eagle	*Harpia harpyja*	NT	SA-CA	20,000–49,999	Decreasing	17,600,000	No	Continent
Ornate hawk-eagle	*Spizaetus ornatus*	NT	SA-CA	13,300–33,300	Decreasing	20,200,000	No	Continent
Grey-bellied goshawk	*Accipiter poliogaster*	NT	SA	1000–10,000	Increasing	11,000,000	No	Continent
Semicollared hawk	*Accipiter collaris*	NT	SA	1500–7000	Stable	1,900,000	No	Continent
Rufous crab hawk	*Buteogallus aequinoctialis*	NT	SA	1,000	Decreasing	5,860,000	No	Continent
Black solitary eagle	*Buteogallus solitarius*	NT	SA-CA	1000–2499	Decreasing	13,500,000	No	Continent
Mantled hawk	*Pseudastur polionotus*	NT	SA	2500–9999	Decreasing	2,080,000	No	Continent

(continued)

Table 16.1 (continued)

Common name	Scientific name	Status	Region	Population size[a]	Population trend	Area[b]	Country endemic[c]	Landmass type
Cuban black hawk	*Buteogallus gundlachii*	NT	CRB	10,000	Decreasing	15,000	Yes	Island
Andean condor	*Vultur gryphus*	NT	SA	6,700	Decreasing	8,520,000	No	Continent

Region indicates the three main continental and insular mass in the Neotropical realm: South America (SA), Central America (CA), and the Caribbean (CRB)
[a]Estimated number of reproductive individuals for the entire population
[b]Estimated area (km^2) for the breeding range of the species
[c]Those species with distribution restricted to a single country irrespective of the size of the distribution range

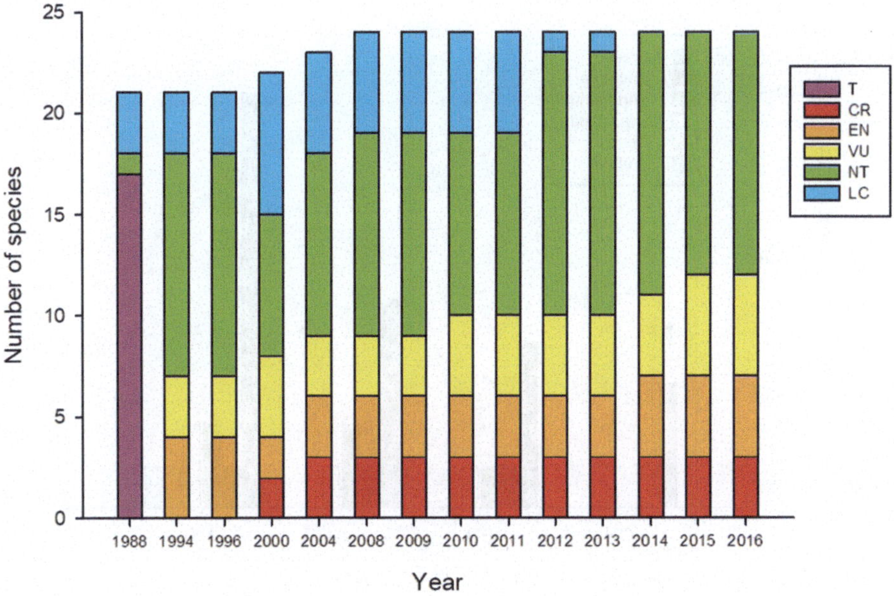

Fig. 16.1 Variation of the conservation status for the 24 raptor species currently categorized as of conservation concern in the Neotropics (LC least concern, NT near threatened, VU vulnerable, EN endangered, CR critically endangered) from 1988 to 2016. The threatened (T) category was applied only for 1988 categorization with the remaining categories being employed by IUCN since 1994

During the period 1994–2016, the conservation status of eight raptor species has been upgraded from lower to higher conservation status. Only one has followed a reverse path, the rufous-tailed hawk (*Buteo ventralis*). However, it is now considered as a vulnerable species after being categorized as of least concern during the period 2000–2004.

Species-Habitat Relationships

Raptor species categorized as of conservation concern in the Neotropical region largely rely on forested habitats (Fig. 16.2). With the exception of the striated caracaras (*Phalcoboenus australis*) and the Andean condor, the remaining 24 species inhabit, either solely or as part of a major variety of habitats, in forested habitats. This is not surprising due to the large amount of land that this habitat type comprises in the Neotropical realm, from extended Amazonian rainforest in the tropics to the seasonally Dry Chaco forest or Andean temperate forest in southern South America. However, most of forest lands in the Neotropics (85%) are concentrated in the humid tropical domain where savannas and tropical rainforests equally comprise the two most important forest landscapes (Bailey 1998).

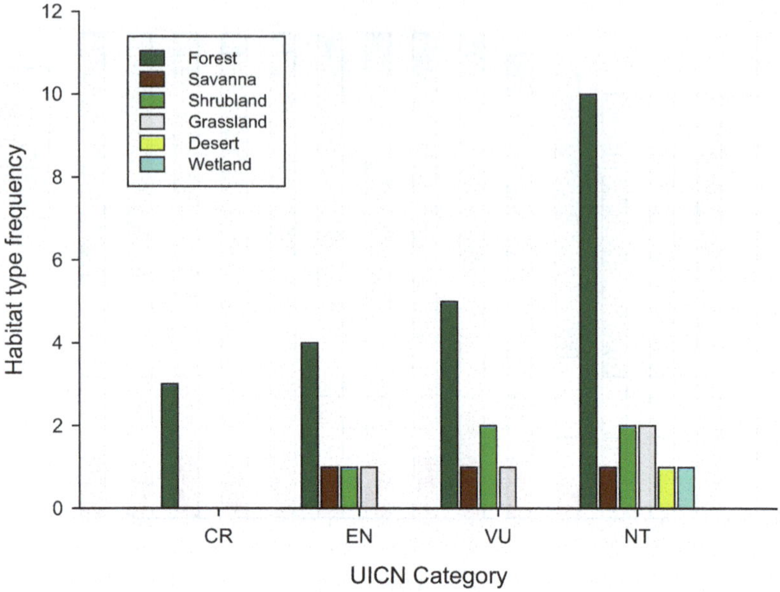

Fig. 16.2 Occurrence of main habitat types in the distribution range of 12 Neotropical raptors categorized either as globally threatened or near threatened according to the IUCN. (Birdlife International 2017)

The three of the most severely threatened Neotropical raptors categorized as Critically Endangered (Table 16.1) are strictly forest dwellers, while more than half of the species categorized as Endangered also inhabit forested landscapes. This pattern is also observed when considering all the 24 species categorized as near threatened or threatened, with 16 of these species (66%) restricted to this single habitat type.

As expected, the diversity of habitats occurring in the species range increases as the species conservation risk decreases. However, this arrangement could be partially due to the fact that species classification criteria take, among several other things, a restricted distribution range as a handicap for species conservation (e.g., species with limited range size are more prone to inhabit a single type of habitat type and under higher risks). There are, however, some exceptions such as the crowned solitary eagle (*Buteogallus coronatus*) (Fig. 16.3). Although this species inhabits a variety of habitats across several countries of southern South America from Brazil, Paraguay, Bolivia, and Argentina, its declining populations are reduced and facing several human-related threats (Sarasola and Maceda 2006; Barbar et al. 2016; Galmes et al. 2017). Also the near-threatened orange-breasted falcon (*Falco deiroleucus*) range occurs in a variety of habitats in Central and South America, from savannas to lowland forests, where its populations show severe declines resulting from different anthropogenic threats. Shrublands (five species), grasslands (four species), and savannas (three species) are also more frequent habitats occurring in the ranges of threatened and near-threatened raptors in the Neotropics.

Fig. 16.3 Dry forests, semiarid savannas, and shrublands are represented across several ecoregions in southern South America, such as the Chaco, Espinal, and Monte Desert biomes. Those are the habitats inhabited by the crowned solitary eagle which is threatened by habitat loss but also for other anthropogenic factors such as direct persecution and electrocution with power lines. (Photo: J. O. Gjershaug)

Major Threats for Neotropical Raptors

Previous assessments of the conservation status of raptors in the Neotropics highlight habitat loss, direct persecution, and environmental contaminants as the main threats for their populations (Thiollay 1985a, b; Bierregaard 1998, Bildstein et al. 1998). Each of these global threats, however, act at different spatial and temporal scales with consequences on raptor populations that sometimes depends on species-specific traits as well as on the degree and extent of such perturbations. There are currently 16 threats, actions, and processes for Neotropical raptors that are included in 9 global threats (Birdlife International 2017).

Habitat Fragmentation and Loss

Development of agriculture activities is one of the most important threats for Neotropical raptors. Seventy-one or 46% of the species among those categorized as either threatened or near threatened are negatively affected by the expansion and intensification of agriculture and by livestock ranching activities (Table 16.2). The ultimate ways in which these two human-related activities threat raptor populations are linked to the destruction of natural habitats by increasing the land areas devoted to ranching and crop implantation at the expense of natural habitats (in this book Chap. 9). Although these land-use changes may affect open habitats like shrublands

Table 16.2 Main threats and number of species affected for each of the 24 Neotropical raptor species categorized either as threatened and near threatened

Activity or process	Particular threat	Number of species
Agriculture and aquaculture	Annual and perennial non-timber crops	17
	Livestock farming and ranching	11
	Wood and pulp plantations	3
	Marine and freshwater aquaculture	1
Biological resource use	Hunting and trapping	15
	Logging and wood harvest	14
Climate change and severe weather	Habitat shifting and alteration	2
Energy production and mining	Mining and quarrying	4
Human intrusions and disturbance	Work and other activities	2
	Recreational activities	1
Invasive species, genes, and diseases	Invasive nonnative/alien species/diseases	4
	Problematic native species/diseases	4
Natural system modifications	Dams and water management/use	2
Residential and commercial development	Housing and urban areas	6
	Commercial and industrial areas	3
Transportation and service corridors	Roads and railroads	3

and natural grasslands, the more notorious effects are observed on forest landscapes. In addition, land-use changes from extensive cattle ranching to intensive agriculture may also threaten some raptor species (Alvarez-López and Kattan 1995).

Comparatively much more attention has been given to the conservation of animal diversity in tropical rainforest than in other forest biomes in the Neotropical region, such as in dry- or semiarid forests. These biomes represent a great portion of landmass in this realm and are also important for biodiversity conservation. The Chaco forests, for example, hold a similar high species richness to that of Amazonian tropical forests (Redford et al. 1990) and experience high rates of deforestation and clearing (Zak et al. 2004). In the Dry Chaco, a total of 15.8 million ha of original habitats were transformed from natural xerophytic forests to crop fields during the period 1976–2012 (Vallejos et al. 2015), with such deforestation particularly noticeable in countries such as Paraguay. For both Amazonia and Chaco regions, direct conversion of forest to croplands seems to be driven by increasing implantation of soybean crops (Grau et al. 2005; Morton et al. 2006). Also deforestation and conversion of semiarid forests of the Espinal biome for cattle ranching and crop implantation significantly diminish the distribution range of crowned solitary eagles in Argentina (Fandiño and Pautasso 2013).

Afforestation of grassland environments for wood and pulp production is another activity that severely modify and affect native habitats, often resulting in less suitable environments for raptors and other open-habitat species. Also the replacement of native forests for stands of non-native species (e.g.,

Eucalyptus sp. and *Pinus* sp.) for forestry purposes results in new habitats with reduced prey availability due to the changes in woody plant density and vegetation structure.

Besides the expansion of agriculture, disappearance or severe modification of native forest through vast areas in the Neotropics has also been caused by logging and wood harvest activities. This activity threatens 58% of the raptor species considered of conservation concern in this region (Table 16.2). Removal of native forest by lumbering operations results in successional stages and forest fragmentation that lead to marked decreases in abundance of forest-dwelling raptor species (Jullien and Thiollay 1996). Up to 42% of forest-interior raptor species have shown significant decreases in their densities through a successional gradient from primary to fragmented forests in French Guiana (Thiollay 1985a), and a similar pattern has been observed in rainforest of northern Andes (Thiollay 2009). Fragmentation and changes in habitat structure also determine shifts in the composition of raptors communities through increasing abundances of savanna and open-habitat raptor species (Thiollay 1984).

Direct Persecution

Direct human perturbation on raptors is mainly related to shooting and hunting and is the second-order threat for Neotropical raptors, affecting no less than 62% of species categorized as of conservation concern. Birds of prey, as other top predators such as felids and large mammal carnivores, usually conflict with human activities mainly with livestock and poultry production. This is particularly true for large eagles, although the ability of raptors to kill livestock is questionable.

In the Neotropics, cases of direct persecution have been documented for several raptor species, such as the harpy eagle (Muñiz-López 2017) and the crowned solitary eagle (Maceda et al. 2003; Sarasola and Maceda 2006; Barbar et al. 2016), while it is suspected but barely documented for many others. The black-and-chestnut eagle (*Spizaetus isidori*; Fig. 16.4) and the Andean condor, for example, are considered as injurious species and shot by local farmers (Zuluaga and Echeverry-Galvis 2016; Cailly Arnulphi et al. 2017), but the extent and demographic impact of human persecution on their populations has not been yet assessed. Illegal shooting on raptors in general occurs in many countries in the Neotropics, despite local regulations that protect birds of prey (Wiley 1986; Bechard and Marquez-Reyes 2003).

Human Disturbance

Urbanization is a widespread form of human disturbance and habitat change occurring worldwide and causing the loss of biodiversity through local extinction processes (McKinney 2002). The development and increase of urban areas at the

Fig. 16.4 The black-and-chestnut eagle is a rainforest-dweller species and also one of the most threatened raptors in the Neotropics. With a narrow distribution restricted to rainforest located in the eastern side of the Andes from Colombia to Argentina, only a few nests for the species have been described up to date. In the picture, an adult individual caring the young eagle at the first discovered nest of the species in the Yungas biome in Argentina. (Photo: R. Aráoz)

expenses of natural habitats force species to move to other sites to avoid the presence of humans and new and unsuitable urban habitats (in this book Chap. 8). This threat affects 25% of Neotropical raptor species of conservation concern (Table 16.2). Furthermore, species that adapt to these modified environments and tolerate human presence often face a variety of mortality factors related to human's infrastructures (e.g., window and vehicle collisions, shooting, etc.) (Hager 2009). In addition, other human activities may also significantly disturb raptors. Hunting game and nongame species has resulted in severe declines of both canopy and subcanopy raptor densities even in primary forest areas, leading some species, like the harpy eagle, to virtually disappear from vast areas as a result either of direct killing (e.g., persecution), disturbance, or reduction of food sources (Thiollay 1984).

Other Sources of Mortality

Noteworthy, some human-related mortality factors affecting wildlife and particularly raptors have received little attention in the Neotropics. Roads and railroads, for example, threaten 12% of globally threatened raptors (Table 16.2), but there are few assessments on this cause of mortality for wildlife in general for much of the region. Most of the studies have been done in the Amazonian region where road killing has been documented affecting several raptor species (Klippel et al. 2015). Furthermore, vehicle collisions have been also recorded for the crowned solitary eagle in central Argentina (Maceda et al. 2003).

Probably more remarkable is the lack of systematic studies on avian electrocution with power lines for great part of the Neotropics (Lehman et al. 2007). This mortality source is one of the major threats that human infrastructures pose for several raptor species around the world (Negro 1999). Recent studies, however, highlight this human-related mortality source as one of the most important for the endangered crowned solitary eagle in central Argentina but also affecting other diurnal raptor species such as the black vulture (*Coragyps atratus*) and the turkey vulture (*Cathartes aura*) (Galmes et al. in press).

Conservation of Migrant Raptors in the Neotropics

The population status and conservation of migratory birds (i.e., species performing seasonal and regular return movements between breeding and non-breeding grounds) are related to direct and indirect factors that limit reproduction and survival of individuals in their breeding areas, along migration routes, and on their non-breeding grounds (Sherry and Holmes 1995; Sillet and Holmes 2002; Newton 2014).

Approximately 30% of the Neotropical raptors are considered as migrant, performing or suspected of performing short- to long-distance migratory movements. Although with representatives in all the four orders, the larger proportion of migrant species in this region belong to Accipitriformes (73%). Among these, one species is considered of conservation concern: the grey-bellied goshawk (*Accipiter poliogaster*). This species is categorized as near threatened and ranges over much of southern South America from Costa Rica to northern Argentina (Thiollay 1994; Fergusson-Lees and Christie 2001). It is considered either as a partial (Bildstein 2006) or full migrant species; however, there is scarce information on its biology and particularly on its migratory patterns which has mostly been inferred from seasonal clusters of individuals in field records in parts of its range. The species is recorded mostly in more pristine habitats, avoiding degraded and fragmented forests (Jullien and Thiollay 1996). The main threat for this bird-eating lowland forest specialist appears to be the acceleration of deforestation in the Amazonia basin (Soares-Filho et al. 2006).

Migratory bird populations are limited by their need of quality habitats in which to maintain high fecundity in summer and high survival in winter (Sherry and Holmes 1995; Newton 2014). Mortality of migratory species in their non-breeding grounds can hence compromise global population viability and demography with carry-over effects on their breeding populations. Although with globally stable populations, at least two long-distance migrant raptors species wintering in the Neotropics have experienced high mortality rates in their non-breeding grounds. Individuals from some osprey (*Pandion haliaetus*) populations migrate from the USA through the Caribbean to Colombia. In their wintering ground, ospreys usually feed on fish raised in aquaculture facilities and are consistently shot by aquaculture farmers. The number of eagles shot may be high, with up to five eagles shot per year/ facility and total numbers of 270 eagles shot/year at some provinces in this country

In the austral summer of 1996, 700 Swainson's hawks (*Buteo swainsoni*) were found killed in the Argentinean pampas by researchers tracking the species' migratory pathways from its breeding grounds in California to austral wintering destinations in southern South America for the first time (Woodbridge et al. 1995). The next austral summer, a total of 5000 hawks were found killed in the same area with an estimated of 20,000 individuals affected (Goldstein et al. 1996). An organophosphate insecticide commercially called monocrotophos was identified as responsible for these massive poisonings. The Swainson's hawk is a highly gregarious species during its non-breeding period (Fig. 16.5), a behavioral trait that may enhance the occurrence of mass mortality incidents also by non-human related causes such as severe weather conditions (Sarasola et al. 2005). As a result of these mortality events and the accurate identification of the product causing the massive poisoning (Goldstein et al. 1999), monocrotophos was banned in Argentina in 1998.

New World Vultures

Vultures are among the largest representatives among birds of prey, and they share as a group some behavioral and ecological traits that are singular and relevant from a conservation perspective. Unlike other raptor species that may consume carrion as an alternative food source (i.e., facultative scavengers), vultures are obligate scavengers and rely exclusively on the availability of carcasses of medium- to large-bodied vertebrates, including livestock, to subsist. Vultures are also exclusive soaring flight birds with high displacement capacity. Other particular but not exclusive traits exhibited by vultures are their social behavior both for foraging and roosting and a markedly delayed sexual maturity.

The New World vultures are included in the order Cathartiformes which are endemic to the Western Hemisphere. With the exception of the California condor (*Gymnogyps californianus*) with populations currently confined to small areas in western North America, all the remaining species range in or are even exclusively in the Neotropics. Among these, the Andean condor (*Vultur gryphus*) is the only categorized as near threatened and hence considered to be of conservation concern

Fig. 16.5 Swainson's hawks in their non-breeding grounds in the Pampas of Argentina. This long-distance migratory species is highly gregarious during the non-breeding period and inhabits agricultural lands of southern South America during its non-breeding period. During the 1990s massive poisoning with organophosphate insecticides in central Argentina killed an estimated of 20,000 hawks. (Photo: J.H. Sarasola)

(Table 16.1). Although with some breeding populations in other mountainous areas of Argentina, the Andean condor distribution is mainly restricted to the Andes from southern Patagonia to Colombia. Major identified threats for this species are secondary poisoning for predator control, poisoning from lead exposure, and direct human persecution. The former has recently drawn attention of conservationist in Argentina because the report of a massive poisoning incident that resulted in 34 condors being killed by a single Carbofuranbaited sheep carcass (Birdlife International 2018). Lead poisoning, however, seems to be widespread through most of the species range (Wiemeyer et al. 2017), and it could be also affecting other endangered raptor species as well (Saggese et al. 2009). In addition, and although not yet assessed in the Neotropics, it is expected that lead poisoning could be also an important threat for other New World vulture species. All individuals of both black and turkey vultures sampled in Virginia (USA), for example, had bone lead levels indicative of chronic exposure to anthropogenic lead from multiple sources (Behmke et al. 2015; for further details on lead effect on raptors, see in this book Chaps. 10 and 11).

New World vultures were sacred and highly symbolic species for many pre-Colombian cultures in Latin American, either associated with death or life, to the point that in some cases it is more common than in other bird species, as displayed in their representation in ritual and religious portrays and sculptures (Sault 2016).

However, after the European colonization and the growth of livestock ranching activities in most parts of the Americas, human perception toward vultures and other scavenger species turned to a more negative one, resulting in active human persecution. Negative perception of people toward vultures is probably associated to their scavenger habits as cases of vultures actively preying on livestock are not well documented in South America.

Although most of these vulture species seem to be human tolerant and more sensitive to environmental food availability than to structural habitat transformations, there are some exceptions. The king vulture (*Sarcoramphus papa*), for example, is reported to not adapt well to human presence, avoiding urban areas and other human-altered landscapes. In addition, this species is strongly associated with primary rainforest resulting in severe populations declines or even local extinctions as a result of deforestation (Jullien and Thiollay 1996). Human disturbance may also determine behavioral changes on key aspects of vulture ecology, such as feeding behavior and habitat use selection (Speziale et al. 2008).

Overall, populations of New World vulture species seem to be far from the current conservation crisis that Old World vultures face that has resulted in massive declines in their populations. However, for most of these Neotropical vulture species, there are important gaps in information about their basic biology and ecology (e.g., food habits, breeding, population dynamic) as well as other threats such as human persecution, chemicals, and diseases, and efforts should be taken on assessing in the Neotropics those threats that may be similar to their Asian or African counterparts.

Endemism and Conservation on Islands

Islands represent a particular paradigm for biodiversity conservation. Some of the most important lessons on the anthropogenic effects on biodiversity belong to the catastrophic consequences of the modern arrival of humans on islands, irrespective of the taxonomic groups we refer to herein. Islands have particular conditions, such as well-defined limits and isolation from other landmasses, that lead to increased speciation processes which determine demographic conditions for these species.

Hundreds of species extinctions and severe reduction or complete loss of singular habitats are the concluding remarks for human impacts on island biodiversity. This was already noted by Charles Darwin when visiting the Malvinas Islands who envisioned the extinction fate of the islands endemic wolf as had previously occurred with the Dodo in Madagascar (Darwin 1871). Throughout a large list of extinct vertebrate species in islands and archipelagoes on all continents, the insular fauna has suffered the most from the impact of human arrival in modern times.

Raptors also seem to have followed this rule as the single case of modern raptor extinction belongs to an island. The Guadalupe caracara (*Caracara lutosa*) was once a common species in the Guadalupe Island (Mexico) (Abbott 1933). The species never had a good reputation among goat herders on the island and that promoted hunting and poisoning campaigns to completely eradicate this raptor from the island arguing the losses caused to goat herding. In the early twentieth century, the Guadalupe caracara was rare and difficult to see (Kaeding 1905), and it is considered to have become extinct in 1906, being one of the few cases in which species extinction was the result of human premeditated actions to that end. Fortunately, other persecuted island inhabitant raptors have had a different end. The striated caracara (*Phalcoboenus australis*), for example, was initially hunted by local farmers on the Malvinas Islands and archipelagoes of southern Argentina and Chile (Thiollay 1994). Although populations were severely reduced in number, and also locally extinct on some islands, the species is no longer intensively persecuted and currently shows stable populations.

Islands are also strongly related with species endemism to the point that vertebrate species richness endemism in islands is globally estimated to be eight times that of mainland regions (Kier et al. 2009). In the Neotropical region, seven raptor species are considered as country endemic, and all of them are of conservation concern, categorized either as threatened or near threatened. All of these country endemic species belong to the order Accipitriformes. Furthermore, five of these species are also island endemics. One exception is the striated caracara whose distribution encompasses several islands in the tip of southern South America. However, this species is not considered as endemic based on the single country distribution criteria (Tierra del Fuego Island is an insular territory geopolitically shared both by Chile and Argentina); it is considered as island endemic for some authors.

Endemics and/or island inhabitant Neotropical raptors comprise a relative small number of species in comparison with that of other regions, such as the Indomalayan and Afrotropical regions (Bildstein et al. 1998), but most of them are of high conservation concern. All the three critically endangered Neotropical raptor species are country endemics and two of them also island endemics. Examples of island-threatened raptors include two Cuban-restricted species, the Cuban kite (*Chondrohierax wilsonii*) and the Gundlach's hawk (*Accipiter gundlachi*). Both are rare and scarce, to the point that Gundlach's hawk was for a time considered to be extinct due to habitat destruction, invasive species, and shooting (Wiley 1986).

The endemic Ridgway's hawk (*Buteo ridgwayi*) is categorized as critically endangered in the Dominican Republic. This hawk is probably the exception among island endemic raptors in the Neotropics with regard to the efforts made to increase the knowledge about its biology and population status (Wiley and Wiley 1981; Thorstrom et al. 2005) but also on the conservation actions undertaken to recover its populations (McClure et al. 2017, in this book Chap. 20). Also the Galapagos hawk (*Buteo galapagoensis*) has attracted attention, and several studies have reported the species food habits, breeding biology, and behavior (Faaborg 1986; Jaramillo et al. 2016).

Conclusions

The number of threatened raptor species in the Neotropics has increased gradually in the last 30 years, and the estimated population trends for most of them do not allow us to visualize a better conservation scenario in the coming years. To the extent that field and scientific research on little studied species improve our capability to obtain better information on population sizes and on the threats that each species faces, it is probable that some species currently considered as of least concern could be prone to be included in riskier categories. Some species, for example, will likely be at risk if actions to stop current global threats, such as deforestation of Amazonian, Chaco, and Espinal forests, are not taken. Most of the globally threatened or near-threatened species in the Neotropics are strongly related with these forested landscapes, which are also one of the habitats suffering from the highest rate of destruction or transformation. In addition, human-related causes of mortality, either through active persecution, electrocution, pollution, or poisoning, need to be adequately assessed, and actions need also to be taken to diminish and prevent such threats.

One of the greatest challenges facing conservationists interested in protecting Neotropical raptors is the scarce information available on basic biology for many of these species. This is true not only for secretive tropical forest-dwelling, island, or range-restricted endemic species but also for wide-ranging species found in a great diversity of habitats and regions but also facing important conservation issues. Although information on species distribution seems to be accurate and continuously improved through and increasing number of researchers and birdwatchers helped with novel tools and citizen science actions, we still lack of reasonable estimates on population sizes, on seasonal movements, as well as on factors limiting and threatening their populations. Such information is basic for better planning species conservation that also needs of information on species habitat preferences and spatial requirements (Thiollay 2008).

Except for some particular cases, our knowledge on the biology of most of the Neotropical raptor species and particularly of those rainforest dwellers is very incomplete and collected almost two decades ago (Bierregaard 1995). For example, information on species food habits is essential for disentangling potential conflicts between raptor species and human economical activities (Sarasola et al. 2010; Zuluaga and Echeverry-Galvis 2016; Aráoz et al. 2017). Also information on nesting and breeding biology, including clutch sizes and breeding success, is unknown for many raptor species. Such baseline information is essential for demographic and viability population analysis aiming to plan management and conservation actions of raptor species with severely reduced populations and close to their extinction vortex.

Fortunately, a growing number of biologists and raptor enthusiastic are gathering valuable field information for many of these species. As a result, some species have been rediscovered after years without records in habitats expected to be suitable for them. That is the case of the white-collared kite (*Leptodon forbesi*) in northeastern Brazil where an increasing number of field records have allowed better estimates for the current distribution of the species (Pereira et al. 2014). For other species, nesting records have been gathered for countries included on their distribution range but never confirmed. This is the case of the black-and-chestnut eagle for which the first nesting record has been recently reported for Argentina (Araoz et al. 2017). Long-term field research projects, such as the Maya project conducted by the Peregrine Fund in Guatemala, have significantly improved our knowledge of several rainforest raptor species in Central America (Whitacre 2012). Efforts like these need to be encouraged and replicated in other regions of the Neotropics.

Even though island systems may be of little significance with regard to the extension or species richness in comparison to the main lands, islands embody singularities and challenges from the perspective of conservation of raptor species in the Neotropics. An example of this is the high proportion of island raptors at risk but also the scarce information on the population status and ecology for most of them.

Despite their cyclically chronic economic crises, many countries in the Neotropics have seen a wide economic expansion in the last decades. This economic growth has come with a wide expansion of environmental impacts to levels previously unknown (e.g., larger than ever deforestation in the Amazon, the Chaco, and other continental forests; large-scale agricultural intensification processes; a wide expansion of human cities and infrastructures like roads, power lines, dams, and power plants; etc.). Although some knowledge has been gathered on the cost of these changes to raptors, it is clear that the future will pose increasing pressures on raptorial birds. A wider interplay between researchers, NGOs, wildlife and nature management agents, and politicians will be needed in the future to tackle the conservation threats imposed by regional development on birds of prey.

References

Abbott CG (1933) Closing history of the Guadalupe Caracara. Condor 35:10–14. https://doi.org/10.2307/1363459

Alvarez-López H, Kattan GH (1995) Notes on the conservation status of resident diurnal raptors of the middle Cauca Valley, Colombia. Bird Conserv Int 5:341–348. https://doi.org/10.1017/S0959270900001088

Aráoz R, Grande JM, López C, Cereghetti J, Vargas FH (2017) The first black-and-chestnut eagle (*Spizaetus isidori*) Nest discovered in Argentina reveals potential human – predator conflicts. J Raptor Res 51:79–82. https://doi.org/10.3356/JRR-16-49.1

Bailey RG (1998) The humid tropical and mountain ecoregions. In: Ecoregions. The ecosystem geography of the oceans and continents. Springer, New York, pp 105–131

Barbar F, Capdevielle A, Encabo M (2016) Direct persecution of crowned eagles (*Buteogallus coronatus*) in Argentina: a new call for their conservation. J Raptor Res 50:115–120. https://doi.org/10.3356/rapt-50-01-115-120.1

Bechard MJ, Marquez-Reyes C (2003) Mortality of wintering ospreys and other birds at aquaculture facilities in Colombia. J Raptor Res 37:292–298

Behmke S, Fallon J, Duerr AE, Lehner A, Buchweitz J, Katzner T (2015) Chronic lead exposure is epidemic in obligate scavenger populations in eastern North America. Environ Int 79:51–55. https://doi.org/10.1016/j.envint.2015.03.010

Bierregaard RO (1995) The biology and conservation status of central and south American Falconiformes: a survey of current knowledge. Bird Conserv Int 5:325–340. https://doi.org/10.1017/S0959270900001076

Bierregaard R (1998) Conservation status of birds of prey in south American tropics. J Raptor Res 32:19–27

Bildstein KL (2006) Migrating raptors of the world: their ecology and conservation. Cornell University Press, New York

Bildstein KL, Schelsky W, Zalles J (1998) Conservation status of tropical raptors. J Raptor Res 32:3–18. https://doi.org/10.7550/rmb.21364

Birdlife International (2017) IUCN red list for birds. [Online.] Available at www.birdlife.org

Birdlife International (2018) 34 Andean condors found dead in Argentina – the poisoning needs to stop. https://www.birdlife.org/worldwide/news/34-andean-condors-found-dead-argentina-poisoning-needs-stop

Cailly Arnulphi VB, Lambertucci SA, Borghi CE (2017) Education can improve the negative perception of a threatened long-lived scavenging bird, the Andean condor. PLoS Oné 12:1–13. https://doi.org/10.1371/journal.pone.0185278

Darwin C (1871) Countries visited during the voyage of H.M.S Beagle round the world. Nature. https://doi.org/10.1038/026169a0

Donázar JA, Cortés-Avizanda A, Fargallo JA, Margalida A, Moleón M, Morales-Reyes Z, Moreno-Opo R, Pérez-García JM, Sánchez-Zapata JA, Zuberogoitia I, Serrano D (2016) Roles of raptors in a changing world: from flagships to providers of key ecosystem services. Ardeola 63:181–234. https://doi.org/10.13157/arla.63.1.2016.rp8

Faaborg J (1986) Reproductive success and survivorship of the Galapagos hawk *Buteo galapagoensis*: potential costs and benefits of cooperative polyandry. Ibis 128:337–347. https://doi.org/10.1111/j.1474-919X.1986.tb02684.x

Fandiño B, Pautasso A (2013) Distribución, historia natural y conservacion de *Harpyhaliaetus coronatus* (Aves: Accipitridae) en el centro-este de Argentina. Natura Neotropicalis 44:41–59

Ferguson-Lees J, Christie DA (2001) Raptors of the world. Christopher Helm, London

Galmes MA, Sarasola JH, Grande JM Vargas FH (2017) Electrocution risk for the endangered crowned solitary eagle and other birds in semiarid landscapes of Central Argentina. Birds Conservation International in press

Goldstein MI, Woodbridge B, Zaccagnini ME, Canavelli SB, Lanusse A (1996) An assessment of mortality of Swainson's hawks on wintering grounds in Argentina. J Raptor Res 30:106–107

Goldstein MI, Lacher TEJ, Woodbridge B, Bechard MJ, Canavelli SB, Zaccagnini ME, Cobb GP, Scollon EJ, Tribolet R, Hooper MJ (1999) Monocrotophos-induced mass mortality of Swainson's hawks in Argentina, 1995–96. Ecotoxicology 8:201–214

Grau HR, Gasparri NI, Aide TM (2005) Agriculture expansion and deforestation in seasonally dry forests of north-West Argentina. Environ Conserv 32:140. https://doi.org/10.1017/S0376892905002092

Hager SB (2009) Human-related threats to urban raptors. J Raptor Res 43:210–226. https://doi.org/10.3356/JRR-08-63.1

del Hoyo J, Elliot A, Sargatal J (1994) Handbook of the birds of the world. Vol. 2. New world vultures to Guineafowls. Lynx Edicions, Barcelona

Jaramillo M, Donaghy-Cannon M, Vargas FH, Parker PG (2016) The diet of the Galapagos hawk (*Buteo galapagoensis*) before and after goat eradication. J Raptor Res 50:33–44

Jullien M, Thiollay J-M (1996) Effects of rain forest disturbance and fragmentation: comparative of the raptor and human-made changes community along natural gradients in French Guiana. J Biogeogr 23:7–25

Kaeding BH (1905) Birds from the west coast of lower California and adjacent island. Condor VII:134–138. https://doi.org/10.2307/1361387

Kier G, Kreft H, Lee TM, Jetz W, Ibisch PL, Nowicki C, Mutke J, Barthlott W (2009) A global assessment of endemism and species richness across island and mainland regions. Proc Natl Acad Sci 106:9322–9327. https://doi.org/10.1073/pnas.0810306106

Klippel AH, Oliveira PV, Britto KB, Freire BF, Moreno MR, Dos Santos AR, Banhos A, Paneto GG (2015) Using DNA barcodes to identify road-killed animals in two Atlantic forest nature reserves, Brazil. PLoS One 10:1–15. https://doi.org/10.1371/journal.pone.0134877

Lehman RN, Kennedy PL, Savidge J a (2007) The state of the art in raptor electrocution research: a global review. Biol Conserv 135:459–474. https://doi.org/10.1016/j.biocon.2006.10.039

Maceda JJ, Sarasola JH, Emilio M, Pessino M (2003) Presas consumidas por el Águila Coronada (*Harpyhaliaetus coronatus*) en el límite sur de su rango de distribución en Argentina. Ornitologia Neotropical 14:419–422

McClure CJW, Rolek BW, Hayes TI, Hayes CD, Thorstrom R, Curti M, Anderson DL (2017) Successful enhancement of Ridgway's hawk populations through recruitment of translocated birds. Condor 117:855–864. https://doi.org/10.1650/CONDOR-17-77.1

McKinney M (2002) Urbanization, biodiversity, and conservation. Bioscience 52:883–890

Morton DC, DeFries RS, Shimabukuro YE, Anderson LO, Arai E, del Bon Espirito-Santo F, Freitas R, Morisette J (2006) Cropland expansion changes deforestation dynamics in the southern Brazilian Amazon. Proc Natl Acad Sci 103:14637–14641. https://doi.org/10.1073/pnas.0606377103

Muñiz-López R (2017) Harpy eagle (*Harpia harpyja*) mortality in Ecuador. Stud Neotropical Fauna Environ 52:81–85. https://doi.org/10.1080/01650521.2016.1276716

Negro JJ (1999) Past and future research on wildlife interactions with power lines. In: Ferrer M, Janss GFE (eds) Birds and power lines: collisions, electrocution and breeding. Quercus, Madrid, pp 21–28

Newton I (2014) Population limitation in migrants. Ibis 146:197–226

Pereira GA, Dantas S d M, Silveira LF, Roda SA, Albano C, Sonntag FA, Leal S, Periquito MC, Malacco GB, Lees AC (2014) Status of the globally threatened forest birds of Northeast Brazil. Pap Avulsos de Zool (São Paulo) 54:177–194. https://doi.org/10.1590/0031-1049.2014.54.14

Redford K, Taber A, Simonetti J (1990) There is more to biodiversity than the tropical rain forests. Conserv Biol 4:328–330

Saggese MD, Quaglia A, Lambertucci SA, BO MS, Sarasola JH, Pereyra-Lobos R, Maceda JJ (2009) Survey of lead toxicosis in free ranging raptors from Central Argentina. In: Watson RT, Fuller M, Pokras M, Hunt WG (eds) Ingestion of lead from spent ammunition: implications for wildlife and humans. The Peregrine Fund, Boise, Idaho, pp 223–231. https://doi.org/10.4080/ilsa.2009.0211

Sarasola JH, Maceda JJ (2006) Past and current evidence of persecution of the endangered crowned eagle *Harpyhaliaetus coronatus* in Argentina. Oryx 40:347–350. https://doi.org/10.1017/S0030605306001013

Sarasola JH, Negro JJ, Salvador V, Maceda JJ (2005) Hailstorms as a cause of mass mortality of Swainson's Hawks in their Wintering grounds. J Wildl Dis 41(3):643–646

Sarasola JH, Santillán MÁ, Galmes MA (2010) Crowned eagles rarely prey on livestock in Central Argentina: persecution is not justified. Endanger Species Res 11:207–2013

Sault N (2016) How hummingbird and vulture mediate between life and death in Latin America. J Ethnobiol 36:783–806. https://doi.org/10.2993/0278-0771-36.4.783

Sherry TW, Holmes RT (1995) Summer versus winter limitations of populations: what are the issues and what the evidences? In: Martin TE, Finch DM (eds) Ecology and management of Neotropical migratory birds. Oxford University Press, Oxford, pp 85–120

Sillett TS, Holmes RT (2002) Variation in survivorship of a migratory songbird throughout its annual cycle. J Animal Ecol 71:296–308

Soares-Filho BS, Nepstad DC, Curran LM, Cerqueira GC, Garcia RA, Ramos CA, Voll E, McDonald A, Lefebvre P, Schlesinger P (2006) Modelling conservation in the Amazon basin. Nature 440:520–523. https://doi.org/10.1038/nature04389

Speziale KL, Lambertucci SA, Olsson O (2008) Disturbance from roads negatively affects Andean condor habitat use. Biol Conserv 141:1765–1772. https://doi.org/10.1016/j.biocon.2008.04.017

Thiollay JM (1984) Raptor community structure of a primary rain forest in French Guiana and effect of human hunting pressure. Raptor Res 18:117–122

Thiollay J-M (1985a) Composition of Falconiform communities along successional gradients from primary rainforests to secondary habitats. ICBP Tech Publ 5:181–190
Thiollay J-M (1985b) Falconiformes of tropical rainforests: a review. ICBP Tech Publ 5:155–165
Thiollay JM (1994) Family Accipitridae (hawks and eagles). In: Handbook of the birds of the world. Vol 2. New world vultures to guineafowl. Lynx Editions, Barcelona, pp 52–205
Thiollay JM (2008) Area requirements for the conservation of rain forest raptors and game birds in French Guiana. Conserv Biol 3:128–137
Thiollay JM (2009) Distributional patterns of raptors along altitudinal gradients in the northern Andes and effects of forest fragmentation. J Trop Ecol 12:535–560
Thorstrom R, Almonte J, Balbuena de la Rosa S, Rodríguez P, Fernández E (2005) Surveys and breeding biology of *Buteo ridgwayi* (Ridgway's hawk) in Los Haitises, Dominican Republic. Caribb J Sci 41:864–869
Trejo A, Ojeda V (2015) Aportes desde la vertiente argentina al conocimiento de las aves rapaces del Bosque Templado Austral. Bol Chil de Ornitología 21:15–28
Udvardy MDF (1975) A classification of the biogeographical provinces of the world. International Union for Conservation of Nature and Natural Resources Occasional Paper 18
Vallejos M, Volante JN, Mosciaro MJ, Vale LM, Bustamante ML, Paruelo JM (2015) Transformation dynamics of the natural cover in the dry Chaco ecoregion: a plot level geo-database from 1976 to 2012. J Arid Environ 123:3–11. https://doi.org/10.1016/j.jaridenv.2014.11.009
Whitacre D (ed) (2012) Neotropical birds of prey: biology and ecology of a forest raptor community. Comstock Publishing Associates, Ithaca
Wiemeyer GM, Perez MA, Torres Bianchini L, Sampietro L, Bravo GF, Jacome NL, Astore V, Lambertucci SA (2017) Repeated conservation threats across the Americas: high levels of blood and bone lead in the Andean condor widen the problem to a continental scale. Environ Pollut 220:672–679. https://doi.org/10.1016/j.envpol.2016.10.025
Wiley JW (1986) Status and conservation of raptors in the West Indies. Birds Prey Bull 3:57–70
Wiley JW, Wiley BN (1981) Breeding season ecology and behavior of Ridgway's hawk (*Buteo ridgwayi*). Condor 83:132–151
Woodbridge B, Finley KK, Seager ST (1995) An investigation of the Swainson's hawk in Argentina. J Raptor Res 29:202–204
Zak MR, Cabido M, Hodgson JG (2004) Do subtropical seasonal forests in the gran Chaco, Argentina, have a future? Biol Conserv 120:589–598. https://doi.org/10.1016/j.biocon.2004.03.034
Zuluaga S, Echeverry-Galvis MA (2016) Domestic fowl in the diet of the black-and-chestnut eagle (*Spizaetus isidori*) in the eastern Andes of Colombia: a potential conflict with humans? Ornitologia Neotropical 27:113–120

Chapter 17
Conservation Threats and Priorities for Raptors Across Asia

Camille B. Concepcion, Keith L. Bildstein, Nigel J. Collar, and Todd E. Katzner

Introduction

With long coastlines and some of the world's most important rivers, mountain ranges, high-altitude plateaus, and islands, Asia is the largest and most populous continent in the world (Lyde 1904; Spencer 1954; Population Reference Bureau 2016). Asia supports all major terrestrial ecosystems and all major climatic types (Galloway et al. 1998; Braimoh and Huang 2015). These include barren ice fields and taigas in North Asia; boreal forests and cold deserts in West, Central, and East Asia; temperate and tropical forests (wet and dry) in East and Southeast Asia; and grasslands in Central and South Asia (Udvardy 1975; Braimoh and Huang 2015). Together, the ecoregions of Asia foster some of the greatest biodiversity on Earth, including six (24%) of 25 global biodiversity hotspots (Myers et al. 2000).

C. B. Concepcion (✉)
Division of Forestry and Natural Resources, West Virginia University, Morgantown, WV, USA

Science Department, College of Natural Sciences and Mathematics, Mindanao State University, Fatima, General Santos City 9500, Philippines

K. L. Bildstein
Acopian Center of Conservation Learning, Hawk Mountain Sanctuary, Orwigsburg, PA, USA

N. J. Collar
BirdLife International, Cambridge, UK

T. E. Katzner
United States Department of Agriculture, Forest Service, Timber and Watershed Laboratory, Parsons, WV, USA

U.S. Geological Survey, Forest and Rangeland Ecosystem Science Center, Boise, ID, USA

Division of Forestry and Natural Resources, West Virginia University, Morgantown, WV, USA

J. H. Sarasola et al. (eds.), *Birds of Prey*,
https://doi.org/10.1007/978-3-319-73745-4_17

Table 17.1 Distribution and conservation status of raptors in the world and in Asia based on BirdLife International (2017). Migratory species include species with a significant portion of populations that make seasonal, cyclical, and predictable movements. Endemic species are species that breed in a single country

	Accipitriformes		Falconiformes		Strigiformes	
	World	Asia	World	Asia	World	Asia
Total	251	105	66	22	241	108
Migratory	165	42	15	10	17	8
Endemic	52	20	8	2	86	52
Extinct	1	0	2	0	5	0
Critically endangered	14	6	0	0	7	2
Endangered	14	5	2	1	9	5
Vulnerable	25	9	4	0	27	13
Near threatened	33	14	9	5	27	21
Least concern	163	70	49	16	161	64
Data deficient	1	1	0	0	5	3

With 47% of the world's threatened and near-threatened raptor species (Table 17.1), Asia is home to 42% of the world's diurnal and nocturnal raptor species, including 30% of all migratory species and 51% of all endemic species. With the exception of work on raptors in tropical Asia (i.e., southeast Asia; Thiollay 1985, 1998; Bildstein 1998), the *Gyps* vultures of South Asia (Prakash et al. 2003, 2012; Gilbert et al. 2004; Shultz et al. 2004; Acharya et al. 2009; Hall et al. 2011, 2015), and a few other country-level studies, the population status of the vast majority of these species has gone largely unstudied. The raptors of South and Southeast Asia provide a useful umbrella for conservation because the survival of their populations continue to be threatened by land-use change, direct human persecution, and environmental contaminants (Bildstein 1998; Thiollay 1998). These threats rarely act in isolation and are directly or indirectly caused by human activities. Raptors are especially vulnerable to these threats because of their life history traits and because they occur at relatively low population densities and in relatively large home ranges (Newton 1979, 1998; Real and Mañosa 1997; Hall et al. 2015).

Threats to Asian Raptors

Of the threats that Asia's raptors face, land-use change is most significant. Agriculture has modified natural landscapes more than any other human activity, and among the most rapidly expanding agricultural industries are oil palm *Elaeis guineensis* and rubber *Hevea brasiliensis* (Clay 2004). The conversion of natural landscapes greatly reduces species richness and diversity (Aratrakorn et al. 2006; Beukema et al. 2007; Fitzherbert et al. 2008; Sodhi et al. 2010). Loss of raptor habitats can also drive spatial and temporal changes in avian populations by affecting territory size and dispersal (Rolstad 1991).

Land-use change also can carry over to loss of habitat of prey species and the consequent reduction in prey abundance. Declines in prey abundance could result in decreases in nesting density and reproductive success of predatory avian populations (Andersson 1978; Newton 1980; Terraube et al. 2011). Decline in prey abundance also can cause conflict with humans if domestic animals are taken as alternative prey (Carrete et al. 2007).

Land-use change magnifies the threat of human persecution as encounters with people increase in fragmented and human-impacted landscapes. Humans have long intentionally destroyed nests and shot and poisoned raptors they consider as pests (Newton 1979, 1998; Bildstein 2001). Human persecution has resulted in raptor population declines by reducing reproductive rates and preadult and adult survival (Whitfield et al. 2004a, b; Tenan et al. 2012) and has caused range retractions and local extinctions of raptor populations (Newton 1998; Smart et al. 2010).

Environmental contaminants and secondary poisoning also severely impact Asian raptor populations (Newton 1979, 1998; Sheffield 1997; Oaks et al. 2004a; Shultz et al. 2004, in this book Chap. 10). Pesticides (Ratcliffe 1970; Mineau et al. 1999; Jagannath et al. 2008) and heavy metals (Solonen and Lodenius 1984; Wallin 1984; Lourenço et al. 2011) are known to reduce survivorship or impair raptor reproductive success. Anticoagulant rodenticides (Stone et al. 2003; Lambert et al. 2007), pharmaceuticals (Oaks et al. 2004a), and lead (Kramer and Redig 1997; Fisher et al. 2006; Gangoso et al. 2009) can weaken or directly kill individuals.

Below we summarize raptor diversity in mainland and oceanic Asia. We then describe the threats that many of these species face and identify priority species and essential steps for their conservation. We close with a section on poorly known species in Asia and opportunities to fill knowledge gaps that will benefit conservation of raptors in this immense but poorly studied continent.

Methods

We define Asia as the area that includes Asiatic Russia in the north, Indonesia in the south (including West Papua, although the avifauna of New Guinea is predominantly Australasian), Japan in the east, and Uzbekistan, Turkmenistan, Afghanistan, and Pakistan in the west (Fig. 17.1). Although the Middle East is, geographically, part of Asia, its avifauna is composed predominantly of European and African species, and we do not consider this area in our treatment of Asian raptors.

Distributions, conservation statuses, threats, raptor habitat, and taxonomy were based on BirdLife International (BirdLife International 2017). BirdLife International provides the most extensive data on the world's bird species and has been compiling data on threatened bird species since 1966. They are the designated Red List Authority for birds (IUCN 2017). BirdLife regularly assesses species at a global level and updates their database based on published information and data collected by 120 partner organizations worldwide (one partner for each country). BirdLife provided us with a database of Asian raptor species including information on Red List category,

Fig. 17.1 Regional Asia. Country boundaries are from the Global Administrative Areas database (www.dabm.org) and, in some regions (e.g., borders of India with Pakistan and China), are not fixed owing to ongoing territorial disputes (Fravel 2015). The Middle East is not included

migratory status, and current population trend. The database also included country of occurrence and land cover types occupied (13 categories). Threats (11 classifications) to globally threatened and near-threatened species were also available.

We summarize the overall Asian raptor diversity and break down the information according to threats, regions, and countries. Regions are also based on BirdLife International (BirdLife International 2017). We look at the overall population trends and population trends per land cover type, again using BirdLife International (2017) data. Subspecies are not differentiated in our summaries. We also consulted Ferguson-Lees and Christie (2001) for distribution and population trends but followed the BirdLife's taxonomic convention.

Results

Asia is home to 235 raptor species (Table 17.1). In total, Asia has 8 Critically Endangered raptor species (Table 17.2), 11 Endangered, 22 Vulnerable, and 40 Near-Threatened species, with a majority of species (64%) listed as Least Concern. Twenty-six percent of Asia's raptors are migratory, and 31% are endemic to a single country (Table 17.1).

Table 17.2 Asia's Critically Endangered and Endangered raptor species based on BirdLife International (2017). Migratory species include species with a significant portion of populations that make seasonal, cyclical, and predictable movements. Endemic species are species that breed in a single country. Y = Yes, N = No, x = present

Species	Common name	Category	Migratory	Endemic	Asian region			
					North	West and Central	East	South and Southeast
Accipitriformes								
Aquila nipalensis	Steppe eagle	Endangered	Y	N	x	x	x	x
Gyps bengalensis	White-rumped vulture	Critically endangered	N	N		x	x	x
Gyps indicus	Indian vulture	Critically endangered	N	N		x		x
Gyps tenuirostris	Slender-billed vulture	Critically endangered	N	N				x
Neophron percnopterus	Egyptian vulture	Endangered	Y	N	x	x	x	x
Nisaetus bartelsi	Javan hawk-eagle	Endangered	N	Y				x
Nisaetus floris	Flores hawk-eagle	Critically endangered	N	Y				x
Nisaetus philippensis	North Philippine hawk-eagle	Endangered	N	Y				x
Nisaetus pinskeri	South Philippine hawk-eagle	Endangered	N	Y				x
Pithecophaga jefferyi	Philippine eagle	Critically endangered	N	Y				x
Sarcogyps calvus	Red-headed vulture	Critically endangered	N	N			x	x
Falconiformes								
Falco cherrug	Saker falcon	Endangered	Y	N	x	x	x	x
Strigiformes								
Bubo blakistoni	Blakiston's fish owl	Endangered	N	N	x		x	
Heteroglaux blewitti	Forest owlet	Critically endangered	N	Y				x
Ninox leventisi	Camiguin boobook	Endangered	N	Y				x
Ninox rumseyi	Cebu boobook	Endangered	N	Y				x
Otus alfredi	Flores scops owl	Endangered	N	Y				x
Otus siaoensis	Siau scops owl	Critically endangered	N	Y				x
Otus thilohoffmanni	Serendib scops owl	Endangered	N	Y				x

Of Asia's raptors, 105 are hawks, eagles, and vultures (Accipitriformes) distributed among 30 genera (Table 17.1). Of these, 19% ($n = 20$) are globally Threatened, and 13% ($n = 13$) are Near-Threatened. Forty percent of Asia's hawks, eagles, and vultures are migratory, and 18% are endemic to a single country. The continent is home to 22 falcon species, in 3 genera. Of these, 5% ($n = 1$) are globally Threatened and 23% ($n = 5$) are Near-Threatened. Forty-five percent are migratory, and 41% are endemic to a single country. Asia is also home to 108 owl species, in 14 genera. Of these, 19% ($n = 20$) are globally Threatened and 19% ($n = 21$) are Near-Threatened. About 7% are migratory and 48% are endemic to a single country.

Overall, Asia's raptors are faring slightly worse than average raptors elsewhere in the world (Table 17.1). A greater proportion of Asia's owls (19%) and migratory species (12%) are Threatened compared to similar species elsewhere in the world (owls, 18%; migratory species, 8%). There is also a greater proportion of Asia's owls (19%) and migratory species (8%) that are Near-Threatened compared to species elsewhere in the world (owls, 11%; migratory species, 6%). Additionally, a greater proportion of Asia's falcons (23%) and endemic species (31%) are Near-Threatened compared with falcons and endemic species elsewhere in the world (falcons, 14%; endemic species, 22%).

Of the threats to Asian raptors, agriculture and aquaculture (e.g., livestock, wood and non-timber crops, and freshwater and marine aquaculture) and biological resource extraction (logging, hunting) are thought to be the greatest threats to Asia's eagle, hawk, vulture, and owl species (Table 17.3). Pollution and biological resource extraction are thought to be the greatest threats to Asian falcons. Secondary threats to these taxa include these same forces as well as changes to land cover, urbanization, and energy production.

Of Asia's 235 raptor species, populations of 57% ($n = 133$) are decreasing (Fig. 17.2). This includes 97% ($n = 33$) of globally Threatened and Near-Threatened

Table 17.3 Summary of ongoing threats to globally Threatened and Near-Threatened diurnal and nocturnal raptor species of Asia. Threats are based on BirdLife International (2017). Percentages represent proportions of Threatened and Near-Threatened species in Asia

	Accipitriformes	Falconiformes	Strigiformes
Agriculture and aquaculture	27 (79%)	2 (33%)	39 (95%)
Biological resource extraction	29 (85%)	5 (83%)	39 (95%)
Climate change and severe weather	3 (9%)	0	0
Energy production and mining	8 (24%)	0	11 (27%)
Human intrusions and disturbance	4 (12%)	0	1 (2%)
Invasive and other problematic species, genes, and diseases	10 (29%)	0	1 (2%)
Natural system modifications	11 (32%)	2 (33%)	12 (29%)
Pollution	16 (47%)	4 (67%)	0
Residential and commercial development	7 (21%)	2 (33%)	15 (5%)
Transportation and service corridors	7 (21%)	0	4 (37%)
Total number of species	34	6	41

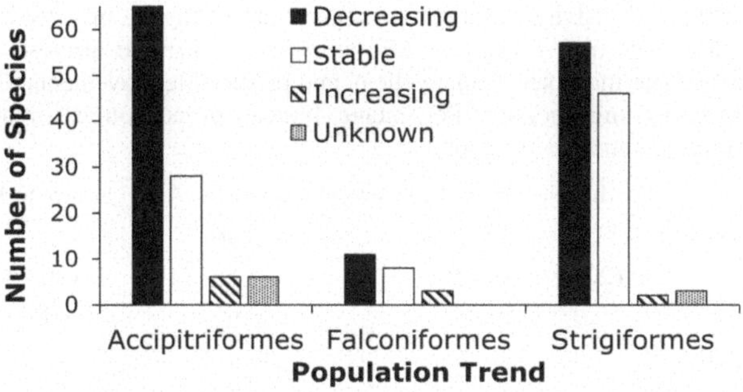

Fig. 17.2 Population trends of Asian raptors. Population trends are based on BirdLife International (2017). Population trend is either directly estimated or suspected based on existing threats and other factors. Population estimates were not taken in the same year, and global population sizes have been estimated between 1999 and 2016

Fig. 17.3 Species status of Asian raptors found in the five most occupied land types. (Conservation status and land types are based on BirdLife International 2017)

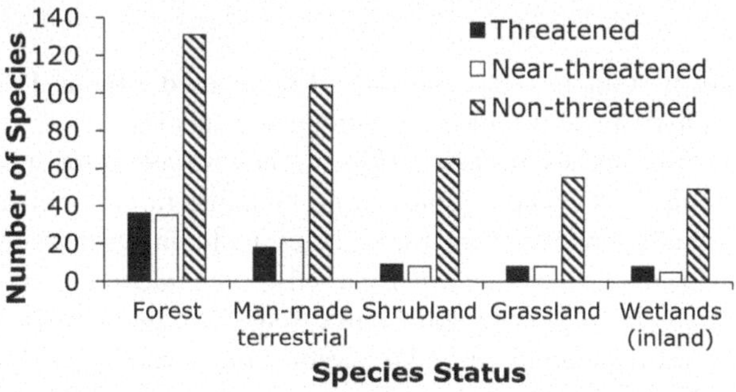

species of hawks, eagles, and vultures, all six globally Threatened and Near-Threatened falcon species, and 88% (*n* = 36) of globally Threatened and Near-Threatened owl species. Among the 150 non-threatened raptor species, populations of 46% (*n* = 32) eagles, hawks, and vultures, 31% (*n* = 5) of falcon species, and 33% (*n* = 21) of owl species are decreasing.

Among the 13 land-use types, forest, man-made or anthropogenic terrestrial (e.g., degraded forests, plantations, arable land, rural gardens, and urban areas), shrubland, grassland, and wetlands were the most commonly occupied by Asian raptors (Fig. 17.3). Forests are most important for 86% (*n* = 202) of Asia's raptor species, and a disproportionately high number (36 of 41) of Asia's globally Threatened raptor species are found in forests. Additionally, although anthropogenic terrestrial land use appears to be acceptable habitat for Least Concern species, 44% (*n* = 18) of globally Threatened and 54% (*n* = 22) of Near-Threatened species are found here as well. Finally, 21% (*n* = 17) of globally Threatened and Near-Threatened species occupy shrubland, 20% (*n* = 16) occupy grassland, and 16% (*n* = 13) occupy wetlands.

At a subcontinental (regional) scale, North Asia (Fig. 17.1) has 61 raptor species, 80% of which migrate (Table 17.4). Although the region has no endemic raptors, its ecosystems are important breeding areas for globally threatened species including Steller's sea eagle *Haliaeetus pelagicus* and Blakiston's fish owl *Bubo blakistoni*.

Table 17.4 Summary of distribution and conservation status of raptors in according to regions of Asia based on BirdLife International (2017). Migratory species include species with a significant portion of populations that make seasonal, cyclical, and predictable movements. Endemic species are species that breed in a single country. Percentages indicate proportion of Threatened and Near-Threatened migratory and endemic species

	North Asia	West and Central Asia	East Asia	South and Southeast Asia
Total	61	59	96	219
Migratory	49 (20%)	44 (25%)	52 (21%)	51 (20%)
Endemic	0	0	0	74 (64%)
Critically endangered	0	2	2	8
Endangered	4	3	4	10
Vulnerable	4	3	5	21
Near threatened	4	7	7	39
Least concern	49	44	78	137
Data deficient	0	0	0	4

North Asia has no Critically Endangered (Table 17.2), four Endangered, four Vulnerable, and four Near-Threatened species.

West and Central Asia (Fig. 17.1) have 59 raptor species, 75% of which migrate (Table 17.4). The region has no endemic raptors but includes important areas for globally threatened and rare species, including eastern imperial eagle *Aquila heliaca*, steppe eagle *Aquila nipalensis*, red-footed falcon *Falco vespertinus*, and saker falcon *Falco cherrug*. West and Central Asia have two Critically Endangered (Table 17.2), three Endangered, three Vulnerable, and seven Near-Threatened species.

East Asia (Fig. 17.1) has 96 raptor species (Table 17.4). There are no endemic species in this region, and 59% of its raptors migrate. All of Asia's migratory owls are in East Asia. The region's notable raptors include steppe eagle, Pallas's fish eagle *Haliaeetus leucoryphus*, saker falcon, and Amur falcon *Falco amurensis*. East Asia has two Critically Endangered (Table 17.2), four Endangered, five Vulnerable, and seven Near-Threatened species.

South and Southeast Asia has the greatest raptor diversity, with 219 species, only 20% of which migrate. Endemism is also highest in this region: 74 species are found only in Asia. South and Southeast Asia also has the highest number of globally Threatened and Near-Threatened species, including 8 Critically Endangered (Table 17.2), 10 Endangered, 21 Vulnerable, and 39 Near-Threatened. Important species in this region include the critically threatened vultures of the Indian subcontinent, as well as two of the most evolutionarily distinct and globally threatened raptors in the world, the forest owlet and the Philippine eagle.

We identified ten Asian countries that have the most Threatened and Near-Threatened species (Table 17.5). Among these countries, eight are in South and Southeast Asia, and the remaining are in North and East Asia. These ten countries have 231 of the 235 raptor species of Asia, including 72 of the 74 Asian endemics. Because of the inclusion of China, Russia, and India, they also include the vast majority of the land mass of the continent. China, Russia, and India also have the

Table 17.5 Summary of distribution and conservation status of raptors in ten Asian countries with the most Threatened and Near-Threatened raptor species based on BirdLife International (2017). Migratory species include species with a significant portion of populations that make seasonal, cyclical, and predictable movements. Endemic species are species that breed in a single country. Percentages indicate proportion of Threatened and Near-Threatened migratory and endemic species

	Indonesia	India	Russian Federation	Philippines	Myanmar	Nepal	Bangladesh	Malaysia	China	Thailand
Total	133	102	102	55	88	83	77	67	94	76
Migratory	24 (4%)	46 (22%)	48 (21%)	14 (0%)	39 (21%)	39 (26%)	33 (27%)	28 (21%)	51 (22%)	32 (19%)
Endemic	39 (54%)	9 (67%)	0	23 (74%)	0	0	0	1 (100%)	0	0
Critically endangered	3	5	0	1	3	3	3	3	2	3
Endangered	3	3	6	4	2	3	3	1	4	1
Vulnerable	12	5	7	6	6	4	4	5	4	5
Near threatened	20	13	8	10	8	9	8	8	6	6
Least concern	92	75	81	34	69	64	59	50	78	61
Data deficient	3	1	0	0	0	0	0	0	0	0

highest number of migratory species (51, 48, and 46 species, respectively). Indonesia and the Philippines have the highest number of endemics – 39 and 23, respectively – 54% and 74% of which are globally Threatened and Near-Threatened. Additionally, Indonesia (29%) and the Philippines (38%) also have the highest proportion of globally Threatened and Near-Threatened species.

Discussion

Overview

Our review indicates that Asia's raptors are faring slightly worse than average raptors elsewhere in the world. Asia's owls and migratory species are at greater risk than similar raptors elsewhere in the world. Additionally, more falcons and endemic species are close to qualifying for a threatened category in the near future than falcons or endemic species elsewhere in the world.

Perhaps not surprisingly, forest is the land cover type most frequently used by Asian raptors, and a vast majority of Threatened and Near-Threatened species are found here. What is surprising is that anthropogenic landscapes such as degraded forests and forest plantations appear to be the second most frequently used land type for Threatened and Near-Threatened species. Forests are also the most used by most of Asia's endemic species and once again artificial landscapes, the second most important. Largely treeless ecosystems, like shrublands and grasslands, and inland wetlands are also frequently used by Asia's raptors. However, only a small proportion of Threatened and Near-Threatened species are found in each of these land cover types. Out of Asia's endemic species, only four endemic owls are found here.

In all likelihood, this does not mean that degraded forests or plantations, for example, are of equal quality to natural landscapes, such as forests. These anthropogenic land-use types may be used but are still of low quality for reproduction or survival (Watson 1992; Donovan and Thompson 2001; Battin 2004; Carrete et al. 2009). It is therefore important for researchers and conservation practitioners to not just determine occupancy of individuals in different land cover types but to evaluate population structure as well, in order to ascertain whether the use of these altered areas is a result of optimized selection or lack of choice (Van Horne 1983).

Threats

Agriculture and aquaculture, biological resource extraction, and pollution are the most common threats to Asia's raptors. Combined, the threats that raptors in Asia face can be grouped into indirect effects and two kinds of direct effects on raptor populations (threats that lead to unintentional fatalities and those that lead to targeted removal of individuals from a population).

Indirect Effects

Indirect effects of land-use change and disturbance are the most significant threat to Asia's raptors. Raptors may be relatively more impacted by habitat change than other birds because they occur at relatively low densities, have relatively large home ranges, and because they are often persecuted by humans (Newton 1979, 1998). For example, land-use change reduces available suitable space and might alter home range sizes (Andersson 1978). If degradation occurs on a large enough scale, it can reduce breeding densities of birds that are forced to occupy larger home ranges (Newton 1980; Amar et al. 2011). Fragmentation additionally creates movement barriers for individuals and affect the movement of individuals among suitable areas (Harrison and Emilio 1999; Bélisle and Desrochers 2002; Harris and Reed 2002).

The effects of land-use change are likely to grow as Asia's economy and infrastructure grows. Those effects are difficult to reverse, especially in tropical areas (Brooks et al. 2006). Agriculture, specifically, has been responsible for converting ~13 million hectares of previously untouched forests to croplands globally every year over the past 30 years (Clay 2004). Agriculture also remains the most widespread occupation for Asian people, and agricultural production has been growing alongside industrial production (Galloway et al. 1998; The Association of Academies of Sciences in Asia 2011). Looking forward, it is estimated that between 2000 and 2050, 10 billion hectares of natural landscapes in the world will be converted to agriculture (Tilman et al. 2001).

In tropical Asia, for example, in the past 40 years, oil palm plantations have expanded faster than any other type of food or industrial agricultural crop (Clay 2004). Malaysia and Indonesia currently have over half of the world's total oil palm plantation area. In these two countries, although preexisting croplands were converted to oil palm plantations, over half of oil palm cultivated areas replaced native forests (Koh and Wilcove 2008). Further, between 1910 and 1940, rubber plantations in Indonesia increased tenfold (Burger and Smit 2001). Seventy percent of all rubber tree plantations are currently in Malaysia, Indonesia, and Thailand (Clay 2004). By the year 2024, the expansion of rubber plantations is expected to continue, and it is estimated that 4.3–8.5 million hectares of forests and swidden land would be needed to answer the global rubber demand (Warren-Thomas et al. 2015).

The expansion of agriculture has also impacted natural grasslands (in this book Chap. 9). Grasslands are heavily used by certain raptor species and are one of the most endangered terrestrial ecosystems (Hoekstra et al. 2005). West and Central Asia, in particular, are a global stronghold for natural grasslands (Hoekstra et al. 2005), and, as a result of the region's low densities of human populations, until the mid-1900s Central Asian ecosystems were largely pristine (Zlotin 2002). However, steppe was targeted by the Virgin Lands project of the USSR, and between 1953 and 1963, an estimated 23–25 million hectares of semidesert grasslands were plowed and converted to agriculture (Kamp et al. 2011; Kraemer et al. 2015) which has had negative consequences for raptor populations (BirdLife International 2001; Sánchez-Zapata et al. 2003).

Direct Effects: Targeted Removal of Individuals

Consumptive uses of a few species, particularly falconry and the pet trade, are thought to drive trajectories of some raptor populations. For example, saker falcons are an Endangered species whose population has been reduced to a tenth of its former abundance (BirdLife International 2001; Kenward 2009). As many as 1000 saker individuals (~8% of global population) were estimated to have been taken illegally from West and Central Asia, particularly Kazakhstan, between 1994 and 1996 (Levin 2011). Small numbers of sakers have always been traded in West and Central Asia, but illegal trading only flourished in 1992 after the fall of the Soviet Union. This geopolitical event opened West and Central Asia's borders and increased foreign access to its saker population (Kenward 2009). In East Asia, especially Mongolia, populations of the Endangered saker falcon are over-trapped (Fox 2001; Zahler et al. 2004). Between 1997 and 2010 alone, an average of 286 saker falcons were traded annually (Dixon et al. 2011). Unfortunately, although local policies are in place to protect saker falcons, it is often difficult to enforce those regulations.

Steppe eagles, imperial eagles, and 27 other species are also sold in markets in China (Yi-Ming et al. 2000; Zhang et al. 2008). Most are taken from the wild during migration (Kenward 2009). More recently, birds are being traded online in South and Southeast Asia. In 2015 alone, over 7500 individuals from 22 diurnal raptor species and 11 owl species were traded online in southeast Asia (Iqbal 2016).

There are also unique forms of consumptive uses in South and Southeast Asia. In India, for example, owls are important in witchcraft and believed to bring gambling luck (Ahmed 2010; Jathar and Rahmani 2013). To answer the demand for witchcraft, between 1992 and 2008, >1000 individuals of 15 species of owls were traded (Ahmed 2010). Another consumptive use occurred in the Doyang Reservoir region of Nagaland, northwestern India, where the mass killing of Amur falcons for human consumption was a major threat (Symes 2012; Dalvi and Haralu 2014). These long-distance migrants form large communal roosts while migrating. During peak migration, as many as 15,000 falcons were taken daily with mist nest, and 120,000–140,000 falcons were killed annually. Fortunately, since 2013, the killings of Amur falcons have been halted as a result of education campaigns and continued population monitoring (NWBCT 2017).

Direct Effects: Unintentional Fatalities

The catastrophic collapse of the *Gyps* population in India is well documented, with an estimated 92% population decline between 1993 and 2000 (Prakash et al. 2003) and further declines since that time. As of 2012 only 0.1–3.2% of original populations remained (Prakash et al. 2012). Similar declines were also recorded for vulture populations in Nepal (Baral et al. 2004; Shultz et al. 2004) and in Pakistan (Gilbert et al. 2004; Oaks et al. 2004b). The decline of *Gyps* vulture was put into spotlight not just for its magnitude but also for the severe economic, cultural, and human health effects that followed (Prakash et al. 2004; Markandya et al. 2008). This cataclysmic phenomenon was attributed to contamination of carcasses with the veterinary drug, diclofenac, that causes vultures to develop visceral gout and acute kidney failure leading to death (Oaks et al. 2004a, b, in this book Chaps. 10, 19, and 20).

In 2006, following the identification of the role of diclofenac, India consequently banned the drug from veterinary use (Taggart et al. 2007). Nepal and Pakistan declared similar bans in 2006 and in 2010, while Bangladesh followed in 2010 (Prakash et al. 2012). This ban and the promotion of the diclofenac alternative, meloxicam (Swarup et al. 2007; Pain et al. 2008), has lowered, but not eliminated, diclofenac's impact (Prakash et al. 2012; Cuthbert et al. 2014, 2016; Galligan et al. 2014). Captive care and breeding of Indian vultures, with a goal of eventual release into the wild, are also currently ongoing (Prakash et al. 2004; Markandya et al. 2008). Vulture restaurants providing clean food likewise are a strategy used to reduce the impacts of diclofenac (Gilbert et al. 2007).

Migratory species may also encounter environmental contaminants in their stopover and wintering areas. For example, Lake Baikal basin, a large steppe area, once had the highest known North Asian concentration of eastern imperial eagles (Ryabtsev and Katzner 2007; Ryabtsev 2011). The population here has declined by at least 80% in the past few decades possibly due to fatalities caused by environmental contaminants in tropical wintering grounds in Southeast Asia (Ueta and Ryabtsev 2001; Ryabtsev and Katzner 2007). Unfortunately, the extent and effects of environmental contaminants are not well studied in the tropics (Lacher and Goldstein 1997). However, this is a highly plausible cause of population decline given the potential environmental impacts of intensified agriculture in Southeast Asian countries (Tinker 1997; Mineau and Whiteside 2013). Another example is the threat of contaminants to Steller's sea eagles that breed on Sakhalin Island and winter in Japan. In summer, Steller's sea eagles typically feed on fish. However, during harsh winters or during periods of overfishing, these birds shift feeding habits and scavenge on sika deer *Cervus nippon* (Shiraki 2001). This has led to increasing incidences of lead poisoning, when birds feed on hunter-killed, but unrecovered, deer or offal (Urosawa 2000; Saito 2009).

Finally, Asia's long-distance migratory species also may face pressures outside of Asia. For example, pallid harriers *Circus macrourus* and red-footed falcons *Falco vespertinus* breed in Central Asia and their populations are complete migrants (Galushin et al. 2003; Panuccio 2007; Brochet et al. 2016; Katzner et al. 2016). On migration, far from their breeding or wintering grounds, these species pass through geographic bottlenecks in Europe where shooting birds of prey is common.

Priority Species

Asia holds a third of the world's Critically Endangered raptors, including two of the most genetically distinct birds of prey, the Philippine eagle and the forest owlet (Jetz et al. 2014). These species, along with the Endangered *Gyps* species, are endemic to the Asian continent and have undergone drastic population declines making them among the highest regional priorities for conservation (BirdLife International 2001).

The Philippine Eagle, whose wild population is estimated at 82–250 pairs (Bueser et al. 2003; BirdLife International 2017), is emblematic of Asia's raptor conservation problems. The species is endemic to the Philippine archipelago (Kennedy 1981; Kennedy et al. 2000) and occupies Philippine tropical forests (Bueser et al. 2003), of which only 3% now remains (Myers et al. 2000). Beyond land-use change, it faces threats from shooting, trade, electrocution, and environmental contaminants. Extensive deforestation also reduced space for the Philippine eagle's prey base (Kennedy 1985; Concepcion et al. 2006) and has increased opportunities for human persecution. Although the Philippine eagle receives legislative protection (Salvador and Ibanez 2006), its current estimated population decline remains at 2–10% per year (Ibañez et al. 2016).

Globally, no other continent has a higher proportion of migratory raptor species than Asia (Bildstein 2006, in this book Chap. 5). Forty percent of all migratory raptors of Asia are endemic to the Asian continent; these endemics include Pallas's fish eagle, Steller's sea eagle, and the Himalayan griffon. Although a majority of Asia's most threatened raptors are not migratory, 15% are full migrants. Migratory raptors of Asia are also faring considerably worse than average migratory raptors elsewhere. Further, compared to their European and African counterparts, Asian populations of cross-continental migrant species are largely unstudied (Kirby et al. 2008). As an example, migrants using Asia's Oceanic and Continental Flyways are poorly known and only recently have studies been published on this flyway (Concepcion et al. 2017; Bildstein 2006). All of these factors make Asia's migratory raptors high priority for conservation and research.

Migratory species face unique threats in their breeding, stopover, and wintering habitats (Webster et al. 2002). For such species, especially species with large home ranges, creation of protected areas for breeding sites alone often is not an efficient conservation approach (BirdLife International 2001; Finch et al. 2017). Comprehensive and actionable management plans would allow identification of these unique threats across all these habitats and provide a framework for addressing those threats.

Priority Regions and Countries

Among regions in Asia, the south and southeast have the highest number of species overall, the highest number of endemic species, and the highest number of globally Threatened and Near-Threatened species. Sixty-two percent of Threatened species and 56% of Near-Threatened are island endemics. As a result, raptors of South and Southeast Asia are more at risk than elsewhere in the continent. Eight of the 10 most speciose Asian countries that also have the most Threatened and Near-Threatened species are also found in this region. This diversity of raptors can be partly attributed to the extent of tropical forests that once covered most of the region (Olson and

Dinerstein 2002; Laurance 2007). Unfortunately, tropical forests are extremely vulnerable landscapes (Brooks et al. 2006), and those in Southeast Asia have experienced, and continue to experience, higher rates of conversion than elsewhere in the world (Laurance 1999; Sodhi et al. 2004). This is to the detriment of South and Southeast Asia's raptors that heavily use these forests.

There is a significant lack of information for tropical Asia's raptors, especially of island endemics (Thiollay 1985; McGowan et al. 1998; BirdLife International 2001). With the exception of work on a few charismatic species such as the Philippine eagle and the Javan hawk-eagle, most of the region's raptors continue to be largely unstudied. South and Southeast Asia are conservation and research priority regions. Protection of the region's raptor species, especially of island endemics, intrinsically has a disproportionately large impact on the global raptor population. There is therefore a need to collect ecologically sound information to enable their protection.

Priority Topics for Research

Without basic information on species biology and demography, it can be difficult to properly direct conservation and management strategies for raptors (Collar 1997; Van Balen et al. 2000; BirdLife International 2001; Mace and Collar 2002; Sutherland et al. 2004; Katzner et al. 2011). For example, Pallas's fish eagle is probably the world's most poorly studied Northern Hemisphere eagle (Katzner and Tingay 2010). It is classified as Vulnerable on the basis of reported population declines (BirdLife International 2001; Ferguson-Lees and Christie 2005). Although its precise distribution is poorly known, the main breeding populations of Pallas's fish eagle are thought to be in China, Mongolia, and India (Ferguson-Lees and Christie 2005; BirdLife International 2017), with lower-density populations in surrounding countries. However, recent field work suggests that Mongolia never was a breeding stronghold for the species (Gilbert et al. 2014), the implication being that conservation action directed there may have little impact for this species. This finding illustrates how a lack of knowledge can confound potential conservation programs.

Apart from research on species biology and demography, there are other overlapping research needs for Asia's Threatened and Near-Threatened species (BirdLife International 2001). There is also a need to identify specific threats to populations and reasons for population declines. For migratory species, this extends to describing the different pressures faced in summer, migratory, and wintering areas (Webster et al. 2002). Likewise, there is a need to study biology, demography, and threats of prey species on which raptors depend (BirdLife International 2001). Finally, laboratory studies may also be important to identify disease and the physiological effects of environmental contaminants (Lacher and Goldstein 1997; BirdLife International 2001).

The effects to raptors of renewable energy and electrocution are also priority topics for research, especially in Asian grassland and treeless ecosystems. As energy development has rapidly grown in West and Central Asia, electrocution has become a growing concern for raptor populations (Kamp et al. 2016). Separate surveys in Kazakhstan documented that as many as 21 raptor species died from electrocution, including eight Threatened and Near-Threatened species such as the saker falcon, the Egyptian vulture *Neophron percnopterus*, and the steppe eagle (Karyakin 2008; Lasch et al. 2010; Levin and Kurkin 2013; Pestov et al. 2015). In East Asia, particularly in Mongolia, older, wood pole electrical structures are being replaced with concrete poles with grounded metal crossarms. The energized parts of these newer, concrete poles are closer to the ground than the older, wood poles, which in turn have increased electrocution of raptors (Harness et al. 2008). As many as 13 species, including 4 migrants, have been documented dying from electrocution in this region (Dixon et al. 2013).

Research elsewhere suggests that new threats may emerge from development of wind and solar energy. Collision with wind turbine blades, for example, can cause trauma or death (Madders and Whitfield 2006; Smith and Dwyer 2016). Trauma and death could also come from solar tower facilities as a result of collision with infrastructure or exposure to concentrated solar energy or solar flux (Smith and Dwyer 2016; Walston et al. 2016). There also are threats of collision, disturbance, and land-use change even before these facilities are completed (Tsoutsos et al. 2005; Smith and Dwyer 2016). Wind energy is particularly interesting because its direct and indirect effects on raptor populations can alter demography (Katzner et al. 2013).

Poorly Known Species

Poorly known species are an important priority for research. Four out of six (67%) of the world's Data Deficient raptor species occur in Asia (Table 17.1; BirdLife International 2017). These are the chestnut-shouldered goshawk *Erythrotriorchis buergersi*, jungle boobook *Ninox theomacha*, the Seram masked owl *Tyto almae*, and the Nicobar scops owl *Otus alius* of South and Southeast Asia. All three owls are island endemics.

Modern and updated identification literature is also key to the quality and reliability of ecological studies (BirdLife International 2001). Establishing identification literature is made more challenging because some Asian species are only now being recognized (e.g., *Spizaetus philippensis* complex, Gamauf et al. 2005; *Ninox philippensis* complex, Rasmussen et al. 2012). Accurate identification of species is obviously crucial to conservation as unrecognized species may be lost due to lack of protection (Gamauf et al. 2005). This is of particular concern in South and Southeast Asia where the range of landscapes available, the isolation of some of these landscapes, and the geography of the region (i.e., islands of varying sizes and distances to the mainland) contribute to a high level of endemism (Gentry 1992; White and Kiff 2000).

Conclusions

Raptors are more at risk in Asia than elsewhere in the world. Our summary of Asia's raptor diversity and human threats supports the need to continue cataloguing the status, threats, and needs and understanding conservation priorities for these species. Holding almost half of all the world's raptor species, and half of all the endemic species, the continued decline and potential loss of raptors in this region are truly a global concern.

One of the things highlighted in our summary is the lack of information on Asia's raptor species. This lack of information is exemplified in the scarce literature available for raptors of Asia. For example, compared to raptors of other tropical regions, tropical Asia has the least published information (Kiff et al. 2007). There also are instances when information is available but unavailable in international abstracting services or unavailable in English (Kiff et al. 2007; Clavero 2010). Similarly, researchers in Asia may have limited access to published information or limited opportunities to publish in English journals (Salager-Meyer 2008).

Lack of information may also mean that actions will not be taken until some species are lost. For example, data from long-term migration monitoring informed of the population decline of several species in North America due to organochlorides (Bednarz et al. 1990). The relatively early warning provided ample time to determine the cause of decline and to respond appropriately (e.g., bald eagle, Grier 1982; Fraser et al. 1996; Watts et al. 2008). This and other types of monitoring are lacking for most, if not all, of Asian species.

Conservation urgency presents opportunities for future work. The establishment of the Asian Raptor Research and Conservation Network (ARRCN) in 1998 has created a community for researchers of Asian raptors (Kiff et al. 2007). The ARRCN is vital in sharing valuable information through organizing symposia, promoting communication among researchers, and disseminating published abstracts. With access to training, mentoring, and funding, it is possible for current and rising raptor researchers in Asia to begin systematically addressing threats and knowledge gaps. Like most things in these contemporary times, international collaboration would be vital in achieving these goals.

Simply put, much work remains to be done to elevate conservation efforts in the continent to at least those of elsewhere in the world. The time to start is now, before many common species decline to the status of conservation priority and before less common species become extinct.

Acknowledgments We thank Tris Allinson who generously compiled the BirdLife data we analyzed. We also thank Tracy Maffett Prado and Medel Silvosa who helped prepare initial data used to organize the manuscript. Statement of author contributions: TEK and CBC organized the manuscript at the invitation of the editors; CBC organized published data and led writing, and all authors contributed to revisions of the manuscript. Any use of trade, firm, or product names is for descriptive purposes only and does not imply endorsement by the US government.

References

Acharya R, Cuthbert R, Baral HS, Shah KB (2009) Rapid population declines of Himalayan griffon *Gyps himalayensis* in upper mustang, Nepal. Bird Conserv Int 19:99–107

Ahmed A (2010) Imperilled custodians of the night: a study on illegal trade trapping and use owls in India. Page (T. India/WWF-India, (ed), New Delhi

Amar A, Davies J, Meek E, Williams J, Knight A, Redpath S (2011) Long-term impact of changes in sheep *Ovis aries* densities on the breeding output of the hen harrier *Circus cyaneus*. J Appl Ecol 48:220–227

Andersson M (1978) Optimal foraging area: size and allocation of search effort. Theor Popul Biol 13:397–409

Aratrakorn S, Thunhikorn S, Donald PF (2006) Changes in bird communities following conversion of lowland forest to oil palm and rubber plantations in southern Thailand. Bird Conserv Int 16:71–82

Baral HS, Giri JB, Virani MZ (2004) On the decline of oriental white-backed vultures *Gyps bengalensis* in lowland Nepal. In: Chancellor RD, Meyburg B-U (eds) Raptors worldwide: proceedings of the 6th world conference on birds of prey and owls. WWGBP and MME/Birdlife Hungary, Berlin/Budapest, pp 215–219

Battin J (2004) When good animals love bad habitats: ecological traps and the conservation of animal populations. Conserv Biol 18:1482–1491

Bednarz J, Klem D, Goodrich LJ, Senner SE (1990) Migration counts of raptors at Hawk Mountain, Pennsylvania, as indicators of population trends, 1934–1986. Auk 107:96–109

Bélisle M, Desrochers A (2002) Gap-crossing decisions by forest birds: an empirical basis for parameterizing spatially-explicit, individual-based models. Landsc Ecol 17:219–231

Beukema H, Danielsen F, Vincent G, Hardiwinoto S, Van Andel J (2007) Plant and bird diversity in rubber agroforests in the lowlands of Sumatra, Indonesia. Agrofor Syst 70:217–242

Bildstein KL (1998) Conservation status of tropical raptors. J Raptor Res 32:3–18

Bildstein KL (2001) Raptors as vermin: a history of human attitudes toward Pennsylvania's birds of prey. Endanger Species Updat 18:124–128

Bildstein KL (2006) Migrating raptors of the world: their ecology and conservation. Cornell University Press, Ithaca

BirdLife International (2001) Threatened birds of Asia: the BirdLife international red data book. BirdLife International, Cambridge, UK

BirdLife International (2017) IUCN red list for birds. http://www.birdlife.org

Braimoh AK, Huang HQ (2015) Land systems vulnerability. In: Braimoh AK, Huang HQ (eds) Vulnerabitliy of land systems in Asia. Wiley, West Sussexx, pp 1–7

Brochet A-L, Van Den Bosschen W, Jbour S, Ndang'ang'a PK, Jones VR, Abdou WALIA, Al-hmoud AR, Asswad NG, Atienza JC, Atrash I, Barbara N, Bensusan K, Bino T, Celada C, Cherkaoui SI, Costa J, Deceuninck B, Etayeb KS, Feltrup-Azafzaf C, Figelj J, Gustin M, Kmecl P, Kocevski V, Korbeti M, Kotrošan D, Mula Laguna J, Lattuada M, Leitão D, Lopes P, López-Jiménez N, Lucić V, Micol T, Moali A, Perlman Y, Piludu N, Portolou D, Putilin K, Quaintenne G, Ramadan-Jaradi G, Ružić M, Sandor A, Sarajli N, Saveljic D, Sheldon RD, Shialis T, Tsiopelas N, Vargas F, Thompson C, Brunner A, Grimmett R, Butchart SHM (2016) Preliminary assessment of the scope and scale of illegal killing and taking of birds in the Mediterranean. Bird Conserv Int 26:1–28

Brooks TM, Mittermeier RA, da Fonseca GAB, Gerlach J, Hoffmann M, Lamoreux JF, Mittermeier CG, Pilgrimd JD, Rodrigues ASL (2006) Global biodiversity conservation priorities. Science 313:58–61

Bueser GL, Bueser KG, Afan DS, Salvador DI, Grier JW, Kennedy RS, Miranda Jr HC (2003) Distribution and nesting density of the Philippine eagle *Pithecophaga jefferyi* on Mindanao Island, Philippines: what do we know after 100 years? Ibis 145:130–135

Burger K, Smit HP (2001) International market responses to the Asian crisis for rubber, cocoa and coffee. In: Gérard F, Ruf F (eds) Agriculture in crisis: people, commodities and natural resources in Indonesia, 1996–2000. Curzon Press, Montpellier, pp 31–48

Carrete M, Grande JM, Tella JL, Sánchez-Zapata JA, Donázar JA, Díaz-Delgado R, Romo A (2007) Habitat, human pressure, and social behavior: Partialling out factors affecting large-scale territory extinction in an endangered vulture. Biological Conservation 136(1):143–154
Carrete M, Tella JL, Blanco G, Bertellotti M (2009) Effects of habitat degradation on the abundance, richness and diversity of raptors across Neotropical biomes. Biol Conserv 142:2002–2011
Clavero M (2010) "Awkward wording. Rephrase": linguistic injustice in ecological journals. Trends Ecol Evol 25:552–553
Clay J (2004) World agriculture and the environment: a commodity-by-commodity guide to impacts and practices. Island Press, Washington, DC
Collar NJ (1997) Species survival versus perpetuation of myth: the case of the Philippine eagle. Oryx 31:4–7
Concepcion CB, Sulapas M, Ibanez JC (2006) Food habits and breeding behavior at a Philippine eagle nest in mount Apo Natural Park, Mindanao, Philippines. Banwa 3:91–105
Concepcion CB, Dumandan PT, Silvosa MR, Bildstein KL, Katzner TE (2017) Species composition, timing and meteorological correlates of autumn open-water crossings by raptors migrating along the east-Asian oceanic flyway. J Raptor Res 51:25–37
Cuthbert RJ, Taggart MA, Prakash V, Chakraborty S, Deori P, Galligan T, Kulkarni M, Ranade S, Saini M, Sharma AK, Shringarpure R, Green RE (2014) Avian scavengers and the threat from veterinary pharmaceuticals. Phil Trans R Soc B 369:369–574
Cuthbert RJ, Taggart MA, Saini M, Sharma A, Das A, Kulkarni MD, Deori P, Ranade S, Shringarpure RN, Galligan TH, Green RE (2016) Continuing mortality of vultures in India associated with illegal veterinary use of diclofenac and a potential threat from nimesulide. Oryx 50:104–112
Dalvi S, Haralu B (2014) Doyang reservoir: a potential IBA in Nagaland. Mistnet 15:24–28
Dixon A, Batbayar N, Purev-Ochir G, Fox N (2011) Developing a sustainable harvest of saker falcons (*Falco cherrug*) for falconry in Mongolia. In: Watson RT, Cade T, Fuller M, Hunt G, Potapov E (eds) Gyrfalcons and ptarmigan in a changing world. The Peregrine Fund, Boise, pp 363–372
Dixon A, Maming R, Gunga A, Purev-Ochir G, Batbayar N (2013) The problem of raptor electrocution in Asia: case studies from Mongolia and China. Bird Conserv Int 23:520–529
Donovan TM, Thompson FR (2001) Modeling the ecological trap hypothesis: a habitat and demographic analysis for migrant songbirds. Ecol Appl 11:871–882
Ferguson-Lees J, Christie DA (2001) Raptors of the world. Houghton Mifflin, New York
Ferguson-Lees J, Christie D (2005) Raptors of the world: a field guide. Princeton University Press, Princeton
Finch T, Butler S, Franco A, Cresswell W (2017) Low migratory connectivity is common in long-distance migrant birds. J Anim Ecol 86:662–673
Fisher IJ, Pain DJ, Thomas VG (2006) A review of lead poisoning from ammunition sources in terrestrial birds. Biol Conserv 131:421–432
Fitzherbert EB, Struebig MJ, Morel A, Danielsen F, Brühl CA, Donald PF, Phalan B (2008) How will oil palm expansion affect biodiversity? Trends Ecol Evol 23:538–545
Fox N (2001) Future trends, captive breeding, trade controls or market forces? In: Potapov E, Banzragch S, Fox N, Barton N (eds) Saker falcon in Mongolia: research and conservation. Mongolian Academy of Sciences, Ulanbaatar, pp 212–214
Fraser JD, Chandler SK, Buehler DA, Seegar JKD (1996) The decline, recovery and future of the bald eagle population of the Chesapeake Bay, U.S.a. In: Meyburg B-U, Chancellor RD (eds) Eagle studies. WWBGBP, Berlin, pp 181–187
Fravel MT (2015) Territorial and maritime boundary disputes in Asia. In: Pekkanen SM, Ravenhill J, Foot R (eds) Oxford handbook of international relations in Asia. Oxford University Press, New York, pp 524–546
Galligan TH, Amano T, Prakash VM, Kulkarni M, Shringarpure R, Prakash N, Ranade S, Green RE, Cuthbert RJ (2014) Have population declines in Egyptian vulture and red-headed vulture in India slowed since the 2006 ban on veterinary diclofenac? Bird Conserv Int 24:272–281
Galloway JN, Ojima DS, Melillo JM (1998) Asian change in the context of global change: an overview. In: Galloway JN, Melillo JM (eds) Asian change in the context of global climate change:

impact of natural and anthropogenic changes in Asia on global biogeochemistry. Cambridge University Press, Cambridge, pp 1–17

Galushin V, Clarke R, Davygora A (2003) International action plan for the pallid harrier (*Circus macrourus*). Birdlife International, Strasbourg

Gamauf A, Gjershaug J-O, Røv N, Kvaløy K, Haring E (2005) Species or subspecies? The dilemma of taxonomic ranking of some south-east Asian hawk-eagles (genus *Spizaetus*). Bird Conserv Int 15:99–117

Gangoso L, Álvarez-Lloret P, Rodríguez-Navarro AAB, Mateo R, Hiraldo F, Donázar JA (2009) Long-term effects of lead poisoning on bone mineralization in vultures exposed to ammunition sources. Environ Pollut 157:569–574

Gentry AH (1992) Tropical forest biodiversity: distributional patterns and their conservational significance. Oikos 63:19–28

Gilbert M, Oaks JL, Virani MZ, Watson RT, Ahmed S, Chaudhry MJI, Arshad M, Mahmood S, Ali A, Khattak RM, Khan AA (2004) The status and decline of vultures in the provinces of Punjab and Sind, Pakistan: a 2003 update. In: Chancellor RD, Meyburg B-U (eds) Raptors worldwide: proceedings of the 6th world conference on birds of prey and owls. WWGBP/MME/Birdlife Hungary, Berlin/Budapest, pp 221–234

Gilbert M, Watson RT, Ahmed S, Asim M, Johnson JA (2007) Vulture restaurants and their role in reducing diclofenac exposure in Asian vultures. Bird Conserv Int 17:63–77

Gilbert M, Tingay R, Jambal L, Sureda N, Gilbert C, Davaasuren B, Sundev G (2014) Distribution and status of the Pallas's fish eagle *Haliaeetus leucoryphus* in Mongolia: a cause for conservation concern? Bird Conserv Int 24:379–388

Grier JW (1982) Ban of DDT and subsequent recovery of reproduction in bald eagles. Science 218:1232–1235

Hall JC, Chhangani AK, Waite TA, Hamilton IM (2011) The impacts of La Niña-induced drought on Indian vulture Gyps indicus populations in western Rajasthan. Bird Conserv Int 2003:1–13

Hall JC, Chhangani AK, Warner TA (2015) Spatial characteristics of nest sites of critically endangered Indian vultures (Gyps indicus) in Rajasthan, India. Indian For 141:1–5

Harness R, Gombobaatar S, Yosef R (2008) Mongolian distribution power lines and raptor electrocutions. In: Rural electric power conference, April 27–30. Institude of Electrical and Electronics Engineer, Washington, DC/Charleston, pp C1.1–C1.6

Harris RJ, Reed JM (2002) Behavioral barriers to non-migratory movements of birds. Ann Zool Fenn 39:275–290

Harrison S, Emilio B (1999) Habitat fragmentation and large-scale conservation: what do we know for sure? Ecography 22:225–232

Hoekstra JM, Boucher TM, Ricketts TH, Roberts C (2005) Confronting a biome crisis: global disparities of habitat loss and protection. Ecol Lett 8:23–29

Ibañez JC, Sumaya AMT, Tampos G, Salvador DJI (2016) Preventing Philippine eagle hunting: what are we missing? J Threat Taxa 8:9505–9511

Iqbal M (2016) Predators become prey! Can Indonesian raptors survive online bird trading? Bird Asia 25:30–35

IUCN (2017) The IUCN red list of threatened species. Version 2016-3. www.iucnredlist.org

Jagannath A, Shore RF, Walker LA, Ferns PN, Gosler AG (2008) Eggshell pigmentation indicates pesticide contamination. J Appl Ecol 45:133–140

Jathar G, Rahmani A (2013) Ecology of the critically endangered Forest owlet *Heteroglaux blewitti*. In: Singaravelan N (ed) Rare animals of India. Bentham Science Publishers, Sharjah, pp 101–112

Jetz W, Thomas GH, Joy JB, Redding DW, Hartmann K, Mooers AO (2014) Global distribution and conservation of evolutionary distinctness in birds. Curr Biol 24:919–930

Kamp J, Urazaliev R, Donald PF, Hölzel N (2011) Post-soviet agricultural change predicts future declines after recent recovery in Eurasian steppe bird populations. Biol Conserv 144:2607–2614

Kamp J, Koshkin MA, Bragina TM, Katzner TE, Milner-Gulland EJ, Schreiber D, Sheldon R, Shmalenko A, Smelansky I, Terraube J, Urazaliev R (2016) Persistent and novel threats to the biodiversity of Kazakhstan's steppes and semi-deserts. Biodivers Conserv 25:2521–2541

Karyakin IV (2008) Lines-killers continue to harvest the mortal crop in Kazakhstan. Raptors Conserv 11:14–21

Katzner TE, Tingay RE (2010) Eagle diversity, ecology and conservation. In: Tingay RE, Katzner TE (eds) The eagle watchers. Cornell University Press, New York, pp 1–25
Katzner TE, Ivy JAR, Bragin EA, Milner-Gulland EJ, Dewoody JA (2011) Conservation implications of inaccurate estimation of cryptic population size. Anim Conserv 14:328–332
Katzner T, Johnson JA, Evans DM, Garner TWJ, Gompper ME, Altwegg R, Branch TA, Gordon IJ, Pettorelli N (2013) Challenges and opportunities for animal conservation from renewable energy development. Anim Conserv 16:367–369
Katzner TE, Bragin EA, Bragin AE, McGrady M, Miller TA, Bildstein KL (2016) Unusual clockwise loop migration lengthens travel distances and increases potential risks for a central Asian, long distance, trans-equatorial migrant, the red-footed falcon *Falco vespertinus*. Bird Study 63:406–412
Kennedy, RS (1977) Notes on the biology and population status of the monkey-eating eagle of the Philippines. Wilson Bull. 89:1–192
Kennedy RS (1985) Conservation research on the Philippine eagle. Natl Geogr Soc Res Rep 18:401–414
Kennedy RS, Gonzales PC, Dickinson EC, Miranda HCJ, Fisher TH (2000) A guide to the birds of the Philippines. Oxford University Press, New York
Kenward RE (2009) Conservation values from falconry. In: Dickson B, Hutton J, Adams WM (eds) Recreational hunting, conservation and rural livelihoods: science and practice. Wiley-Blackwell, London, pp 179–196
Kiff LF, Bijlsma RG, Severinghaus LL, Shergalin J (2007) The raptor literature. In: Bird DM, Bildstein KL (eds) Raptor research and management techniques. Hancock House Publishers, Surrey/Blane, pp 11–46
Kirby JS, Stattersfield AJ, Butchart SHM, Evans MI, Grimmett RF a, Jones VR, O'Sullivan J, Tucker GM, Newton I (2008) Key conservation issues for migratory land- and waterbird species on the world's major flyways. Bird Conserv Int 18:S49–S73
Koh LP, Wilcove DS (2008) Is oil palm agriculture really destroying tropical biodiversity? Conserv Lett 1:60–64
Kraemer R, Prishchepov AV, Müller D, Kuemmerle T, Radeloff VC, Dara A, Terekhov A, Frühauf M (2015) Long-term agricultural land-cover change and potential for cropland expansion in the former virgin lands area of Kazakhstan. Environ Res Lett 10:54012
Kramer JL, Redig PT (1997) Sixteen years of lead poisoning in eagles, 1980–95: an epizootiologic view. J Raptor Res 31:327–332
Lacher TE, Goldstein MI (1997) Tropical ecotoxicology : status and needs. Environ Toxicol Chem 16:100–111
Lambert O, Pouliquen H, Larhantec M, Thorin C, L'Hostis M (2007) Exposure of raptors and waterbirds to anticoagulant rodenticides (difenacoum, bromadiolone, coumatetralyl, coumafen, brodifacoum): epidemiological survey in Loire Atlantique (France). Bull Environ Contam Toxicol 79:91–94
Lasch U, Zerbe S, Lenk M (2010) Electrocution of raptors at power lines in Central Kazakhstan. Waldökologie, Landschaftsforschung Naturschutz 9:95–100
Laurance WF (1999) Reflections on the tropical deforestation crisis. Biol Conserv 91:109–117
Laurance WF (2007) Forest destruction in tropical Asia. Curr Sci 93:1544–1550
Levin A (2011) Illegal trade and decrease in numbers of the Saker falcon in Kazakhstan. Raptors Conserv 23:64–73
Levin AS, Kurkin GA (2013) The scope of death of eagles on powerlines in western Kazakhstan. Raptors Conserv 27:240–244
Lourenço R, Tavares PC, Del Mar Delgado M, Rabaça JE, Penteriani V (2011) Superpredation increases mercury levels in a generalist top predator, the eagle owl. Ecotoxicology 20:635–642
Lyde LW (1904) A geography of Asia including the east indies. Adam and Charles Black, London
Mace GM, Collar NJ (2002) Priority-setting in species conservation. In: Norris K, Pain DJ (eds) Conserving bird biodiversity: general principles and their application, Conservation biology series no. 7. Cambridge University Press, Cambridge, UK, pp 61–73
Madders M, Whitfield DP (2006) Upland raptors and the assessment of wind farm impacts. Ibis 148:43–56

Markandya A, Taylor T, Longo A, Murty MN, Murty S, Dhavala K (2008) Counting the cost of vulture decline-an appraisal of the human health and other benefits of vultures in India. Ecol Econ 67:194–204
McGowan P, Gillman M, Dodd M (1998) Assessing the status of poorly known species: lessons from partridges and pheasants of Southeast Asia. Biol Conserv 83:1–7
Mineau P, Whiteside M (2013) Pesticide acute toxicity is a better correlate of U.S. grassland bird declines than agricultural intensification. PLoS One 8:e57457
Mineau P, Fletcher MR, Glaser LC, Thomas NJ, Brassard C, Wilson LK, Lyon LA, Henny CJ, Bollinger T, Porter SL (1999) Poisoning of raptors with organophosphorus and carbamate pesticides with emphasis on Canada, U.S. and U.K. J Raptor Res 33:1–37
Myers N, Mittermeier RA, Mittermeier CG, da Fonseca GAB, Kent J (2000) Biodiversity hotspots for conservation priorities. Nature 403:853–858
Newton I (1979) Population ecology of raptors. T & AD Poyser, Berkhamsted
Newton I (1980) The role of food in limiting bird numbers. Ardrea 68:11–30
Newton I (1998) Population limitation in birds. Academic Press, London
NWBCT (2017) Amur falcon conservation in Nagaland, India. https://medium.com/@NWBCT/amur-falcon-conservation-in-nagaland-india-29259f793ef#.3kyrrndpx
Oaks JL, Gilbert M, Virani MZ, Watson RT, Meteyer CU, a Rideout B, Shivaprasad HL, Ahmed S, Chaudhry MJI, Arshad M, Mahmood S, Ali A, Khan AA (2004a) Diclofenac residues as the cause of vulture population decline in Pakistan. Nature 427:630–633
Oaks JL, Meteyer CU, Rideout BA, Shivaprasad HL, Gilbert M, Virani M, Watson RT, Khan AA (2004b) Diagnostic investigation of vulture mortality: the anti-inflammatory drug diclofenac is associated with visceral gout. In: Chancellor RD, Meyburg B-U (eds) Raptors worldwide: proceedings of the 6th world conference on birds of prey and owls. WWGBP/MME/Birdlife Hungary, Berlin/Budapest, pp 241–243
Olson DM, Dinerstein E (2002) The global 200: priority ecoregions for global conservation. Ann Mo Bot Gard 89:199–224
Pain DJ, Bowden CGR, Cunningham AA, Das D, Gilbert M, Jakati RD, Jhala Y, Khan AA, Naidoo V, Rahmani A, Ranade SP, Baral HS, Senacha KR, Saravanan S, Shah N, Swan G, Swarup D, Taggart MA, Watson RT, Virani MZ, Wolter K, Green RE (2008) The race to prevent the extinction of south Asian vultures. Bird Conserv Int 18:S30–S48
Panuccio M (2007) Protection of migratory raptors in the Mediterranean. Sustain Mediterr 35:13–14
Pestov MV, Saraev FA, Terentiev VA, Nurmuhambetov ZE (2015) The project outcome "assessment of the impact of medium voltage power lines on avifauna in Mangistau region (Kazakhstan)". Raptors Conserv 31:64–74
Population Reference Bureau (2016) 2016 world population data sheet. http://www.prb.org/pdf16/prb-wpds2016-web-2016.pdf
Prakash V, Pain DJ, Cunningham AA, Donald PF, Prakash N, Verma A, Gargi R, Sivakumar S, Rahmani AR (2003) Catastrophic collapse of Indian white-backed *Gyps bengalensis* and long-billed *Gyps indicus* vulture populations. Biol Conserv 109:381–390
Prakash V, Pain DJ, Shultz S, Cunningham A (2004) Saving Asia's *Gyps* vultures: the "vulture rescue" team' s conservation programme. In: Chancellor RD, Meyburg B-U (eds) Raptors worldwide: proceedings of the 6th world conference on birds of prey and owls. WWGBP/MME/Birdlife Hungary, Berlin/Budapest, pp 245–255
Prakash V, Bishwakarma MC, Chaudhary A, Cuthbert R, Dave R, Kulkarni M, Kumar S, Paudel K, Ranade S, Shringarpure R, Green RE (2012) The population decline of *Gyps* vultures in India and Nepal has slowed since veterinary use of diclofenac was banned. PLoS One 7:e49118
Rasmussen PC, Allen DNS, Collar NJ, Demeulemeester B, Hutchinson RO, Jakosalem PGC, Kennedy RS, Lambert FR, Paguntalan LM (2012) Vocal divergence and new species in the Philippine hawk owl *Ninox philippensis* complex. Forktail 28:1–20
Ratcliffe DA (1970) Changes attributable to pesticides in egg breakage frequency and eggshell thickness in some British birds. J Appl Ecol 7:67–115
Real J, Mañosa S (1997) Demography and conservation of western European Bonelli's eagle *Hieraaetus fasciatus* populations. Biol Conserv 79:59–66

Rolstad J (1991) Consequences of forest fragmentation for the dynamics of bird populations: conceptual issues and the evidence. Biol J Linn Soc 42:149–163

Ryabtsev VV (2011) Easternmost eastern imperial eagle population in crisis. Acta Zool Bulg 63:29–36

Ryabtsev VV, Katzner TE (2007) Severe declines of eastern imperial eagle *Aquila heliaca* populations in the Baikal region, Russia: a modern and historical perspective. Bird Conserv Int 17:197–209

Saito K (2009) Lead poisoning of Steller's sea eagle (*Haliaeetus pelagicus*) and white-tailed eagle (*Haliaeetus albicilla*) caused by the ingestion of lead bullets and slugs, in Hokkaido, Japan. In: Ingestion of lead from spent ammunition: implications for wildlife and humans. The Peregrine Fund, Boise, pp 302–309

Salager-Meyer F (2008) Scientific publishing in developing countries: challenges for the future. J Engl Acad Purp 7:121–132

Salvador DI, Ibanez JC (2006) Ecology and conservation of Philippine eagles. Ornithol Sci 5:171–176

Sánchez-Zapata JA, Carrete M, Gravilov A, Sklyarenko S, Ceballos O, Donázar JA, Hiraldo F (2003) Land use changes and raptor conservation in steppe habitats of eastern Kazakhstan. Biol Conserv 111:71–77

Sheffield S (1997) Owls as biomonitors of environmental contamination, pp 383–398. In: Duncan, James R.; Johnson, David H.; Nicholls, Thomas H., eds. Biology and conservation of owls of the Northern Hemisphere: 2nd International symposium. Gen. Tech. Rep. NC-190. St. Paul, MN: U.S. Dept. of Agriculture, Forest Service, North Central Forest Experiment Station

Shiraki S (2001) Foraging habitats of Steller's sea-eagles during the wintering season in Hokkaido, Japan. J Raptor Res 35:91–97

Shultz S, Baral HS, Charman S, Cunningham AA, Das D, Ghalsasi GR, Goudar MS, Green RE, Jones A, Nighot P, Pain DJ, Prakash V (2004) Diclofenac poisoning is widespread in declining vulture populations across the Indian subcontinent. Proc R Soc B 271:S458–S460

Smart J, Amar A, Sim IMW, Etheridge B, Cameron D, Christie G, Wilson JD (2010) Illegal killing slows population recovery of a re-introduced raptor of high conservation concern – the red kite *Milvus milvus*. Biol Conserv 143:1278–1286

Smith JA, Dwyer JF (2016) Avian interactions with renewable energy infrastructure: an update. Condor 118:411–423

Sodhi NS, Koh LP, Brook BW, Ng PKL (2004) Southeast Asian biodiversity: an impending disaster. Trends Ecol Evol 19:654–660

Sodhi NS, Koh LP, Clements R, Wanger TC, Hill JK, Hamer KC, Clough Y, Tscharntke T, Posa MRC, Lee TM (2010) Conserving southeast Asian forest biodiversity in human-modified landscapes. Biol Conserv 143:2375–2384

Solonen T, Lodenius M (1984) Mercury in Finnish sparrowhawks *Accipiter nisus*. Ornis Fenn 61:58–63

Spencer JE (1954) Asia, east by south. Wiley, New York

Stone WB, Okoniewski JC, Stedelin JR (2003) Anticoagulant rodenticides and raptors: recent findings from New York, 1998–2001. Bull Environ Contam Toxicol 70:34–40

Sutherland WJ, Pullin AS, Dolman PM, Knight TM (2004) The need for evidence-based conservation. Trends Ecol Evol 19:305–308

Swarup D, Patra RC, Prakash V, Cuthbert R, Das D, Avari P, Pain DJ, Green RE, Sharma AK, Saini M, Das D, Taggart M (2007) Safety of meloxicam to critically endangered *Gyps* vultures and other scavenging birds in India. Anim Conserv 10:192–198

Symes CT (2012) Amur falcon *Falco amurensis* slaughter in Nagaland, India. Gabar 23:69–73

Taggart MA, Senacha KR, Green RE, Jhala YV, Raghavan B, Rahmani AR, Cuthbert R, Pain DJ, Meharg AA (2007) Diclofenac residues in carcasses of domestic ungulates available to vultures in India. Environ Int 33:759–765

Tenan S, Adrover J, Muñoz Navarro A, Sergio F, Tavecchia G (2012) Demographic consequences of poison-related mortality in a threatened bird of prey. PLoS One 7:e49187

Terraube J, Arroyo B, Madders M, Mougeot F (2011) Diet specialisation and foraging efficiency under fluctuating vole abundance: a comparison between generalist and specialist avian predators. Oikos 120:234–244

The Association of Academies of Sciences in Asia (2011) Towards a sustainable Asia: environment and climate change. Springer/Science Press, Heidelberg/Beijing
Thiollay J-M (1985) The tropical rainforest raptors: state of knowledge, world situation and conservation strategy. ICBP Tech Publ 5:223–225
Thiollay JM (1998) Current status and conservation of falconiformes in tropical Asia. J Raptor Res 32:40–55
Tilman D, Fargione J, Wolff B, D'Antonio C, Dobson A, Howarth R, Schindler D, Schlesinger WH, Simberloff D, Swackhamer D (2001) Forecasting agriculturally driven global environmental change. Science 292:281–284
Tinker PB (1997) The environmental implications of intensified land use in developing countries. Philos Trans R Soc Lond Ser B Biol Sci 352:1023–1033
Tsoutsos T, Frantzeskaki N, Gekas V (2005) Environmental impacts from the solar energy technologies. Energy Policy 33:289–296
Udvardy MDF (1975) A classification of the biogeographical provinces of the world. Occasional. International Union of Conservation of Nature and Natural Resources, Morges
Ueta M, Ryabtsev VV (2001) Migration routes of four juvenile imperial eagles *Aquila heliaca* from the Baikal region of eastern Russia. Bird Conserv Int 11:93–99
Urosawa NK (2000) Lead poisoning in Steller's sea eagles and White-Tailed Sea eagles. In: Ueta M, McGrady MJ (eds) First symposium on Steller's and White-Tailed Sea eagles in East Asia. Wild Bird Society of Japan, Tokyo, pp 107–109
Van Balen S, Nijman V, Prins HHT (2000) The Javan hawk-eagle: misconceptions about rareness and threat. Biol Conserv 96:297–304
Van Horne B (1983) Density as a misleading indicator of habitat quality. J Wildl Manag 47:893–901
Wallin K (1984) Decrease and recovery patterns of some raptors in relation to the introduction and ban of alkyl-mercury and DDT in Sweden. Ambio 13:263–265
Walston LJ, Rollins KE, LaGory KE, Smith KP, Meyers SA (2016) A preliminary assessment of avian mortality at utility-scale solar energy facilities in the United States. Renew Energy 92:405–414
Warren-Thomas E, Dolman PM, Edwards DP (2015) Increasing demand for natural rubber necessitates a robust sustainability initiative to mitigate impacts on tropical biodiversity. Conserv Lett 8:230–241
Watson J (1992) Nesting ecology of the Seychelles kestrel *Falco araea* on Mahé, Seychelles. Ibis 134:259–267
Watts BD, Therres GD, Byrd MA (2008) Recovery of the Chesapeake Bay bald eagle nesting population. J Wildl Manag 72:152–158
Webster MS, Marra PP, Haig SM, Bensch S, Holmes RT (2002) Links between worlds: Unraveling migratory connectivity. Trends Ecol Evol 17:76–83
White CM, Kiff LF (2000) Biodiversity, island raptors and species concepts. In: Chancellor RD, Meyburg B-U (eds) Raptors at risk. World Working Group on Birds of Prey/Hancock House, Blaine/Washington, DC, pp 633–652
Whitfield DP, Fielding AH, Mcleod DRA, Haworth PF (2004a) Modelling the effects of persecution on the population dynamics of golden eagles in Scotland. Biol Conserv 119:319–333
Whitfield DP, Fielding AH, McLeod DRA, Haworth PF (2004b) The effects of persecution on age of breeding and territory occupation in golden eagles in Scotland. Biol Conserv 118:249–259
Yi-Ming L, Zenxiang G, Xinhai L, Sung W, Niemelä J (2000) Illegal wildlife trade in the Himalayan region of China. Biodivers Conserv 9:901–918
Zahler P, Lhagvasuren B, Reading RP, Wingard JR, Amgalanbaatar S, Gombobaatar S, Barton N, Onon Y (2004) Illegal and unsustainable wildlife hunting and trade in Mongolia. Mong J Biol Sci 2:23–31
Zhang L, Hua N, Sun S (2008) Wildlife trade, consumption and conservation awareness in Southwest China. Biodivers Conserv 17:1493–1516
Zlotin R (2002) Biodiversity and productivity of ecosystems. In: Shahgedanova M (ed) The physical geography of northern Eurasia. Oxford University Press, Oxford, UK, pp 169–190

Chapter 18
Conservation and Ecology of African Raptors

Arjun Amar, Ralph Buij, Jessleena Suri, Petra Sumasgutner, and Munir Z. Virani

Introduction

Africa represents one of the most important continents for both endemic resident and migratory raptors (Brown et al. 1982). Sub-Saharan Africa alone supports breeding populations of over 20% of all raptor species globally, while also hosting more than 20 regular Palearctic migratory raptors (Ferguson-Lees and Christie 2001). However, over the last several decades, numbers and diversity of raptors in many regions have declined dramatically, principally because of an increase in human populations across the continent, which has resulted in habitat loss, agricultural intensification and the increased use of poisons. Africa is experiencing rapid growth, both economically (www.worldbank.org) and in terms of human numbers (United Nations 2017), and these changes are having considerable impacts on raptor populations in Africa. Such pressures are likely to increase significantly in the next few decades, as the human population in sub-Saharan Africa grows to over 2 billion

A. Amar (✉)
FitzPatrick Institute of African Ornithology, DST-NRF Centre of Excellence,
University of Cape Town, Cape Town, South Africa
e-mail: arjun.amar@uct.ac.za

R. Buij
Animal Ecology Department, Alterra Wageningen UR, Wageningen, The Netherlands

J. Suri · P. Sumasgutner
FitzPatrick Institute of African Ornithology, DST-NRF Centre of Excellence,
University of Cape Town, Cape Town, South Africa

Centre for Statistics in Ecology, the Environment and Conservation, Department of Statistical Science, University of Cape Town, Cape Town, South Africa

M. Z. Virani
The Peregrine Fund, Boise, ID, USA

Department of Zoology, National Museums of Kenya, Nairobi, Kenya

J. H. Sarasola et al. (eds.), *Birds of Prey*,
https://doi.org/10.1007/978-3-319-73745-4_18

by 2050 (United Nations 2017). Additionally, Africa, together with Asia, is urbanising faster than any other regions in the world. For example, Nigeria alone is projected to add 212 million urban dwellers by 2050 (Baloye and Palamuleni 2015). This rapid urbanisation poses challenges for sustainable development and public health and will also have large impacts on the continent's wildlife.

Given these development, Africa's protected area network is an increasingly vital stronghold for many raptor species (Amar and Cloete 2017; Herremans and Herremans-Tonnoeyr 2000), especially for the larger, more vulnerable species that are generally the first to vanish from human-dominated landscapes (Brown 1991; Thiollay 2006). But the capacity of the existing protected area network to support the long-term persistence of remaining, increasingly small and fragmented populations of large raptors has been questioned (Murn et al. 2016). Many large raptors are especially vulnerable to pressures operating beyond protected area boundaries when they forage beyond the realms of protected areas (van Eeden et al. 2017; Krüger et al. 2014), which may also explain their significant declines inside protected areas in West (Thiollay 2006), East (Virani et al. 2011) and Southern Africa (Amar and Cloete 2017). Future and increasing threats to raptors in Africa include the predicted growth in Africa's energy sector from oil and mining exploration, construction of additional power lines and wind turbines, as well as the other infrastructure associated with these developments (Edwards et al. 2014). The impacts of climate change on raptors are yet to be understood, although changing rainfall patterns in Africa are likely to have important effects on the population dynamics of a range of species (Martin et al. 2014b; Wichmann et al. 2003). Going forward, efforts to mitigate the magnitude of raptor declines will need commitment from African governments and the conservation communities to ensure the survival of species that are on the verge of collapse (Ogada et al. 2015).

There is a long history of raptor research in Africa, and some of the most famous raptor studies have been conducted in Africa. For example, Leslie Brown conducted pioneering studies on raptors in Kenya and elsewhere on the continent (Brown 1953; Brown and Brown 1979; Brown and Cade 1972); Valerie Gargett's study on Verreaux's Eagles (*Aquila verreauxii*) in Zimbabwe is long regarded as one of the classic long-term population studies (Gargett et al. 1995); and Peter Steyn has produced several books documenting his long-term and diverse raptor research in Southern Africa (Steyn 1973, 1982). However, despite the wealth of raptor research in Africa, basic ecological knowledge of many species remains extremely limited (Virani and Watson 1998).This underpins the need to continually and consistently develop the next generation of African raptor biologists.

In this chapter we discuss the importance of Africa for raptor species, in terms of the diversity of both resident and migrant species hosted in the varied habitats of the continent. Although the continent is important to migrant raptors from Europe to the Far East, in this chapter, we consider primarily those species breeding on the mainland and south of the Sahara. We also discuss the importance of these raptor species in providing vital ecosystem services within Africa. We examine what monitoring of raptor species has occurred in Africa to allow changes in abundance to be measured and thus trends to be determined. We then explore how the global conservation status of African raptors has changed over the last three decades and discuss some of the key threats faced by African raptors that may contribute to these changes in their conservation status. We then look into some of the research that has been conducted on African raptors in order to identify strengths and gaps in our knowledge of these species, as well as important areas of research for the future in order to prevent declines in raptor populations. For this, we undertake a systematic review of the literature in order to document the current scientific knowledge of raptors species in Africa. Lastly, we describe some of the ongoing research on raptors in Africa and our hope for the future.

Importance of Africa for Raptors

Sub-Saharan Africa is host to many unique raptor species, from some of the largest and most powerful eagles in the world such as the African crowned eagle, to some of the smallest falcons like the African pygmy falcon (*Polihierax semitorquatus*). The continent's savannas also host several monotypic species, such as the martial eagle (Fig. 18.1), which is the only member of the *Polemaetus* genus, and even a monophyletic genus, the secretarybird (*Sagittarius serpentarius*), which is the only member of the Sagittariidae family. The variety of African raptors range from food generalists such as augur buzzard (*Buteo augur*) and tawny eagle (*Aquila rapax*), to food specialists such as the bat hawk (*Macheiramphus alcinus*), six snake eagles, three *Gyps* vultures and Verreaux's eagle (but see Murgatroyd et al. 2016a).

In terms of migrants, Africa is the non-breeding ground for a large number of raptor species from Eurasia that fly into Africa during the boreal winter. These include falcons, harriers, buzzards, hawks, kites, eagles and vultures from Europe to pallid

Fig. 18.1 Adult martial eagle with prey. The species has declined in many countries, both inside and outside of protected areas. (Photo by Munir Virani – The Peregrine Fund)

harriers (*Circus macrourus*), steppe eagles (*Aquila nipalensis*) and Amur falcons (*Falco amurensis*) from Asia. Some of these migrants undergo the most incredible journeys that are now being revealed thanks to recent advances in tracking technologies. For example, by tagging Amur falcons, a team led by Bernd Ulrich and Christina Meyburg recently found that these raptors cover a distance of some 14,500 km from their breeding grounds in Mongolia and China to Southern Africa each year (Meyburg et al. 2017) and do so by flying 2500–3000 km across the Indian Ocean nonstop in 2–3 days!

Because Africa is such a large continent and spans both the northern and southern hemisphere, it also hosts a substantial number of raptors that are intra-African migrants. Unlike Palearctic migrants who winter in Africa to avoid the harsh boreal winters and track shifting prey availability with the dry and rainy seasons in Africa, the movements of intra-African migrants appear to be entirely driven by regional rainfall patterns and resulting changes in the activity, abundance and accessibility of prey. The clearest examples of migration have therefore been reported from areas with a single and well-defined rainy season, in West and Southern Africa (Thiollay 1978a, Brown et al. 1982). Conversely, in regions where conditions are more stable throughout the year (the tropical forests) or where rainfall is more variable and occurs twice yearly (equatorial savannas), migrations are absent or at best far less predictable (Smeenk 1974). Some intra-African migrants, such as African swallow-tailed kite (*Chelictinia riocourii*), grasshopper buzzard (*Butastur rufipennis*), Beaudouin's snake eagle (*Circaetus beaudouini*) and red-necked buzzard (*Buteo auguralis*), are typically migratory throughout most of their distribution range. Unlike most Palearctic migrants, the movements of these and other African raptors are very poorly understood and highly variable even within species. For example, yellow-billed kites (*Milvus aegyptius*), shikras (*Accipiter badius*) and Wahlberg's eagles (*Hieraaetus wahlbergi*) are either sedentary, short- or long-distance migrants, nomadic or partial migrants in different regions of Africa (Brown 1971; Smeenk 1974; Thiollay 1978b). Recent efforts to fit GPS satellite tags on some of these species are allowing us to better understand their movements and the drivers of these movements. In one of the few satellite tagging studies conducted on an African migratory raptor to date, Wahlberg's eagles were shown to migrate to Sudan, Chad, Nigeria and Cameroon after breeding in South Africa and Namibia (Meyburg et al. 1995), thus tracking the rains throughout their annual cycle. Disentangling the patterns and drivers of the movement of these intra-African migrants is vital for understanding potential impacts of climate change on these species.

Ecosystem Services of Raptors in Africa

Raptors in Africa provide vital ecosystems services. For example, the ecosystem services that vultures provide fall into three of the four categories suggested by the Millennium Ecosystem Assessment (Duraiappah et al. 2005): their iconic circling gives away the location of carcass – which would have benefitted hungry

Fig. 18.2 Hooded vultures help to clean up carcasses from the environment and therefore prevent the spread of pathogens. (Photo by Ralph Buij)

scavenging cavemen and in modern times alerts farmers and anti-poaching teams to otherwise difficult to detect carcasses. By removing carcasses from the landscape, vultures also control the spread of infectious disease (Fig. 18.2). Indeed, when vulture numbers crashed in India, the increased number of decomposing carcasses supported a growing feral dog population which resulted in a marked increase in rabies infections and deaths which was estimated to have cost the country around $34 billion (Markandya et al. 2008). Research in Africa has shown that when vultures are absent, carcasses take up to three times longer to decompose, thereby creating a hub for pathogens that would have otherwise been rapidly removed. At lingering carcasses, the number and physical interactions of mammalian scavengers also increase (Ogada et al. 2012) thereby potentially contributing to spread of disease between mammals. Similar to Asia, the continued decline in African vultures may thus come with consequences to human health (Ogada et al. 2012; Markandya et al. 2008). Finally, vultures are a considerable cultural asset in Africa including more recently, being a photographic attraction at vulture restaurants (see below).

Other raptor species in Africa likely play important roles in the ecosystem. In many regions raptors are amongst the most common top predators and are likely to shape the species assemblages of birds and mammals (Ritchie et al. 2013), as well as their behaviours (Shultz and Noë 2002; Willems and Hill 2009). Within some areas in Africa, the loss of large raptors from urban systems is thought to have contributed to the increase in some species that have become nuisance species for humans (Atkins et al. 2017), and experimentally introducing raptors back into the system influences both the number of individuals and the behaviours (vigilance) of these nuisance species (Atkins et al. 2017). This has also been shown in other places around the world in relation to pest bird species (Kross et al. 2012) and in Africa also probably extends to controlling other non-avian agricultural pests.

Small mammals, queleas and locusts represent major agricultural and social pests in some regions of Africa (Fiedler 1994; Mundy and Jarvis 1989) and attract a

Table 18.1 Important avian predators of major agricultural pest species in sub-Saharan Africa

Raptor species	Pest (crop)	References
Black kite (PAM), shikra, tawny eagle, steppe eagle (PM), Wahlberg's eagle, African harrier-hawk, lanner falcon, peregrine falcon, gabar goshawk, steppe buzzard (PM), augur buzzard, dark chanting goshawk, red-necked falcon	Red-billed quelea (sorghum, millet, rice, wheat)	Keith and Bruggers (1998), Ogada et al. (2015), Thiollay (1975, 1978a, 1989), Thomsett (1987)
Eurasian marsh harrier (PM), pallid harrier (PM), booted eagle (PM), steppe eagle (PM), tawny eagle, black-winged kite, long-crested eagle, augur buzzard, steppe buzzard (PM), jackal buzzard, common kestrel (PAM)	Rodents (rice, sorghum, maize, sugarcane)	Arroyo and King (1995), Brown (1970, 1955), Buij et al. (2012), Buij and Croes (2013), Van Gulck et al. (1998), Norgarb and Lasbrey (1953), Smeenk (1974), Tarboton (1978), Virani (1999)
Lesser kestrel (PM), common kestrel (PM), lanner falcon, African swallow-tailed kite, steppe eagle (PM), Montagu's harrier (PM), pallid harrier (PM), black kite (PAM), Amur falcon (PM), red-footed falcon (PM), African cuckoo-hawk, grasshopper buzzard	Locusts (including desert locusts) and grasshoppers (sorghum, millet)	Anderson et al. (1999), Augiron et al. (2015), Buij et al. (2012, 2013a, c), Kopij (2009), Sánchez-Zapata et al. (2007), Steedman (1988), Thiollay (1989), Trierweiler and Koks (2009)

The crops in which the raptors were seen foraging on the pest species are indicated in parentheses
PM Palearctic migrant, *PAM* Palearctic and African migrants

variety of raptors (Table 18.1). Although in Africa, a definitive experimental test of their role in limiting agricultural pests has yet to be undertaken, the results of natural experiments with nest boxes elsewhere suggest that raptors probably play a role in controlling the numbers of some of these prey species, such as rodents (Paz et al. 2013). Studies using experimentally manipulated predation levels in East Africa indicated faster rodent population growth rates and higher rodent peak population sizes in the absence of raptors compared to situations where raptor presence was stimulated (e.g. through perches; Vibe-Petersen et al. 2006). There is also some evidence from West Africa that raptors, like other acridivorous birds, can suppress grasshopper or locust abundance, which may help to reduce agricultural damages (Mullié and Guèye 2010). Similarly, red-billed queleas (*Quelea quelea*) in grain-producing regions of Africa are commonly preyed on by a range of sedentary and migratory raptors, which often gather in large numbers at quelea breeding colonies when the nestlings fledge (Smeenk 1974; Thiollay 1978b). Given that raptors, depending on their size, can consume between 4 and 20 fledgling queleas per day, a congregation of several hundred raptors may consume up to ten thousand queleas at single breeding colonies or up to a quarter of their offspring (Thiollay 1975). However, whether such predation can influence overall population size, and the amount of damage they cause, is unknown. On the other hand, the attraction of raptors to concentrations of agricultural pest species may make them vulnerable to exposure of pesticides and secondary poisoning, especially avicides and rodenticides (Keith and Bruggers 1998; McWilliam and Cheke 2004; Thomsett 1987).

Fig. 18.3 Ecosystem services in action – vultures feeding at a carcass on the African savannah. Raptor spectacles such as these contribute to the popularity of avitourism in many regions of Africa. (Photo by Munir Virani – The Peregrine Fund)

Raptors can also be hugely valuable in terms of the tourism revenue they generate. With birding being one of the world's fastest growing hobbies, avitourism has seen a steady increase in Africa too (Biggs et al. 2011; Conradie et al. 2013). For example, in South Africa alone, the avitourism industry is estimated to be worth over ZAR 1 billion ($73 million; Nicolaides 2013). Part of this involves birding as a by-product of other nature-based tourism, while most involves dedicated trips to see certain species. Within this, raptors rank high amongst the kind of charismatic species that motivate avitourists, and Africa's diverse raptor community represents a large attraction. With a wide array of endemic African species (many of which are extremely range restricted, such as black harriers and Taita falcons), large aggregations of migrant species (e.g. roosting sites of species like African swallow-tailed kites, lesser kestrels (*Falco naumanni*) and Amur falcons) and somewhat unique African sights like several species of vulture scavenging on megafauna carcasses (Fig. 18.3), it is easy to imagine how Africa could be considered a raptorphile's paradise. For example, although a major tourism attraction in its own right, South Africa's Kruger National Park is well known to boast a high variety of raptor species that can be seen relatively easily, and potential to see localised species like Dickinson's kestrels and magnificent martial eagles can generate a lot of buy-in from avitourists. The same is true of countries such as Kenya, Ethiopia, Ghana and Botswana where avitourism is on the rise. For example, African Fish Eagle Photography Tours at Lake Naivasha rakes in nearly US$ 250,000 per year for local communities (Virani pers. obs.).

Vulture restaurants provide an interesting opportunity for photographers. The first established vulture restaurant was opened in the Drakensberg, South Africa, in 1966 (Piper 2004). Since then the number of restaurants has grown to over 200 in South Africa. Beside the value for tourism, supplementary feeding may also contribute positively to vultures' breeding success and survival (Cortés-Avizanda et al. 2016; Ewen et al. 2015; Moreno-Opo et al. 2015), although it may have its downsides (e.g. Cortés-Avizanda et al. 2009), but such issues for African vultures remain poorly studied. The success of a vulture restaurant ultimately relies on the motivation and dedication of the people who run it, and in South Africa, 80% of active vulture restaurants are run by self-motivated members of the public (i.e. commercial pig farmers, Selinske et al. 2015). These issues are currently the focus of a PhD study (Christian Brink, FitzPatrick Institute, University of Cape Town).

In many parts of the world, raptors have also been used as indicators of environmental health. Because raptors occupy higher trophic levels, they can act as bio-accumulators and bio-magnifiers of environmental toxins, and raptors have proven particularly useful as indicators of environmental pollutants (Gomez-Ramirez et al. 2014). The best examples include the declines of many bird- and fish-eating raptors caused by DDT and PCB contamination in the 1960s and 1970s (Newton 1979), which led to extensive investigations into their effects on human health, culminating in the widespread ban of these products. Within Africa, there have been a few studies exploring pollutants in African raptors, most from Southern Africa (García-Heras et al. 2017; Garcia-Heras et al. 2017a; Van Wyk et al. 2001). However, generally speaking, raptors in Africa represent an untapped resource by which to measure environmental quality. This is particularly true given the heavy exploitation of natural resources and associated risks of pollution that occurs in Africa, for example, in the mining industry, and the heavy and widespread (mis)use of pesticides in agriculture (Maumbe and Swinton 2003; Ngowi et al. 2007; Williamson et al. 2008). As such, raptors may in future offer a valuable means by which to assess the levels of sustainability of such activities and to monitor pollution also for the benefit of human health.

Population Monitoring of Raptors in Africa

Levels of biological monitoring in a country are usually linked to the country's economic status (Amano and Sutherland 2013). Africa remains an economically underdeveloped region compared with many other regions of the world. Most countries in Africa are classified by the UN as being low income countries, with a few being classified as middle income (e.g. South Africa). Thus, despite many regions in Africa designated as biodiversity hotspots (Larsen et al. 2011) and Important Bird Areas (IBAs) (Fishpool and Evans 2001), the levels of biological monitoring are relatively low. In most of Africa, there is little systematic monitoring of avian biodiversity, including raptors, unlike in more developed countries and regions (e.g. BBS schemes in the UK, https://www.bto.org/volunteer-surveys/bbs) and the USA (Silvertown 2009), or the common bird indexes across Europe (Schmeller et al. 2012).

In some countries and regions however, African raptor populations have received some degree of monitoring. Some of this monitoring has been at the level of species, with repeat nesting surveys being undertaken (Table 18.2). In other areas, changes in abundance of the entire raptor assemblage have been attempted through repeats of raptor road transects over time (Table 18.3). South Africa is the exception to the rule, because it has the Southern African Bird Atlas Project 1 and 2. This country-wide citizen science atlas project has allowed reporting rate changes of some raptor

Table 18.2 Details of species-specific studies undertaking repeat or long-term monitoring of African raptors to detect population trends

Species	Country/region	Methods	Time period (first; second)	Population trend	Source
Augur buzzard	Kenya (Lake Naivasha	Nesting surveys	1995; 2014	−47%	Eichenwald et al. (in prep.)
Martial eagle	South Africa (Kalahari)	Nesting surveys	1993; 2011/2012	−43.8%	Amar et al. (2016)
Black sparrowhawk	South Africa (Cape Town)	Nesting surveys	2001–2012	+276%	Martin et al. (2014a)
Peregrine falcon	South Africa (Cape Town)	Nesting surveys	1997–2010	+500%	Altwegg et al. (2014)
Hooded vultures	Senegal (Dakar)	Roost counts	1969–1971; 2016	−85%	Mullié et al. (2017)
Bearded vulture	Southern Africa	Nesting surveys	1960–1999; 2000–2012	−32 to 51%	Krüger et al. (2013)
White-backed vulture	South Africa (Kimberley)	Nesting surveys	2001; 2014	−23%	Murn et al. (2013)
Cape vulture	South Africa (northern provinces)	Colony counts	1985; 2000; 2013	Largely stable	Benson (2014)
Cape vulture	Southern Africa	Colony counts	2010–2015	Largely stable	Wolter et al. (2016)

species to be explored at a level that is unprecedented in an African context (Amar and Cloete 2017; Hofmeyr et al. 2014). There are hopes that the success of this project can be expanded to the rest of the continent, and there has already been growing involvement from citizen scientists in the rest of Southern Africa, as well as in Nigeria and Kenya, paving the way for a future African Bird Atlas Project (Underhill pers. comm.).

We have attempted to compile the published studies on raptor monitoring across Africa, both at the species level (Table 18.2) and at the assemblage level (Table 18.3). These studies are only those studies where comparisons have been made over time and may not be exhaustive. From the relatively few surveys that have examined changes in raptor abundance in different African regions, the overall trend is for declines, with a few exceptions, for example, some urban-dwelling species (Altwegg et al. 2014; Martin et al. 2014a). For several species these declines have been catastrophic (Thiollay 2006), and in general declines appear to be worse outside of protected areas than inside them (Amar and Cloete 2017; Thiollay 2006; Virani et al. 2011).

Several of these surveys were instrumental in alerting the conservation community to the large declines of African vulture species. The widespread decline of this group of raptors has resulted in many describing the situation as an 'African Vulture Crisis' (Krüger et al. 2013; Ogada et al. 2015), akin (but driven by different causes) to the crisis seen for this group of birds in Asia (Oaks et al. 2004). The decline of raptors and vultures, in particular, has raised major concerns over the impacts this loss may have to ecosystems and even to human health (Markandya et al. 2008).

Table 18.3 Repeat surveys which monitor the assemblages of raptor species in Africa

Study	Country/ region	Species	Methods	Time period 1	Time period 2	Source
Southern African Bird Atlas Project 1 and 2	South Africa	All raptors	Atlas surveys	1987–1992	2007–ongoing	Amar et al. (2016), Amar and Cloete (2017), Hofmeyr et al. (2014)
West African raptor surveys	Burkina Faso, Mali, Niger, Cameroon	All diurnal raptors	Road counts	1969–1973	2000–2004	Thiollay (2007)
Botswana raptor surveys	Northern Botswana	All diurnal raptors	Road counts	1991–1995	2015–2016	Herremans and Herremans-tonnoeyr (2000), Garbett et al. (in review)
Masai Mara road counts	Kenya	6 species of large raptor	Road counts	1976 & 1988	2003–2005	Virani et al. (2011)
Ugandan road counts	Uganda	All diurnal raptors	Road counts	2009–2015	ongoing	Pomeroy et al. (2016)

These surveys are only those that involve repeats and therefore where change can be estimated

Additional to these resurveys, there are also other forms of baseline monitoring. One of the most important projects is the African Raptor Databank (ARDB). This project was launched in 2012 with support from The Peregrine Fund and has grown rapidly since. The ambitious goal has been to map all African raptors on the continent with the support of hundreds of raptor observers across the continent. In 2017, the ARDB completed a 5-year data collection period. Nearly 180,000 records have been assimilated. The ARDB is managed by Rob Davies (Habitat Info Wales, UK) and makes use of the latest spatial data technologies to acquire, manage, analyse and disseminate raptor distribution data. The project was conceived to address the issue of habitat loss for many species; however, with the imminent threat facing vultures from poisoning, the purpose and objectives of the ARDB were adapted to enable the ARDB to serve a future role as a monitoring tool. The ARDB has already provided benefits, for example, improved vulture range maps have been created and have helped with the IUCN uplisting of African vultures and have informed the international vulture Multi-species Action Plan (MsAP). The ARDB has now adapted the technologies to roll out this model worldwide with a global databank and mobile phone application, with the aim to add different language modules over time.

Conservation Status of African Raptors

Given that the existing, albeit limited, raptor monitoring from Africa has shown a general tendency of population declines (Tables 18.2 and 18.3), we might expect this to be coupled with a deterioration in the global conservation status of African

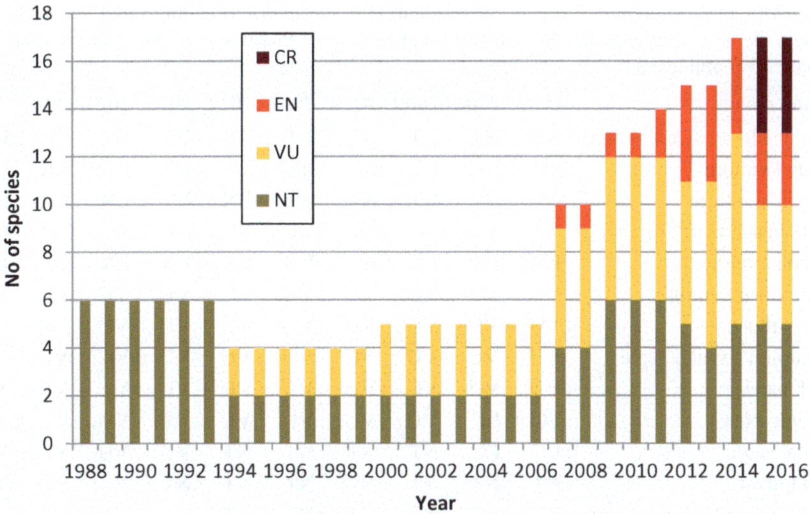

Fig. 18.4 Change in the global conservation status of 66 species of diurnal African raptors (defined as species breeding south of the Sahara) between 1988 and 2016. NT near threatened, VU vulnerable, EN endangered, CR critically endangered. In total 17 species were listed on the Red List in 2016, representing 26% of all species examined

raptors. We used information from the BirdLife Data zones (http://datazone.birdlife.org/home) to explore this question for 66 species of diurnal African raptors, which we define as being resident species breeding south of the Sahara (Appendix A). We explored the number of these species that were listed on the IUCN Red List as being extinction prone at the four different levels (i.e. near threatened, vulnerable, endangered and critically endangered) and how this had changed since 1988.

The growing pressures on Africa's raptors are becoming increasingly apparent in their population trends and their conservation status. Currently 17 species out of the 66 explored are on the Red List, which represents 26% of African species (Fig. 18.4). Examining how this has changed over time, it is clear that the conservation status of diurnal raptors appeared to decline rapidly in the mid-2000s, with the first vulnerable species being listed in 1994, the first endangered species being listed in 2007 and the first critically endangered species being listed in 2015.

From the 17 species that were listed in 2016 on the IUCN's Red List, 5 species were listed as being near threatened, 5 species were listed as being vulnerable, 3 species were listed as being endangered and 4 species were listed as being critically endangered (Table 18.4). From this analysis, it is also clear that there is a steady worsening of the conservation status for threatened species; between 2007 and 2016, no less than 6 of 17 threatened species moved into an increasingly worse conservation status.

Examining the conservation status of the different groups of raptors also points to some interesting patterns, although for most groups, sample sizes prevent a more formal analysis of these patterns. Nevertheless, it is striking that eight out of nine

Table 18.4 Table showing all species of resident African raptors (breeding south of the Sahara) currently listed as being prone to extinction and how their threat status has changed each year between 2007 and 2016

Common name	2007	2008	2009	2010	2011	2012	2013	2014	2015	2016
African white-backed vulture	NT	NT	NT	NT	NT	EN	EN	EN	CR	CR
Hooded vulture	LC	LC	LC	LC	EN	EN	EN	EN	CR	CR
Rüppell's vulture	NT	NT	NT	NT	NT	EN	EN	EN	CR	CR
White-headed vulture	VU	VU	VU	VU	VU	VU	VU	VU	CR	CR
Egyptian vulture	EN	EN	EN	EN	EN	EN	EN	EN	EN	EN
Lappet-faced vulture	VU	VU	VU	VU	VU	VU	VU	VU	EN	EN
Cape vulture	VU	VU	VU	VU	VU	VU	VU	VU	EN	EN
Beaudouin's snake eagle	VU	VU	VU	VU	VU	VU	VU	VU	VU	VU
Black harrier	VU	VU	VU	VU	VU	VU	VU	VU	VU	VU
Secretarybird	LC	LC	VU	VU	VU	VU	VU	VU	VU	VU
Martial eagle	LC	LC	NT	NT	NT	NT	VU	VU	VU	VU
Taita falcon	NT	NT	NT	NT	NT	NT	NT	VU	VU	VU
Southern banded snake eagle	NT	NT	NT	NT	NT	NT	NT	NT	NT	NT
Bateleur	LC	LC	NT	NT	NT	NT	NT	NT	NT	NT
African crowned eagle	LC	LC	LC	LC	LC	NT	NT	NT	NT	NT
Bearded vulture	LC	LC	LC	LC	LC	LC	LC	NT	NT	NT
Mountain buzzard	NR	NR	NR	NR	NR	NR	NR	NT	NT	NT

NR not evaluated, *LC* least concern, *NT* near threatened, *VU* vulnerable, *EN* endangered, *CR* critically endangered

African vulture species (89%) are listed as being extinction prone and all the species that are listed in the two highest threat categories are all vultures, which further emphasises the perilous conservation state of that group of Africa raptors (Ogada et al. 2015). For eagles, 5 out of 17 species (29%) are listed, but the proportion of threatened species is apparently higher for larger eagle species (3/6 = 50%) and snake eagle species (2/6 = 33%) than for the smaller eagle species (0/5 = 0%). Buzzards (1/6 = 17%), falcons (1/11 = 9%), hawks (0/16 = 0%) and kites (0/3 = 0%) also appear to have less threatened species than the other groups.

For many species there is a need for their conservation status to be re-evaluated, but this can only be achieved through rigorous monitoring which for most of Africa is generally lacking (Amar and Cloete 2017). Many species that are currently not listed as being globally threatened species according to IUCN's Red Data Book but that are regionally threatened are in desperate need of population studies to accurately establish their status and effects of human-caused habitat alteration and persecution; these include a variety of species from different genera such as palm-nut vulture (*Gypohierax angolensis*), red-necked buzzard (*Buteo auguralis*), long-crested eagle (*Lophaetus occipitalis*), African hawk-eagle (*Aquila spilogaster*), Cassin's hawk eagle (*Aquila africana*), and African marsh harrier (*Circus ranivorus*).

Scientific Knowledge on African Raptors

We undertook a systematic literature review using Web of Science (https://webofknowledge.com/) to compile a database of published studies on African raptors. In a sense, this review was intended as a repeat of that carried out by Virani and Watson (1998) in their review of ecological knowledge of raptors in the East African tropics and Western Indian Ocean Islands but covered a more extensive geographical area. However, our review was limited to diurnal raptors of the orders Accipitriformes and Falconiformes (i.e. owls were excluded) from mainland sub-Saharan Africa only, and we also made no attempt to specifically distinguish the levels of knowledge for each study, as was done by Virani and Watson (1998).

Our literature searches were conducted on September 15, 2016 (except Cape vulture which was done on June 15, 2017). For each individual search, the search terms used were the species' common and scientific names. To ensure that the maximum amount of studies were captured in our results, we also included any alternative scientific and common names, using the *Illustrated Checklist of the Birds of the World* (del Hoyo et al. 2014). For cosmopolitan and migrant species which are known to occur outside of Africa, we conducted an additional search for the word 'Africa' and combined this with the results from the search on the species names to limit the results to research originating from Africa only. For example, for common kestrel, the following search was conducted: 'Common-Kestrel OR Rock-Kestrel OR Falco-tinnunculus OR Falco-rupicolus; Africa; COMBINE (AND)'.

We exported our results and selected only studies from journals that were currently ISI-listed. We then read the titles and abstracts of each paper to determine the location of the study, whether it was relevant or not (i.e. dealing with the species in question and conducted in Africa), and the broad subject matter of the paper. Where the required information could not be determined from the title or abstract alone, the paper was sourced and read in order to extract these details.

Our search, which covered 67 species (Appendix A – this includes all the 66 species from the previous section, plus Barbary Falcon) between 1949 and 2016, produced a total of 555 papers published. One hundred and thirteen papers contained information on multiple raptor species (range of raptor species per paper 2–11, mean ± 1 S.D. = 2.99 ± 1.58). We considered information on each species as a different 'study' for the purposes of this review, and therefore the 555 papers contained 780 'studies'.

The research was published in 137 journals. The top 10 journals with the most number of studies were *Ostrich* ($n = 230$), followed by *Ibis* (73), Ardea (40), *Journal of Raptor Research* (31), *Bird Conservation International* (30), *South African Journal of Wildlife Research* (27), *African Journal of Ecology* (22), *Biological Conservation* (18), *PLoS ONE* (12) and *Behavioural Ecology and Sociobiology* (10).

There was considerable variation in the number of studies per species (Fig. 18.5). The average (± 1 S.D.) number of studies per species was 11.6 ± 14.7. From the 67 species examined, 24 could be described as relatively 'well-studied' having 12 or more studies, with the remaining 41 species having <10 studies. It was interesting

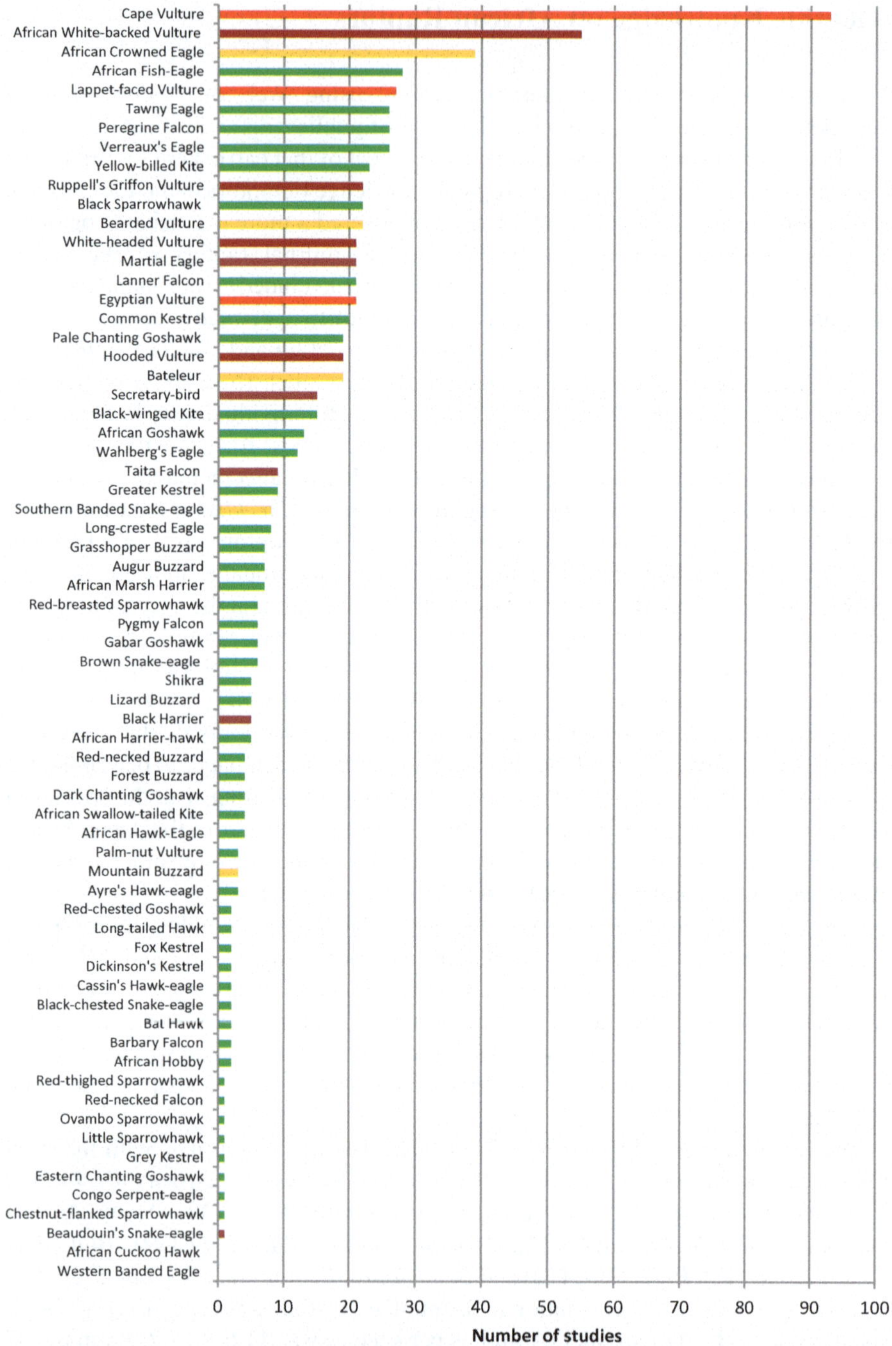

Fig. 18.5 Numbers of 'studies' from a systematic search of Web of Science for 67 diurnal raptor species that are resident breeders in sub-Saharan Africa. Species bars are colour-coded by conservation status according to the IUCN Red List. (see Table 18.4)

to note that of the 17 species listed as extinction prone (Table 18.3), 12 fell above the average and could be placed into the category of being 'well-studied' (Fig. 18.5). This included eight of the nine African vulture species. Indeed, Cape vulture (*Gyps coprotheres*; $n = 93$) and African white-backed vulture (*Gyps africanus*; $n = 55$) stood out as having many more studies than the other raptor species. This result suggests that research is generally being directed towards species of conservation concern or that declines are reported most often in well-studied species which are then more likely to become uplisted. Indeed, only five species that are currently on the IUCN Red List could be classified as relatively poorly studied (with less than 12 studies): the Taita falcon (*Falco fasciinucha*), southern banded snake eagle (*Circaetus fasciolatus*), black harrier (*Circus maurus*), mountain buzzard (*Buteo oreophilus*) and Beaudouin's snake eagle. The latter two species were particularly poorly studied with only three and one study, respectively. Two species had no studies published on them at all (or at least none that were in ISI journals); these were the African cuckoo-hawk (*Aviceda cuculoides*) and the western banded snake eagle (*Circaetus cinerascens*).

Changes in African Raptor Research over Time

There has been a general and consistent increase in the numbers of studies on African raptors produced each year (Fig. 18.6), from an average of only around 1 study per year in the 1960s to an average of over 25 studies per year in the 2010s. The slope suggests an increase of around five studies per year per decade over this 60-year period. A similar trend was seen when we examined the number of papers produced per year. Thus, these results suggest an increase in the number of studies over time and are not simply a consequence of more multispecies papers over time (Fig. 18.6).

Research by Region

The increasing rate of publication on African raptors suggests that African raptors are studied more intensively than ever before. Still knowledge of many species remains almost poor, for example, because much research is performed on a rather small number of species in a relatively small geographic area. To examine this further, we identified the location of the study for each study where possible. This was possible for 730 out of the 780 studies in our database. These were then classified by broad geographic region (North, South, East or West Africa). In some cases, where the study was not specific to a geographic region but was relevant to the species wherever it occurred (e.g. genetics study or general review), the study was described as being relevant to all locations where the species was present. Thus, the total number of studies in the different geographical regions can be greater than the total number of all studies.

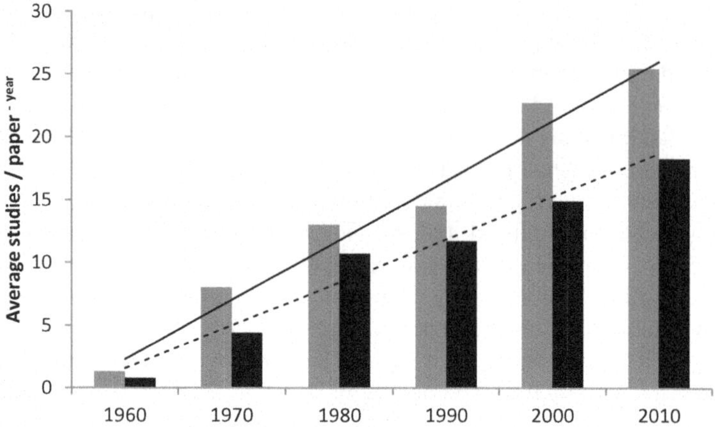

Fig. 18.6 Average number of studies (grey, solid line) or papers (black, dash line) produced per year on 67 diurnal resident sub-Saharan African raptor species from a systematic search using Web of Science. Lines show a linear regression line. (The 2010 data only includes those studies from 2010 to 2016)

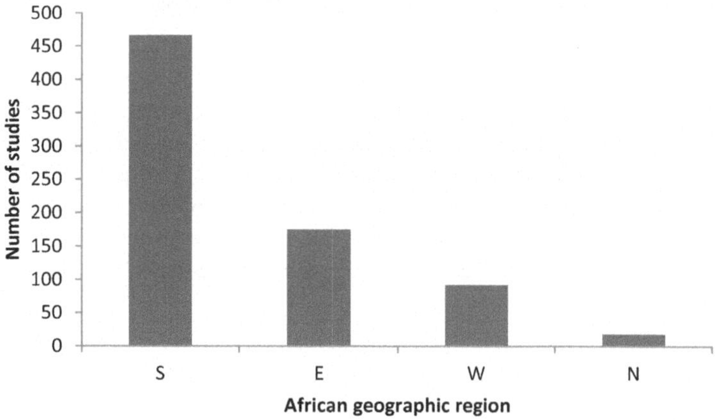

Fig. 18.7 Total number of studies by geographic region in Africa. The data comes from a systematic Web of Science search for 67 diurnal raptor species that are resident breeders in sub-Saharan Africa. Each study can have information on a species from different geographic regions, thus the total number of studies across all regions can equal more than the total number of studies

Our results showed a striking disparity in the numbers of studies from the different regions (Fig. 18.7). There were more studies in Southern Africa ($n = 466$, 62%) than in all the other regions combined. East Africa was the next best studied region with 23% of studies ($n = 175$), followed by West Africa (92 studies, 12%), with North Africa particularly poorly studied ($n = 18$, 2%), although this last result will partly reflect that this study considered species of African raptor that were resident south of the Sahara.

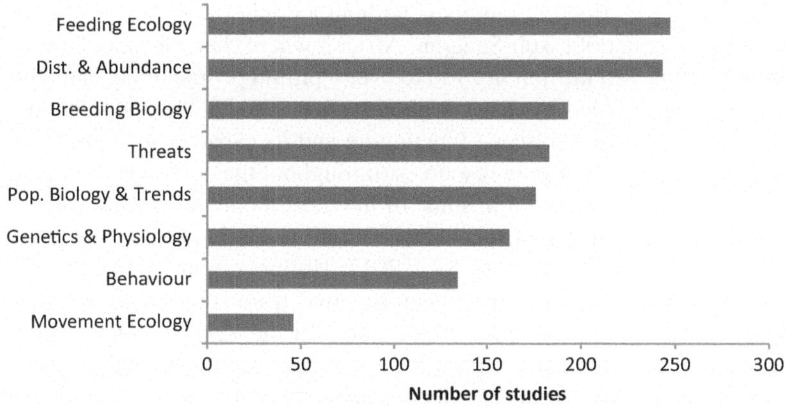

Fig. 18.8 Number of studies focused on different topics for raptors in Africa. Data comes from a systematic Web of Science search for 67 diurnal raptor species that are resident breeders in sub-Saharan Africa. Each study can be included in multiple topics and thus the total numbers in this figure (1384) sum to more than the 780 studies considered

Research by Topic

Studies were also classified into eight broad categories depending on their subject matter. We used the four categories originally devised by Virani and Watson's (1998) review of ecological knowledge of raptors in the East African tropics and Western Indian Ocean Islands, which were (1) distribution and abundance, (2) population biology and trends, (3) breeding biology and (4) feeding ecology. Additionally, we included four other categories: (5) genetics and physiology (papers related to genetics and evolution, ecophysiology and ecotoxicology, disease ecology and parasites and morphology and individual condition), (6) movement ecology, (7) threats and conservation and (8) behaviour. A given paper could fall into multiple categories (like the previous geographic regions analysis).

From our review, we found there were most studies on feeding ecology (247) and distribution and abundance (243) (Fig. 18.8). Movement ecology stood out for having the least number of studies on that topic (46), although that result is not surprising given the fact that studies on movement ecology have only really advanced in recent years with the advent of animal tracking technology. Indeed, from the 46 studies on movement ecology, 28 of these were published since the year 2000.

Threats to African Raptors

Many of the threats faced by raptors in Africa are similar to those faced by raptors around the globe, although the intensity of some threats appears to be more severe here than anywhere else, leading to unprecedented declines in some species such as vultures (Krüger et al. 2013; Ogada et al. 2015; Thiollay 2006).

The major factor affecting raptors is without a doubt the strong human population increase throughout sub-Saharan Africa, where human populations have increased from ca. 180 million in 1950 to over a billion people today and which are expected to have doubled by 2050 (United Nations 2017). This population explosion has greatly altered landscape functioning and its capacity to support wildlife, including raptors and their prey resources, throughout the continent. Although raptor declines have also occurred in some of the most iconic wilderness areas on the continent, such as Kenya's Maasai Mara (Virani et al. 2011) and South Africa's Kruger National Park (Amar and Cloete 2017; Hofmeyr et al. 2014), the strongest and most widespread declines have been reported from rural areas, where former wildlands with few people have made way for densely populated farmlands (Thiollay 2006). The human footprint is most pronounced just south of the Sahara such as in West Africa's savannas, which used to support some of the richest raptor communities on the continent, many of which are now greatly impoverished (Thiollay 2006). Of course, tGyps vultures and Verreaux'shese developments have not affected all raptors equally, further underpinning the need for research on a species-by-species basis. For example, tree-dotted farmlands in the Sahel may still provide adequate habitat for nesting intra-African migrants such as grasshopper buzzards and African swallow-tailed kites which are adapted to exploit unpredictable, seasonal and easily accessible prey, although scarcity of food in farmlands may lower fledgling condition (Buij et al. 2013b, c). Rodent specialists such as black-winged kites (*Elanus caeruleus*) and some Palearctic migrants (e.g. booted eagle, steppe eagle) may benefit locally from rodent-rich cultivation (Buij and Croes 2013; Herremans and Herremans-tonnoeyr 2000) and acridivorous Montagu's harrier from woodland conversion to grasshopper-rich shrublands (Augiron et al. 2015). We also now know that even species that were thought of as specialists, such as the Verreaux's eagle, can adapt to certain levels of agriculture intensification (Murgatroyd et al. 2016a, c). Where raptors are not persecuted, some adaptable species are even able to make use of abundant resources inside city boundaries, although there might be hidden costs of urban living which could undermine their long-term health and persistence in urban environments (Suri et al. 2017). Examples of raptors reaching high densities in cities are peregrine falcons (*Falco peregrinus;* Altwegg et al. 2014), black sparrowhawks (Martin et al. 2014a; Rose et al. 2017), African crowned eagles (McPherson et al. 2016a), yellow-billed kites and hooded vultures (*Necrosyrtes monachus;* Brown 1970).

However, agricultural expansion and intensification at the cost of natural habitats appear to deprive the majority of Africa's breeding raptors of their food resources and nesting habitats. Agricultural development projects aimed at boosting food production have decreased the productivity of unique and restricted habitats, such as West African floodplains, to the detriment of once rich raptor communities (Thiollay 2001), and similar plans are likely to further expose Africa's pristine forest and savannas to such developments (Edwards et al. 2014; Weng et al. 2013). Importantly, the use of pesticides and other poisons associated with cultivation probably kills many raptors, although most deaths remain undocumented. Quelea control operations are still common and known to cause mass casualties amongst raptors (Keith and Bruggers 1998), while rodenticides and insecticides cause direct mortalities from secondary poisoning or depress seasonally important food resources. Pesticides, often hazardous ones banned from western countries, are cheaply and widely available in local markets throughout sub-Saharan Africa and widely abused to kill wildlife or pollute the environment (Craig 2017; Odino 2011; Ogada 2014; Santangeli et al. 2016), while their heavy use in some cash crops (e.g. cotton) may depress raptor abundance (Buij et al. 2013b). Shockingly, a recent study identified the presence of DDE (the metabolite of DDT) in the chicks and adults of black harriers breeding in the South West of South Africa (Garcia-Heras et al. 2017a), although the exact cause and route of this pollutant are not yet known. Pesticides are also used for fishing and thus contaminate waterbodies, while declining fish stocks are intensively utilised, further weakening the prey base for African fish eagles (*Haliaeetus vocifer*) and ospreys. Rural populations not only supplement their diet with small to medium-sized vertebrates, thus reducing raptor prey availability, but also kill raptors for consumption (Whytock et al. 2014). Carbofuran, other carbamate or organophosphate insecticides and rodenticides are used to poison raptors for consumption, using poisoned baits and at waterholes during the dry season by digging small depressions close to larger waterholes (Ogada 2014; Saidu and Buij 2013). In addition, many farmers keep poultry and small livestock, and retaliatory persecution targets a range of raptors, from hawks to eagles, for their (often alleged) role in livestock predation (Brown 1991; Nyirenda et al. 2017). Other threats, some hardly documented, lurk in farmlands; for example, 18 raptor species were reported to have drowned in farm reservoirs in Southern Africa (Anderson 2000).

Electrocution and collision with power lines increasingly threaten raptors throughout the continent (Jenkins et al. 2010). Energy infrastructure can be problematic for raptor in Africa. In South Africa, at least 14 diurnal raptor species have been electrocuted on power facilities (Boshoff et al. 2011; Ledger and Annegarn 1981), and one power line in Sudan may have caused the death of sufficient migratory Egyptian vultures to partially explain their population declines (Angelov et al. 2013). A novel and increasing threat to African raptors stems from the global need for renewable energy. Wind turbines are often considered as a viable option for 'green' energy production. However, concerns over the impacts on bird populations are increasing, as collisions and displacement from home ranges are evident at operational wind energy projects worldwide (Madders and Whitfield 2006; Smallwood and Thelander 2008). Collision risk is associated with species-specific flight behaviour, topographic factors and seasonal abundances, but exact mechanisms of collision susceptibility still remain unclear (Barrios and Rodríguez 2004; de Lucas et al. 2008; Reid et al. 2015; Rushworth and Krüger 2014). It is generally agreed that the best mitigation methods are to place wind turbine developments in locations that pose minimal risk (Reid et al. 2015) and, at wind developments with multiple raptor mortalities, a few specific turbines can cause disproportionate number of deaths (Ralston Paton et al. 2017). Current research is focused on building species-specific risk models that have already been undertaken for bearded vultures in the Southern African range (Reid et al. 2015).

Climate change may reduce the breeding suitability of some areas for some raptors whose prey availability may be limited both by rainfall and by temperature (Buij et al. 2013b; Garcia-Heras et al. 2016; Gargett et al. 1995; Hustler and Howells 1988, 1990); some studies suggest that the impacts on raptor population dynamics and survival caused by climate change will be especially great in the most arid savannas (Wichmann et al. 2003).

The Demise of Vultures and Large Eagles

Africa's vultures and large eagles have suffered some of the largest declines and are currently deserving of most conservation attention. The vulture decline and disappearance of their breeding colonies, first reported in West Africa using repeat road surveys (Thiollay 2006), but later elsewhere in Africa (e.g. Ogada et al. 2015; Virani et al. 2011), highlighted that of all large African raptors, vultures seem to be suffering disproportionally and for reasons that were not necessarily shared by other raptors. Throughout Africa, vultures are being poisoned on an unprecedented scale, initially mostly as 'by-catch' resulting from poisoned baits targeted at mammalian carnivores and, more recently, because they reveal the whereabouts of poached animals, particularly elephants, to wildlife officials (Ogada et al. 2016; Ogada 2014). In West and South Africa, deliberate targeting of vultures for belief-based use is also likely to have played a key role in their recent declines. Vulture carcasses sold on markets throughout the region are evidence of this large-scale persecution (Buij et al. 2016), and although this threat may not be novel (Rondeau and Thiollay 2004), the profitability and extent of the trade have certainly increased. All parts of vultures are used, even eggs, faeces and nests, but heads of *Gyps* and lappet-faced vultures are especially valued for their purported medicinal properties (Nikolaus 2011). In West Africa, tree-nesting lappet-faced and white-backed vultures are probably more vulnerable to nest harvesting or disturbance by humans than Rüppell's vulture (Wacher et al. 2013), with most remaining nest sites of the latter being on steep, inaccessible cliffs.

Additional issues that may have also contributed to vulture declines include habitat loss, wind turbine collisions and electand evaluate the availabilityrocutions, loss of food resources including wild herbivores, nest disturbance even inside national parks and better sanitation in human settlements, including more rapid disposal of carcasses (Ogada et al. 2015; Rondeau and Thiollay 2004). Other threats such as lead ingestion from the game hunting industry (Naidoo et al. 2017) remains poorly studied but is the focus of current research by PhD student Beckie Garbett and Raptors Botswana, an NGO, in Botswana (Garbett et al. 2018).

In addition to vultures, the wide-ranging and similarly long-lived, slow-reproducing large African eagles are perhaps the most sensitive to human disturbance and persecution – accounts of their demise over vast areas are often anecdotal. Where quantitative information is available, it shows that declining eagle populations are the norm both inside and outside protected areas (Amar and Cloete 2017; Thiollay 2007; Thomsett 2015). For example, four eagle species decreased by

86–93% in ca. three decades in central West Africa, and seven were not even recorded outside protected areas (Thiollay 2006). In this region, a few eagles (such as Wahlberg's, tawny and long-crested eagles) continue to breed in small numbers where woodland fragments harbour food and nest sites, sometimes in cultivated fields and where rocky ledges protect them from disturbance. Some species, including large eagles such as South Africa's crowned and Verreaux's eagles, may be fairly adaptable and even persist in urban or intensive agricultural habitats where they are not persecuted (McPherson et al. 2016a; Murgatroyd et al. 2016c). Others, such as martial eagle, breed only within national parks or in sparsely populated rangelands and forage widely over settlements and cultivation, where high mortality rates due to persecution lead to declining populations (Brown 1991; Van Eeden et al. 2017).

Ongoing or Recent Studies of African Raptors

Here we list a number of ongoing raptor studies in Africa. We do not attempt to make this list exhaustive and merely seek to demonstrate the types of studies that are currently being undertaken. We group these studies taxonomically for ease of presentation.

Martial Eagles in Protected Areas

Large and widespread declines in martial eagle populations have been recorded across West and South Africa (Amar and Cloete 2017; Thiollay 2006). These trends, together with other anecdotal declines elsewhere, resulted in the species being uplisted in 2013 to vulnerable. Amar and Cloete (2017) revealed that within South Africa, these declines had also occurred in some of the key protected areas, including the supposed stronghold of Kruger National Park (KNP). Rowen van Eeden has just recently completed his PhD exploring the causes of these declines. This and ongoing research suggested that there may be problems with productivity, with only around 0.20 young per pair being produced each year. Adults and juveniles were tracked using GPS devices, which showed that territorial eagles held home ranges averaging ca. 108 km^2, preferentially using areas that had higher tree cover and were classed as dense bush rather than open bush or grassland (Van Eeden et al. 2017). Relatively high mortality of wide-ranging 'floaters' due to anthropogenic causes (three of four eagles) raises concerns for the species' persistence (Van Eeden et al. 2017). In July 2016, another field study was initiated on the species in the Maasai Mara ecosystem in southwest Kenya to investigate its demography, diet and habitat use (Hatfield et al. in prep.). Other research on the species has focused on the population in the Karoo in South Africa which nests on the power line network there (Machange et al. 2005); Jessie Berndt's MSc research suggested that this population may comprise around 150 pairs and may thus represent an important component of the population.

African Crowned Eagles in Urban Environments

In the city of Durban, South Africa alone, a population of over 30 pairs of breeding crowned eagles resides within the Metropolitan green space system (McPherson et al. 2016a). This urban population has been the focus of a recent PhD by Shane McPherson. Crowned eagles take advantage of alien *Eucalyptus* trees as nesting trees and the abundant and diverse food supply of urban wildlife, including rock hyrax (*Procavia capensis*) as primary prey, hadada ibis, forest antelopes and vervet monkeys (*Chlorocebus pygerythrus*). Human-wildlife conflicts can occur because juvenile eagles can predate on pet cats and small dogs, but domestic stock comprised 6% of the identifiable prey at nests, and contrary to popular belief, no dogs and few cats were delivered to the nest in this study (McPherson et al. 2016b). Another study on the species in Nairobi, Kenya, and the surrounding region is currently being conducted by MSc student Washington Wachira.

Verreaux's Eagles in Transformed Landscapes

Verreaux's eagles are generally regarded as mountain and hyrax specialist. However, recent findings from a PhD by Megan Murgatroyd have demonstrated that the agricultural Sandveld region in South Africa can not only sustain a productive population (Murgatroyd et al. 2016c) but also reveal adaptability in diet (Murgatroyd et al. 2016a). Similar adaptability could be shown for a peri-urban population (Kailen Padayachee MSc thesis). Both habitat types can sustain a population provided that there is some suitable breeding habitat (i.e. rocky outcrops with cliffs) within the agricultural or urbanised landscape. The productivity of these populations is contributed to by the provision of a diverse prey base, an environment conducive to good nesting success and a landscape favourable for low-energy flying. Data from high-resolution GPS tags were used to identify different flight behaviours (Murgatroyd et al. 2016b) and evaluate the availability of lift in the topographically contrasting habitat. In a next step, these high-resolution tracking data will be used to identify topographical and meteorological correlates of orographic and thermal soaring so that the spatial distribution of lift driving soaring flights can be predicted across the landscape, ultimately to predict collision risk with wind turbines, which is a project called 'VERA', Verreaux's Eagle Risk Assessment (http://blackeagleproject.blogspot.co.za/p/about-black-eagle-project.html).

African Fish Eagles at Lake Naivasha

Building on Dr Leslie Brown's studies of the late 1960s, a range of research organisations have collected annual population data on the African fish eagle of Lake Naivasha since 1994 (Harper et al. 2002). This population in Kenya's Great Rift

Valley is regarded as the densest population in Africa with eagle pairs as close as 300 m apart (Virani et al. in prep.). This aquatic species can act as an important indicator of the quality of aquatic ecosystems where anthropogenic developments such as geothermal energy, invasive species, fisheries, unplanned human settlements, water abstraction and horticultural activities have impacted the lake's once pristine ecology. Recent research has found that African fish eagle populations are highly correlated with the lake's water level that influences fish populations and in turn productivity (Virani et al. in prep.). Over the last 10 years, studies have diversified to understand population density in relation to land use (Kapila 2009), heavy metal contamination (Mutia et al. 2012) and pesticide concentrations in the eagles, water and sediments (Gudka 2012; Otieno et al. 2012). Despite anthropogenic pressures on the lake, the population of Naivasha's African fish eagles appears remarkably resilient at just around 230 individuals, similar to what Leslie Brown recorded in 1971 (Brown and Hopcraft 1973), and thriving on elevated populations of the common carp (*Cyprinus carpio*).

Augur Buzzards in Kenya's Rift Valley

The augur buzzard (*Buteo augur*) population in Kenya's Rift Valley provides important information about how rapidly changing agricultural landscapes affect a common species. The species was intensively studied in the mid-1990s (Virani and Harper 2009). They adapt well to cultivated land and dense human habitation (Brown et al. 1982). In Kenya's Lake Naivasha region, the species nests on both cliffs and trees (Virani 1999). Its diet consists mainly of rodents, the yellow-toothed mole-rat (henceforth mole-rat) *Tachyoryctes splendens* which, because of its high densities in cultivated areas, has become the augur buzzard's principal food source (Virani and Harper 2009). In the southern parts of the Lake Naivasha area, there has been an increase in human and livestock population growth with resultant impacts on the ecosystem and its associated flora and fauna (Harper et al. 2011). Since the mid-1990s, the augur buzzard population has declined by nearly 50%, although the exact reasons are not yet known, but suspected causes include overgrazing by livestock and conversion of Acacia woodland pastures and agricultural fields to intensive horticultural activities (Eichenwald et al. in prep.). The region is also a hub for the development of geothermal energy resulting in a labyrinth of ill-designed power lines that have caused numerous fatalities from electrocutions of sit-and-wait raptors such as augur buzzards (Virani pers. obs.).

Black Sparrowhawks in Cape Town

The black sparrowhawk (*Accipiter melanoleucus*) population in Cape Town, South Africa, has been the focus of an ongoing long-term study by researchers at the FitzPatrick Institute. This urban population has been monitored since 2000 shortly after the black sparrowhawk colonised the Cape Peninsula (Martin et al. 2014a, b). The species can be characterised as an urban adapter, exploiting nesting opportunities in alien tree plantations and the high prey abundance of pigeons and doves (Suri et al. 2017). The species apparently suffers no negative effects of urbanisation in terms of either productivity (Rose et al. 2017), survival (Sumasgutner et al. in prep.) or individual health (Suri et al. 2017). However, there is potentially a higher infection prevalence with *Knemidokoptes* mites in urban environments, as recorded in Cape Town and Durban (Van Velden et al. 2017). The research on this species has also focused on understanding the evolution and maintenance of the two morphs (light or dark adult morph) of this polymorphic species (Amar et al. 2013). The morphs in South Africa show clinal variation following rainfall patterns (Amar et al. 2014) and solar radiation intensity (Tate et al. 2016a). The two morphs occur in the same population but differ in their blood parasite infection intensity (Lei et al. 2013) and their habitat selection within their home ranges (Tate and Amar 2017). Dark morphs were found to have better hunting success in darker light conditions (Tate et al. 2016a). Additionally, mixed-morph pairs (consisting of two opposite morphs) show higher productivity (Tate et al. 2016b) and higher offspring survival (Sumasgutner et al. 2016) than pairs consisting of the same morph. Understanding why these mixed morph pairs produce more and better quality offspring is the focus on a PhD currently being undertaken by Carina Nebel.

Black Harrier research in Southern Africa

The black harrier is endemic to southwestern South Africa, where it breeds in indigenous vegetation, essentially along the South African coast within the fynbos biome and inland within the Karoo biome (Van Der Merwe 1981). The species has been the focus of a long-term study by Rob Simmons and was the focus of a recently completed PhD by Sophie-Marie Garcia-Heras. Due to the rapid increase of agriculture and urbanisation in the region, black harriers' breeding habitats have been reduced by 50% since the 1950s (Curtis et al. 2004), and many nesting areas are now surrounded by agricultural or urbanised land, increasing the potential exposure to contaminants and pollutants. The species global population has been estimated at <1000 breeding birds. The species has an extended laying period over 8 months (Garcia-Heras et al. 2017b) and is a small mammal specialist (64.4% and 78.2% of prey and consumed biomass, respectively) with birds and reptiles as alternative prey. Recently, eco-toxicological and eco-physiological research have assessed environmental contamination; shockingly PCBs and DDTs were detected in 79% and 84% of sampled individuals (Garcia-Heras et al. 2017a), which impacted their individual health (García-Heras et al. 2017).

Population Ecology of Cape Peninsula Peregrines

Andrew Jenkins has been studying peregrine falcons (*Falco peregrinus*) in the Greater Cape Town area since the late 1980s. This study has focused on individual colour ringing of adults and their young produced to monitor survival, dispersal and breeding success of these marked birds in relation to a variety of potential biotic and abiotic drivers (Jenkins 1994, 2000a, b; Jenkins and Benn 1998). The study has now accumulated over two decades of detailed demographic data and has documented a more than fivefold increase in the size of the population over that time, including a major expansion of breeding birds into the deep urban areas of the city of Cape Town where nest boxes are provided (Altwegg et al. 2014).

Ecology and Breeding System of the African Pygmy Falcon

The African pygmy falcon (*Polihierax semitorquatus*) depends on weaver structures for nesting. In Southern Africa this species breeds exclusively in massive sociable weaver (*Philetairus socius*) colonies. Yet despite the conspicuousness of this system, the falcon remained largely unstudied (but see Maclean 1970), until a comprehensive investigation by Robert Thomson was initiated in 2011 in the Kalahari, Northern Cape, South Africa. There, over 250 sociable weaver colonies host up to 34 active pygmy falcon nests/territories (Lowney et al. 2017), with approximately 35% of weaver colonies showing signs of past falcon occupation. The study also reveals an interesting breeding system of delayed offspring dispersal, particularly by male young, but also potential polyandry of multiple unrelated males within breeding groups.

Pan-African Vulture Studies

Over the past 5 years (2013–2017), the status, distribution and threats to seven threatened African vulture species (bearded, Egyptian, lappet-faced, white-headed, white-backed, Rüppell's, hooded) have been investigated in the Horn of Africa (including Ethiopia, Djibouti and Somaliland). Fifteen tagged Egyptian vultures provided insights into the migration bottlenecks and stopovers sites along the Red Sea flyway (Buechley et al. in review-a), as well as the home range characteristics and habitat use in the Middle East and East Africa (Buechley et al. in review-b). Future and ongoing work focuses on addressing threats to vultures from secondary poisoning and electrocutions, evaluating the status of the bearded vulture in Ethiopia and studying the movement ecology of white-backed and Rüppell's vultures.

Various studies have recently been initiated of one of the most abundant yet threatened raptors in Africa north of the equator, the hooded vulture, and notably its breeding ecology, movements, distribution, threats and trends in the Gambia, Burkina Faso, Nigeria, Ghana, Senegal and Guinea-Bissau. These showed amongst others that numbers remain high in some areas (Henriques et al. 2017), although strong long-term declines (>85% in 47 years) occurred in Dakar (Mullié et al. 2017). Most of these studies are ongoing, including a Pan-African study to investigate home range sizes and patterns of human commensalism (Bildstein pers. comm.) and a study on the breeding biology of the species in South Africa (Thompson et al. 2017a, b).

Recent studies are shedding more light on the trade in vultures for belief-based use and consumption in West and Central Africa. These showed that the trade in body parts of raptors is widespread with an annual trade of hundreds of vultures in West Africa alone (Buij et al. 2016), where younger people and low income earners are more likely to participate in the trade than older, more affluent people (Atuo et al. 2015). Apart from being traded for belief-based use, vultures are also consumed for their meat in West and Central Africa. In Central African forests, palm-nut vultures and other raptors are hunted more frequently for their meat than previously assumed, because remains are often discarded in hunting camps rather than ending up at markets (Whytock et al. 2014).

Future Research and Goals for Raptor Research in Africa

Capacity Building and Training of the Future African Raptor Biologists

Developing future leadership will be critical for successful raptor conservation in Africa. Qualified and experienced individuals are the essential driving force for conservation progress. Furthermore, good science, especially knowledge and understanding of biotic processes and systems, combined with good leadership skills is the recipe for effective conservation around the world. Since the 1990s, raptor biologists in Africa have made great strides in increasing knowledge about little known raptors. The current cohort of African raptor biologists is constantly striving to improve the situation for a species or remove threats, protect a landscape, or enable humans to co-exist with wildlife and nature. Leaders at every level understand that the world is changing, and changing rapidly. If we want to remain effective and continue increasing knowledge about African raptors, the most valuable investment we can make is the development of our leaders today to ensure great leadership tomorrow. As stated previously, Africa is an important continent for raptor species globally. Substantial work on little-known raptors has been achieved in South Africa and to a large extent in East Africa, yet much still remains to be learned about raptors in Central and West Africa, particularly the tropical forest-dwelling species. Raptor species are being up listed on the IUCN Red Data List on an annual basis, with the situation for African vultures being particularly severe. Africa's landscape is fast changing, and the continent's biodiversity is a fundamental basis of its development and underpins the wellbeing of current and future generations. For the great majority of Africans, biodiversity represents their only lifeline that can no longer be ignored. The reality is that biodiversity in Africa is threatened by the needs of the continent's human population that is rapidly increasing at an unsustainable rate. In addition, a bustling economy aimed at developing the continent's infrastructure is sweeping across sub-Saharan Africa. Energy needs for Africa will double over the next decade with huge investments earmarked for the development of energy infrastructure. Consumer spending is set to rise 80 percent by 2020, and Africa now has a fast-growing middle class, expected to increase from 60 million to 100 million people by 2025. This will have far reaching consequences for the survival of Africa's biodiversity and especially raptors. In 2015, Africa's trade with China had topped the US$300 billion mark. For conservation and development to be sustainable in Africa, there is an urgent need to understand how development affects biodiversity and what measures are required for mitigation. Raptors, as far-ranging, easily seen top predators, are excellent 'flagships for conservation'. Understanding their population dynamics, distribution and abundance and ecological requirements in a rapidly changing continental landscape and creating an appreciation for the ecosystem services that they provide will be key towards ensuring their survival for future generations. Most importantly, there will be a critical need to develop continental-wide local capacity in the field of raptor biology, conservation and management, as well as harnessing the use of modern technology to achieve conservation goals.

Raptor populations in Africa are being impacted by myriad threats. These include habitat loss as a result of rapidly growing human populations, rampant poisoning, persecution and developmental impacts such as the growth of the energy sector. Conservation organisations and institutions of higher learning in Africa are struggling to develop new approaches to protect the continent's remaining natural heritage. The challenge is to design strategies that not only ensure the long-term viability of species and ecosystems but also ones that are politically and economically acceptable to local communities and governments. Developing the local capacity needed to gather scientific information and translating that to the policy at grass root level will require in-country champions to bear the burden of that responsibility with passion, commitment and a burning desire to change the future for the benefit of raptors and their habitats. There will also be a need to develop capacity in the fields of raptor management, rehabilitation and restoration.

Realistically, the capacity development goal for the short term should be to test and establish training and educational procedures that will lead to the effective development of passionate raptor biologists and conservation leaders from African countries where conservation is little appreciated and that will lead to a greatly expanded programme in succeeding years. Building partnerships will be key towards the success of this goal. More importantly securing funding, for example, through an endowment for this initiative will be vital for long-term sustainability. This could be achieved through an alliance for the development of African raptor champions so that a specific and recognised group is set up to oversee and streamline a funding disbursement mechanism with priority goals, activities and indicators of success. It is our desire that by 2050, each African country will have a local PhD level raptor biologist to champion raptor conservation in that country and to contribute to the wider conservation in each region.

Appendix A

List of 67 breeding resident diurnal African raptors and their IUCN Red List status

Common name	Scientific name	IUCN Red List status
African crowned eagle	*Stephanoaetus coronatus*	NT
African cuckoo-hawk	*Aviceda cuculoides*	LC
African fish eagle	*Haliaeetus vocifer*	LC
African goshawk	*Accipiter tachiro*	LC
African harrier-hawk	*Polyboroides typus*	LC
African hawk-eagle	*Aquila spilogaster*	LC
African hobby	*Falco cuvierii*	LC
African marsh harrier	*Circus ranivorus*	LC
African swallow-tailed kite	*Chelictinia riocourii*	LC
African white-backed vulture	*Gyps africanus*	CR
Augur buzzard	*Buteo augur*	LC
Ayres's hawk-eagle	*Hieraaetus ayresii*	LC
Barbary falcon	*Falco pelegrinoides*	Not listed
Bat hawk	*Macheiramphus alcinus*	LC
Bateleur	*Terathopius ecaudatus*	NT
Bearded vulture	*Gypaetus barbatus*	NT
Beaudouin's snake eagle	*Circaetus beaudouini*	VU
Black eagle	*Aquila verreauxii*	LC
Black harrier	*Circus maurus*	VU
Black kite	*Milvus migrans*	LC
Black sparrowhawk	*Accipiter melanoleucus*	LC
Black-chested snake eagle	*Circaetus pectoralis*	LC
Black-winged kite	*Elanus caeruleus*	LC
Brown snake eagle	*Circaetus cinereus*	LC
Cape vulture	*Gyps coprotheres*	LC
Cassin's hawk-eagle	*Aquila africana*	LC
Chestnut-flanked sparrowhawk	*Accipiter castanilius*	LC
Common kestrel	*Falco tinnunculus*	LC
Congo serpent eagle	*Dryotriorchis spectabilis*	LC
Dark chanting goshawk	*Melierax metabates*	LC
Dickinson's kestrel	*Falco dickinsoni*	LC
Eastern chanting goshawk	*Melierax poliopterus*	LC
Egyptian vulture	*Neophron percnopterus*	EN
Forest buzzard	*Buteo trizonatus*	LC
Fox kestrel	*Falco alopex*	LC
Gabar goshawk	*Micronisus gabar*	LC
Grasshopper buzzard	*Butastur rufipennis*	LC
Greater kestrel	*Falco rupicoloides*	LC
Grey kestrel	*Falco ardosiaceus*	LC
Hooded vulture	*Necrosyrtes monachus*	CR

Common name	Scientific name	IUCN Red List status
Lanner falcon	*Falco biarmicus*	LC
Lappet-faced vulture	*Torgos tracheliotus*	EN
Little sparrowhawk	*Accipiter minullus*	LC
Lizard buzzard	*Kaupifalco monogrammicus*	LC
Long-crested eagle	*Lophaetus occipitalis*	LC
Long-tailed hawk	*Urotriorchis macrourus*	LC
Martial eagle	*Polemaetus bellicosus*	VU
Mountain buzzard	*Buteo oreophilus*	LC
Ovambo sparrowhawk	*Accipiter ovampensis*	LC
Pale chanting goshawk	*Melierax canorus*	LC
Palm-nut vulture	*Gypohierax angolensis*	LC
Peregrine falcon	*Falco peregrinus*	LC
Pygmy falcon	*Polihierax semitorquatus*	LC
Red-breasted sparrowhawk	*Accipiter rufiventris*	LC
Red-chested goshawk	*Accipiter toussenelii*	LC
Red-necked buzzard	*Buteo auguralis*	LC
Red-necked falcon	*Falco chicquera*	LC
Red-thighed sparrowhawk	*Accipiter erythropus*	LC
Ruppell's griffon/vulture	*Gyps rueppellii*	CR
Secretarybird	*Saggitarius serpentarius*	VU
Shikra	*Accipiter badius*	LC
Southern banded snake eagle	*Circaetus fasciolatus*	NT
Taita falcon	*Falco fasciinucha*	VU
Tawny eagle	*Aquila rapax*	LC
Wahlberg's eagle	*Hieraaetus wahlbergi*	LC
Western banded snake eagle	*Circaetus cinerascens*	LC
White-headed vulture	*Trigonoceps occipitalis*	CR

NT near threatened, *VU* vulnerable, *EN* endangered, *CR* critically endangered

References

Altwegg R, Jenkins A, Abadi F (2014) Nestboxes and immigration drive the growth of an urban Peregrine Falcon Falco peregrinus population. Ibis 156:107–115

Amano T, Sutherland WJ (2013) Four barriers to the global understanding of biodiversity conservation: wealth, language, geographical location and security. Proc R Soc B Biol Sci 280:20122649–20122649

Amar A, Cloete D (2017) Quantifying the decline of the Martial Eagle Polemaetus bellicosus in South Africa. Bird Conserv Int 43:1–12

Amar A, Koeslag A, Curtis O (2013) Plumage polymorphism in a newly colonized black sparrowhawk population: classification, temporal stability and inheritance patterns. J Zool 289:60–67

Amar A, Koeslag A, Malan G, Brown M, Wreford E (2014) Clinal variation in the morph ratio of Black Sparrowhawks Accipiter melanoleucus in South Africa and its correlation with environmental variables. Ibis 156:627–638

Amar A, Cloete D, Whittington M (2016) Using independent nest survey data to validate changes in reporting rates of Martial Eagles between the Southern African Bird Atlas Project 1 and 2. Ostrich 87:1–5

Anderson MD (2000) Raptor conservation in the Northern Cape Province, South Africa. Ostrich 71:25–32
Anderson PC, Kok OB, Erasmus BH (1999) Diet, body mass and condition of Lesser Kestrels Falco naumanni in South Africa. Ostrich 70:112–116
Angelov I, Hashim I, Oppel S (2013) Persistent electrocution mortality of Egyptian Vultures Neophron percnopterus over 28 years in East Africa. Bird Conserv Int 23:1–6
Arroyo BE, King JR (1995) Observations on the ecology of Montagu's and Marsh Harriers wintering in north-west Senegal. Ostrich 66:37–40
Atkins A, Redpath SM, Little RM, Amar A (2017) Experimentally manipulating the landscape of fear to manage problem animals. J Wildl Manag 81:610–616
Atuo FA, Timothy JO, Peter UA (2015) An assessment of socio-economic drivers of avian body parts trade in West African rainforests. Biol Conserv 191:614–622
Augiron S, Gangloff B, Brodier S, Chevreux F, Blanc JF, Pilard P, Coly A, Sonko A, Schlaich A, Bretagnolle V, Villers A (2015) Winter spatial distribution of threatened acridivorous avian predators: implications for their conservation in a changing landscape. J Arid Environ 113:145–153
Baloye DO, Palamuleni LG (2015) A comparative land use-based analysis of noise pollution levels in selected urban centers of Nigeria. Int J Environ Res Public Health 12:12225–12246
Barrios L, Rodríguez A (2004) Behavioural and environmental correlates of soaring-bird mortality an an-shore wind turbines. J Appl Ecol 41:72–81
Benson P (2014) A survey of Cape Vulture breeding colonies in South Africa's northern provinces (Transvaal Region)- an update 2013. Ornithol Obs 6:31–36
Biggs D, Turpie J, Fabricius C, Spenceley A (2011) The value of avitourism for conservation and job creation-An analysis from South Africa. Conserv Soc 9:80
Boshoff AF, Minnie JC, Tambling CJ, Michael MD (2011) The impact of power line-related mortality on the Cape Vulture Gyps coprotheres in a part of its range, with an emphasis on electrocution. Bird Conserv Int 21:311–327
Brown LH (1953) On the biology of the large birds of prey of the Embu district, Kenya Colony. Ibis 95:74–114
Brown LH (1955) Supplementary notes on the biology of the large birds of prey of Embu district, Kenya Colony. Ibis 97:38–64
Brown LH (1970) African birds of prey: a study of Africa's 89 diurnal raptors and 31 owls. Collins, London
Brown LH (1971) African birds of prey. Houghton Mifflin, Boston
Brown CJ (1991) Declining martial Polemaetus bellicosus and tawny Aquila rapax eagle populations and causes of mortality on farmlands in central Namibia. Biol Conserv 56:49–62
Brown LH, Brown B (1979) The behaviour of the Black Sparrowhawk Accipiter melanoleucus. Ardea 67:77–95
Brown LH, Cade TJ (1972) Age classes and population dynamics of the Bateleur and African Fish Eagle. Ostrich 43:1–16
Brown LH, Hopcraft JBD (1973) Population structure and dynamics in the African fish eagle Haliaeetus vocifer (Daudin) at Lake Naivasha, Kenya. Afr J Ecol 11:255–269
Brown LH, Urban EK, Newman K, Safford R, Hawkins F (1982) The birds of Africa, vol 1. Academic Press, London
Buechley ER, McGrady MJ, Şekercioğlu ÇH (in press a) Satellite tracking a wide-ranging endangered vulture species to target conservation actions in the Middle East and East Africa. Biodivers Conserv https://doi.org/10.1007/s10531-018-1538-6
Buechley ER, Oppel S, Beatty WS, Nikolov SC, Dobrev V, Arkumarev V, Saravia V, Bougain C, Bounas A, Kret E, Skartsi T, Aktay L, Aghababyan K, Frehner E, Sekercioglu ÇH (forthcoming-b) Identifying important stopover sites and flyway corridors for a soaring migrant across three continents. J Avian Biol
Buij R, Croes BM (2013) Raptor habitat use in the Lake Chad Basin: Insights into the effect of flood-plain transformation on Afrotropical and Palearctic raptors. Bird Conserv Int 23:199–213

Buij R, van der Goes D, de Iongh HH, Gagare S, Haccou P, Komdeur J, de Snoo G (2012) Interspecific and intraspecific differences in habitat use and their conservation implications for Palaearctic harriers on Sahelian wintering grounds. Ibis 154:96–110

Buij R, Croes BM, Gort G, Komdeur J (2013a) The role of breeding range, diet, mobility and body size in associations of raptor communities and land-use in a West African savanna. Biol Conserv 166:231–246

Buij R, Croes BM, Komdeur J (2013b) Biogeographical and anthropogenic determinants of landscape-scale patterns of raptors in West African savannas. Biodivers Conserv 22:1623–1646

Buij R, Kortekaas K, van Krimpen RRD, van Wijk R, van der Zanden S, de Iongh HH, Heitkönig IMA, de Snoo GR, Komdeur J (2013c) Breeding performance of the Grasshopper Buzzard (Butastur rufipennis) in a natural and a human-modified West African Savanna. Condor 115:47–57

Buij R, Nikolaus G, Whytock R, Ingram DJ, Ogada D (2016) Trade of threatened vultures and other raptors for fetish and bushmeat in West and Central Africa. Oryx 50:606–616

Conradie N, van Zyl C, Strasheim A (2013) What inspires birders to migrate South towards Africa? A quantitative measure of international avitourist motivation. S Afr Bus Rev 17:128–167

Cortés-Avizanda A, Carrete M, Serrano D, Donázar JA (2009) Carcasses increase the probability of predation of ground-nesting birds: a caveat regarding the conservation value of vulture restaurants. Anim Conserv 12:85–88

Cortés-Avizanda A, Blanco G, Devault TL, Markandya A, Virani MZ, Brandt J, Donázar JA (2016) Supplementary feeding and endangered avian scavengers: benefits, caveats, and controversies. Front Ecol Environ 14:191–199

Craig C (2017) Assessing the anthropogenic threats to vultures in the communal farmlands of Namibia. University of Cape Town, Cape Town

Curtis O, Simmons RE, Jenkins AR (2004) Black Harrier Circus maurus of the Fynbos biome , South Africa: a threatened specialist or an adaptable survivor? Bird Conserv Int 14:233–245

Del Hoyo J, Collar NJ, Christie DA, Elliot A, Fishpool LDC (2014). Illustrated checklist of the birds of the world, Lynx Edicions, vol 1, Barcelona

Kapila S (2009) An assessment of the health of the Lake Naivasha Ramsar site with special consideration of the population and breeding dynamics of the African Fish Eagle, Haliaeetus vocifer. MSc Thesis, University College London (UCL), United Kingdom

de Lucas M, Janss GFE, Whitfield DP, Ferrer M (2008) Collision fatality of raptors in wind farms does not depend on raptor abundance. J Appl Ecol 45:1695–1703

Duraiappah A, Naeem S, Agardy T, Ash N (2005) Ecosystems and human well-being: biodiversity synthesis; a report of the Millennium Ecosystem Assessment. World Resources Institute, Washington, DC

Edwards DP, Sloan S, Weng L, Dirks P, Sayer J, Laurance WF (2014) Mining and the African environment. Conserv Lett 7:302–311

Eichenwald AJ, Amar A, Tyrrell P, Buechley ER, Virani M (in prep.) Is anthropogenic development responsible for declines in local abundance of a raptor of least concern?

Ewen JG, Walker L, Canessa S, Groombridge JJ (2015) Improving supplementary feeding in species conservation. Conserv Biol 29:341–349

Ferguson-Lees J, Christie DA (2001) Raptors of the world (Boston, Houghton Mifflin). Houghton Mifflin, Boston

Fiedler LA (1994) Rodent pest management in Eastern Africa. Plant Production and Protection paper No. 123. FAO, Rome.

Fishpool LDC, Evans MI (2001) Important Bird Areas in African and associated islands: priority sites for conservation. Pisces Publications and BirdLife International, Newbury/Cambridge, UK

Garbett, R., Maude, G., Hancock, P., Kenny, D., Reading, R., & Amar, A. (2018). Association between hunting and elevated blood lead levels in the critically endangered African white-backed vulture Gyps africanus. Science of The Total Environment. https://doi.org/10.1016/j.scitotenv.2018.02.220

Garbett R, Herremans M, Maude G, Reading R, Amar A In review. Raptor population trends in northern Botswana: a re-survey of road transects after 20 years. Biological Conservation

Garcia-Heras M-S, Arroyo B, Mougeot F, Amar A, Simmons RE (2016) Does timing of breeding matter less where the grass is greener? Seasonal declines in breeding performance differ between regions in an endangered endemic raptor. Nat Conserv 15:23–45

García-Heras MS, Arroyo B, Simmons RE, Camarero PR, Mateo R, García JT, Mougeot F (2017) Pollutants and diet influence carotenoid levels and integument coloration in nestlings of an endangered raptor. Sci Total Environ 603:299–307

Garcia-Heras M, Arroyo B, Simmons RE, Camarero PR, Mateo R, Mougeot F (2017a) Blood concentrations of PCBs and DDTs in an avian predator endemic to southern Africa : Associations with habitat , electrical transformers and diet. Environ Pollut 232:440–449

Garcia-Heras MS, Mougeot F, Arroyo B, Avery G, Avery M, Simmons RE (2017b) Is the Black Harrier Circus maurus a specialist predator? Assessing the diet of a threatened raptor species endemic to southern Africa. Ostrich 88(1):1–8

Gargett V, Gargett E, Damania D (1995) The influence of rainfall on black eagle breeding over 31 years in the Matobo Hills, Zimbabwe. Ostrich 66:114–121

Gomez-Ramirez P, Shore RF, van den Brink NW, van Hattum B, Bustnes JO, Duke G, Fritsch C, Garcia-Fernandez AJ, Helander BO, Jaspers V, Krone O, Marinez-Lopez E, Mateo R, Movalli P, Sonne C (2014) An overview of existing raptor contaminant monitoring activities in Europe. Environ Int 67:12–21

Gudka M (2012) Assessment of pesticide concentrations in environmental and biological parameters from two Kenyan Rift Valley Lakes. MSc Thesis, Percy Fitzpatrick Institute of the Department of Zoology, University of Cape Town, South Africa

Hatfield S. et al. (in prep.). An ecological study of the Martial Eagle *Polemaeetus bellicosus* in the Greater Masai Mara region, Kenya

Harper DM, Harper MM, Virani MZ, Smart A, Childress RB, Adatia R, Henderson I, Chege B (2002) Population fluctuations and their causes in the African Fish Eagle, (Haliaeetus vocifer (Daudin)) at Lake Naivasha, Kenya. Hydrobiologia 488:171–180

Harper DM, Morrison EHJ, Macharia MM, Mavuti KM, Upton C (2011) Lake Naivasha, Kenya: ecology, society and future. Freshwater Rev 4:89–114

Henriques M, Lecoq M, Monteiro H, Regalla A, Granadeiro JP, Catry P (2017) Status of birds of prey in Guinea-Bissau: first assessment based on road surveys. Ostrich 88:101–111

Herremans M, Herremans-tonnoeyr D (2000) Land use and the conservation status of raptors in Botswana. Biol Conserv 94:31–41

Hofmeyr SD, Symes CT, Underhill LG (2014) Secretarybird Sagittarius serpentarius population trends and ecology: insights from South African citizen science data. PLoS One 9:e96772

Hustler K, Howells W (1988) The effect of primary production on breeding success and habitat selection in the African Hawk-Eagle Author (s). Condor 90:583–587

Hustler K, Howells W (1990) The influence of primary production on a raptor community in Hwange National Park, Zimbabwe. J Trop Ecol 6:343–354

Jenkins AR (1994) The influence of habitat on the distribution and abundance of Peregrine and Lanner Falcons in South Africa. Ostrich 65:281–290

Jenkins A (2000a) Factors affecting breeding success of peregrine and lanner falcons in South Africa. Ostrich 71:385–392

Jenkins AR (2000b) Hunting mode and success of African Peregrines Falco peregrinus minor: does nesting habitat quality affect foraging efficiency? Ibis (London 1859) 142:235–246

Jenkins AR, Benn GA (1998) Home range size and habitat requirements of Peregrine falcons on the Cape Peninsula, South Africa. J Raptor Res 32:90–99

Jenkins AR, Smallie JJ, Diamond M (2010) Avian collisions with power lines: a global review of causes and mitigation with a South African perspective. Bird Conserv Int 20:263–278

Keith JO, Bruggers RL (1998) Review of hazards to raptors from pest control in Sahelian Africa. J Raptor Res 32:151–158

Kopij G (2009) Seasonal variation in the diet of the Amur kestrel (Falco amurensis) in its winter quarter in Lesotho. Afr J Ecol 48:546–550

Kross SM, Tylianakis JM, Nelson XJ (2012) Effects of introducing threatened falcons into vineyards on abundance of passeriformes and bird damage to grapes. Conserv Biol 26:142–149

Krüger SC, Allan DG, Jenkins AR, Amar A (2013) Trends in territory occupancy, distribution and density of the Bearded Vulture Gypaetus barbatus meridionalis in southern Africa. Bird Conserv Int 24:162–177

Krüger S, Reid T, Amar A (2014) Differential range use between age classes of Southern African bearded vultures Gypaetus barbatus. PLoS One 9:1–18

Ledger JA, Annegarn HJ (1981) Electrocution hazards to the cape vulture Gyps coprotheres in South Africa. Biol Conserv 20:15–24

Lei B, Amar A, Koeslag A, Gous TA, Tate GJ (2013) Differential haemoparasite intensity between black sparrowhawk (Accipiter melanoleucus) morphs suggests an adaptive function for polymorphism. PLoS One 8:e81607

Lowney A, Green K, Ngomane BP, Thomson RL (2017) Mortal combat: intraspecific killing by an African Pygmy-Falcon (*Polihierax semitorquatus*) to acquire new mate and territory. J Raptor Res 51:89–91

Machange RW, Jenkins AR, Navarro RA (2005) Eagles as indicators of ecosystem health: is the distribution of Martial Eagle nests in the Karoo , South Africa , influenced by variations in land-use and rangeland quality ? J Arid Environ 63:223–243

Maclean GL (1970) The pygmy falcon Polihierax semitorquatus. Koedoe 13:1–22

Madders M, Whitfield DP (2006) Upland raptors and the assessment of wind farm impacts. Ibis 148:43–56

Markandya A, Taylor T, Longo A, Murty MN, Murty S, Dhavala K (2008) Counting the cost of vulture decline—an appraisal of the human health and other benefits of vultures in India. Ecol Econ 67:194–204

Martin RO, Koeslag A, Curtis O, Amar A (2014a) Fidelity at the frontier: divorce and dispersal in a newly colonized raptor population. Anim Behav 93:59–68

Martin RO, Sebele L, Koeslag A, Curtis O, Abadi F, Amar A (2014b) Phenological shifts assist colonisation of a novel environment in a range-expanding raptor. Oikos 123:1457–1468

Maumbe BM, Swinton SM (2003) Hidden health costs of pesticide use in Zimbabwe's smallholder cotton growers. Soc Sci Med 57:1559–1571

McPherson SC, Brown M, Downs CT (2016a) Crowned eagle nest sites in an urban landscape: requirements of a large eagle in the Durban Metropolitan Open Space System. Landsc Urban Plan 146:43–50

McPherson SC, Brown M, Downs CT (2016b) Diet of the crowned eagle (Stephanoaetus coronatus) in an urban landscape: potential for human-wildlife conflict? Urban Ecosyst 19:383–396

McWilliam AN, Cheke RA (2004) A review of the impacts of control operations against the red-billed quelea (Quelea quelea) on non-target organisms. Environ Conserv 31:130–137

Meyburg BU, Meyburg C, Pretorius R (2017) Year-round satellite tracking of Amur Falcon (Falco amurensis) reveals the longest migration of any raptor species across the open sea. In: From avian tracking to population processes, British Ornithologists' Union Annual Conference University of Warwick

Meyburg B-U, Mendelsohn JM, Ellis DH, Smith DG, Meyburg C, Kemp AC (1995) Year-round movements of a Wahlberg's Eagle *Aquzla Wahlbergz* tracked by satellite. Ostrich 66:135–140

Mittermeier RA, Turner WR, Larsen FW, Brooks TM, Gascon C (2011) Global biodiversity conservation: the critical role of hotspots. In: Biodiversity hotspots. Springer, Berlin, Heidelberg, pp 3–22

Moreno-Opo R, Trujillano A, Arredondo Á, González LM, Margalida A (2015) Manipulating size, amount and appearance of food inputs to optimize supplementary feeding programs for European vultures. Biol Conserv 181:27–35

Mullié WC, Guèye Y (2010) Does bird predation enhance the impact of green muscle® (Metarhizium acridum) used for grasshopper control? J Orthoptera Res 19:139–155

Mullié WC, Couzi FX, Diop MS, Piot B, Peters T, Reynaud PA, Thiollay JM (2017) The decline of an urban Hooded Vulture Necrosyrtes monachus population in Dakar, Senegal, over 50 years§. Ostrich 88:131–138

Mundy PJ, Jarvis MJF (1989) Africa's feathered locust. Baobab Books, Harare

Murgatroyd M, Avery G, Underhill LG, Amar A (2016a) Adaptability of a specialist predator: the effects of land use on diet diversification and breeding performance of Verreaux's eagles. J Avian Biol 47:834–845

Murgatroyd M, Underhill LG, Bouten W, Amar A (2016b) Ranging behaviour of Verreaux's eagles during the pre-breeding period determined through the use of high temporal resolution tracking. PLoS One 11:1–18

Murgatroyd M, Underhill LG, Rodrigues L, Amar A (2016c) The influence of agricultural transformation on the breeding performance of a top predator : Verreaux's Eagles in contrasting land use areas. Condor 118:238–252

Murn C, Combrink L, Ronaldson GS, Thompson C, Botha A (2013) Population estimates of three vulture species in Kruger National Park, South Africa. Ostrich 84:1–9

Murn C, Mundy P, Virani MZ, Borello WD, Holloway GJ, Thiollay J (2016) Using Africa's protected area network to estimate the global population of a threatened and declining species: a case study of the Critically Endangered White-headed Vulture Trigonoceps occipitalis. Ecol Evol 6:1092–1103

Mutia TM, Virani MZ, Moturi WN, Muyela B, Mavura WJ, Lalah JO (2012) Copper, lead and cadmium concentrations in surface water, sediment and fish, C. Carpio, samples from Lake Naivasha: effect of recent anthropogenic activities. Environ Earth Sci 67:1121–1130

Naidoo V, Wolter K, Botha CJ (2017) Lead ingestion as a potential contributing factor to the decline in vulture populations in southern Africa. Environ Res 152:150–156

Newton I (1979) Population ecology of raptors. T & AD Poyser, Berkhamsted

Ngowi AVF, Mbise TJ, Ijani ASM, London L, Ajayi OC (2007) Smallholder vegetable farmers in Northern Tanzania: pesticides use practices, perceptions, cost and health effects. Crop Prot 26:1617–1624

Nicolaides A (2013) Promoting Avitourism as a special niche area of Ecotourism in South Africa. African J Hosp Tour Leis 2:1–16

Nikolaus G (2011) The fetish culture in West Africa: an ancient tradition as a threat to endangered bird life? In: Schuchmann KL (ed) Tropical Vertebrates in a Changing World. Zoologisches Forschungsmuseum Alexander Koenig, Bonn, Germany, pp 145–150

Norgarb C, Lasbrey J (1953) Jackal buzzards. Ostrich 24:33–36

Nyirenda VR, Musonda F, Kambole S, Tembo S (2017) Peasant farmer-raptor conflicts around Chembe Bird Sanctuary, Zambia, Central Africa: poultry predation, ethno- biology, land use practices and conservation. Anim Biodivers Conserv 40:121–132

Oaks JL, Gilbert M, Virani MZ, Watson RT, Meteyer CU, Rideout BA, Shivaprasad HL, Ahmed S, Iqbal Chaudhry MJ, Arshad M, Mahmood S, Ali A, Ahmed Khan A (2004) Diclofenac residues as the cause of vulture population decline in Pakistan. Nature 427:630–633

Odino M. (2011) Measuring the conservation threat that deliberate poisoning poses to birds in Kenya: the case of pesticide hunting with Furadan in the Bunyala Rice Irrigation Scheme. In: Richards N (ed) Carbofuran and Wildlife Poisoning: Global Perspectives and Forensic Approaches. Wiley, Chichester, UK, pp 53–69. https://doi.org/10.1002/9781119998532

Ogada DL (2014) The power of poison: pesticide poisoning of Africa's wildlife. Ann N Y Acad Sci 1322:1–20

Ogada DL, Torchin ME, Kinnaird MF, Ezenwa VO (2012) Effects of vulture declines on facultative scavengers and potential implications for mammalian disease transmission. Conserv Biol 26:453–460

Ogada D, Shaw P, Beyers RL, Buij R, Murn C, Thiollay JM, Beale CM, Holdo RM, Pomeroy D, Baker N, Krüger SC, Botha A, Virani MZ, Monadjem A, Sinclair ARE (2015) Another continental vulture crisis: Africa's vultures collapsing toward extinction. Conserv Lett 9:89–97

Ogada D, Botha A, Shaw P (2016) Ivory poachers and poison: drivers of Africa's declining vulture populations. Oryx 50:593–596

Otieno PO, Schramm KW, Pfister G, Lalah JO, Ojwach SO, Virani M (2012) Spatial distribution and temporal trend in concentration of carbofuran, diazinon and chlorpyrifos ethyl residues in sediment and water in Lake Naivasha, Kenya. Bull Environ Contam Toxicol 88:526–532

Paz A, Jareño D, Arroyo L, Viñuela J, Arroyo B, Mougeot F, Luque-Larena JJ, Fargallo JA (2013) Avian predators as a biological control system of common vole (Microtus arvalis) populations in north-western Spain: experimental set-up and preliminary results. Pest Manag Sci 69:444–450
Piper SE (2004) Vulture restaurants. In: Monadjem A, Anderson MD, Piper SE, Boshoff AF (eds) The vultures of Southern Africa – ouo vadis ? Vulture Study Group, Johannesburg
Pomeroy D, Shaw P, Opige M, Kaphu G, Ogada DL, Virani MZ (2016) Vulture populations in Uganda : using road survey data to measure both densities and encounter rates within protected and unprotected areas. Bird Conserv Int 25:399–414
Ralston Paton S, Smallie J, Pearson A, Ramalho R. 2017. Wind energy's impacts on birds in South Africa: A preliminary review of the results of operational monitoring at the first wind farms of the Renewable Energy Independent Power Producer Procurement Programme in South Africa. BirdLife South Africa Occasional Report Series No. 2. BirdLife South Africa, Johannesburg, South Africa.
Reid T, Krüger S, Whitfield DP, Amar A (2015) Using spatial analyses of bearded vulture movements in southern Africa to inform wind turbine placement. J Appl Ecol 52:881–892
Ritchie EG, Elmhagen B, Glen AS, Letnic M, Ludwig G, McDonald R (2013) Ecosystem restoration with teeth: what role for predators? Trends Ecol Evol 27:265–271
Rondeau G, Thiollay JM (2004) West African vulture decline. Vulture News 51:13–33
Rose S, Sumasgutner P, Koeslag A, Amar A (2017) Does seasonal decline in breeding performance differ for an African raptor across an urbanization gradient ? Front Ecol Evol 5:1–9
Rushworth I, Krüger S (2014) Wind farms threaten southern Africa's cliff-nesting vultures. Ostrich 85:13–23
Saidu Y, Buij R (2013) Traditional medicine trade in vulture parts in northern Nigeria. Vulture News 65:4–14
Sánchez-Zapata JA, Donázar JA, Delgado A, Forero MG, Ceballos O, Hiraldo F (2007) Desert locust outbreaks in the Sahel: resource competition, predation and ecological effects of pest control. J Appl Ecol 44:323–329
Santangeli A, Arkumarev V, Rust N, Girardello M (2016) Understanding , quantifying and mapping the use of poison by commercial farmers in Namibia – implications for scavengers â€™ conservation and ecosystem health. Biol Conserv 204:205–2011
Schmeller D, Henle K, Loyau A, Besnard A, Henry P-Y (2012) Bird-monitoring in Europe – a first overview of practices, motivations and aims. Nat Conserv 2:41–57
Selinske MJ, Coetzee J, Purnell K, Knight AT (2015) Understanding the motivations, satisfaction, and retention of landowners in private land conservation programs. Conserv Lett 8:282–289
Shultz S, Noë R (2002) The consequences of crowned eagle central-place foraging on predation risk in monkeys. Proc R Soc B 269:1797–1802
Silvertown J (2009) A new dawn for citizen science. Trends Ecol Evol 24:467–471
Smallwood KS, Thelander C (2008) Bird mortality in the Altamont Pass wind resource area, California. J Wildl Manag 72:215–223
Smeenk C (1974) Comparative ecological studies of some East-African birds of prey. Ardea 62:1–97
Steedman A (1988) Locust handbook. Overseas Development Natural Resources Institute, London
Steyn P (1973) Eagle days: a study of African eagles at the nest. Sable Press, Sandton
Steyn P (1982) Birds of prey of southern Africa. David Phillip, Cape Town
Sumasgutner P, Koeslag A, Amar A (in prep.). Senescence in the city: Do urban black sparrowhawks age faster?
Sumasgutner P, Tate GJ, Koeslag A, Amar A (2016) Family morph matters: factors determining survival and recruitment in a long-lived polymorphic raptor. J Anim Ecol 85:1043–1055
Suri J, Sumasgutner P, Hellard É, Koeslag A, Amar A (2017) Stability in prey abundance may buffer Black Sparrowhawks Accipiter melanoleucus from health impacts of urbanization. Ibis 159:38–54
Tarboton WR (1978) Hunting and the energy budget of the Black-shouldered Kite. Condor 80:88–91

Tate GJ, Amar A (2017) Morph specific foraging behavior by a polymorphic raptor under variable light conditions. Sci Rep 7: 9161

Tate GJ, Bishop JM, Amar A (2016a) Differential foraging success across a light level spectrum explains the maintenance and spatial structure of colour morphs in a polymorphic bird. Ecol Lett 19:679–686

Tate GJ, Sumasgutner P, Koeslag A, Amar A (2016b) Pair complementarity influences reproductive output in the polymorphic Black Sparrowhawk Accipiter melanoleucus. J Avian Bio 48: 387–398

Thiollay JM (1975) Exemple de prédation naturelle sur une population nicheuse de Quelea quelea (L) au Mali. Rev Ecol Terre Vie 29:31–54

Thiollay JM (1978a) Les migrations de rapaces en Afrique occi- dentale: adaptations écologiques aux fluctuations saisonnières de production des écosystèmes. Terre la Vie 32:89–133

Thiollay JM (1978b) Population structure and seasonal fluctuations of the Falconiformes in Uganda National Parks. Afr J Ecol 16:145–151

Thiollay JM (1989) Distribution and ecology of Palearctic birds of prey wintering in West and Central Africa. In: Meyburg B-U, Chancellor R (eds) Raptors in the modern world. World Working Group on Birds of Prey/Hancock House, Berlin/Blaine, pp 95–107

Thiollay JM (2001) Long-term changes of raptor populations in Northern Cameroon. J Raptor Res 353:173–118

Thiollay JM (2006) The decline of raptors in West Africa: long-term assessment and the role of protected areas. Ibis (London 1859) 148:240–254

Thiollay JM (2007) Raptor declines in West Africa: comparisons between protected, buffer and cultivated areas. Oryx 41:322–329

Thompson LJ, Davies JP, Bildstein KL, Downs CT (2017a) Removal (and attempted removal) of material from a Hooded Vulture Necrosyrtes monachus nest by a starling and a Hooded Vulture. Ostrich 88:183–187

Thompson LJ, Davies JP, Gudehus M, Botha AJ, Bildstein KL, Murn C, Downs CT (2017b) Visitors to nests of Hooded Vultures Necrosyrtes monachus in northeastern South Africa§. Ostrich 88:155–162

Thomsett S (1987) Raptor deaths as a result of poisoning quelea in Kenya. Gabar 2:33–38

Thomsett S (2015) Eagle Hill, Kenya: changes over 60 years. Scopus 34:24–30

Trierweiler C, Koks BJ (2009) Montagu's harrier Circus pygargus. Living on the edge: Wetlands and birds in a changing Sahel, vol 312. KNNV Publishing, Zeist, Netherlands, p 327

United Nations Department of Economic and Social Affairs and Population Division (2017) World population prospects: the 2017 revision, Key findings and advance tables http://www.un.org/en/development/desa/population/

Van Der Merwe F (1981) Review of the status and biology of the Black Harrier. Ostrich 52:193–207

van Eeden R, Whitfield DP, Botha A, Amar, A (2017) Ranging behaviour and habitat preferences of the martial eagle: Implications for the conservation of a declining apex predator. PloS one 12(3):e017395

Van Gulck T, Stoks R, Verhagen R, Sabuni CA, Mwanjabe P, Leirs H (1998) LeirsShort-term effects of avian predation variation on population size and local survival of the multimammate rat, Mastomys natalensis(Rodentia, Muridae). Mammalia 62:329–339

Van Velden JL, Koeslag A, Curtis O, Gous T, Amar A (2017) Negative effect of mite (Knemidokoptes) infection on reproductive output in an African raptor. Auk 134:498–508

Van Wyk E, Bouwman H, Van Der Bank H, Verdoorn GH, Hofmann D (2001) Persistent organochlorine pesticides detected in blood and tissue samples of vultures from different localities in South Africa. Comp Biochem Physiol C Toxicol Pharmacol 129:243–264

Vibe-Petersen S, Leirs H, De Bruyn L (2006) Effects of predation and dispersal on Mastomys natalensis population dynamics in Tanzanian maize fields. J Anim Ecol 75:213–220

Virani M. et al. (in prep.). Factors affecting the population dynamics of the African fish eagle Haliaeetus vocifer at Lake Naivasha, Kenya: a twenty year study

Virani MZ (1999) The breeding ecology and behaviour of the augur buzzard Buteo augur in relation to different land-uses in the southern Lake Naivasha area, Kenya. University of Leicester, UK, Leicester

Virani MZ, Harper DM (2009) Factors influencing the breeding performance of the Augur Buzzard Buteo augur in southern Lake Naivasha, Rift Valley, Kenya. Ostrich 80:9–17

Virani MZ, Watson RT (1998) Raptors in the East African tropics and western Indian Ocean islands: state of ecological knowledge and conservation status vation strategies locally, regionally, and globally cies" because their large home ranges and low eas encompassing viable Population. J Raptor Res 32: 28–39

Virani MZ, Kendall C, Njoroge P, Thomsett S (2011) Major declines in the abundance of vultures and other scavenging raptors in and around the Masai Mara ecosystem, Kenya. Biol Conserv 144:746–752

Wacher T, Newby J, Houdou I, Harouna A, Rabeil T (2013) Vulture observations in the Sahelian zones of Chad and Niger. Bull ABC 20:186–199

Weng L, Boedhihartono AK, Dirks PHGM, Dixon J, Lubis MI, Sayer JA (2013) Mineral industries, growth corridors and agricultural development in Africa. Glob Food Sec 2:195–202

Whytock RC, Buij R, Virani MZ, Morgan BJ (2014) Do large birds experience previously undetected levels of hunting pressure in the forests of Central and West Africa? Oryx 50:1–8

Wichmann MC, Jeltsch F, Dean WRJ, Moloney KA, Wissel C (2003) Implication of climate change for the persistence of raptors in arid savanna. Oikos 1:186–202

Willems EP, Hill RA (2009) Predator-specific landscapes of fear and resource distribution: effects on spatial range use. Ecology 90:546–555

Williamson S, Ball A, Pretty J (2008) Trends in pesticide use and drivers for safer pest management in four African countries. Crop Prot 27:1327–1334

Wolter K, Neser W, Hirschauer MT, Camiña A (2016) Cape Vulture Gyps coprotheres breeding status in southern Africa: monitoring results from 2010 to 2014. Ostrich 6525:1–5

Chapter 19
Old World Vultures in a Changing Environment

Antoni Margalida and Darcy Ogada

Introduction

Carrion is a pulsed food resource of unpredictable occurrence in space and time that offers a high nutritive biomass, but is not globally widespread across all habitats and territories, and can be considered free because it does not require a large physical investment derived from predation (Ostfeld and Keesing 2000; DeVault et al. 2003; Selva and Fortuna 2007; Barton et al. 2013; Moleón et al. 2014a). As a result of these features, organisms feeding on this resource have developed morphological and behavioural adaptations to optimise its exploitation (Hertel 1994; DeVault et al. 2003; Moreno-Opo et al. 2015a, 2016), establishing guilds of species as in the case of carrion-eating birds (Selva and Fortuna 2007). Among terrestrial vertebrates, only vultures (families Accipitridae and Cathartidae) have evolved into obligate scavengers (DeVault et al. 2003; Beasley et al. 2015; Moleón et al. 2014b). Scavengers exploit carrion at different levels of intensity, and depending on their degree of carrion consumption and their own adaptive traits, they are defined as either obligate or facultative (DeVault et al. 2003; Wilson and Wolkovich 2011). As a result of the unpredictability in the availability of carcasses, the proportion of obligate scavenger species are scarce in comparison with facultative species that

A. Margalida (✉)
Department of Animal Science (Division of Wildlife), Faculty of Life Sciences and Engineering, University of Lleida, Lleida, Spain

Division of Conservation Biology, Institute of Ecology and Evolution, University of Bern, Bern, Switzerland
e-mail: antoni.margalida@iee.unibe.ch

D. Ogada
The Peregrine Fund, Boise, ID, USA

Ornithology Section, National Museums of Kenya, Nairobi, Kenya

J. H. Sarasola et al. (eds.), *Birds of Prey*,
https://doi.org/10.1007/978-3-319-73745-4_19

scavenge at variable rates but that can subsist on other food resources in the absence of carrion (DeVault et al. 2003; Beasley et al. 2015; Mateo-Tomás et al. 2015).

Worldwide there are 23 obligate scavenger species. Of the 16 Old World vultures, 81% are globally threatened or near-threatened. Most of these species are declining, particularly in Asia and Africa, mainly as a consequence of anthropogenic activities such as the illegal use of poisons, landscape transformation, health policies and ingestion of toxic veterinary drugs (Ogada et al. 2012a). These threats persist and continue to increase, despite vultures' importance to mankind. Vultures have traditionally provided important ecosystem services such as controlling diseases and pests, recycling nutrients and providing cultural inspiration and recreational value (Moleón et al. 2014a).

Vulture declines have far-reaching consequences, especially when considering the ecosystem services they provide to humans. It has been estimated that Spanish vultures remove >8000 tons of livestock carcasses per year, which prevents the release of greenhouse gases and provides economic savings estimated at 1.5 million Euros (Margalida and Colomer 2012; Morales-Reyes et al. 2015). Similarly, in developing countries such as India and Kenya, vultures likely play an outsized role in limiting disease transmission at carcasses (Markandya et al. 2008; Ogada et al. 2012b). In India, vulture declines have been linked to a subsequent increase in feral dogs, resulting in a rise in rabies transmission with an estimated $34 billion in healthcare costs between 1993 and 2006 (Markandya et al. 2008).

Vultures have biological adaptations that greatly complicate their conservation. Most significant among these is their need to range over vast landscapes in order to obtain food, which necessitates that conservation efforts be initiated across international boundaries (Margalida et al. 2013a, b; Lambertucci et al. 2014). In addition, they have evolved a life history strategy that emphasises long life and slow reproduction. Individuals do not begin breeding until *c.* 5 years of age, most species lay a single egg, and one breeding cycle can last up to 2 years in some species (see reviews in Mundy et al. 1992; Donázar 1993; Houston 2001). This ensures that once populations decline, they will take a long time to recover and the financial resources needed to restore depleted populations will be significant.

Perhaps not surprisingly, the conservation of Old World vultures varies widely by region. The European region is home to four vulture species of which three are threatened or near-threatened. Vultures were extirpated from several European countries by human activities in the nineteenth and twentieth centuries, but Eurasian griffon (*Gyps fulvus*) and cinereous vultures (*Aegypius monachus*) have been successfully reintroduced in recent decades to France; and bearded vultures (*Gypaetus barbatus*) have been reintroduced to Austria, Switzerland and Italy. Since 1993, the EU and various national governments have invested significant financial resources in the conservation of vultures – including at least 76 LIFE projects related to these species – and between 1993 and 2014 spent € 121.9 million, of which € 59.7 million came from European funds. During this period, Spain, home to 90% of all European vultures, invested € 72.8 million (€ 30.8 million received from the EU) on 38 projects related to vulture conservation (Margalida and Oliva-Vidal 2017).

In Asia, conservation efforts have become critically important after the Asian vulture crisis of the 1990s. Currently, Saving Asia's Vultures from Extinction (SAVE) coordinates a consortium of regional and international organisations that work to conserve and to advocate and fundraise for activities that sustain South Asia's vultures (www.save-vultures.org). Conservation activities include an intensive (and expensive) captive breeding programme.

In Africa, four species of vulture have recently been uplisted to critically endangered (hooded *Necrosyrtes monachus*, white-headed *Trigonoceps occipitalis*, white-backed *Gyps africanus* and Rüppell's *Gyps rueppelli*) and two species to endangered (Cape *Gyps coprotheres* and lappet-faced *Torgos tracheliotus*) as a result of severe continental declines (Ogada et al. 2016a; BirdLife International 2017). This has triggered new interest (and funding) for studying and monitoring populations (e.g. in Ethiopia, Ghana, Guinea Bissau, Mozambique, Zambia and Burkina Faso); despite this, the majority of African countries still lack even the most basic information on their vulture populations. Across all regions, conservation efforts are still menaced by anthropogenic activities that are jeopardising the viability of many species.

Here we describe several global or regional threats that, taken together, provide an overview of the main factors affecting the conservation of Old World vultures and, by extension, the important ecosystem services these species provide. We give examples of conservation measures that have helped to improve their management. Finally, we suggest what is required to ensure the long-term survival of Old World vultures.

Diclofenac, the Asian Crisis and Lessons Not Learned

In the 1980s, three *Gyps* vulture species endemic to South Asia were the most abundant large raptors in the world, but during the 1990s, their populations were reduced by more than 99% (Prakash 1999; Prakash et al. 2003, 2007) and all are now listed as critically endangered (BirdLife International 2017). The non-steroidal anti-inflammatory drug (NSAID) diclofenac was identified in 2004 as the primary cause of the rapid declines in Pakistan, India and Nepal (Oaks et al. 2004). Low-cost veterinary diclofenac-based products were being widely administered to livestock, some of which subsequently died, and due to cultural practices, their carcasses were widely available to scavengers. Sufficient residues remained in the carcasses of treated animals to cause acute renal failure and rapid death of vultures feeding on them (Taggart et al. 2007). Controlled experiments subsequently demonstrated that many species of *Gyps* vultures are susceptible to diclofenac poisoning including Oriental white-backed (*Gyps bengalensis*), African white-backed, Cape griffon and Eurasian griffon vultures (Oaks et al. 2004; Swan et al. 2006a, b; Naidoo et al. 2009). Since death rates of vultures are low under normal conditions, additional mortalities caused by veterinary drugs can have a profound impact on populations. Models indicated that contamination of just 0.3–0.7% of ungulate carcasses containing lethal levels of diclofenac was sufficient to cause vulture populations to decline at ~50% per year, as observed for one species in India and Pakistan (Green et al. 2004).

The Government of India enacted a ban on the production, importation and sale of veterinary diclofenac products in 2006. Similar measures were quickly taken in Pakistan and Nepal and then in Bangladesh in 2010 (www.save-vultures.org). This action was further supported by the identification of meloxicam as a suitable alternative drug that has proven safe for *Gyps* vultures (Swan et al. 2006a, b; Swarup et al. 2007; Naidoo et al. 2008). Since 2010, diclofenac prevalence in dead cattle has reduced by 70% (Cuthbert et al. 2014), and population declines in South Asia have slowed and may have reversed in some areas (Prakash et al. 2012; Chaudrhy et al. 2012). However, current usage is still sufficient to eradicate vulture populations (Green et al. 2006), and this has been attributed to human-use diclofenac, available in larger 30 ml vials, which makes it very convenient and cheap for its illegal use as a veterinary dose (Bowden 2015). These large 'multi-dose' vials for human use were finally banned by the Indian government in July 2015 (Bowden 2015). In the meantime, a number of other NSAIDs (e.g. carprofen, flunixin and ketoprofen) have been shown to cause mortality in *Gyps* vultures (Cuthbert et al. 2007), although the likelihood of these causing widespread mortality varies (Cuthbert et al. 2014); it illustrates the potential minefield among insufficiently tested pharmaceutical drugs and their cost to susceptible populations.

Despite this history, the government of Spain authorised the marketing of diclofenac as a veterinary pharmaceutical for use in cattle, pigs and horses in 2013 (Margalida et al. 2014b). Spain is important for the global conservation of avian scavengers, as it holds >90% of the European population of vultures, the entire population of the globally threatened Spanish imperial eagle (*Aquila adalberti*) and important numbers of red kites (*Milvus milvus*), which may also be at risk. Spain has a history of providing carcasses for wild scavengers to aid in their conservation. Spanish law (RD 1632/2011) facilitates this by allowing carcasses of farm animals to be left in the field in some protected areas or to be taken to 'muladares' (vulture feeding stations) to provide food for wildlife, including vultures (Margalida et al. 2012).

By law, diclofenac should only be administered under veterinary supervision and should not to be given to animals that are likely to enter the natural food chain. However, there remains a real risk that carcasses containing diclofenac residues will be consumed by vultures. Spain has an important livestock industry, with around 25 million pigs and 5.7 million cattle, and diclofenac is licensed for use in many clinical conditions that occur in these animals (Margalida et al. 2014d). Several studies have demonstrated that Eurasian griffon vultures are indeed accessing antibiotic residues in livestock carcasses provided at feeding stations (Casas-Díaz et al. 2016), and modelling of a simulated vulture population predicted that the potential rate of decline of the Spanish population of Eurasian griffon vultures due to diclofenac would be 0.9–7.7% per year (Green et al. 2016).

It is clear, then, that the risk to avian scavengers from veterinary use of diclofenac and other NSAIDs has not yet been evaluated adequately, and given the current global status of vultures, the precautionary principle ought to be evoked to ban diclofenac for veterinary use in the EU and other vulture range states, as has already occurred in South Africa, Iran and several Asian countries.

Additional trials evaluating the long-term impact of meloxicam and other veterinary drugs on scavengers, as well as reducing the availability of diclofenac in livestock carcasses, are needed to ensure the recovery of critically endangered *Gyps* vultures. In addition, veterinary treatment should be applied cautiously to reduce the prevalence of other anti-inflammatories, antibiotics and antiparasitic agents that currently contaminate carcasses (Donázar et al. 2009a; Casas-Díaz et al. 2016; Blanco et al. 2017a, b). Livestock management has important implications for the ecosystem services provided by scavengers (Dupont et al. 2012; Moleón et al. 2014a), as well as for the associated economic, human health and environmental effects (Margalida and Colomer 2012; Morales-Reyes et al. 2015, 2017).

Poisoning and Environmental Contaminants

Among raptors, vultures are particularly vulnerable to contaminants due to their reliance on carrion (Chaps. 10 and 11 in this book). As a result of their feeding habits, large numbers of individuals can be poisoned at a single carcass. In most cases, illegal poisoned baits are used to attract carnivores, provoking collateral mortality of facultative and obligate scavengers (Hernández and Margalida 2008, 2009a; Ogada et al. 2012a; Margalida 2012; Mateo-Tomás et al. 2012; Richards 2012). Although the use of poisons to manage carnivore populations has been banned in many countries, poisoning continues to be a common illegal tool used in all regions to manage game species and/or to protect livestock (Hernández and Margalida 2008, 2009a; Ogada et al. 2012a; Tingay et al. 2012; Berny et al. 2015; Bowden 2015). Accordingly, intentional poisoning of carnivores by humans is likely the most widespread cause of vulture mortality worldwide (Donázar 1993; Margalida 2012; Ogada et al. 2012a). For example, in Spain 8000 cases of illegal poisoning were reported during the period 1990–2010 with victims including 53 bearded vultures, 366 Egyptian vultures (*Neophron percnopterus*), 759 cinereous vultures and 2877 Eurasian griffon vultures (Margalida 2012).

The majority of poisons used are highly toxic pesticides. Many are carbamate pesticides that kill rapidly, and their detection post-mortem can be difficult, particularly in developing countries where preservation of suspected poisoned carcasses and access to expensive laboratory testing equipment are difficult (Ogada 2014). The most notorious wildlife poisons include carbofuran (trade name Furadan) and aldicarb (trade name Temik), and even strychnine is still used (Guitart et al. 2010; Ogada 2014; Botha et al. 2015). Although many of these pesticides have been banned particularly in the EU, their continued misuse to poison wildlife persists (e.g. Berny et al. 2015; Ruiz-Suárez et al. 2015), illustrating the inherent difficulties involved in regulating pesticides.

Widespread availability, low cost, ease of use and effectiveness have ensured that pesticide poisoning remains the generalized method used to predators and by association vultures. Worryingly, there is a growing use of poisons to intentionally harvest vultures, particularly in Africa (Ogada 2014). Vultures are routinely harvested for their body parts using pesticides, which are then sold for fetish, throughout West and Central Africa and in southern Africa (Beilis and Esterhuizen 2005; Mander et al. 2007; Saidu and Buij 2013; Buij et al. 2016). This method of harvest represents obvious, but as yet untested, potential hazards for consumers who may be ingesting or inhaling powders procured from vulture parts (Mander et al. 2007). Since 2012, there has been a dramatic rise in the intentional poisoning of vultures at the scene of poaching incidents, mostly involving elephants. Vultures are targeted by poachers because they rapidly discover poaching incidents where their overhead circling alerts the authorities to the location of these illicit activities (Ogada et al. 2016b).

Given the long-standing and global use of pesticides and other synthetic poisons to illicitly kill predators and other wildlife, innovative solutions are clearly paramount. In southern Spain, detection dogs have been trained to recover carcasses of poisoned wildlife; combined with state-of-the-art forensic analysis capabilities, these efforts have been successful in prosecuting wildlife poisoners (Fajardo et al. 2012). The detection dogs and their handlers also undertake regular patrols in known or suspected poisoning hotspot areas that alongside the prosecution of poisoners have been proven effective in decreasing and deterring vulture poisoning incidents in Spain (Fajardo et al. 2012). In southern Africa and in Kenya, poison response trainings have empowered rangers, police officers and community group members to recognise and rapidly respond to poisoning incidents. While the underlying causes of vulture poisonings still persist, these trainings have been shown to be effective in reducing the probability of extinction for critically endangered vultures (Murn and Botha 2017) and even in preventing poisoning incidents (Ogada pers. observ). It is clear that eliminating illegal wildlife poisoning is a complex task that must involve legal, educational, economic and punitive measures.

Ingestion of pellets or fragments from lead bullets poses another significant threat to scavengers (Hunt et al. 2006; Kelly et al. 2011; Lambertucci et al. 2011). Upon impact, lead bullets often fragment and become lodged in muscle and soft tissue where they become available to scavengers that consume viscera or muscle tissue from field-processed and unrecovered big game. Sublethal exposure to heavy metals may affect bone mineralisation (Gangoso et al. 2009), reduce muscle and fat concentrations (Carpenter et al. 2003), cause organ damage and internal lesions (Pattee et al. 1981) and reduce hatching success (Steidl et al. 1991). Lead exposure linked to hunting has been documented for threatened species such as Egyptian vultures (Gangoso et al. 2009) and bearded vultures (Hernández and Margalida 2009b; Berny et al. 2015), and it has been suspected as a source of contamination in endangered *Gyps* vultures in Africa (Naidoo et al. 2017).

Given the ongoing and perilous situation involving lead poisoning of California condors (*Gymnogyps californianus*) in the New World, legislation to minimise exposure to lead should be urgently implemented or improved, depending on the country, to reduce its effects on large predatory and scavenging birds.

Bovine Spongiform Encephalopathy in Europe

In 2001, a new European health crisis emerged with an outbreak of the so-called 'mad cow' disease. Acquired from cattle infected by bovine spongiform encephalopathy (BSE), it affects the brain and nervous system of humans and animals and leads to the degeneration of brain tissue, giving it a sponge-like appearance (EFSA 2017). The source of the disease is cattle feed prepared from BSE-infected animal tissues, such as the brain and spinal cord (EFSA 2017).

The outbreak led to the passing of sanitary legislation by the European Union (Regulation EC 1774/2002) that greatly restricted the use of animal by-products not intended for human consumption. Henceforth, all carcasses of domestic animals had to be collected from farms and transformed or destroyed at authorised facilities; as a result, many vulture feeding stations associated with stock farms closed down almost overnight. In Spain, it was estimated that 80% of all sheep and goat carcasses and 100% of all cow carcasses were being disposed of as per the new law (Donázar et al. 2009a, b; Margalida et al. 2010). Between 2005 and 2008 in the Navarre and Aragon regions of northern Spain, it is believed that 80% of all feeding stations were forced to close (Cortés-Avizanda et al. 2010; Margalida et al. 2014c).

The economic impact was also significant, as businesses had to invest in the collection of animal carcasses that were once destined to be consumed by vultures and other wild scavengers. In Spain, the cost of collecting a carcass was estimated at around 20 € per animal (Donázar and Margalida 2009; Donázar et al. 2009a), and a similar figure was estimated in France (Boumellasa 2004). Additionally, the industrial transformation of carcasses costs between 66 and 96 € per tonne (burning and recycling of animal remains) (Boumellasa 2004; Donázar et al. 2009a). Further environmental damage included the energetic costs of these industrial activities, for example, in the form of emission of greenhouse gases (Morales-Reyes et al. 2015).

As the collection of dead cattle became more widespread during the last decade, the behavioural and demographic effects on vultures appeared suddenly. For example, reports of vultures (mostly griffon vultures) attacking and killing cattle increased exponentially (Margalida et al. 2011a, b, 2014a).

The new regulations provoked an annual decrease in breeding success, reduced population growth, increased mortality of young age classes, changes in the diet and behaviour of Eurasian griffon vultures and a reduction on egg quality (Donázar et al. 2009a, b, 2010; Zuberogoitia et al. 2010; Margalida et al. 2011a, 2014a, c; Hernández et al. 2018).

It was patent that the restrictions imposed by this new EU legislation, which deprived scavenger populations of essential resources, flagrantly contradicted member states' obligations to conserve these threatened species (Tella 2001; Donázar et al. 2009b; Margalida et al. 2010). Fortunately, recommendations made by scientists, conservationists and conservation managers recently led to new EU guidelines allowing farmers to abandon dead animals in the field and/or at feeding stations (Margalida et al. 2012). This example illustrates how scientific arguments can trigger positive political action and help to reconcile conservation challenges and human activities (Sutherland et al. 2004; Margalida et al. 2012).

Collision and Electrocution Due to Energy Infrastructure

The worldwide increase in the production and development of both traditional and alternative power sources (Northrup and Wittenmeyer 2012) constitutes a serious and growing threat to avian scavengers (Chaps. 12 and 13 in this book). Among the most important risks, mortality from electrocution and collisions with power lines and wind farms has been documented for numerous raptor species (Ledger and Annegarn 1981; González et al. 2007; Margalida et al. 2008; Lehman et al. 2010; Guil et al. 2011, 2015), and collision with wind turbines is now the main cause of death for griffon vultures in Spain (see Tellería 2009; Carrete et al. 2009). Impacts of this mortality may be severe enough to affect long-term population size (Bevanger 1998) and population dynamics (Ledger and Annegarn 1980; Leshem 1985; Nikolaus 1984; Angelov et al. 2013). What is clear is that the energy sector represents an increasing threat to vultures (and other birds and bats) due to its rapid, global growth, with Asia, led by China and India, having the most cumulative installations (GWEC 2015).

In Europe, among all species studied, vultures are among those most frequently killed by collision with spinning turbine blades (Carrete et al. 2012). Even slight increases in mortality due to poorly placed wind turbines can significantly affect populations and potentially accelerate extinction of sensitive or endangered species such as the Egyptian vulture (Carrete et al. 2009). Wind energy infrastructure in Africa may jeopardise the existence of local or regional populations of several species (Drewitt and Langston 2006; Jenkins et al. 2010; Bellebaum et al. 2013). In southern Africa, where bearded vultures are now critically endangered in the region, with only about 100 pairs remaining, two wind farms are planned for Lesotho within this species' core habitat (Reid et al. 2015).

In Europe, effective guidelines for wind farm establishment have been carried out to place them far from sensitive species (Carrete et al. 2012), while in Africa, apart from South Africa, the development of national regulations and guidelines in regard to the establishment of wind farms lags far behind the current pace of development (Reid et al. 2015; Ogada pers. obs.). In addition, national or regional sensitivity mapping to guide the placement of wind farms is urgently needed throughout most of Africa and Asia. Sensitivity maps that are easily accessible to developers are a critical component for guiding the placement of wind turbines to avoid migratory flyways and other sensitive areas for vultures and other birds.

Frequently overlooked is the substantial threat posed to vultures and other large birds, particularly in developing regions, from the installation of hundreds of thousands of kilometres of new power lines that are necessary to transport power from these often remote wind-generating facilities to consumers.

Mitigation measures for reducing electrocution and collisions with power lines include reviewing the placement of new lines, removing the earth wire, fitting lines with markers or changing the pylon design. These have been developed in several countries, particularly in Europe, North America and South Africa (Lehman et al. 2007). In Spain, several efforts have identified mortality hotspots and conducted pylon and line modifications (Tintó et al. 2010; Guil et al. 2011). In South Africa, collision rates of some species such as cranes and bustards were partially reduced after attaching bird-flight diverters on ground wires (Anderson 2002). Regrettably, the actual effectiveness of these measures for vultures is unknown, although for other raptor species some reductions in mortality have been reported.

The Future of Old World Vultures in a Changing World

Populations of obligate scavengers have declined significantly over the last several decades across the globe, mainly due to a suite of anthropogenic factors. In particular, habitat loss and human persecution have played a prominent role in vulture declines in many regions; however, unintentional poisoning has emerged as one of the greatest threats to avian scavengers globally (Oaks et al. 2004; Ogada et al. 2012a).

At the same time, the conservation needs of vultures represent a uniquely complex mix of factors. Apart from the often confounding influences of economics, politics, conservation science, human health, industrial development and veterinary practices as illustrated here, vultures' biology and life history strategy further complicates their management and conservation. For example, few efficient supranational management strategies exist, which inevitably limits the effectiveness of any conservation strategy (see Margalida et al. 2013a, b).

In order to reverse the negative trend for Old World vultures, innovative and collaborative management solutions are needed. For example, the Endangered Wildlife Trust in South Africa has developed a strategic partnership with their national power provider, Eskom, to minimise the negative impacts associated with electrical infrastructure on vultures and other wildlife through risk and impact assessments, training and research (www.ewt.org.za). More such partnerships will be needed with the energy sector, particularly in developing regions where limited capacity and access to resources favour such collaborations.

Other possible interventions include the creation of large feeding stations, which may involve the private sector (as suppliers) and the public sector (as managers). While feeding stations are an attractive solution from a management perspective, they must also be carefully evaluated in terms of their conservation effectiveness and appropriateness based on local conditions (Cortés-Avizanda et al. 2010, 2016; Moreno-Opo et al. 2015a, b).

Cultural services provided by avian scavengers offer a growing economic benefit associated with vulture and eagle viewing and photography, which is an emerging and powerful conservation tool, provided that a part of the profits are allocated to avian scavenger management programmes and also benefit local communities (Becker et al. 2005; Piper 2005; Donázar et al. 2009a).

Throughout the regions inhabited by Old World vultures, the challenges to their conservation are shaped by local conditions and cultural practices. Similarly, any conservation solutions must inevitably involve local people and communities who share the same landscapes and benefit from the ecological services provided by vultures.

The future of Old World vultures depends on developing long-term, collaborative research and conservation strategies that recognise the economic, political, social and cultural realities inherent in the various regions where these birds persist.

Acknowledgements The authors acknowledge all people involved in different scientific and conservation projects related with vulture conservation that provided most of the data at which this chapter was based. D. O. acknowledges N. Richards for the assistance with references. A. M. was supported by a Ramón y Cajal research contract by the Ministry of Economy and Competitiveness (RYC-2012-11867).

References

Anderson MD (2002) Karoo large terrestrial bird powerline project, report no. 1. Eskom, Johannesburg. Unpublished report

Angelov I, Hashim I, Oppel S (2013) Persistent electrocution mortality of Egyptian vultures *Neophron percnopterus* over 28 years in East Africa. Bird Conserv Int 23:1–6

Barton PS, Cunningham SA, Lindenmayer DB, Manning AD (2013) The role of carrion in maintaining biodiversity and ecological processes in terrestrial ecosystems. Oecologia 171:761–772

Beasley JC, Olson ZH, DeVault TL (2015) Ecological role of vertebrate scavengers. In: Benbow ME, Tomberlin J, Tarone A (eds) Carrion ecology, evolution, and their application. CRC Press, Boca Raton, pp 107–128

Becker N, Inbar M, Bahat O, Choresh Y, Ben–Noon G, Yaffe O (2005) Estimating the economic value of viewing griffon vultures *Gyps fulvus*: a travel cost model study at Gamla Nature Reserve, Israel. *Oryx* 39:429–434

Beilis N, Esterhuizen J (2005) The potential impact on Cape Griffon *Gyps coprotheres* populations due to the trade in traditional medicine in Maseru, Lesotho. Vulture News 53:15–19

Bellebaum J, Korner-Nievergelt F, Dürr T, Mammen U (2013) Wind turbine fatalities approach a level of concern in a raptor population. J Nat Conserv 21:394–400

Berny P, Vilagines L, Cugnasse JM, Mastain O, Chollet JY, Joncour G, Razin M (2015) VIGILANCE POISON: illegal poisoning and lead intoxication are the main factors affecting avian scavenger survival in the Pyrenees (France). Ecotoxicol Environ Saf 118:71–82

Bevanger K (1998) Biological and conservation aspects of bird mortality caused by electricity power lines: a review. Biol Conserv 86:67–76

BirdLife International (2017) IUCN Red List for birds. Downloaded from http://www.birdlife.org on 26 Dec 2017

Blanco G, Junza A, Barrón D (2017a) Food safety in scavenger conservation: diet-associated exposure to livestock pharmaceuticals and opportunists mycoses in threatened Cinereous and Egyptian vultures. Ecotoxicol Environ Saf 135:292–301

Blanco G, Junza A, Barrón D (2017b) Occurrence of veterinary pharmaceuticals in golden eagle nestlings: Unnoticed scavenging on livestock carcasses and other potential exposure routes. Sci Total Environ 586:355–361

Botha CJ, Coetser H, Labuschagne L, Basson A (2015) Confirmed organophosphorus and carbamate pesticide poisonings in South African wildlife (2009–2014). J S Afr Vet Assoc 86:1–4

Boumellassa H. (2004) Rapaces nécrophages: concilier conservation de l'espece et minimisation des dispenses, vers un renforcement du lien Agriculture-Environment. D.E.A. Economie

de l'Environment et gestion des Resources Naturelles. École Doctorale d'Economie et Mathématiques de Paris-Ouest. Université de Paris X, Nanterre.
Bowden C (2015) Asian vulture crisis-40 million gone, but have we reached a turning point? Curr Conserv 9:4–10
Buij R, Nikolaus G, Whytock R, Ingram DJ, Ogada D (2016) Trade of threatened vultures and other raptors for fetish and bushmeat in West and Central Africa. Oryx 50:606–616
Carrete M, Sánchez-Zapata JA, Benítez JR, Lobón M, Donázar JA (2009) Large scale risk-assessment of wind-farms on population viability of a globally endangered long-lived raptor. Biol Conserv 142:2954–2961
Carrete M, Sánchez-Zapata JA, Benítez JR, Lobón M, Montoya F, Donázar JA (2012) Mortality at wind–farms is positively correlated to large–scale distribution and aggregation in griffon vultures. Biol Conserv 145:102–108
Carpenter, JW, Pattee OH, Fritts SH, Rattner BA., Wiemeyer SN., Royle JA., Smith, MR (2003) Experimental lead poisoning in turkey vultures (Cathartes aura). J Wildl Dis 39:96–104
Casas-Díaz E, Cristofol C, Cuenca R, Agustín S, Carneiro M, Marco I, Lavín S, Margalida A (2016) Determination of antibiotic residues in plasma of two populations of Eurasian griffon vulture (*Gyps fulvus*) in Spain. Sci Total Environ 557:620–626
Chaudhry MJI, Ogada DL, Malik RN, Virani MZ, Giovanni MD (2012) First evidence that populations of the critically endangered Long-billed Vulture *Gyps indicus* in Pakistan have increased following the ban on diclofenac in South Asia. Bird Conserv Int 22:389–397
Cortés-Avizanda A, Carrete M, Donázar JA (2010) Managing supplementary feeding for avian scavengers: guidelines for optimal design using ecological criteria. Biol Conserv 143:1707–1715
Cortés-Avizanda A, Blanco G, DeVault TL, Markandya A, Virani MZ, Brandt J, Donázar JA (2016) Supplementary feeding and endangered avian scavengers: benefits, caveats, and controversies. Front Ecol Environ 14:191–199
Cuthbert R, Parry-Jones J, Green RE, Pain DJ (2007) NSAIDs and scavenging birds: potential impacts beyond Asia's critically endangered vultures. Biol Lett 3:91–94
Cuthbert RJ, Taggart MA, Prakash V, Chakraborty SS, Deori P, Galligan T, Kulkarni M, Ranade S, Saini M, Sharma AK, Shringarpure R (2014) Avian scavengers and the threat from veterinary pharmaceuticals. Philos Trans Royal Soc B 369:20130574
DeVault TL, Rhodes OE Jr, Shivik JA (2003) Scavenging by vertebrates: behavioral, ecological, and evolutionary perspectives on an important energy transfer pathway in terrestrial ecosystems. Oikos 102:225–234
Donázar JA (1993) Los buitres ibéricos: biología y conservación. J.M. Reyero Editor, Madrid
Donázar JA, Margalida A (2009) European vultures on a knife-edge: the consequences of changes in animal husbandry and sanitary legislation. In: Donázar JA, Margalida A, Campión D (eds) Vultures, feeding stations and sanitary legislation: a conflict and its consequences from the perspective of conservation biology, Munibe 29 (Suplemento). Sociedad de Ciencias Aranzadi, San Sebastián, pp 526–551
Donázar JA, Margalida A, Campión D (eds) (2009a) Vultures, feeding stations and sanitary legislation: a conflict and its consequences from the perspective of Conservation Biology, Munibe 29 (Suplemento). Sociedad de Ciencias Aranzadi, San Sebastian
Donázar JA, Margalida A, Carrete M, Sánchez-Zapata JA (2009b) Too sanitary for vultures. Science 326:664
Donázar JA, Cortés-Avizanda A, Carrete M. (2010) Dietary shifts in two vultures after the demise of supplementary feeding stations: consequences of the EU sanitary legislation. European J. Wildl. Res. 56, 613–621
Drewitt AL, Langston RHW (2006) Assessing the impacts of wind farms on birds. Ibis 148:29–42
Dupont H, Mihoub JB, Bobbe S, Sarrazin F (2012) Modelling carcass disposal practices: implications for the management of an ecological service provided by vultures. J Appl Ecol 49:404–411
European Food Safety Authority (EFSA) (2017) Bovine Spongiform Encephalopathy (BSE). Downloaded from https://www.efsa.europa.eu on 26 Dec 2017

Fajardo I, Ruiz A, Zorrilla I, Valero A, Fernández I, Sáez E, Molino FM, Olivares J (2012) Use of specialised canine units to detect poisoned baits and recover forensic evidence in Andalucía (Southern Spain). In: Richards NL (ed) Carbofuran and wildlife poisoning: global perspectives and forensic approaches. Wiley, Chichester, pp 147–155

Gangoso L, Álvarez-Lloret P, Rodríguez-Navarro A a B, Mateo R, Hiraldo F, Donázar JA (2009) Long-term effects of lead poisoning on bone mineralization in vultures exposed to ammunition sources. Environ Pollut 157:569–574

González LM, Margalida A, Mañosa S, Sánchez R, Oria J (2007) Causes and spatio-temporal variations of non-natural mortality in the globally endangered Spanish imperial eagle (*Aquila adalberti*) during a recovery period. Oryx 41:495–502

Green RE, Taggart MA, Das D, Pain DJ, Kumar CS, Cunningham AA, Cuthbert R (2004) Collapse of Asian vulture populations: risk of mortality from residues of the veterinary drug diclofenac in carcasses of treated cattle. J Appl Ecol 43:949–956

Green RE, Taggart MA, Das D, Pain DJ, Sashi Kumar C, Cunningham AA, Cuthbert R (2006) Collapse of Asian vulture populations: risk of mortality from residues of the veterinary drug diclofenac in carcasses of treated cattle. J Appl Ecol 43:949–956

Green RE, Donázar JA, Sánchez-Zapata JA, Margalida A (2016) The threat to Eurasian griffon vultures *Gyps fulvus* in Spain from veterinary use of the drug diclofenac. J Appl Ecol 53:993–1003

Guil F, Fernández-Olalla M, Moreno-Opo R, Mosqueda I, García ME, Aranda A, Arredondo A, Guzmán J, Oria J, González LM, Margalida A (2011) Minimising mortality in endangered raptors due to power lines: the importance of spatial aggregation to optimize the application of mitigation measures. PLoS One 6:e28212

Guil F, Colomeer MA, Moreno-Opo R, Margalida A (2015) Space-time trends in Spanish bird electrocution rates from alternative information sources. Glob Ecol Conserv 3:379–388

GWEC (2015). Global Wind Report. Annual Market Update 2015. Global Wind Energy Council. Downloaded from htpps://www.gwec.net on 30 Dec 2017

Guitart R, Sachana M, Caloni F, Croubels S, Vandenbroucke V, Berny P (2010) Animal poisoning in Europe. Part 3: wildlife. Vet J 183:260–265

Hernández M, Margalida A (2008) Pesticide abuse in Europe: effects on the Cinereous vulture (*Aegypius monachus*) population in Spain. Ecotoxicology 17:264–272

Hernández M, Margalida A (2009a) Poison-related mortality effects in the endangered Egyptian vulture (*Neophron percnopterus*) population in Spain: conservation measures. Eur J Wildl Res 55:415–423

Hernández M, Margalida A (2009b) Assessing the risk of lead exposure for the conservation of the endangered Pyrenean bearded vulture (*Gypaetus barbatus*) population. Environ Res 109:837–842

Hernández M, Colomer MA, Pizarro M, Margalida A (2018) Changes in eggshell thickness and ultrastructure in the Bearded vulture (*Gypaetus barbatus*) Pyrenean population: a long-term analysis. Sci Total Environ 624:713–721

Hertel F (1994) Diversity in body size and feeding morphology within past and present vulture assemblages. Ecology 75:1074–1084

Houston D (2001) Condors & vultures. Colin Baxter Photography, Grantown-on-Spey

Hunt WG, Burnham W, Parish CN, Burnham KK, Mutch B, Oaks JL (2006) Bullet fragments in deer remains: implications for lead exposure in avian scavengers. Wildl Soc Bull 34:167–170

Jenkins AR, Smallie JJ, Diamond M (2010) Avian collisions with power lines: a global review of causes and mitigation with a south African perspective. Bird Conserv Int 20:263–278

Kelly TR, Bloom PH, Torres SG, Hernandez YZ, Poppenga RH, Boyce WM, Johnson CK (2011) Impact of the California lead ammunition ban on reducing lead exposure in golden eagles and turkey vultures. PLoS One 6:e17656

Lambertucci SA, Donázar JA, Delgado A, Jiménez B, Sáez M, Sánchez-Zapata JA, Hiraldo F (2011) Widening the problem of lead poisoning to a South-American top scavenger: lead concentrations in feathers of wild Andean condors. Biol Conserv 144:1464–1471

Lambertucci SA, Alarcón PAE, Hiraldo F, Sánchez-Zapata JA, Blanco G, Donázar JA (2014) Apex scavenger movements call for transboundary conservation policies. Biol Conserv 170:145–150

Ledger JA, Annegarn HJ (1981) Electrocution hazards to the cape vulture *Gyps coprotheres* in South Africa. Biol Conserv 20:15–24

Lehman RN, Kennedy PL, Savidge JA (2007) The state of the art in raptor electrocution research: a global review. Biol Conserv 136:159–174

Leshem Y (1985) Griffon vultures in Israel: electrocution and other reasons for a declining population. Vulture News 13:14–20

Mander M, Diederichs N, Ntuli L, Mavundla K, Williams V, McKean S (2007) Survey of the trade in vultures for the traditional health industry in South Africa. FutureWorks, Durban

Margalida A (2012) Baits, budget cuts: a deadly mix. Science 338:192–192

Margalida A, Colomer MA (2012) Modelling the effects of sanitary policies on European vulture conservation. Sci Rep 2:753

Margalida A, Oliva-Vidal P (2017) The shadow of diclofenac hangs over European vultures. Nat Ecol Evol 1:1050

Margalida A, Heredia R, Razin M, Hernández M (2008) Sources of variation in mortality of the Bearded vulture *Gypaetus barbatus* in Europe. Bird Conserv Int 18:1–10

Margalida A, Donázar JA, Carrete M, Sánchez Zapata JA (2010) Sanitary versus environmental policies: fitting together two pieces of the puzzle of European vulture conservation. J Appl Ecol 47:931–935

Margalida A, Campión D, Donázar JA (2011a) European vultures' altered behaviour. Nature 480:457

Margalida A, Colomer MA, Sanuy D (2011b) Can wild ungulate carcasses provide enough biomass to maintain avian scavenger populations? An empirical assessment using a bio-inspired computational model. PLoS One 6:e20248

Margalida A, Carrete M, Sánchez-Zapata JA, Donázar JA (2012) Good news for European vultures. Science 335:284–284

Margalida A, Arlettaz R, Donázar JA (2013a) Poisoning and lead ammunition: further international agreements are needed to preserve vultures and the crucial sanitary service they provide. Environ Sci Technol 47:5522–5523

Margalida A, Carrete M, Hegglin D, Serrano D, Arenas R, Donázar JA (2013b) Uneven large-scale movement patterns in wild and reintroduced pre-adult bearded vultures: conservation implications. PLoS One 8:e65857

Margalida A, Campión D, Donázar JA (2014a) Vultures vs livestock: conservation relationships in an emerging conflict between humans and wildlife. Oryx 48:172–176

Margalida A, Sánchez-Zapata JA, Blanco G, Hiraldo F, Donázar JA (2014b) Diclofenac approval as a threat to Spanish vultures. Conserv Biol 28:631–632

Margalida A, Colomer MÀ, Oro D (2014c) Man-induced activities modify demographic parameters in a long-lived species: effects of poisoning and health policies. Ecol Appl 24:436–444

Margalida A, Bogliani G, Bowden C, Donázar JA, Genero F, Gilbert M, Karesh W, Kock R, Lubroth J, Manteca X, Naidoo V, Neimanis A, Sánchez-Zapata JA, Taggart M, Vaarten J, Yon L, Kuiken T, Green RE (2014d) One health approach to use of pharmaceuticals. Science 346:1296–1298

Markandya A, Taylor T, Longo A, Murty MN, Murty S, Dhavala K (2008) Counting the cost of vulture decline – an appraisal of the human health and other benefits of vultures in India. Ecol Econ 67:194–204

Mateo-Tomás P, Olea PP, Sánchez-Barbudo IS, Mateo R (2012) Alleviating human-wildlife conflicts: identifying the causes and mapping the risk of illegal poisoning of wild faun. J Appl Ecol 49:376–385

Mateo-Tomás P, Olea PP, Moleón M, Vicente J, Botella F, Selva N, Viñuela J, Sánchez-Zapata JA (2015) From regional to global patterns in vertebrate scavenger communities subsidized by big game hunting. Divers Distrib 21:913–924

Moleón M, Sánchez-Zapata JA, Margalida A, Carrete M, Donázar JA, Owen-Smith N (2014a) Hominids and scavengers: the evolution of interactions and ecosystem services. Bioscience 64:394–403

Moleón M, Sánchez-Zapata JA, Selva N, Donázar JA, Owen-Smith N (2014b) Inter-specific interactions linking predation and scavenging in terrestrial vertebrate assemblages. Biol Rev 89:1042–1054

Morales-Reyes Z, Pérez-García JM, Moleón M, Botella F, Carrete M, Lazcano C, Moreno-Opo R, Margalida A, Donázar JA, Sánchez-Zapata JA (2015) Supplanting ecosystem services provided by scavengers raises greenhouse gas emissions. Sci Rep 5:7811

Morales-Reyes Z, Pérez-García JM, Moleón M, Botella F, Carrete M, Donázar JA, Cortés-Avizanda A, Arrondo E, Moreno-Opo R, Margalida A, Sánchez-Zapata JA (2017) Evaluation of the network of protection areas for the feeding of scavengers (PAFs) in Spain: from biodiversity conservation to greenhouse gas emission savings. J Appl Ecol 54:1120–1129

Moreno-Opo R, Trujillano A, Arredondo Á, González LM, Margalida A (2015a) Manipulating size, amount and appearance of food inputs to optimize supplementary feeding programs for European vultures. Biol Conserv 181:27–35

Moreno-Opo R, Trujillano A, Margalida A (2015b) Optimization of supplementary feeding programs for European vultures depend on environmental and management factors. Ecosphere 6:127

Moreno-Opo R, Trujillano A, Margalida A (2016) Behavioural coexistence and feeding efficiency drive niche partitioning at carcasses within the guild of European avian scavengers. Behav Ecol 27:1041–1052

Mundy PJ, Butchart D, Ledger J, Piper S (1992) The vultures of Africa. Academic Press, London

Murn C, Botha A (2017) A clear and present danger: impacts of poisoning on a vulture population and the effect of poison response activities. Oryx. https://doi.org/10.1017/S0030605316001137

Naidoo V, Wolter K, Cromarty AD, Bartels P, Bekker L, McGaw L, Taggart MA, Cuthbert R, Swan GE (2008) The pharmacokinetics of meloxicam in vultures. J Vet Pharmacol Ther 31:128–134

Naidoo V, Wolter K, Cuthbert R, Duncan N (2009) Veterinary diclofenac threatens Africa's endangered vulture species. Regul Toxicol Pharmacol 53:205–208

Naidoo V, Wolter K, Botha CJ (2017) Lead ingestion as a potential contributing factor to the decline in vulture populations in southern Africa. Environ Res 152:150–156

Nikolaus G (1984) Large numbers of birds killed by electric power line. Scopus 8:42

Northrup JM, Wittenmeyer G (2012) Characterising the impacts of emerging energy development on wildlife, with an eye towards mitigation. Ecol Lett 16: 112–125

Oaks JL, Gilbert M, Virani MZ, Watson RT, Meteyer CU, Rideout BA, Shivaprasad HL, Ahmed S, Chaudhry MJI, Arshad M, Mahmood S (2004) Diclofenac residues as the cause of vulture population decline in Pakistan. Nature 427:630–633

Ogada DL (2014) The power of poison: pesticide poisoning of Africa's wildlife. Ann N Y Acad Sci 1322:1–20

Ogada DL, Keesing F, Virani MZ (2012a) Dropping dead: causes and consequences of vulture population declines worldwide. Ann N Y Acad Sci 1249:57–71

Ogada DL, Torchin ME, Kinnaird MF, Ezenwa VO (2012b) Effects of vulture declines on facultative scavengers and potential implications for mammalian disease transmission. Conserv Biol 26:453–460

Ogada D, Shaw P, Beyers RL, Buij R, Murn C, Thiollay JM, Beale CM, Holdo RM, Pomeroy D, Baker N, Krüger SC, Botha A, Virani MZ, Monadjem A, Sinclair ARE (2016a) Another continental vulture crisis: Africa's vultures collapsing toward extinction. Conserv Lett 9:89–97

Ogada D, Botha A, Shaw P (2016b) Ivory poachers and poisons; drivers of Africa's declining vulture populations. Oryx 50:593–596

Ostfeld RS, Keesing F. (2000) Pulsed resources and community dynamics of consumers in terrestrial ecosystems. Trends Ecol Evol. 15:232–237

Pattee OH, Wiemeyer SN, Mulhern BM, Sileo L, Carpenter JW (1981) Experimental Lead-Shot Poisoning in Bald Eagles. J Wildl Manage 45:806–810

Piper SE (2005) Supplementary feeding programmes: how necessary are they for the maintenance of numerous and healthy vulture populations. In: Conservation and management of vulture populations. Thessaloniki, Greece pp 41–50

Prakash V (1999) Status of vultures in Keoladeo National Park, Bharatpur, Rajasthan, with special reference to population crash in Gyps species. J Bombay Nat Hist Soc 96:365–378

Prakash V, Pain DJ, Cunningham AA, Donald PF, Prakash N, Verma A, Gargi R, Sivakumar S, Rahmani AR (2003) Catastrophic collapse of Indian white-backed *Gyps bengalensis* and long-billed *Gyps indicus* vulture populations. Biol Conserv 109:381–390

Prakash V, Green RE, Pain DJ, Ranade SP, Saravanan S, Prakash N, Venkitachalam R, Cuthbert R, Rahmani AR, Cunningham AA (2007) Recent changes in populations of resident Gyps vultures in India. J Bombay Nat Hist Soc 104:129–135

Prakash V, Bishwakarma MC, Chaudhary A, Cuthbert R, Dave R, Kulkarni M, Kumar S, Paudel K, Ranade S, Shringarpure R, Green RE (2012) The population decline of Gyps vultures in India and Nepal has slowed since veterinary use of diclofenac was banned. PLoS One 7:e49118

Reid T, Kruger S, Whitfield DP, Amar A (2015) Using spatial analyses of bearded vulture movements in southern Africa to inform wind turbine placement. J Appl Ecol 52:881–892

Richards N (ed) (2012) Carbofuran and wildlife poisoning: global perspectives and forensic approaches. Wiley, Chichester

Ruiz-Suárez N, Boada LD, Henríquez-Hernández LA, González-Moreo F, Suárez-Pérez A, Camacho M, Zumbado M, Almeida-González M, del Mar Travieso-Aja M, Luzardo OP (2015) Continued implication of the banned pesticides carbofuran and aldicarb in the poisoning of domestic and wild animals of the Canary Islands (Spain). Sci Total Environ 505:1093–1099

Saidu Y, Buij R (2013) Traditional medicine trade in vulture parts in northern Nigeria. Vulture News 65:4–14

Selva N, Fortuna MA (2007) The nested structure of a scavenger community. Proc R Soc Lond B 274:1101–1108

Sutherland WJ, Pullin AS, Dolman PM, Knight TM (2004) The need for evidence-based conservation. Trends Ecol Evol 19:305–308

Steidl RJ, Griffin CR, Niles LJ (1991) Contaminant levels of osprey eggs and prey reflect regional differences in reproductive success. J Wildl Manage 55:601–608

Swan GE, Cuthbert R, Quevedo M, Green RE, Pain DJ, Bartels P, Cunningham AA, Duncan N, Meharg AA, Oaks JL, Parry-Jones J (2006a) Toxicity of diclofenac to Gyps vultures. Biol Lett 2:279–282

Swan G, Naidoo V, Cuthbert R, Green RE, Pain DJ, Swarup D, Prakash V, Taggart M, Bekker L, Das D, Diekmann J (2006b) Removing the threat of diclofenac to critically endangered Asian vultures. PLoS Biol 4:e66

Swarup DPRC, Patra RC, Prakash V, Cuthbert R, Das D, Avari P, Pain DJ, Green RE, Sharma AK, Saini M, Taggart M (2007) Safety of meloxicam to critically endangered Gyps vultures and other scavenging birds in India. Anim Conserv 10:192–198

Taggart MA, Senacha KR, Green RE, Jhala YV, Raghavan B, Rahmani AR, Cuthbert R, Pain DJ, Meharg AA (2007) Diclofenac residues in carcasses of domestic ungulates available to vultures in India. Environ Int 33:759–765

Tella JL (2001) Action is needed now, or BSE crisis could wipe out endangered birds of prey. Nature 410:408

Tellería JL (2009) Potential impacts of wind farms on migratory birds crossing Spain. Bird Conserv Int 19:131–136

Tingay RE, McAdam D, Taylor MJ (2012) Perspectives on wildlife poisoning by carbofuran in the United Kingdom and Republic of Ireland–with a particular focus on Scotland. In: Richards N (ed) *Carbofuran and Wildlife Poisoning: Global Perspectives and Forensic Approaches*. Wiley, Chichester, pp 171–188

Tintó A, Real J, Mañosa S (2010) Predicting and correcting electrocution of birds in Mediterranean areas. J Wildl Manag 74:1852–1862

Wilson EE, Wolkovich EM (2011) Scavenging: how carnivores and carrion structure communities. Trends Ecol Evol 26:129–135

Zuberogoitia I, Martínez JE, Margalida A, Gómez I, Azkona A, Martínez JA (2010) Reduced food availability induces behavioural changes in Griffon vulture. Ornis Fennica 87:52–60

Chapter 20
Raptor Conservation in Practice

Richard T. Watson

Introduction

You only have to look at a falcon or an eagle to understand the human fascination for these birds. They are stunningly beautiful, powerful, charismatic, and truly one of nature's greatest works of art. For most, that is reason enough to invest time, money, and effort in conserving them for the enjoyment of future generations. More importantly, however, raptors serve a role in maintaining ecosystem structure and function and evolutionary processes in biomes from forests to deserts and grasslands. Because of their position at the top of the food chain, and their tenuous, predatory lifestyle, birds of prey are often sensitive to changes in the environment, particularly the presence of toxins, but other changes also. In many cases they have been the bellwether of contamination, disease, and other changes that also can affect human health and welfare (Figs. 20.1, 20.2, 20.3, 20.4, 20.5, and 20.6).

The Peregrine Fund has identified several practices that have proven to be consistently important for successful raptor conservation. These include a scientific understanding of a species' population status and factors limiting its distribution and abundance. The latter usually requires quantitative studies on the species' population ecology and behavior which typically precede conservation interventions. If the species is in decline, the cause must be resolved before conservation can be successful, but, in some cases, experimental restoration may be initiated even before the cause is known so that reintroduced birds can be studied to learn what factors limit their survival or reproduction.

A well-defined and achievable conservation goal helps to focus effort on conservation actions that will make a measurable difference. Such interventions must be adapted to the goal, and methods have usually involved restoration through captive breeding and release of critically endangered species, habitat protection,

R. T. Watson (✉)
The Peregrine Fund, Boise, ID, USA
e-mail: rwatson@peregrinefund.org

J. H. Sarasola et al. (eds.), *Birds of Prey*,
https://doi.org/10.1007/978-3-319-73745-4_20

Fig. 20.1 Vultures in Africa are threatened by poisoning of animal carcasses upon which they feed. Poisoning includes deliberate killing of vultures by poachers to prevent them from revealing poachers' crime scenes, retaliation to livestock depredation, and killing vultures for body parts used in sorcery. (Photo credit: Munir Virani/The Peregrine Fund)

Fig. 20.2 Biologists draw blood from a California condor to test for lead poisoning from ingested fragments of lead bullets used to harvest game and other animals. (Photo credit: The Peregrine Fund)

Fig. 20.3 The presence of a juvenile harpy eagle reveals successful reproduction in Panama, but human encroachment on the forest results in shooting of adult eagles and population decline. (Photo credit: José de Jesús Vargas-González/The Peregrine Fund)

Fig. 20.4 The critically endangered Ridgway's hawk is threatened by human persecution, but perceptions changed when the value of the hawk for controlling predators such as snakes and rats was made clear and when incentives to protect them were provided. (Photo credit: David Anderson/ The Peregrine Fund)

Fig. 20.5 The Ridgway's hawk is critically endangered and threatened by human persecution as well as reproductive failure due to *Philornis* botfly infestation of nestlings. (Photo credit: Russell Thorstrom/The Peregrine Fund)

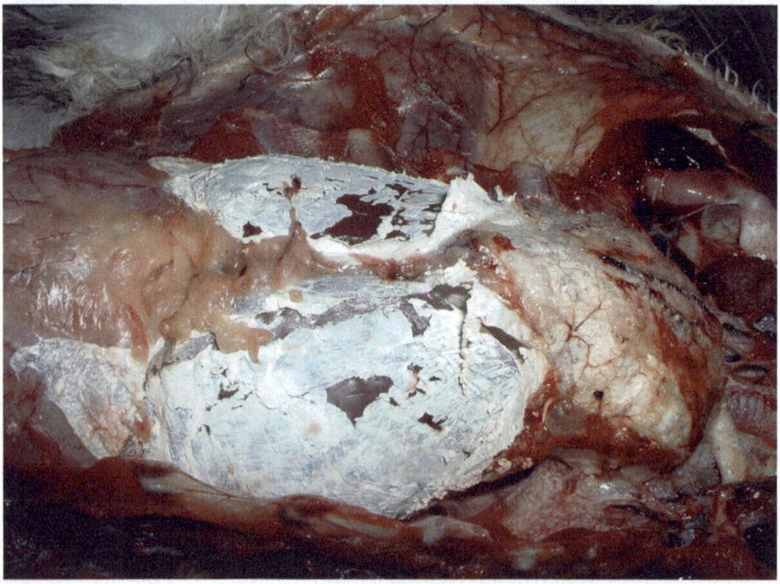

Fig. 20.6 Diclofenac, a veterinary pharmaceutical, drove at least three vulture species close to extinction in South Asia. Discovery by The Peregrine Fund of the drug's fatal effect on vulture kidney function, seen as white deposits on internal body organs, led to its ban in veterinary use and saved the species. (Photo credit: Martin Gilbert/The Peregrine Fund)

and public awareness to reduce shooting and other anthropogenic effects such as use and misuse of pesticides, poisons, and pharmaceutical drugs.

Sufficient funding over time is one of the most important resources needed, but talent and skill are just as important and often overlooked. Talent includes people with the knowledge, experience, passion, drive, and determination to be successful. Developing skill locally is important for success, but finding the right mix of knowledge and passion is often difficult, especially in developing nations where conservation is considered a luxury by the rural poor and well-educated individuals prefer desk jobs in the city near the comfort of home.

Other useful elements for successful conservation include organizational support, a method for annually evaluating results, and a strategy that includes adaptive management of the project as new information is gained. We have found that working collaboratively with people who may be responsible for the species' demise produces better, long-term results than confrontation and litigation, and a cooperative philosophy also appeals to land owners and financial donors whose support is often needed. Laws can be either beneficial or detrimental to conservation results, even if they were intended to be beneficial, and should be introduced sparingly and preferably only after voluntary compliance has been tried first.

These elements will be described and explained in the context of five critically endangered species and one vulnerable species: peregrine falcon recovery, successfully completed; California condor recovery, a successful project that is ongoing and still responding to new information; the Asian *Gyps* vulture population crash, which presents some new challenges; Madagascar fish eagle conservation, to illustrate conservation success in one of the world's poorest nations; and harpy eagle conservation which aims to avoid species endangerment before more expensive interventions are needed. Climate change is the conservation challenge of the twenty-first century, and its effects are most measurable in the arctic home of the gyrfalcon where The Peregrine Fund has started a project to collect empirical evidence of its impact on predator-prey interactions and the potential for behavioral or ecological resilience of the species and conservation actions in the future. Understanding species rarity and what it means to species' persistence is helpful for prioritizing conservation actions, especially for species that may have persisted in rarity for hundreds of years. Assessment of species phylogenetic relations is important for conservation decision-making because actions typically center on the existence of species in need and the timing of separation of sister taxa can influence whether conservation resources are made available. I conclude with a description of the value and importance of student training and support to build local capacity worldwide for raptor conservation and research.

Peregrine Falcon Recovery from DDT

The decline of the peregrine falcon (*Falco peregrinus*) from the 1950s through 1970s warned us that all was not well with the pesticide DDT. Research showed that DDT and its derivatives broke down very slowly in the environment, accumulated in the food chain becoming more concentrated with each link, and reached such high levels in top predators that it interfered with their calcium metabolism and caused eggshell thinning in birds. For over two decades, peregrine falcons throughout much of the world failed to hatch eggs and raise young because of DDT, and their populations dwindled (Hickey 1969; Ratcliffe 1980). By 1972 when DDT was banned in the United States, *F. p. anatum* was extirpated from the Eastern United States (Berger et al. 1969), the migratory *F. p. anatum* and *F. p. tundrius* from the arctic and subarctic range of the species were much diminished in abundance (Kiff 1988), and the species had been listed as endangered. As a result of banning DDT, species restoration efforts, and natural repopulation, peregrine falcons in the United States rebounded, and *F. p. tundrius* and *F. p. anatum* were delisted in 1994 and 1999, respectively (Heinrich 2011). Field monitoring and genetic methods to assess population stability and size over time both indicate a stable or growing population (Johnson et al. 2011).

The peregrine's near demise encouraged research on DDT's effects in humans and revealed that DDT binds to fats in the body and could reach levels seven times higher in breast milk than in blood. Recent studies have shown that childhood exposure to DDT and its derivatives from 1945 through the 1960s is linked to a fivefold increased risk of breast cancer in women who are now in their 50s and 60s (Cohn et al. 2007). At the time DDT was banned, insects were already becoming resistant because of its widespread overuse. Its potential utility today, under carefully regulated conditions for mosquito control inside homes (WHO 2006) is, in part, because the pesticide was removed from widespread agricultural use in the 1970s. Despite its negative human health effects, since 2000 DDT has once again been used to aid in control of malaria in quantities from 3500 to 6200 tons/year, mainly in India, other Southeast Asian countries, and Africa (WHO 2011). DDT's characteristics have not changed; it is still a persistent organochlorine pesticide that binds to fats, affects calcium metabolism, and concentrates in breast milk. The lessons we learned from the decline of peregrines and other species still need to be considered.

Recovery of the peregrine in North America was successful for many reasons, but most agree that banning DDT from widespread use was essential. The much lower levels of DDT and its metabolites in the peregrine's prey and in turn in the falcons themselves made the species' recovery possible. Had the level of use in the 1950s–1960s continued, the recovery would have failed. The second reason for success was the release of nearly 7000 peregrines in a collaborative program of captive breeding and release across North America.

The Peregrine Fund was created in direct response to the imminent extinction of the peregrine falcon in the United States and the need to save the species, both for the enjoyment of future generations and for their functional place as top predators in ecosystems across the continent. Founded in 1970 by Tom Cade, then Professor at Cornell University, the goal was to learn how to breed falcons in captivity and how to release them to the wild so that they survived and bred successfully in numbers sufficient to overcome natural mortality and create a growing population. The remarkable efforts of thousands of people and numerous organizations, including The Peregrine Fund, who collaborated on this goal over about 35 years of work, are told in the book *Return of the Peregrine: A North American Saga of Tenacity and Teamwork* (Cade and Burnham 2003).

Many critics questioned the technical effectiveness of captive breeding and release as a conservation method, as well as their relative benefits compared to more holistic approaches such as habitat preservation and ecosystem management. Time and experience have shown that the suite of actions that include captive breeding and release, and others such as assisted dispersal, have proven effective in restoring many endangered species to unoccupied range. Their value as tools for conservation will only increase as more species become threatened by habitat fragmentation and degradation. These hands-on, species-specific methods lie at one end of a continuum of actions needed to conserve biodiversity, with ecosystem management at the other end (Cade and Temple 1995; Cade 2000). They are neither better nor worse than other methods of conservation; they are simply applied for different purposes. Once a species becomes threatened, conservation usually requires special attention to its particular needs for restoration. More holistic actions focused on ecosystems and habitats are essential for maintaining biological diversity, preventing additional species from becoming vulnerable to extinction and providing the ecological scope for the recovery of endangered species (Cade and Burnham 2003).

California Condor Restoration and Lead Poisoning

Lead poisoning from ingestion of lead pellets, bullets, or fragments of either found in hunter-harvested animal carcasses is a threat to avian scavengers (in this book Chap. 11), especially the California condor (Cade 2007; Hunt et al. 2009a) but also bald and golden eagles (Stauber et al. 2010; Cruz-Martinez et al. 2012) in the United States and at least 60 other raptors, game birds, water birds, and other species worldwide (Pain et al. 2009). Lead poisoning is the leading cause of death in the reintroduced Arizona-Utah population of condors (Green et al. 2009; Parish et al. 2009), is one of the leading causes in California (Finkelstein et al. 2012), and remains the single most important obstacle to the recovery of the species. Without intensive intervention to treat lead-poisoned condors, this species would not be sustained in the wild (Finkelstein et al. 2012). Annual trapping, testing, and treatment of condors exposed to lead (Fig. 20.2) help to reduce direct mortality, but the effects of lead poisoning and its treatment continue to reduce reproductive rates and curb population viability. Asking hunters and other shooting sportsmen to switch to non-lead ammunition would solve the problem and appears to be working in Arizona where the state game and fish agency has invested in a voluntary, incentive-based program (Sieg et al. 2009). Efforts to make the switch have largely been thwarted elsewhere by incomplete regulations and by misinformation from ammunition industry and gun rights lobbyists who create fear among gun users that the gravity of lead poisoning in wildlife is a fabrication of anti-gun and anti-hunting activists. In contrast, condor conservationists support hunting, recognizing that hunters play an important role in helping condor recovery by increasing the food supply available to condors (Cade 2007), and they help regulate deer populations in the absence of many other large predators. The discovery of the effects of spent ammunition as a source of lead poisoning in condors and other wildlife led to research that also implicates this source of exposure on human health (Hunt et al. 2009b; Iqbal et al. 2009). The impacts can be significant, at least in subsistence hunter communities (Johansen et al. 2006; Tsuji et al. 2008). This problem would be relatively easy to solve, given political will and honesty.

Like the peregrine recovery effort that began in the 1970s in North America, California condor recovery began in the 1980s with much discussion and controversy, eventually leading to the last of the condors being removed from the wild in 1987 for captive breeding and reintroduction, with a founder population of just 22 individual birds. The Peregrine Fund joined the effort in 1993, building the largest condor breeding facility in the country, housing up to 20 pairs of condors, and annually producing most condors for release at sites from California to Arizona and Mexico and presumably elsewhere in the future. The Peregrine Fund also established the field program to restore condors in Northern Arizona under Section 10(j) "experimental and nonessential" rules of the Endangered Species Act (ESA), while restoration in California occurred with the full protection of the ESA. Unlike the peregrine recovery, however, the cause of condor decline was not well-established until after restoration releases began in the late 1990s. By 2002 condors tracked with satellite telemetry in Arizona were found to be dying from lead poisoning in unsustainable numbers, and by 2005 research had identified the source of lead exposure in condors. It came from the fragmented remains of spent bullets used to harvest big game, small game, varmint, and other animals (Hunt et al. 2009a, b).

The Peregrine Fund endorsed and supported a voluntary switch by hunters to non-lead ammunition in Arizona and Utah, with information and incentives from the state fish and game agencies that began in 2005. In California, the state passed legislation in 2008 that banned lead from hunting ammunition used in the condor's range and later expanded the ban statewide to take full effect by 2019. By 2016, neither strategy has yet proven sufficiently effective to prevent condor fatalities from lead poisoning, which still occurs at unsustainable rates in Arizona and Utah (where condors are also now nesting) and in California. The species survives in the wild only with annual trapping and testing for lead in the blood during the big game hunting season and chelation therapy of birds with high blood lead levels, plus replacement of those that die in the wild with captive-raised condors. In both restoration sites, additional strategies are being considered for removing lead from the condors' food supply, as breeding and release efforts that are still needed for the species to persist in the wild are not sustainable forever. Ultimately, species restoration can only be considered successful once a population can exist without human intervention.

Asian *Gyps* Vulture Mortality from Pharmaceuticals

The rapid population decline of two of the world's most abundant raptors, the Oriental white-backed vulture (*Gyps bengalensis*) and the long-billed vulture (*Gyps indicus*), was first noticed in India in the mid-1990s (Prakash 1999, in this book Chap. 19). By the early 2000s, similar catastrophic declines were detected in neighboring Pakistan and Nepal (Baral et al. 2004; Gilbert et al. 2002, 2006), and three once-common species were listed as Critically Endangered (BirdLife International 2000). Initial theories focused on pathogenic disease and previously unreported neck-drooping behavior (Cunningham et al. 2003). Ecological, behavioral, and forensic research by The Peregrine Fund and partners ultimately found that the nonsteroidal anti-inflammatory pharmaceutical, diclofenac, was responsible for vulture mortality in Pakistan (Oaks et al. 2004) and subsequently confirmed by others in India and Nepal (Green et al. 2004; Schultz et al. 2004) and that neck-drooping behavior had alternative explanations (Gilbert et al. 2007a; Watson et al. 2008a). Diclofenac, long used in human medicine, had found its way into the veterinary market as a safe, inexpensive, and very popular drug for palliative care of pain and inflammation in livestock in southern Asia. Residues of the drug in livestock that died within days after treatment were sufficient to cause kidney failure (Fig. 20.6) and death in *Gyps* vultures that fed upon their carcasses, then a common form of carcass disposal (Watson et al. 2004). This was the first evidence of a major ecological impact on a large vertebrate species caused by a pharmaceutical product. The mortality of adult vultures had a rapid and profound impact on the population, leading to a rate of population decline of about 30% per year (Gilbert et al. 2006).

In 2004, a series of meetings were held with government officials to inform them of this discovery. Although vulture feeding stations were tried as a method to reduce diclofenac exposure, they were of limited success (Gilbert et al. 2007b). Extensive lobbying efforts led in 2006 to a ban on the manufacture and sale of veterinary formulations of diclofenac in India, Pakistan, and Nepal (Pain et al. 2008; Oaks and Watson 2011). Vulture population surveys through 2012 showed that this quick action had stopped the catastrophic rate of decline of vultures (Chaudhry et al. 2012), even though diclofenac was available in formulations for human use and sometimes illegally used in veterinary medicine. More recently, the recovery of vultures has been jeopardized once again by the availability of alternative, synonymous drugs, though the impact may be less severe as people now bury or burn livestock carcasses more frequently than in former times when vultures were abundant.

The population crash of Asian *Gyps* vultures revealed that even trace amounts of pharmaceutical residue in the diet can be fatal to nontargeted and untested wildlife species. It showed that we cannot be complacent in our testing and use of drugs, pesticides, and poisons, and that once a solvable problem is discovered, a swift and unequivocal response from those responsible can be achieved with potentially dramatic results.

The Peregrine Fund played a critical role in saving South Asian vulture species from extinction by leading the team of ecologists and veterinarians in the forensic research to discover and prove the cause of mortality behind the catastrophic vulture decline. Once the cause was understood, we shared the knowledge quickly and efficiently. Our scientific peers first heard the news at the World Conference of Birds of Prey and Owls in Budapest in May 2003, which led to the rapid confirmation of diclofenac's role in India and Nepal the same year. We also published our research in the journal *Nature* within a year (Oaks et al. 2004). Government officials from India, Pakistan, and Nepal were presented with the evidence and conservation recommendations at the Kathmandu Summit on South Asian Vultures in February 2004.

Our recommendations to ban the veterinary use of diclofenac as quickly as possible, and to put sufficient numbers of each affected vulture species into captivity for safekeeping and potential restoration, were adopted with speed and resolution by each of the countries. A consortium of organizations working for vulture conservation across South Asia was created to coordinate efforts and introduce additional strategies such as the creation of vulture safe zones (Bowden et al. 2016).

However, aside from monitoring the state of breeding populations in India, Pakistan, and Nepal, our role in conservation ended as other national and international organizations took up the challenge in an effort that continues to this day. Diclofenac has been banned for veterinary use for over a decade, and our monitoring showed that populations of long-billed vultures stabilized quickly following the ban, but we have not seen a rebound of vulture numbers that we expected. Although vulture-safe alternatives to diclofenac were found and made available, other relatives of diclofenac, such as aceclofenac, that are just as fatal to vultures found their way into the market without testing (Galligan et al. 2016). Food abundance for vultures is not as it was prior to the crash because now livestock owners burn or bury the carcasses of dead animals, so vulture populations may never again be as abundant as they once were. But the lack of evidence for increasing populations is a concern that indicates actions needed to restore South Asian vultures cannot be ended yet, and vigilance will always be required.

African Vultures and Ending Wildlife Poisoning

The poisoning of predators to control livestock depredation may be the most globally widespread cause of unintentional poisoning of avian scavengers, especially vultures but also eagles and other species (in this book Chap. 10). Using poison to kill predators is grossly inefficient, not to mention ethically deplorable, because so many nontarget species are also killed. Among diurnal species, avian scavengers are more efficient than mammals at finding carrion and die in larger numbers from poisoning generally aimed at mammalian predators. Bateleurs in South Africa detected 67% of experimentally observed meat baits and were the first animal to find the baits 94% of the time (Watson 2000). Vultures in East Africa entirely consumed 84% of experimentally placed carcasses before any mammalian scavengers appeared (Houston 1974). Facultative scavengers, such as hyenas and especially lions, use the activity of vultures to detect carrion (Schaller 1972), but avian scavengers more than compensate for this competition by arriving rapidly in large numbers (Fig. 20.1) and consuming food quickly, leaving them vulnerable to poisoning intended for mammalian predators.

Predator poisoning has probably occurred for centuries and continues to be common, especially in Africa and Europe, where poisoning is used to avenge or deter the killing of livestock and game animals (e.g., rabbits, pheasants, and partridges) where hunters believe carnivore depredation reduces their hunting success (Ogada et al. 2011). Many types of poison are used including the compound 1080, strychnine, and others, but the misuse of the pesticide carbofuran is the most devastating recent example from Africa, where this low-cost and readily available pesticide is misused to kill predators like hyenas and lions.

Carbofuran has been especially popular as a poison in Kenya. Initially imported into the country in the 1960s for use as a pesticide in rice paddies, carbofuran's misuse first occurred in the mid-1990s when it was applied undiluted to kill ducks and other water birds that were harvested for human consumption (Odino and Ogada 2008; Otieno et al. 2011; Lalah et al. 2011), presumably at some risk to the consumers. Carbofuran has been poisoning wildlife for about 40 years, causing wildlife mortality from both misuse (i.e., baiting and intentional poisoning), as well as legal applications of the compound to crops (Richards 2012). Although the distinction between use and misuse is important because each elicits a different response and solution, ultimately carbofuran has the unpleasant distinction of being so hazardous to wildlife that it simply cannot be effectively regulated or managed without incidental wildlife mortality (Richards 2012). Researchers in the United States and Canada concluded that the sole condition under which carbofuran can be safely applied is if an area is already entirely devoid of wildlife (Mineau et al. 2012). As a result, it has been banned for any use in the United States, and there is a movement to ban it worldwide that has met resistance primarily from the manufacturers. A swift and unequivocal response from those responsible should be required.

The Peregrine Fund collaborated with researchers and organizations across Africa to collate data and publish results on vulture population declines in Africa. The results were alarming. Vulture populations of several species have crashed by 80% or more over three generations due primarily to the retaliatory poisoning of predators (Ogada et al. 2015). We worked together to reevaluate the IUCN conservation status of African raptors, leading to up-listing the status of four species to Critically Endangered and two species to Endangered in 2015 (BirdLife International 2015). Action to save these species is urgently needed. In 2016 we began a campaign to end wildlife poisoning in Kenya, and by working with other nongovernmental organizations across Africa, we hope to scale up actions as quickly as possible to the East African region and ultimately the African continent. Using Northern Kenya's landscape as a pilot, The Peregrine Fund and partners have identified a three-pronged approach to stop wildlife poisoning by:

1. Building better bomas within communities living alongside livestock predators through improved cost-shared fencing options. Bomas are livestock corrals which are often poorly built, allowing lions and other predators to attack livestock at night.
2. Training a network of rapid response teams to be the first responders to poisoning incidents. These teams of specially trained rangers and field scouts will improve the recovery of evidence, eliminate further poisoning at the scene, and educate communities about the hazard to themselves and wildlife of using poisons.
3. Engaging with communities and government to seek effective and practical alternatives and policies to stop wildlife poisoning in retaliation to livestock depredation.

In addition, we will monitor vulture populations at known breeding sites, monitor movements and survival of satellite-tagged vultures to identify vulture poisoning rates and hotspots, and detect change in response to this conservation strategy as a means of measuring success.

Ridgway's Hawk, Harpy Eagle, and Combatting Persecution

Human persecution refers to the intentional killing or disturbance of raptors through actions such as shooting, harassment, and deliberate poisoning and has probably occurred for centuries due to human prejudice, ignorance, superstition, wantonness, and retaliation (Curti and Valdez 2009). For species that normally reproduce slowly, such as harpy eagles (*Harpia harpyja*) in the neotropics (Fig. 20.3), Philippine eagles (*Pithecophaga jefferyi*) in Asia, and crowned eagles (*Stephanoaetus coronatus*) in Africa, even a slightly elevated rate of adult mortality can have a large negative effect on the population and cause extirpation even in advance of forest habitat loss (Vargas-González and Vargas 2011; Salvador and Ibañez 2006; Schultz 2002).

Critically endangered Ridgway's hawks (*Buteo ridgwayi*) in Dominican Republic (Figs. 20.4 and 20.5) are limited to a national park in a tiny portion of their former range because people shoot them for fear they will take their chickens, often mistaking them for the more abundant and larger red-tailed hawk (*Buteo jamaicensis*) which does take free-range chickens (Thorstrom et al. 2005; Woolaver 2011). Such persecution is not unusual, and it can be devastating for some species. Shooting raptors was encouraged in the United States through the 1950s but has diminished substantially today since bounties were stopped, and the public perception of value in wildlife has become widespread in the generally more economically advantaged population. Persecution continues worldwide where public value of raptors in the ecosystem is not widely perceived or where raptors are believed to kill small livestock, such as chickens. Raising public awareness of the cultural and aesthetic values of raptors in society and their value in ecosystem function and evolutionary processes is increasingly important as human populations increase in density, dominate landscapes, and introduce new forms of anthropogenic mortality on raptor populations (Curti and Valdez 2009).

The Peregrine Fund engages in public education at multiple levels and locations, from our interpretive center at our World Center for Birds of Prey headquarters in Boise, Idaho, to village presentations and door-to-door conversations with families adjacent to our restoration sites for harpy eagles in Panama, Ridgway's hawks in Dominican Republic, and many other sites. The beneficial impact of education is hard to measure unless metrics are established in advance and baseline data collected before and after education events. The long-term benefits of education often require frequent reinforcement over several years, even decades, before a positive outlook on raptors and wildlife in general becomes the paradigm for a community. Measuring impact and attitude adjustment among communities of people is essential to demonstrate results and justify the ongoing, long-term funding needed to sustain the effort.

In 1998, The Peregrine Fund experimentally released the first captive-bred harpy eagles in Soberania National Park in Panama. Shortly after, two of these eagles were shot and killed by local hunters. Realizing that captive breeding and release alone would not be enough to restore this species, we temporarily stopped releases to focus on changing people's perception of and attitude toward harpy eagles. We developed an education program to mitigate human persecution of raptors by promoting positive attitudes and values toward all birds of prey among communities in Panama (Curti and Valdez 2009). The program had six objectives:

1. Use the harpy eagle as a flagship species to increase general knowledge about conservation issues, focusing on the needs and existing knowledge of rural, indigenous, and *campesino* communities near where captive-bred harpy eagles were released or where wild populations of this species still occurred.
2. Develop pride and understanding among communities in areas where we aimed to restore harpy eagles or where wild populations of harpy eagles still exist by hiring and training community members in raptor release, tracking, and monitoring techniques.
3. Create and distribute easy-to-use educational materials for schools, and run corresponding teacher training workshops with the Ministry of Education's support.
4. Create an exchange of scientific and cultural knowledge between biologists and members of indigenous communities.
5. Develop and apply age-specific and culturally appropriate educational activities for both children and adults within target communities.
6. Develop and utilize ongoing evaluations to measure the effectiveness of the program, and make necessary modifications through a constant adaptive management process.

The program was successful in changing attitudes and behaviors and showed that when armed with a set of clearly defined goals, an identifiable target audience, and a wide variety of well-developed presentations and activities, an effective environmental education program can be accomplished with limited resources and personnel (Curti and Valdez 2009).

Offering alternatives to retaliation against raptors for depredation of small livestock, and incentives to adopt the alternatives, has proven essential among poor communities that have no financial resources and are rarely exposed to the benefits of wildlife. Examples include providing free chicken coops to communities in Dominican Republic to protect Ridgway's hawks from retaliation in and around Los Haitises National Park, the hawk's last stronghold, and around Puntacana Resort, where we have restored the species on land privately protected for tourism. Ridgway's hawks take chickens that are less than a week old (older chicks are too large for the diminutive hawk), so protecting chickens up to 2 weeks of age in chicken coops prevents depredation from hawks and, as collateral benefit, from other predators such as rats and snakes.

Through experiences such as these, we have learned the importance of starting an education campaign simultaneously with conservation efforts. Ideally, the education effort should be the first step in any conservation program. However, as long as there is wildlife to save, it is never too late to begin implementing an education component (Curti and Valdez 2009).

Madagascar Fish Eagle, Harpy Eagle, and Reversing Habitat Loss

As human demand for resources has grown, so has our impact on species dependent on resource-rich habitats from forests to grasslands. Habitat loss and fragmentation can be blamed for declines in forest-dependent species worldwide where forest has been clear-cut and remaining stands left fragmented, e.g., Philippine eagle (Salvador and Ibañez 2006), harpy eagle (Vargas-González and Vargas 2011), and grassland-dependent species such as aplomado falcons in the Southern United States and Northern Mexico where grasslands have been plowed under for agricultural development (Macías-Duarte et al. 2009, 2016). Orange-breasted falcons have become isolated by habitat loss in the northern extent of their range in Belize/Guatemala, with the next known extant population to the south occurring in Panama (Berry et al. 2010). Lack of prey due to overhunting or changes in land use is thought to have contributed to large-scale declines of vultures in West Africa, Southeast Asia, and Europe (Ogada et al. 2011). In Africa, species common in the landscape less than two decades ago, such as augur buzzards, have lost foraging ground as a result of horticultural expansion, and their breeding population is down by 60% in a southern Lake Naivasha study area (Virani and Harper 2009; Buechley 2011). Owls have also been affected with populations of the endangered East African endemic Sokoke Scops Owl in Kenya's Arabuko-Sokoke Forest having declined by 25% over the last two decades (Virani et al. 2010). The African crowned eagle and other forest-dependent raptors are not only faced with rapidly fragmenting forest cover but also a new threat of being persecuted and eaten by humans in places like Cameroon's Ebo Forest (Whytock et al. 2018). Changing land-use patterns, livestock overgrazing, and the intensification of agriculture resulting from Africa's rapidly growing human population and economy have reduced the distributional ranges of large raptors such as the martial eagle, African hawk eagle, and secretary bird. Many species are now only abundant in the larger protected areas.

The Peregrine Fund began working with indigenous communities in Madagascar in 1993 following an initial 3 years of research into the population ecology of the critically endangered Madagascar fish eagle. By working with and employing local people, we also developed an important level of trust and understanding with the local communities (Watson and Rabarisoa 2000). Following the 1993 invasion of the fish eagles' lakes stronghold by migrant fisherman from other parts of Madagascar, we joined with the local indigenous communities to help pioneer the model of community-based habitat conservation in Madagascar. Our premise was simple: …if there are enough fish in the lakes and trees in the forest for local fishing communities, there are enough of these critical resources to sustain fish eagles (Watson et al. 2008b). Two years of engagement with local communities by our professional sociologist in 1996–1997 (Razandrizanakanirina 1997) led to a project to implement a new national law (1996) decreed to decentralize control of natural resources from government by empowering local communities. We helped communities create official community associations that had authority to sign agreements on behalf of the community and to establish, monitor, and enforce resource extraction rules, such as fishing and timber extraction limits. The associations were also responsible for development aid, such as tree nurseries for commercial fruit crops and indigenous tree reforestation projects, fuel-efficient fish-drying ovens, and beehives for honey production, among many others. Implementation of the law was slow and at times convoluted and set back when violated by members of the communities and migrant fisherman alike. Ultimately, after 20 years of persistent effort by us and the communities, in 2015 three community conservation areas became official contributors to the country's national protected areas system, protecting 200,000 ha (nearly 500,000 acres) of biologically rich forest, wetlands, and grassland and saving critical habitat for the Madagascar fish eagle and other species.

The harpy eagle is endangered in Central America where The Peregrine Fund has been learning about the species' population ecology in a long-term study since 2000 in the lowland forests of Darién, Panama. The Chocó-Darién ecoregion is a global priority conservation area because it supports one of the world's most diverse assemblages of lowland tropical forest plants and animals, with exceptional species richness and endemism among birds, reptiles, amphibians, butterflies, and plants. This forest also holds the largest known population of harpy eagles in Central America as testament to its intact state. A combination of Emberá, Wounaan, and Kuna indigenous communities and Spanish-origin *colono-campesino* groups inhabit Darién and are having an influence on the ecosystem.

The human population in Darién is increasing at nearly 2% per year due, in part, to illegal land invaders and growth among the indigenous communities. These groups have limited opportunities for high school and college education, access to training, and technology for sustainable use of soil and forest resources, yet their cultural traditions are being changed and lost as they attempt to enter the twenty-first century economy. Land is community owned and farmed by indigenous Emberá and Wounaan communities. Slash-and-burn agriculture is resulting in deforestation, soil and nutrient loss, and biodiversity loss, as well as causing negative impacts on the local human population. Following an external cultural affinity to cattle ranching, Emberá and Wounaan indigenous people are now also clearing the forest around their communities to raise small herds of cattle. In 2015–2016 over 192,000 hectares of Darién were under pasture for raising cattle.

Marketing of land by *colono-campesinos* is the most important reason for land invasions in Darién. Land invaders cut the forest and create pastures to increase the price of the area in order to sell to the highest bidder. Associated negative impacts are illegal hunting, increased frequency of forest fires, contamination of water resources, loss of forest, and release of atmospheric carbon, among others. Nearly all forest fires in Darién are caused by humans and are initiated in agricultural and pastureland, escaping into adjacent forest. In 2016, at least 5000 ha of pristine forest and 45,000 ha of secondary forest were lost to forest fires in Darién, exacerbated by a strong drought from an El Niño climate event. It is predicted that forest fires will increase in frequency and intensity under future climate scenarios. Actions taken now to limit forest clearing and control fires are essential to conserve unique Chocó-Darién biodiversity and help limit impacts from climate change.

So far, the lack of roads in Darien represented a barrier of protection for the conservation of the biodiversity of Darién-Choco ecoregion as a whole. At present, there is a gravel highway to within 100 km of the Colombian border. While most indigenous inhabitants of the Darién travel by river, the highway has opened the region to loggers, cattle ranchers, and *colono-campesinos* land invaders. This new wave of human migration is threatening both the indigenous people and the natural ecosystem.

Bushmeat extraction is a threat to peccaries, agoutis, iguanas, guans, deer, and tapir that are illegally hunted and traded in communities of Darién. In a week one person can kill and sell five to ten wild animals to earn money for food and school supplies for children. Over the last 15 years, 31 harpy eagles are known to have been killed. As numbers of people living in and around the forest increase, hunting pressure increases, and hunters move further into remote forest in search of bushmeat.

Ecotourism could play an important role as an incentive for forest protection. However, indigenous communities in Darién have little capacity to manage ecotourism and benefit from this source of revenue. Tourist management plans that integrate interpretation of natural resources, biodiversity conservation, and equitable distribution of revenue among indigenous communities are needed.

The Peregrine Fund has been engaged with several Emberá-Wounaan communities for almost two decades since 2000 and has developed their mutual trust and confidence. In 2017 we began a pilot project to demonstrate that the kind of community-based conservation that has been successful for us in Madagascar can be replicated in Panama and potentially elsewhere. We aim to show that, with guidance, finance, and incentives, communities can address most of the threats described above, while implementing sustainable actions to protect the forest and help indigenous human communities become more resilient to climate change impacts, such as longer dry seasons, drier forest, and more frequent fires. Specifically, this project will improve forest and biodiversity conservation by ending forest clearing from uncontrolled cattle ranching and agricultural encroachment, reducing uncontrolled fires from forest clearing, reducing bushmeat extraction, and promoting alternative and sustainable sources of revenue.

Gyrfalcon and Climate Change

Responding to climate change may be one of the world's most important endeavors of the twenty-first century. Climate change models predict large changes in species' distributions and reduction in abundance of species from gyrfalcons (*Falco rusticolus*) in the arctic (Huntley and Green 2011; Booms et al. 2011) to Sokoke scops owls (*Otus ireneae*) in East Africa (Monadjem et al. 2012). Biotic interactions such as predator-prey, inter- and intraspecific competition, and phenology add a level of complexity that is difficult to model and in urgent need of empirical research (Watson and Hunt 2011). From the 2011 conference on Gyrfalcons and Ptarmigan in a Changing World (Watson et al. 2011), we learned that climate change is measurable, and it is happening more rapidly in the arctic than anywhere else on earth. As a result, the tree line is rising in both altitude and latitude, bird phenology is changing, and the tundra habitat of the gyrfalcon is shrinking (Bachelet 2011). Prey population cycles have been disrupted in the most recent decade, eliminating the periodic abundance of food on which gyrfalcons depend to assure successful reproduction (Gilg et al. 2011; Mossop 2011). There is evidence that gyrfalcon populations and breeding are diminished in the south of their range (Mossop 2011; Burnham and Burnham 2011; Mechnikova et al. 2011).

Climate models have become more sophisticated in recent years as our understanding of climate science has improved, and predictions are now more certain than ever before (Bachelet 2011). Using assumption scenarios to fill in the unknown parameters, ecologists use these models to forecast the ecological future for arctic habitats and species that depend on them. These models forecast a 50–60% loss of suitable habitat for gyrfalcons and their prey by 2080 (Booms et al. 2011; Huntley and Green 2011), a time span equivalent to my children's lifetime. Top predators such as the gyrfalcon are often sensitive to environmental change and can serve as early indicators of threat and as models for conservation intervention. Gyrfalcons and their principal prey, ptarmigan, are widely distributed and far-ranging species in the arctic ecosystem and are therefore good candidates for measuring, understanding, and mitigating current and predicted changes in their landscape. Coordinated research on gyrfalcon ecology across the species' global range would illuminate the species' ecological changes as they occur and inform conservation decisions for this and probably many other species also by seeking areas of ecological or behavioral vulnerability or resilience to climate change effects.

The Peregrine Fund began investigating climate change impacts on raptors by convening the 2011 conference on Gyrfalcons and Ptarmigan in a Changing World as a first step to understanding how we might address the threat of climate change. What emerged from this fascinating conference was an urgent need to develop coordinated research into the ecological and behavioral vulnerability and resilience to climate change of gyrfalcons throughout their arctic range and seek measurable actions to conserve them in the tundra landscape (Newton 2011). As a result, we now operate a research and conservation program focused on the effects of climate change on the gyrfalcon. In partnership with Alaska's department of fish and game and others, we expanded research on gyrfalcon ecology at a site where a decade of baseline data had already been collected on a large number of breeding territories that were logistically feasible to study. Early results were both exciting and encouraging. For example, gyrfalcons showed prey-switching in their diet, from ptarmigan to ground squirrels, between 2014 and 2015. This implies greater flexibility in foraging strategy than previously believed which could improve their resilience to climate change effects.

We have developed a manual of methods and terminologies for use by arctic raptor researchers. The manual provides the tools to ask relevant research questions and design a study to answer them, and it standardizes research methods so that gyrfalcon biologists collect comparable data with similar methods (Anderson et al. 2017). This allows researchers to combine data sets into meta-analyses at larger spatial scales than previously possible.

In partnership with Habitat Info, Inc., we developed a "bank vault" depository for data on arctic raptors called the Polar Raptor Data Bank (PRDB, https://gis.habitatinfo.com/tpf/). It is a secure storage site for data that safeguards the legacy of scientists who wish to protect their data in perpetuity. Once enough users are registered and data entered, it will facilitate research on arctic raptors over the long term and is intended to open communication and collaboration among biologists specializing on arctic raptors. The PRDB utilizes concepts first developed for the African Raptor Data Bank (http://www.habitatinfo.com/african-raptor-databank/) and, once combined, will expand to a Global Raptor Data Bank that serves both as a secure data vault and a center for global collaboration.

We started the Tundra Conservation Network in 2014 by assembling an international body of scientists to work together to understand the effects of climate change on the status, distribution, and population dynamics of tundra-obligate wildlife and plant species and the ecosystem- and community-level processes of which these species are an integral part. The network possesses a comprehensive knowledge base because each partner holds expertise in one or more components of the tundra food web, including vegetation ecology, predator-prey population ecology, plant-herbivore interactions, and global climate change research and modeling. Because our partners work in all eight arctic countries, combining our expertise provides a comprehensive, arctic-wide synthesis of the effects of climate change on tundra-obligate species. Research is still in its infancy, so we encourage future collaborations by contacting our arctic program director.

Issues of Natural Rarity and Population Isolation

Not all rare and potentially endangered species have become so due to recent anthropogenic influence. Some species have existed in comparatively low numbers for long periods, apparently without harm, though in the rapidly changing environments of today's world, that could change quickly. The Madagascar fish eagle (*Haliaeetus vociferoides*) is one of the world's rarest birds of prey and at significant risk of extinction. In the most recent census, only 287 individuals were recorded with an estimated total breeding population of about 120 pairs (Razafimanjato et al. 2013), an estimate that remained unchanged from the first global census in 1995, 10 years before (Rabarosoa et al. 1997). Levels of population genetic diversity are extremely low when compared with sister species, the African fish eagle (*Haliaeetus vocifer*), white-tailed sea eagle (*Haliaeetus albicilla*), and bald eagle (*Haliaeetus leucocephalus*) (Johnson et al. 2009). A Bayesian coalescent-based method showed that Madagascar fish eagles have maintained a small effective population size for hundreds to thousands of years and that its low level of neutral genetic diversity is not the result of a recent bottleneck (Johnson et al. 2009). The species has persisted in rarity for a very long time, so efforts to prevent species' extinction may not benefit from hands-on restoration but rather should place priority on maintaining habitat and reducing human persecution (Watson and Rabarisoa 2000). As feasible, actions to expand the species' range and abundance would help the very small population persist in the event of stochastic environmental impacts such as hurricane, fire, disease, or climate change that would threaten a small and confined population.

The Peregrine Fund focused first on understanding the distribution and abundance of the Madagascar fish eagle, conducting detailed baseline surveys in the 1990s throughout the species' range along the west coast and up to 100 km inland. We studied their behavioral ecology to understand the limitations of their abundance and used GIS models to predict areas of suitable habitat. We tested methods such as "sibling rescue" as a tool to increase population size and reduce extinction risk, but GIS models indicated that habitat may be limiting and that habitat conservation was the most important conservation action to take. As a result, since the mid-1990s, we have focused over 20 years of effort on community-based conservation of the ecotone between wetland (for fish, their principle prey) and forest (for nest sites, though they also occasionally nest on rocky cliffs). Our success is measured in both the area protected (almost 500,000 acres) and the stability of the fish eagle population, which we measure every 5 years as part of our conservation strategy. Because suitable habitat is limited and may have been limiting the eagle's abundance for centuries, the Madagascar fish eagle may forever be vulnerable to anthropogenic influence. It must, therefore, be forever monitored with a commitment to take further actions if evidence of decline is detected.

Historical records and recent surveys suggest that the orange-breasted falcon (*Falco deiroleucus*) has been extirpated from much of Central America and Southern Mexico and that its range is contracting in South America, despite large areas of apparently suitable habitat. The species has probably always been rare and local because of its ecology and specialized habitat requirements, but the surveys indicate that the small population of orange-breasted falcons in Belize and Guatemala is isolated (Thorstrom et al. 2002) and now appears to be in decline in Belize (Berry et al. 2010). Territory occupancy in Belize (12 pairs) declined from 83% (1992–1997) to 54% (2003–2009), and occupancy in 2009 was only half the mean in the prior decade. Mean annual production of fledglings per territorial pair declined 35% from 0.77 to 0.50. Mean annual population productivity, which measures the combined effect of occupancy and fecundity, declined 57% from 0.90 to 0.38 (Berry et al. 2010). In contrast, neither occupancy nor fledging success in Guatemala (seven pairs) declined over the same time period. Berry et al. (2010) proposed that factors contributing to this decline include conflict with humans and competition for nest sites and depredation by increasing numbers of black vultures (*Coragyps atratus*). They suggest management solutions including genetic restoration and the creation of safe harbors for nesting.

In Belize, The Peregrine Fund is monitoring occupancy and breeding productivity at nest sites of the orange-breasted falcon. Evidence suggests reduced fecundity among this small, remnant Central American population, which may be related to inbreeding depression. We are experimentally introducing captive-bred birds of Panamanian origin to test whether inbreeding depression might be reducing the falcon's productivity. Introduced birds have the potential to increase genetic variation and improve fitness characters related to fecundity, such as fertility, hatchability, and fledgling survival (Swinnerton et al. 2004).

Role of Phylogenetic Studies to Inform Conservation Decisions

Recent phylogenetic research on eagles has clarified relationships between species and genera and, in general, identified more species than previously recognized based on non-molecular methods (Lerner and Mindell 2005). Many of the new species are of small population size and restricted geographic range, and consequently, their discovery has increased the number of species at risk. This number may further increase as more presumed subspecies are included in phylogenetic analyses. For example, serpent eagles (*Spilornis* spp.) are generally in jeopardy wherever they occur worldwide. They often have small population size and limited distribution confined to islands and occupy habitats under threat, such as tropical forest. The phylogeography, population status, and ecology of many of these species are not well understood which seriously hampers efforts for their conservation. Ferguson-Lees and Christie (2001) proposed changes to the taxonomy of the serpent eagle group based on traditional morphological and behavioral methods in which they assigned species status to seven distinctive island subspecies of the crested serpent eagle *Spilornis cheela.* In contrast, the IUCN and BirdLife International continue to recognize this species complex, with its 23 subspecies, as a single species (*S. cheela*). If the Ferguson-Lees and Christie (2001) division of *Spilornis cheela* to form seven new species is accepted, then four of the species would be considered Data Deficient (*S. abbotti*, *S. asturinus*, *S. sipora*, and *S. natunensis*) and one Critically Endangered (*S. baweanus*) (GRIN 2017). These proposed taxonomic changes have not been confirmed or widely accepted and urgently need to be examined with a molecular phylogenetic study before rare and endangered island forms disappear forever. Depending on the results, conservation intervention may be justified to prevent the extinction of island endemic forms. Arguments about taxonomy may seem removed from effective conservation, but most conservation is about prioritizing effort on behalf of full species and is rarely concerned about conserving subspecies. If elevating these subspecies to the rank of species is genetically defensible, additional resources and protection may become more readily available.

Conversely, until recently the Cape Verde Kite (*Milvus milvus fasciicauda*) was considered to be one of the rarest birds of prey in the world and at significant risk of extinction. The Peregrine Fund collected field samples from kites on the Cape Verde islands, and Johnson et al. (2005) included them in a phylogenetic analysis based on three mitochondrial genes. They compared a sampling of kites in the genus *Milvus*, including a broad geographical sampling of black kites (*Milvus migrans*), red kites (*Milvus milvus*), Cape Verde kite museum specimens collected between 1897 and 1924, and the kites trapped on the Cape Verde Islands in 2002. They found that the historical Cape Verde kites, including the type specimen, were non-monophyletic and scattered within a larger red kite clade. The recently trapped kites from the Cape Verde Islands were all phylogenetically identified as Black Kites. Our results suggested that the traditionally recognized Cape Verde Kite was not a distinctive evolutionary unit, and the case for species status was not supported. Urgent conservation action on its behalf was therefore not justified. Interestingly, the study did reveal support for recognition of at least one clade of Yellow-billed Kites from South Africa and Madagascar, traditionally considered a Black Kite subspecies, as a distinctive phylogenetic species. Fortunately, this species is common and widespread and not in need of conservation intervention.

Investing in Future Conservation Leaders

Qualified and experienced individuals are the essential, driving force for conservation progress and initiatives. Science, especially knowledge and understanding of biotic processes and systems, is the anchor for effective conservation. In all The Peregrine Fund's programs, we strive to ensure sustainability of effort by supporting and training local individuals, so they are in a position to continue the raptor research and conservation that we start. Our goal is to make a lasting contribution to global biodiversity conservation and science by developing a raptor biologist in every country in the world through training and support. By the end of 2016, students receiving support and training from The Peregrine Fund since beginning work in 1970 have earned 109 advanced degrees, including 28 PhD and 81 MSc or equivalent degrees. Among a sample of 48 graduates of our international programs training, 83% have gone on to professional careers in wildlife conservation. In addition to those whom we employ to run our programs, such as Dr. Hernan Vargas, Director of Neotropical Programs; Dr. Munir Virani, Africa Programs Director; and Professor Lily-Arison René de Roland, National Director for Madagascar, other former students are influencing raptor conservation in positions with governmental and nongovernmental organizations such as the species conservation coordinator for Conservation International, terrestrial program coordinator for World Wildlife Fund, African vulture conservation manager for BirdLife International, wildlife conservation biologist for the Environment Agency—Abu Dhabi, and conservation biologist for the Ramsar Convention on Wetlands. Considering both direct and indirect influence on conservation outcomes, The Peregrine Fund's strategy to prioritize work on saving raptor species from extinction while building capacity for raptor conservation and research in every country in which we work has been transformative.

Acknowledgments Many Peregrine Fund staff contributed ideas and text to this chapter, which is a condensed compilation of conservation actions and research that have often led to success over the organization's nearly 50-year lifespan. I particularly wish to acknowledge valuable contributions by David Anderson, Marta Curti, Thomas Hayes, Chris McClure, Darcy Ogada, Lily Arison René de Roland, Amy Siedenstrang, Russell Thorstrom, Hernan Vargas, José Vargas, and Munir Virani.

References

Anderson DL, McClure CJW, Franke A (2017) Applied raptor ecology: Essentials from gyrfalcon research. The Peregrine Fund, Boise

Bachelet D (2011) Conservation strategies in a changing world. In: Watson RT, Cade TJ, Fuller M, Hunt G, Potapov E (eds) Gyrfalcons and ptarmigan in a changing world, vol II. The Peregrine Fund, Boise, pp 385–398. https://doi.org/10.4080/gpcw.2011.0317

Baral HS, Giri JB, Virani M (2004) On the decline of oriental white-backed vultures Gyps bengalensis in lowland Nepal. In: Chancellor RD, Meyburg B-U (eds) Raptors worldwide. World Working Group on Birds of Prey and Owls/MME/BirdLife Hungary, Berlin/Budapest, pp 215–219

Berger DD, Sindelar CR Jr, Gamble KE (1969) The status of the breeding peregrines of the eastern United States. In: Hickey JJ (ed) Peregrine falcon populations: their biology and decline. University of Wisconsin Press, Madison, pp 165–173

Berry RB, Benkman CW, Muela A, Seminario Y, Curti M (2010) Isolation and decline of a population of the Orange-breasted falcon. Condor 112(3):479–489. https://doi.org/10.1525/cond.2010.100012

BirdLife International (2000) Threatened birds of the world. Lynx Edicions and BirdLife, Barcelona/Cambridge, UK

BirdLife International (2015) Global IUCN Red List for birds – 2015 changes. https://globally-threatened-birdforums.birdlife.org/2015/10/global-iucn-red-list-for-birds-2015-changes/

Booms T, Lindgren M, Huettmann F (2011) Linking Alaska's predicted climate, gyrfalcon, and ptarmigan distributions in space and time: a unique 200-year perspective. In: Watson RT, Cade TJ, Fuller M, Hunt G, Potapov E (eds) Gyrfalcons and ptarmigan in a changing world, vol I. The Peregrine Fund, Boise, pp 177–190. https://doi.org/10.4080/gpcw.2011.0116

Bowden CGR, Galligan TH, Prakash V, Paudel K, Cuthbert R, Green RE (2016) An overview of recent advances and priorities for vulture conservation in the South Asia region. Regional symposium on vulture conservation in Asia. IUCN Pakistan, Karachi

Buechley E. (2011) Augur Buzzard declines at Lake Naivasha. Swara July: 54–56

Burnham KK, Burnham WA (2011) Ecology and biology of gyrfalcons in Greenland. Pages 1-20 in Watson RT, Cade TJ, Fuller, M, Hunt G, Potapov E (Eds) Gyrfalcons and ptarmigan in a changing world, Volume II. The Peregrine fund, Boise

Cade TJ (2007) Exposure of California condors to lead from spent ammunition. J Wildl Manag 71(7):2125–2133. https://doi.org/10.2193/2007-084

Cade TJ, Burnham WA (2003) Return of the peregrine. A North American Saga of tenacity and teamwork. The Peregrine Fund, Boise

Cade TJ (2000) Progress in translocation of diurnal raptors. Pages 342-372 in Raptors at risk (Chandellor RD, Meyburg B-U, Eds) World Working Group on Birds of Prey, Berlin / Hancock House, Blaine

Cade TJ, Temple SA (1995) Management of threatened bird species: evaluation of the hands-on approach. Ibis 137 (Supplement):S161–172

Chaudhry MJI, Ogada DL, Malik RN, Virani MZ, Giovanni MD (2012) First evidence that populations of the critically endangered long-billed vulture *Gyps indicus* in Pakistan have increased following the ban of the toxic veterinary drug diclofenac in South Asia. Bird Conserv Int 22:389–397

Cohn BA, Wolff MS, Cirillo PM, Sholtz RI (2007) DDT and breast cancer in young women: new data on the significance of age at exposure. Environ Health Perspect 115(10):1406–1414. https://doi.org/10.1289/ehp.10260

Cruz-Martinez L, Redig PT, Deen J (2012) Lead from spent ammunition: a source of exposure and poisoning in bald eagles. Hum-wildl interact 6(1):94–104

Cunningham AA, Prakash V, Pain DJ, Ghalsasi GR, Wells AH, Kolte GN, Nighot P, Goudar MS, Kshirgar S, Rahmani A (2003) Indian vultures: victims of a disease epidemic? Anim Conserv 6:189–197

Curti M, Valdez U (2009) Incorporating community education in the strategy for harpy eagle conservation in Panama. J Environ Educ 40(4):3–15

Ferguson-Lees J, Christie DA (2001) Raptors of the world. Christopher Helm, London

Finkelstein ME, Doak DF, George D, Burnett J, Brandt J, Church M, Grantham J, Smith DR (2012) Lead poisoning and the deceptive recovery of the critically endangered California condor. Proc. Natl. Acad. Sci. U.S.A 109(28):11449–11454. https://doi.org/10.1073/pnas.1203141109

Galligan TH, Taggart MA, Cuthbert RJ, Svobodova D, Chipangura J, Alderson D, Prakash VM, Naidoo V (2016) Metabolism of aceclofenac in cattle to vulture-killing diclofenac. Conserv Biol. https://doi.org/10.1111/cobi.12711

Gilbert M, Virani M, Watson RT, Oaks JL, Benson PC, Khan AA, Ahmed S, Chaudhry MJI, Arshad M, Mahmood S, Shah QA (2002) Breeding and mortality of oriental white-backed vulture *Gyps bengalensis* in Punjab Province, Pakistan. Bird Conserv Int 12:311–326

Gilbert M, Watson RT, Virani MZ, Oaks JL, Ahmed S, Chaudhry MJI, Arshad M, Mahmood S, Ali A, Khan AA (2006) Rapid population declines and mortality clusters in three oriental white-backed vulture *Gyps bengalensis* colonies due to diclofenac poisoning. Oryx 40:388–399

Gilbert M, Watson RT, Virani MZ (2007a) Neck-drooping posture in oriental white-backed vultures (*Gyps bengalensis*): an unsuccessful predictor of mortality and its probable role in thermoregulation. J Raptor Res 41:35–40

Gilbert M, Watson RT, Ahmed S, Asim M, Johnson JA (2007b) Vulture restaurants and their role in reducing diclofenac exposure in Asian vultures. Bird Conserv Int 17:63–77

Gilg O, Sittler B, Hanski I (2011) Will collared lemmings and their predators be the first vertebrates to "fall over the cliff" in Greenland due to global climate changes? Extended abstract. In: Watson RT, Hunt G, Cade TJ, Fuller M, Potapov E (eds) Gyrfalcons and ptarmigan in a changing world, vol I. The Peregrine Fund, Boise, pp 131–132. https://doi.org/10.4080/gpcw.2011.0113

Global Raptor Information Network (GRIN) (2017) Species account: Bawean Serpent eagle *Spilornis baweanus*. Downloaded from http://www.globalraptors.org on 1 Jan 2017

Green RE, Newton I, Shultz S, Cunningham AA, Gilbert M, Pain DJ, Prakash V (2004) Diclofenac poisoning as a cause of population declines across the Indian subcontinent. J Appl Ecol 41:793–800

Green RE, Hunt WG, Parish CN, Newton I (2009) Effectiveness of action to reduce exposure of free-ranging California condors in Arizona and Utah to lead from spent ammunition. In: Watson RT, Fuller M, Pokras M, Hunt WG (eds) Ingestion of lead from spent ammunition: implications for wildlife and humans. The Peregrine Fund, Boise, pp 240–253. https://doi.org/10.4080/ilsa.2009.0218

Heinrich WR (2011) Peregrine falcon recovery in the continental United States, 1974–1999, with notes on related programs of the Peregrine Fund. In: Sielicki J, Mizera T (eds) Peregrine falcon populations: status and perspectives in the 21st century. Turul Publishing/Poznañ University of Life Sciences Press, Warsaw/Poznañ, pp 431–444

Hickey JJ (ed) (1969) Peregrine falcon populations: their biology and decline. University of Wisconsin Press, Madison

Houston DC (1974) Food searching behaviour in griffon vultures. E Afr Wildl J 12:63–77

Hunt WG, Parish CN, Orr K, Aguilar RF (2009a) Lead poisoning and the reintroduction of the California condor in Northern Arizona. J Avian Med Surg 23(2):145–150

Hunt WG, Burnham WA, Parish CN, Burnham KK, Mutch B, Oaks JL (2009b) Bullet fragments in deer remains: implications for lead exposure in scavengers. In: Watson RT, Fuller M, Pokras M, Hunt WG (eds) Ingestion of lead from spent ammunition: implications for wildlife and humans. The Peregrine Fund, Boise, pp 254–258. https://doi.org/10.4080/ilsa.2009.0123

Huntley B, Green RE (2011) Bioclimatic models of the distributions of gyrfalcons and ptarmigan. In: Watson RT, Cade TJ, Fuller M, Hunt G, Potapov E (eds) Gyrfalcons and ptarmigan in a changing world, vol II. The Peregrine Fund, Boise, pp 329–338. https://doi.org/10.4080/gpcw.2011.0312

Iqbal S, Blumenthal W, Kennedy C, Yip FY, Pickard S, Flanders WD, Loringer K, Kruger K, Caldwell KL, Brown MJ (2009) Hunting with lead: association between blood lead levels and wild game consumption. Environ Res. https://doi.org/10.1016/j.envres.2009.08.007

Johansen P, Pedersen HS, Asmund G, Riget F (2006) Lead shot from hunting as a source of lead in human blood. Environ Pollut 142:93–97

Johnson JA, Watson RT, Mindell DP (2005) Prioritizing species conservation: does the Cape Verde kite exist? Proc R Soc B 272:1365-1372

Johnson JA, Tingay RE, Culver M, Hailer F, Clarke ML, Mindell DP (2009) Long-term survival despite low genetic diversity in the critically endangered Madagascar fish-eagle. Mol Ecol 18:54–63. https://doi.org/10.1111/j.1365-294X.2008.04012.x

Johnson JA, Talbot S, Sage G, Burnham KK, Brown J, Maechtle T, Seegar WS, Yates M, Anderson B, Mindell DP (2011) The use of genetics for the management of a recovering population: temporal assessment of migratory Peregrine falcons in North America. PLoS One 2(11):e14042. https://doi.org/10.1371/journal.pone.0014042

Kiff LF (1988) Changes in the status of the peregrine in North America: an overview. In: Cade TJ, Enderson JH, Thelander CG, White CM (eds) Peregrine falcon populations: their management and recovery. The Peregrine Fund, Boise, pp 123–139

Lalah JO, Otieno PO, Richards NL, Odino M, Frank L, Cotterill A, Dolrenry S, Hazzah L, Ogada D, Virani M, Martins DJ (2011) A chronicling of long-standing carbofuran use and its menace to wildlife in Kenya. In: Richards N (ed) Carbofuran and wildlife poisoning: global perspectives and forensic approaches. Wiley, New York, pp 39–98. https://doi.org/10.1002/9781119998532.ch3

Lerner HRL, Mindell DP (2005) Phylogeny of eagles, old world vultures, and other Accipitridae based on nuclear and mitochondrial DNA. Mol Phylogenet Evol 37:327–346

Macías-Duarte A, Montoya A, Méndez-González CE, Rodríguez-Salazar JR, Hunt WG, Krannitz PG (2009) Factors influencing habitat use by migratory grassland birds in Chihuahua, Mexico. Auk 126(4):896–905. https://doi.org/10.1525/auk.2009.08251

Macías-Duarte A, Montoya A, Rodriguez-Salazar JR, Hunt G (2016) The imminent disappearance of the aplomado falcon from the Chihuahuan Desert. Journal of Raptor Research 50(2):211–216

Mechnikova S, Romanov M, Kudryavtsev N (2011) Change in numbers and nesting ecology of the gyrfalcon in the Yamal peninsula, Russia, from 1981 to 2010. In: Watson RT, Cade TJ, Fuller M, Hunt G, Potapov E (eds) Gyrfalcons and ptarmigan in a changing world, vol II. The Peregrine Fund, Boise, pp 205–212. https://doi.org/10.4080/gpcw.2011.0220

Mineau P, Lyon L, McMillin S (2012) Impacts of carbofuran on birds in Canada and the United States. In: Richards N (ed) Carbofuran and wildlife poisoning: global perspectives and forensic approaches, 1st edn. Wiley, Chichester, pp 208–250

Monadjem A, Virani MZ, Jackson C, Reside A (2012) Rapid decline and shift in the future distribution predicted for the endangered Sokoke Scops owl *Otus ireneae* due to climate change. Bird Conserv Int. https://doi.org/10.1017/S0959270912000330

Mossop DH (2011) Long-term studies of willow ptarmigan and gyrfalcon in the Yukon territory: a collapsing 10-year cycle and its apparent effect on the top predator. In: Watson RT, Cade TJ, Fuller M, Hunt G, Potapov E (eds) Gyrfalcons and ptarmigan in a changing world, vol I. The Peregrine Fund, Boise, pp 323–336. https://doi.org/10.4080/gpcw.2011.0206

Newton I (2011) Conference summary. In: Watson RT, Cade TJ, Fuller M, Hunt G, Potapov E (eds) Gyrfalcons and ptarmigan in a changing world, vol I. The Peregrine Fund, Boise, pp 5–10. https://doi.org/10.4080/gpcw.2011.0318

Oaks JL, Watson RT (2011) South Asian vultures in crisis: environmental contamination with a pharmaceutical. In: Elliot J et al (eds) Wildlife ecotoxicology: forensic approaches. Springer, New York/Dordrecht/Heidelberg/London, pp 413–441

Oaks JL, Gilbert M, Virani M, Watson RT, Meteyer CU, Rideout BA, Shivaprasad HL, Ahmed S, Chaudhry MJI, Arshad M, Mahmood S, Ali A, Khan AA (2004) Diclofenac residues as the cause of vulture population decline in Pakistan. Nature 427:630–633

Ogada D, Keesing F, Virani M (2011) Dropping dead: causes and consequences of vulture population declines worldwide. Ann N Y Acad Sci 2011:1–15. https://doi.org/10.1111/j.1749-6632.2011.06293.x

Ogada D, Shaw P, Beyers RL, Buij R, Murn C, Thiollay J-M, Beale CM, Holdo RM, Pomeroy D, Baker N, Krüger SC, Botha A, Virani M, Monadjem A, Sinclair ARE (2015) Another Continental Vulture Crisis: Africa's Vultures Collapsing toward Extinction. Conservation Letters 2015. https://doi.org/10.1111/conl.12182

Odino M and Ogada D (2008) Furadan use in Kenya: a survey of the distributors and end-users of toxic Carbofuran (Furadan) in pastoralist and rice growing areas. Zoology Department, Ornithology Section, National Museums of Kenya, Nairobi

Otieno PO, Lalah JO, Virani M, Jondiko IO, Schramm K-W (2011) Carbofuran use and abuse in Kenya: residues in soils, plants, water courses and the African white-backed vultures (*Gyps africanus*) found dead. Environmentalist 31:382–393

Pain DJ, Bowden CGR, Cunningham AA, Cuthbert R, Das D, Gilbert M, Jakati RD, Jhala Y, Khan AA, Naidoo V, Oaks JL, Parry-Jones J, Prakash V, Rahmani A, Ranade SP, Baral HS, Senacha KR, Saravanan S, Shah N, Swan G, Swarup D, Taggart MA, Watson RT, Virani M, Wolter K, Green RE (2008) The race to prevent the extinction of South Asian vultures. Bird Conserv Int 18:S30–S48. https://doi.org/10.1017/S0959270908000324

Pain DJ, Fisher IJ, Thomas VG (2009) A global update of lead poisoning in terrestrial birds from ammunition sources. In: Watson RT, Fuller M, Pokras M, Hunt WG (eds) Ingestion of lead from spent ammunition: implications for wildlife and humans. The Peregrine Fund, Boise, pp 99–118. https://doi.org/10.4080/ilsa.2009.0108

Parish CN, Hunt WG, Feltes E, Sieg R, Orr K (2009) Lead exposure among a reintroduced population of California condors in Northern Arizona and Southern Utah. In: Watson RT, Fuller M, Pokras M, Hunt WG (eds) Ingestion of lead from spent ammunition: implications for wildlife and humans. The Peregrine Fund, Boise, pp 259–264. https://doi.org/10.4080/ilsa.2009.0217

Prakash V (1999) Status of vultures in Keoladeo National Park, Bharatpur, Rajasthan, with special reference to population crash in Gyps species. J Bombay Nat Hist Soc 96:365–378

Rabarosoa R, Watson RT, Thorstrom R, Berkelman J (1997) Status of the Madagascar fish eagle *Haliaeetus vociferoides* in 1995. Ostrich 68:8–12

Ratcliffe DA (1980) The Peregrine falcon. Buteo Books, Vermillion

Razafimanjato G, Sam TS, Rakotondratsima M, Rene de Roland L-A, Thorstrom R (2013) Population status of the Madagascar fish eagle *Haliaeetus vociferoides* in 2005–2006. Bird Conserv Int 24:88–99. https://doi.org/10.1017/S0959270913000038

Razandrizanakanirina D (1997) Community-based wetland conservation project: mission notes. In: Watson RT (ed) Madagascar wetlands conservation project, progress report III. The Peregrine Fund, Boise, pp 105–120

Richards N (ed) (2012) Carbofuran and wildlife poisoning: global perspectives and forensic approaches, 1st edn. John Wiley & Sons, Inc., Chichester

Salvador DJI, Ibañez JC (2006) Ecology and conservation of Philippine eagles. Ornithol Sci 5:171–176

Schaller GB (1972) The Serengeti lion. University of Chicago Press, Chicago

Schultz S (2002) Population density, breeding chronology and diet of crowned eagles *Stephanoaetus coronatus* in Taï National Park, Ivory Coast. Ibis 144:135–138

Schultz S, Baral HS, Charman S, Cunningham AA, Das D, Ghalsasi GR, Goudar MS, Green RE, Jones A, Nighot P, Pain DJ, Prakash V (2004) Diclofenac poisoning is widespread in declining vulture populations across the Indian subcontinent. Proc Roy Soc Lond B (Supplement) 271(Suppl 6):S458–S460

Sieg R, Sullivan K, Parish CN (2009) Voluntary lead reduction efforts within the Northern Arizona range of the California condor. In: Watson RT, Fuller M, Pokras M, Hunt WG (eds) Ingestion of lead from spent ammunition: implications for wildlife and humans. The Peregrine Fund, Boise, pp 341–349. https://doi.org/10.4080/ilsa.2009.0309

Stauber E, Finch N, Talcott PA, Gay JM (2010) Lead poisoning of bald (*Haliaeetus leucocephalus*) and golden (*Aquila chrysaetos*) eagles in the U.S. inland Pacific northwest region—an 18 year retrospective study: 1991–2008. J Avian Med Surg 24:279–287

Swinnerton KJ, Groombridge JJ, Jones CG, Burn RW, Mungroo Y (2004) Inbreeding depression and founder diversity among captive and free-living populations of the endangered pink pigeon *Columba mayeri*. Anim Conserv 7:353–364

Thorstrom R, Watson RT, Baker A, Ayers S, Anderson D (2002) Preliminary ground and aerial surveys for Orange-breasted falcons in Central America. J Raptor Res 36(1):39–44

Thorstrom R, Almonte J, Balbuena de la Rosa S, Rodríguez P, Fernández E (2005) Surveys and breeding biology of *Buteo ridgwayi* (Ridgway's hawk) in Los Haitises, Dominican Republic. Caribb J Sci 41:864–869

Tsuji LJS, Wainman BC, Martin ID, Sutherland C, Weber J-P, Dumas P, Nieboer E (2008) Lead shot contribution to blood lead of first nations people: the use of lead isotopes to identify the source of exposure. Sci Total Environ 405:180–185

Vargas-González J d J, Vargas H (2011) Nesting density of harpy eagles in Darien with population size estimates for Panama. J Raptor Res 45(3):199–210

Virani M, Harper DM (2009) Factors influencing the breeding performance of the augur buzzard *Buteo augur* in southern Lake Naivasha, Rift Valley, Kenya. Ostrich 80(1):9–17. https://doi.org/10.2989/OSTRICH.2009.80.1.2.760

Virani M, Njoroge P, Gordon I (2010) Disconcerting trends in populations of the endangered Sokoke Scops owl *Otus ireneae* in the Arabuko-Sokoke Forest, Kenya. Ostrich 81(2):155–158. https://doi.org/10.2989/00306525.2010.488429

Watson RT (2000) Flight, foraging and food of the Bateleur *Terathopius ecaudatus*: an aerodynamically specialized, opportunistic forager. In: Chancellor RD, Meyburg B-U (eds) Raptors at Risk. WWGBP/Hancock House, Berlin, pp 65–75

Watson RT, Hunt G (2011) Preface. In: Watson RT, Cade TJ, Fuller M, Hunt G, Potapov E (eds) Gyrfalcons and ptarmigan in a changing world, vol I. The Peregrine Fund, Boise, pp 1–3. http://peregrinefund.org/subsites/conference-gyr/proceedings/99-Preface.pdf

Watson RT, Rabarisoa R (2000) Sakalava fishermen and Madagascar fish eagles: enhancing traditional conservation rules to control resource abuse that threatens a key breeding area for an endangered eagle. Ostrich 71:2–10

Watson RT, Gilbert M, Oaks JL, Virani M (2004) The collapse of vulture populations in South Asia. Biodiversity 5:3–7

Watson RT, Gilbert M, Virani M (2008a) Neck-drooping posture of oriental white-backed vultures (*Gyps bengalensis*) in close proximity to human observers. J Raptor Res 42:66–67

Watson RT, Rene de Roland LA, Rabearivony J, Thorstrom R (2008b) Community-based wetland conservation protects endangered species in Madagascar: Lessons from science to conservation. Proc 16th Biodiversity Symposium, Philippine Wildlife Conservation Society, Manila, Philippines. Banwa 4(1):81–95

Watson RT, Cade T, Hunt WG, Fuller M, Potapov E (eds) (2011) Gyrfalcons and ptarmigan in a changing world, vol I and II. The Peregrine Fund, Boise. http://www.peregrinefund.org/subsites/conference-gyr/proceedings/index.html

Whytock RC, Morgan BJ, Awa T, Bekokon Z, Abwe EA, Buij R, Virani M, Vickery JA, Bunnefeld N (2018) Quantifying the scale and socioeconomic drivers of bird hunting in Central African forest communities. Biological Conservation 218:18–25

Woolaver LG Jr (2011) Ecology and conservation genetics of Ridgway's Hawk *Buteo ridgwayi*. Ph.D. dissertation, York University, Toronto, Ontario. p 291

World Health Organization (2006) Indoor residual spraying: WHO position statement. Global Malaria Programme, Geneva. http://www.who.int/malaria. Last Accessed 10 Nov 2008

World Health Organization (2011) Global insecticide use for vector-borne disease control: A ten year assessment, 2000–2009. 5th edn. World Health Organization, Geneva. http://whqlibdoc.who.int/publications/2011/9789241502153_eng.pdf. Last Accessed 18 Sept 2012

Raptor Species Index

J. H. Sarasola et al. (eds.), *Birds of Prey*,
https://doi.org/10.1007/978-3-319-73745-4

Word-Topics Index

J. H. Sarasola et al. (eds.), *Birds of Prey*,
https://doi.org/10.1007/978-3-319-73745-4

J

K

The manufacturer's authorised representative in the EU is Springer Nature Customer Service Centre GmbH, Europaplatz 3, 69115 Heidelberg, Germany. If you have any concerns regarding our products, please contact ProductSafety@springernature.com

Printed and bound by CPI Group (UK) Ltd, Croydon, CR0 4YY
25/09/2026
02222853-0012